JN416614

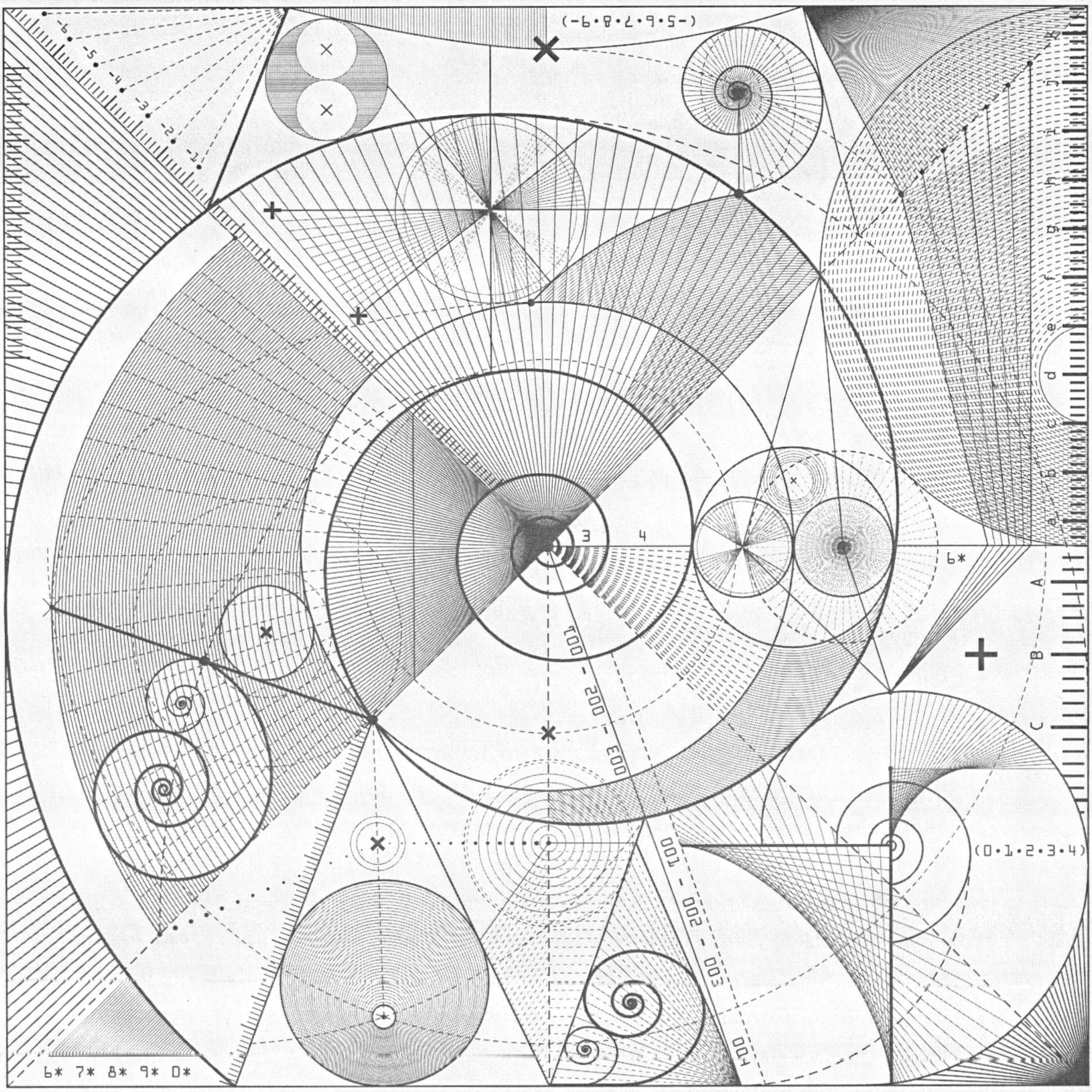

CALCULUS

Man Sik Min · Hyeong Chul Jeong · Hyejung Lee

한티미디어

PREFACE

We wrote this calculus textbook for students seriously considering a career in STEM: Science, Technology, Engineering, Mathematics. Having taught calculus for over thirty years at universities, We have gained invaluable insights into making the material more accessible and less daunting through the use of simple, direct language. Thus, even for students with weak mathematical foundations or for those who are approaching calculus for the very first time, We believe that this textbook can facilitate their comprehension. Numerous examples and problem sets were designed so that the major concepts of calclus could be absorbed, then solved, and finally applied. In addition to single variable calculus, We cover vector calculus, partial derivatives, and multiple integrals, as these are the minimum requirements necessary to survive a university STEM curriculum.

The contents of this book are as follows. Chapter 1 introduces the definitions of functions and deals with inverse and composite functions. Specifically, exponential, log functions, and trigonometric functions—all crucial parts in science and engineering—are thoroughly examined. Chapter 2 deals with the limits of functions and the concept of continuity. Special attention is paid to the limits of transcendental functions and the limits of infinity, as well as the mean value theorem through continuity. Chapter 3 introduces derivatives and deals with tangent, normal line equations, and various differential calculations using slope. Chapter 4 examines the close relationship between mathematics and its practical applications in real life via local maximum, local minimum, and optimization problems. Chapter 5 introduces indefinite integrals and definite integrals, focusing on the relationship between the derivative and integral through the fundamental theorem of calculus. Chapter 6 introduces the application of definite integrals. Mainly examined are the areas between the curves, the volume of revolution, the movement of curves, the length of arcs on a plane, and improper integrals. This chapter expands the concept of derivatives or integrals from single variable calculus to solve problems related to maximum or minimum values of multivariate functions or volumes of solids, helping students understand physical phenomena using vector functions.

Chapter 7 deals with differential equations, Chapter 8 deals with series. Chapter 9 deals with the concept of polar coordinates and parametric calculus. Chapter 10 introduces vectors, straight lines, and planes using the dot product and the cross product of vectors.

We received a lot of help from Chung Gu Hyun in writing this book. In reviewing the text, examples, and exercises, they offered creative solutions and suggestions, providing student perspectives that allowed me to anticipate potential difficulties in the material. Among the many examples that come to mind, while calculating the limit of a trigonometric function graph, instead of obtaining the derivative first, Gu Hyun suggested an easier method of obtaining the graph by transforming the given trigonometric function. It was through his diligence that, in the process of substitution integration, he organized the solution himself to solve the first-order linear differential equation, thereby greatly improving the quality of this book.

In addition, Gu Hyun's careful review of the text and scope of this book, correcting formulas as well as English grammar, allowed me to devote my energies to the mathematical concepts and overarching whole of the project. Finally, I would like to extend my special gratitude to my publisher, Hanti Media. I hope that my textbook can help students develop a better understanding and appreciation of calculus and the benefits it will have in their professional lives.

2022. 03. 17

Authors

CONTENTS

QUESTION

CHAPTER 03 Derivatives

CHAPTER 04 Application of Derivatives

CHAPTER 05 Integration

CONTENTS

ANSWERS AND EXPLANATIONS

CONTENTS

CHAPTER 08 Sequence and series

CHAPTER 09 Parametric Equations and Polar Coordinates

CHAPTER 10 Vectors and the Geometry of Space

CALCULUS

CHAPTER 01

Pre-Calculus

1.1 Relations and Functions

A relation is a set of ordered pairs (x,y). The domain of a relation is the set of all x-values from the ordered pairs, and the range of a relation is the set of all y-values from the ordered pairs. A mapping diagram shows how each element of the domain is paired with each element of the range.

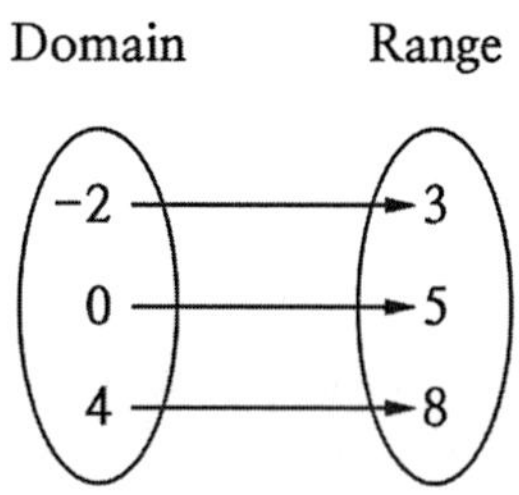

{(−2,3),(0,5),(4,8)}

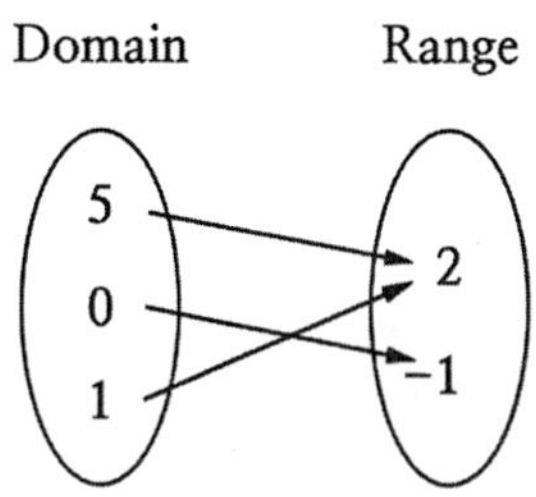

{(5,2),(0,−1),(1,2)}

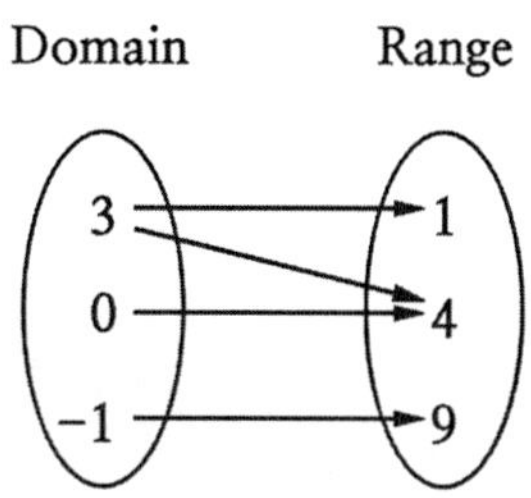

{(3,1),(3,4),(0,4),(−1,9)}

Figure 1.1

A function is a special type of relation in which each element of the domain is paired with exactly one element of the range. The first two relations above are functions, but the third relation is not a function because the 3 in the domain is paired with both 1 and 4 in the range.

A relation is a function if and only if no vertical line intersects its graph more than once.

Example 1

State the domain and range of each relation. Is the relation a function?

a.

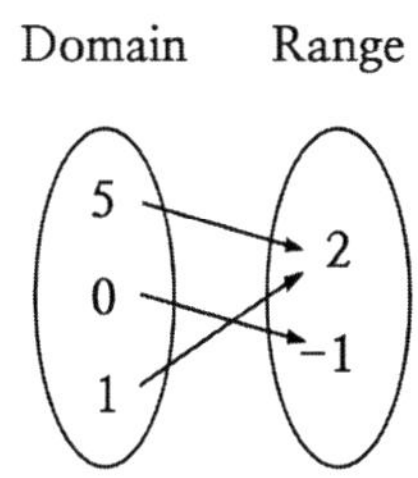

b.

x	y
-3	2
1	3
5	3

c.

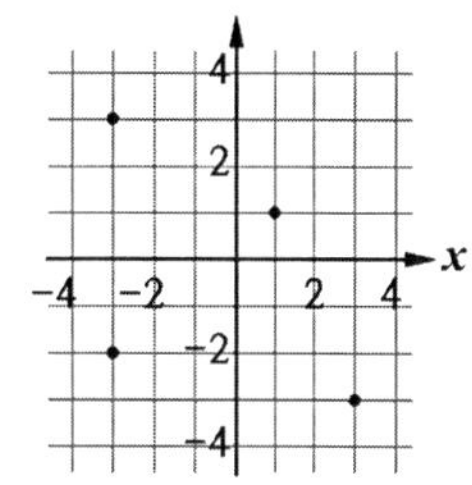

Figure 1.2

Solution

a. The domain is {5,0,1}, The range is {2,−1}, This relation is a function.

b. The domain is {−3,1,3}, The range is {2,3}, This relation is a function.

c. The domain is {−3,1,3} The range is {3,1,−2,−3}, This relation is not a function, since the -3 in the domain is paired with both 3 and -2 in the range.

1.2 Definition of Slope

The slope m of the line passing through points (x_1, y_1) and (x_2, y_2) is given by

$m = \dfrac{y_2 - y_1}{x_2 - x_1}$, where $x_1 \neq x_2$.

Example 2

If a line passing through the points $(t, 3)$ and (1,8) has a slope of $\dfrac{5}{3}$, what is the value of t?

Solution

$$m = \frac{8-3}{1-t} = \frac{5}{3} \Rightarrow \frac{5}{1-t} = \frac{5}{3}$$

$$\Rightarrow 1 - t = 3$$

$$\Rightarrow t = -2$$

1.3 Linear Functions

A function f is a linear function if it can be written in the form $f(x) = mx + b$, where m and b are constants.

The Graph of a linear function is a nonvertical straight line with slope m and y-intercept b.

Example 3

The function f is linear. If $f(0) = 3$ and $f(-2) = 7$, what is the value of $f(3)$?

Solution

$f(x) = mx + b$
$f(0) = m(0) + b = 3 \Rightarrow b = 3$
$\Rightarrow f(x) = mx + 3$
$f(-2) = m(-2) + 3 = 7$
$\Rightarrow -2m = 4 \Rightarrow m = -2$
$f(x) = -2x + 3$
$f(3) = -2(3) + 3 = -3$

1.4 Equations of Lines

Slope-Intercept Form: $y = mx + b$, where m is the slope and b is the y-intercept

Point-Slope Form: $y - y_1 = m(x - x_1)$, where m is the slope and (x_1, y_1) is a point on the line.

Standard Form: $Ax + By = C$, where A, B, and C are integers.

Vertical Line: $x = a$, where a is the x-intercept. the slope is undefined.

Horizontal Line: $y = b$, where b is the y-intercept. The slope is zero.

Example 4

Find an equation of a line if the x-intercept is 2 and the y-intercept is 5.

Solution

If the x-intercept is 2, the line passes through $(2,0)$.
If the y-intercept is 5, the line passes through $(0,5)$.

$$m=\frac{5-0}{0-2}=-\frac{5}{2}$$

$$y=-\frac{5}{2}x+5$$

1.5 Parallel and Perpendicular Lines

Two different lines with equations $y=m_1x+b_1$ and $y=m_2x+b_2$ are parallel if $m_1=m_2$ (the slopes are equal).
They are perpendicular if $m_1 \cdot m_2=-1$ (the slopes are negative reciprocals).

Example 5

a. Find an equation of the line which passes through the point $(3,-1)$ and is parallel to $2x-y=10$

b. Find an equation of the line which passes through the point $(2,5)$ and is perpendicular to $x-2y=-6$

Solution

a. First solve the equation of the given line for y.

$y=2x-10$

Then the slope of the line parallel to this line is 2.

$y=2x+b$

$-1=2(3)+b$

$b=-7$

$y=2x-7$

b. First solve the equation of the given line for y.

$y=\frac{1}{2}x+3$

Then the slope of the line perpendicular to this line is -2.

$y=-2x+b$

$5=-2(2)+b$

$b=9$

$y=-2x+9$

1.6 Quadratic Functions

A quadratic function is function that can be written in the form $f(x)=ax^2+bx+c$ $(a\neq 0)$.

The graph of a quadratic function is a parabola .

The line of symmetry for a quadratic function is $x=-\frac{b}{2a}$, and the vertex is $(-\frac{b}{2a},f(-\frac{b}{2a}))$.

If $a<0$, f has a maximum value at the vertex, that is when $x=-\frac{b}{2a}$.

If $a>0$, f has a minimum value at the vertex, that is when $x=-\frac{b}{2a}$.

Solutions, Roots, and Zeros of Functions

The solution of a function is the value(s) of x for which $f(x)=0$. Solutions of functions are also called roots or zeros On a graph, the solution of the function is the x -intercept(s).

Discriminant and Roots of a Quadratic Function

The solutions of the quadratic equation $ax^2+bx+c=0$ are given by the quadratic formula:

$x=\frac{-b\pm\sqrt{b^2-4ac}}{2a}$. The expression b^2-4ac is called the discriminant.

1. If the discriminant $b^2 - 4ac > 0$, then there are two real roots, and the graph crosses the x-axis twice.

2. If the discriminant $b^2 - 4ac = 0$, then there is one real root, and the graph is tangent to the x-axis.

3. If $b^2 - 4ac > 0$, then there are no real roots, and the graph does not cross the x-axis.

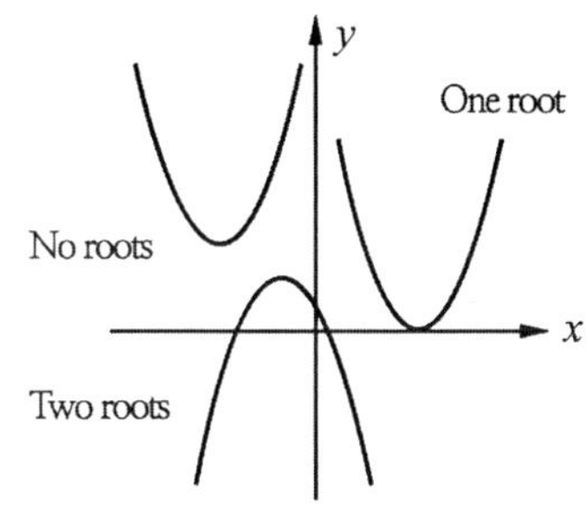

Figure 1.3

Example 6

a. Find the line of symmetry, minimum value, and the vertex of the graph $f(x) = 2x^2 - 4x + 5$.

b. If k is a positive integer and $x^2 + 4x + k = 0$ has two real roots, what is one ossible value of k?

Solution

a. The line of symmetry is $x = -\frac{b}{2a} = -\frac{-4}{2(2)} = \frac{4}{4} = 1$

The minimum occurs when $x = 1$.

$f(-\frac{b}{2a}) = f(1) = 2(1)^2 - 4(1) + 5 = 3$

The minimum value of f is 3.

The vertex is (1,3).

b. If a quadratic function has two real roots, $b^2 - 4ac > 0$.

$4^2 - 4(1)k > 0 \Rightarrow 16 - 4k > 0 \Rightarrow 16 > 4k \Rightarrow 4 > k$

Since k is a positive integer, the possible values of k are 1, 2 and 3.

1.7 Laws of Exponents

Laws		Examples
Products of Powers	$a^m \cdot a^n = a^{m+n}$	$2^5 \cdot 2^3 = 2^{5+3} = 2^8$
Quotient of Powers	$\frac{a^m}{a^n} = a^{m-n}$	$\frac{2^7}{2^4} = 2^{7-4} = 2^3$
Power of a Power	$(a^m)^n = a^{mn}$	$(2^3)^4 = 2^{3\times4} = 2^{12}$
power of a Quotient	$(\frac{a}{b})^n = \frac{a^n}{b^n}$	$(2 \cdot 3)^4 = 2^4 \cdot 3^4$
Negative Exponent	$a^{-n} = \frac{1}{a^n} \quad (a \neq 0)$	$2^{-3} = \frac{1}{2^3} = \frac{1}{8}$
	$(\frac{a}{b})^{-n} = (\frac{b}{a})^n \ (ab \neq 0)$	$(\frac{2}{3})^{-3} = (\frac{3}{2})^3 = \frac{3^3}{2^3} = \frac{27}{8}$
Zero Exponent	$a^0 = 1 \ (a \neq 0)$	$3^0 = 1, \ (2ab^2)^0 = 1$

Example 7

Simplify each expression.

a. $(-2t^3)^2(s^2t)^{-3}$ b. $(\frac{x}{2y})^0(\frac{x^{-1}}{y^2})^2(x^2y)^{-1}$

Solution

a. $(-2t^3)^2(s^2t)^{-3} = (-2)^2(t^3)^2(s^2)^{-3}(t)^{-3} = 4 \cdot t^6 \cdot s^{-6} \cdot t^{-3} = 4 \cdot t^{6-3} \cdot s^{-6}$

$= 4 \cdot t^3 \cdot (\frac{1}{s^6}) = \frac{4t^3}{s^6}$

b. $(\frac{x}{2y})^0(\frac{x^{-1}}{y^2})^2(x^2y)^{-1} = 1 \cdot (\frac{x^{-2}}{y^4})(x^{-2}y^{-1}) = \frac{x^{-2}x^{-2}y^{-1}}{y^4} = x^{-4}y^{-1-4}$

$= x^{-4}y^{-5} = \frac{1}{x^4y^5}$

1.8 Special Products and Factoring Polynomials

Difference of Two Squares: $a^2 - b^2 = (a+b)(a-b)$

Sum of Two cubes: $a^3 + b^3 = (a+b)(a^2 - ab + b^2)$

Difference of Two Cubes: $a^3 - b^3 = (a-b)(a^2 + ab + b^2)$

Perfect Square Trinomials: $(a^2 + 2ab + b^2) = (a+b)^2$

$(a^2 - 2ab + b^2) = (a-b)^2$

Cubes of Binomials: $(a^3 + 3a^2b + 3ab^2 + b^3) = (a+b)^3$

$(a^3 - 3a^2b + 3ab^2 - b^3) = (a-b)^3$

Example 8

a. $6x^3 - 15x^2 - 9x$ b. $x^2 - 8x + 16$ c. $x^4 - 81$

d. $2x^3 - 16$ e. $3x^3 - 6x^2 + 5x - 10$

Solution

a) $6x^3 - 15x^2 - 9x = 3x(2x^2 - 5x - 3) = 3x(2x+1)(x-3)$

b) $x^2 - 8x + 16 = x^2 - 8x + 16 = (x)^2 - 2(x)(4) + 4^2 = (x-4)^2$

c) $x^4 - 81 = (x^2+9)(x^2-9) = (x^2+9)(x-3)(x+3)$

d) $2x^3 - 16 = 2(x^3 - 8) = 2(x-2)(x^2+2x+4)$

e) $3x^3 - 6x^2 + 5x - 10 = (3x^3 - 6x^2) + (5x - 10) = 3x^2(x-2) + 5(x-2)$

$= (3x^2+5)(x-2)$

1.9 Composition of Functions

Given the two functions f and g, the composite function, denoted by $f \circ g$ is defined as $(f \circ g)(x) = f(g(x))$.
In order for a value of x to be in the domain of $f \circ g$, two conditions must be satisfied:

1) x must be in the domain of g, and
2) $g(x)$ must be in the domain of f.

Example 9

If $f(x) = x^2 - 1$ and $g(x) = \sqrt{x+2}$, find the following:

a. $f(g(3))$ b. $g(f(x))$

Solution

a. $g(3) = \sqrt{3+2} = \sqrt{5}$
   $f(g(3)) = f(\sqrt{5}) = (\sqrt{5})^2 - 1 = 4$

b. $g(f(x)) = g(x^2-1) = \sqrt{(x^2-1)+2} = \sqrt{x^2+1}$

1.10 Inverse Functions

The functions f and g are inverse funtions if
$f(g(x)) = x$ for all x-values in the domain of g and
$g(f(x)) = x$ for all x-values in the domain of f.

The inverse of a function f is usually denoted by f^{-1}, which is read "f inverse."

Thus for the definition of inverse functions we can write
$f^{-1}(f(x)) = x$ and $f(f^{-1}(x)) = x$.

The graph of a function f and the graph of its inverse f^{-1} are symmetric about the line $y = x$.

The graph of f contains the point (a,b)
if and only if the graph of f^{-1} contains the point (b,a).

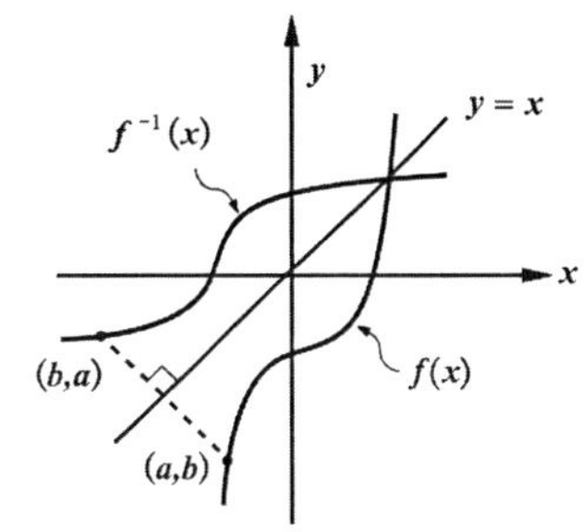

Figure 1.4

Finding an Inverse Function

1. Replace $f(x)$ with y.
2. Interchange x and y.
3. Solve for y.

Example 10

Find the inverse of each function.

a. $(-3,2)$, $(1,5)$, $(4,-7)$

b. $f(x)=-2x+3$

Solution

a. $(2,-3),(5,1),(-7,4)$

b. $f(x)=-2x+3$

$y=-2x+3$

$x=-2y+3$

$y=\dfrac{-x+3}{2}$

The inverse of $f(x)=-2x+3$ is $f^{-1}(x)=\dfrac{-x+3}{2}$

1.11 Even and Odd Functions

The function $f(x)$ is even if $f(-x)=f(x)$.
The graph of an even function is symmetric about they y-axis.

The function $f(x)$ is odd if $f(-x)=-f(x)$.
The graph of an odd function is symmetric about the origin.

Example 11

Determine Whether each function is even, odd, or neither

a. $f(x) = x^3 - 2x$ b. $f(x) = x^4 - 5x^2$ c. $f(x) = x^2 + 4x$

Solution

a. $f(-x) = (-x)^3 - 2(-x) = -x^3 + 2x = -(x^3 - 2x) = -f(x)$
The function is odd because $f(-x) = -f(x)$.
The graph is symmetric about the origin.

b. $f(-x) = (-x)^4 - 5(-x)^2 = x^4 - 5x^2 = f(x)$.
The function is odd because $f(-x) = -f(x)$.
The graph is symmetric about the y-axis.

c. $f(-x) = (-x)^2 + 4(-x) = x^2 - 4x \neq f(x)$
The function is neither even nor odd because $f(-x) \neq f(x)$ and
$f^{-1}(x) = \dfrac{-x+3}{2}$

1.12 Special Functions

The greatest integer function is defined as $f(x) = [x]$ =the greatest integer less than or equal to x.

When the domain of a function is divided into several parts and a different function rule is applied to each part, the function is called a piecewise-defined function.

Example 12

a. If $f(x) = [x]$, find $f(-1.5)$, $f(0.5)$,and $f(2.9)$.

b. Find a complete graph of $f(x) = [x]$.

c. Find a complete graph of the piecewise function $f(x) = \begin{cases} 3 - x & \text{if} x \leq -1 \\ x^2 + 1 & \text{if} x > -1 \end{cases}$

Solution

a. $f(-1.5) = [-1.5] = -2$

$f(0.5) = [0.5] = 0$
$f(2.9) = [2.9] = 2$

b. To graph the greatest integer function $f(x) = [x]$, find several ordered pairs for the values $-1 \le x \le 1$.

x	-1	-0.5	0	0.5	1	1.5	2
$[x]$	-1	-1	0	0	1	1	2

For $-1 \le x < 0$, $[x] = -1$.
For $0 \le x < 1$, $[x] = 0$.
For $1 \le x < 2$, $[x] = 1$ and so on.

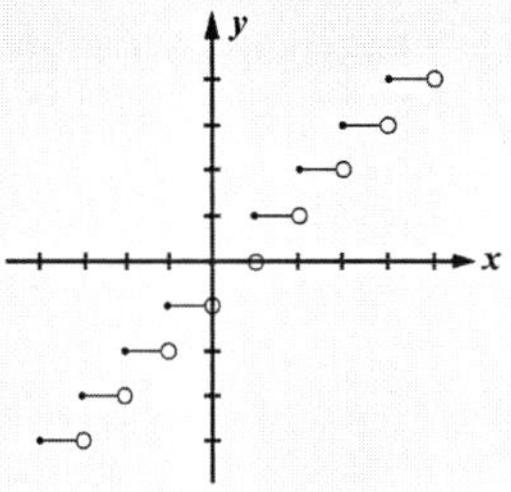

Figure 1.5

c.

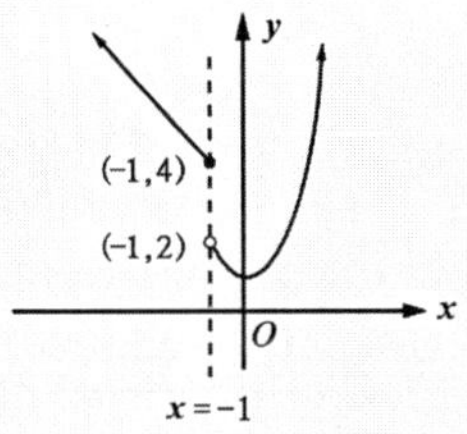

Figure 1.6

1.13 Distance Formula

The distance between two points with coordinates (x_1, y_1) and (x_2, y_2) is given by $d = \sqrt{(x_2 - x_1)^2 + (y_2 - y_1)^2}$.

Example 13

Find the value of h if the distance between the points $(3, 10)$ and $(h, 4)$ is 10.

Solution

$$d = \sqrt{(h-3)^2 + (4-10)^2} = 10$$
$$(h-3)^2 + 36 = 100$$
$$(h-3)^2 = 64$$
$$h - 3 = \pm 8$$
$$h = 11 \text{ or } h = -5$$

1.14 Midpoint Formula

The midpoint of a line segment with endpoints (x_1,y_1) and (x_2,y_2) is the point with coordinates $M(\frac{x_1+x_2}{2},\frac{y_1+y_2}{2})$.

Example 14

If a line segment has its midpoint at the coordinates $(5,12)$ and one endpoint at the coordinates $(11,4)$, what are the coordinates of the other endpoint?

Solution

Let (x,y) be the coordinates of the other endpoint.

Then $(5,12)=(\frac{11+x}{2},\frac{4+y}{2})$. $\Rightarrow 5=\frac{11+x}{2}$ and $12=\frac{4+y}{2}$ $\Rightarrow x=-1$ and $y=20$

The coordinates of the other endpoint are $(-1,20)$.

1.15 Circles

The equation of a circle with center (h,k) and radius r units if $(x-h)^2+(y-k)^2=r^2$

Example 15

Find the center and radius of a circle with equation $x^2+y^2+10x-6y-47=0$.

Solution

$(x^2+10x+\square)+(y^2-6y+\square)=47$

$(x^2+10x+25)+(y^2-6y+9)=47+25+9$

$(x+5)^2+(y-3)^2=9^2$

The center of the circle is $(-5,3)$ and the radius is 9.

1.16 Parabolas

A parabola whose equation is $y = a(x-h)^2 + k$ opens upward if $a > 0$ and downward if $a < 0$. it has a vertex $V(h,k)$ and an axis of symmetry $x = h$.

A parabola whose equation is $y = a(x-k)^2 + h$ opens to the right if $a > 0$ and to the left if $a < 0$. It has a vertex $V(h,k)$ and an axis of symmetry $y = k$.

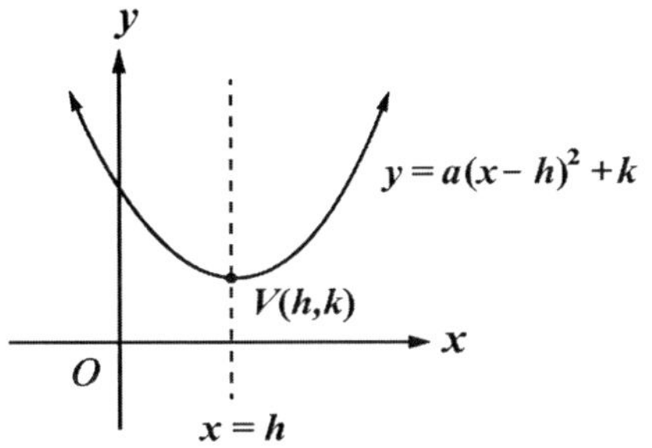

Figure 1.7

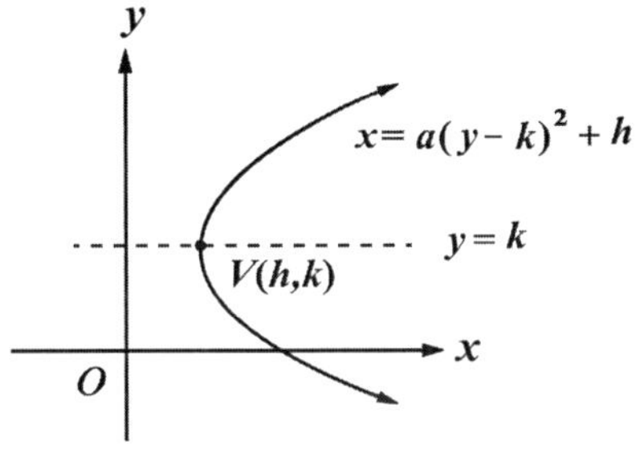

Figure 1.8

Example 16

Find the vertex and axis of symmetry of the parabola $y^2 - 4x - 10y + 5 = 0$

Solution

$4x - 5 = y^2 - 10y$

$4x - 5 + 25 = y^2 - 10y + 25$

$4(x+5) = (y-5)^2$

$x + 5 = \frac{1}{4}(y-5)^2$ or $x = \frac{1}{4}(y-5)^2 - 5$

The vertex is $V(-5,5)$ and the axis of symmetry is $y = 5$.

1.17 Ellipses

The standard form for the equation of an ellipse with center(h,k) and a horizontal major axis is

$$\frac{(x-h)^2}{a^2}+\frac{(y-k)^2}{b^2}=1, \text{ where } a^2 > b^2.$$

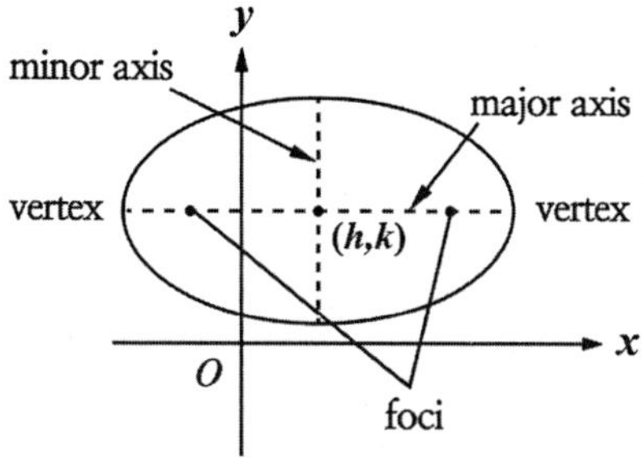

Figure 1.9

The standard form for the equation of an ellipse with center(h,k) and a verical major axis is

$$\frac{(x-h)^2}{b^2}+\frac{(y-k)^2}{a^2}=1, \text{ where } a^2 > b^2.$$

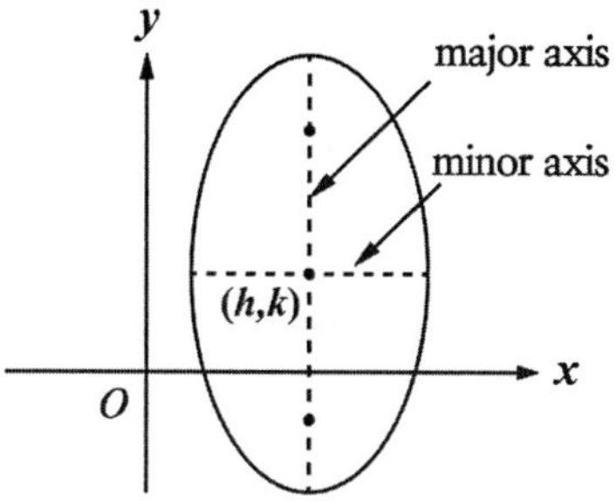

Figure 1.10

The length of major axis of an ellipse is 2a. The length of the minor axis of ellipse is 2b.
The area of an ellipse=πab.

If the center is $(0,0)$ the equation of an ellipse with a horizontal major axis is

$$\frac{x^2}{a^2}+\frac{y^2}{b^2}=1, \text{ where } a^2 > b^2.$$

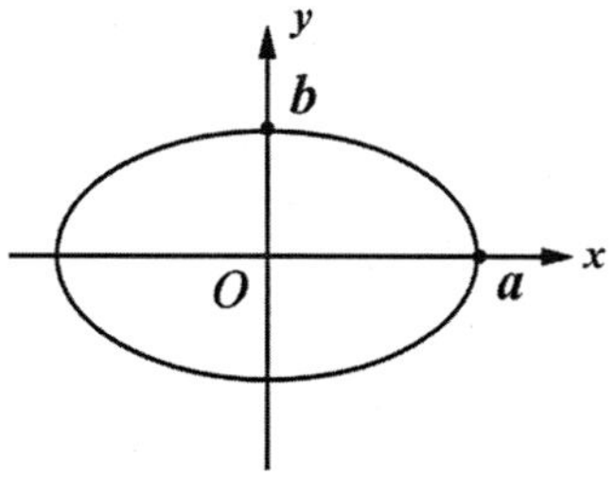

Figure 1.11

If the center is $(0,0)$ the equation of an ellipse with a vertical major axis is

$$\frac{x^2}{b^2}+\frac{y^2}{a^2}=1, \text{ where } a^2 > b^2.$$

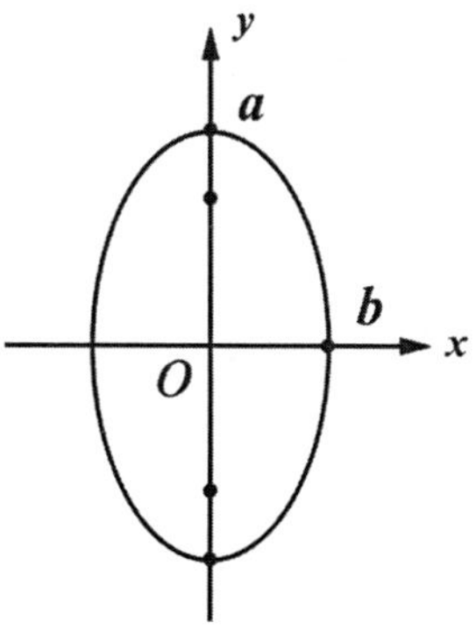

Figure 1.12

Example 17

What is the length of the major axis of the ellipse whose equation is $12x^2+4y^2=48$?

Solution

$$\frac{12x^2}{48}+\frac{4y^2}{48}=\frac{48}{48}$$

$$\frac{x^2}{4}+\frac{y^2}{12}=1$$

So, $a=\sqrt{12}$ and $b=2$

The length of the major axis is $2a=2\cdot\sqrt{12}=2\cdot2\sqrt{3}=4\sqrt{3}$.

1.18 Hyperbolas

For a hyperbola with center (h,k) that opens to the left and right, the standard form of the equation is $\frac{(x-h)^2}{a^2}-\frac{(y-k)^2}{b^2}=1$, where a is the distance from the center to a vertex.

If the center is $(0,0)$ the standard form of the equation is $\frac{x^2}{a^2}-\frac{y^2}{b^2}=1$.

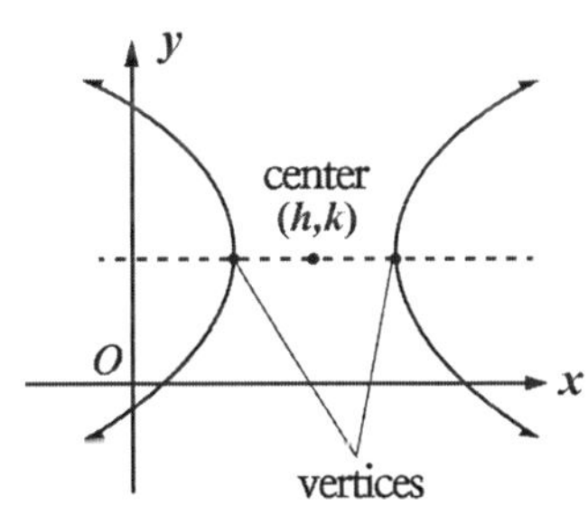

Figure 1.13

For a hyperbola with center (h,k) that opens upward and downward, the standard form of the equation is $\frac{(y-k)^2}{a^2}-\frac{(x-h)^2}{b^2}=1$, where a is the distance from the center to a vertex.

If the center is $(0,0)$ the standard form of the equation is $\frac{y^2}{a^2}-\frac{x^2}{b^2}=1$.

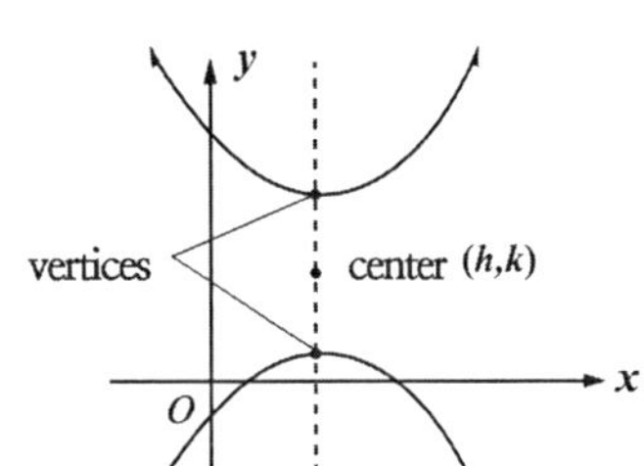

Figure 1.14

Example 18

Find the coordinates of the center of the hyperbola whose equation is
$x^2 - 4y^2 + 10x + 24y - 27 = 0$.

Solution

$(x^2+10x+25)-4(y^2-6y+9)$	To complete the square rearrange terms
$=27+25-36$	and add $(\frac{10}{2})^2$ and $-4 \cdot (\frac{-6}{2})^2$ to both sides.
$(x+5)^2-4(y-3)^2=16$	Complete the square.
$\frac{(x+5)^2}{16}-\frac{(y-3)^2}{4}=1$	Standard form for the equation of a hyperbola

Therefore the center of the hyperbola is (-5, 3).

1.19 Definition of Logarithm

If b and x are positive and $b \neq 1$, $\log_b x = y$ if and only if $x = b^y$. An equation of the form $y = \log_b x$ is called a logarithmic function. The domain of a logarithmic function is $(0, \infty)$ and the range is $(-\infty, \infty)$.

Example 19

a. Simplify $\log_3 9\sqrt{3}$.

b. Solve $\log_5 x = -2$.

c. Evaluate $\log_7 7^3$.

d. Evaluate $4^{\log_4 9}$.

Solution

a. Let $x = \log_3 9\sqrt{3}$

$3^x = 9\sqrt{3}$

$3^x = 3^2 \cdot 3^{\frac{1}{2}}$

$x = \frac{5}{2}$, $\log_3 9\sqrt{3} = \frac{5}{2}$

b. $\log_5 x = -2$

$x = 5^{-2}$

$x = \frac{1}{25}$

c. $\log_7 7^3 = 3$

d. $4^{\log_4 9} = 9$

1.20 Properties of Logarithms

For all positive numbers M, N, and b, where $b \neq 1$:

1. $\log_b MN = \log_b M + \log_b N$
2. $\log_b \frac{M}{N} = \log_b M - \log_b N$
3. $\log_b M^p = p \cdot \log_b M$
4. $b^{\log_b N} = N$
5. $\log_b b = 1$
6. $\log_b 1 = 0$
7. $\log_b x = \frac{\log_a x}{\log_a b}$

Example 20

Solve $\log_2(x+2) + \log_2(x-2) = 5$.

Solution

$$\log_2(x+2) + \log_2(x-2) = 5$$
$$\log_2(x+2) \cdot (x-2) = 5$$
$$\log_2(x^2-4) = 5$$
$$x^2 - 4 = 2^5$$
$$x^2 = 36$$
$$x = 6$$

1.21 Common and Natural Logarithms

Common Logarithms

The function $y = \log_{10} x$ is called the common logarithmic function. It is usually written without the subscript 10, so $\log_{10} x$ is written as $\log x$.

Natural Logarithms

The function $y = \log_e x$ is called the natural logarithmic function. It is usually written without the subscript e, and is written as $\ln x$. The number e is an irrational number whose value is approximately 2.718.

Example 21

a. Find the value of $\log_7 52$.

b. Solve $5^x=38$.

Solution

a. $\log_7 52 = \dfrac{\log 52}{\log 7} = \dfrac{1.716}{845} = 2.03$

b. $5^x = 38$

$\ln 5^x = 38$

$x \ln 5 = \ln 38$

$x = \dfrac{\ln 38}{\ln 5}$

$x = 2.26$

1.22 Trigonometric Functions of Acute Angles

The trigonometric functions of any angle $0° < \theta < 90°$ are defined as follows:

$$\sin\theta = \frac{\textit{opposite side}}{\textit{hypotenuse}} = \frac{x}{r} \qquad \csc\theta = \frac{1}{\sin\theta} = \frac{r}{y}(y \neq 0)$$

$$\cos\theta = \frac{\textit{adjacent side}}{\textit{hypotenuse}} = \frac{x}{r} \qquad \sec\theta = \frac{1}{\cos\theta} = \frac{r}{x}(x \neq 0)$$

$$\tan\theta = \frac{\textit{opposite side}}{\textit{adjacent side}} = \frac{y}{x} \qquad \cot\theta = \frac{1}{\tan\theta} = \frac{x}{y}(y \neq 0)$$

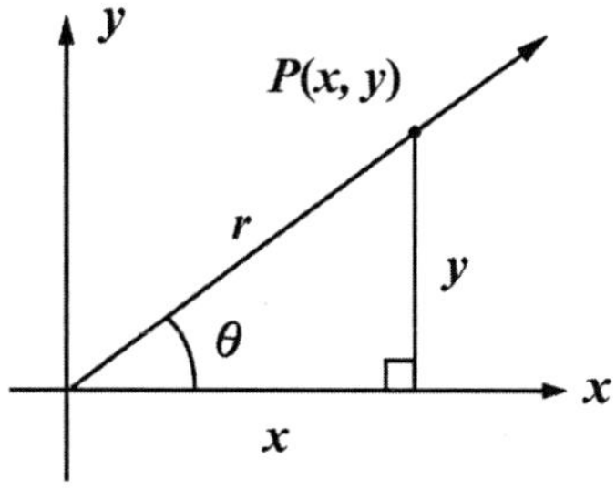

Figure 1.15

Example 22

In the figure at the right, what is the value of x?

Solution

$\sin\theta = \dfrac{opposite}{hypotenuse}$

$\sin 38^\circ = \dfrac{15}{x} \Rightarrow \sin 38^\circ \cdot x = 15$

$x = \dfrac{15}{\sin 38^\circ} = \dfrac{15}{0.616} = 24.36$

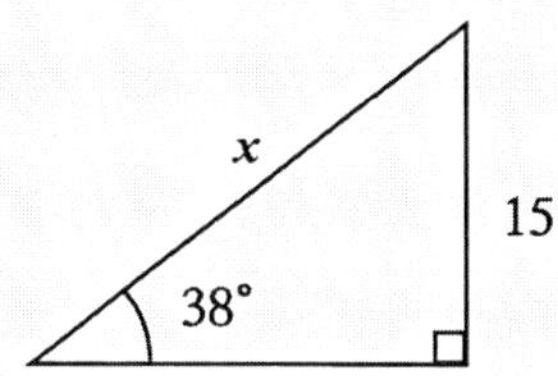

Figure 1.16

1.23 Trigonometric Functions and the Unit Circle

The unit circle is a circle with center (0,0) and radius 1, whose equation is $x^2+y^2=1$. Let $P(x,y)$ be a point on the unit circle on the terminal side of θ. Then the six trigonometric functions for any angle θ are defined as follows:

$\sin\theta = \dfrac{y}{r} = \dfrac{y}{1} = y$ $\qquad \csc\theta = \dfrac{1}{y}\,(y \neq 0)$

$\cos\theta = \dfrac{x}{r} = \dfrac{x}{1} = x$ $\qquad \sec\theta = \dfrac{1}{x}\,(x \neq 0)$

$\tan\theta = \dfrac{y}{x}\,(x \neq 0)$ $\qquad \cot\theta = \dfrac{x}{y}\,(y \neq 0)$

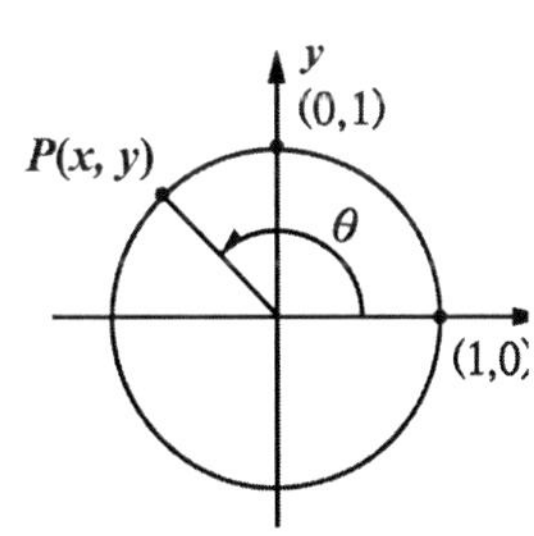

Figure 1.17

$\cos\theta$ is the x coordinate of the point $P(x,y)$ and $\sin\theta$ is the y coordinate of the point $P(x,y)$, each associated with angle θ. Therefore $\cos\theta$ is positive in quadrants I and IV, and $\sin\theta$ is positive in quadrants I and II.

The reference angle associated with θ is the acute angle formed by the x-axis and the terminal side of the angle θ. A reference angle can be used to evaluate trigonometric functions for angles greater than $90°$.

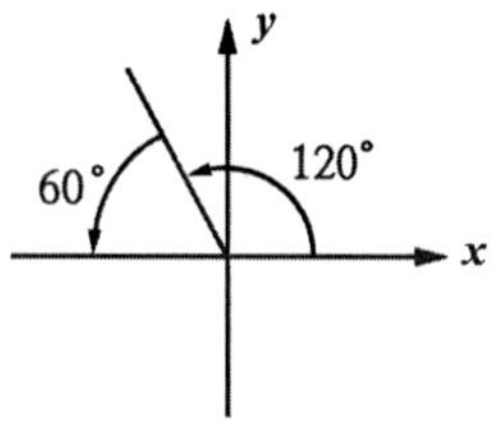

The reference angle for $120°$ is $60°$

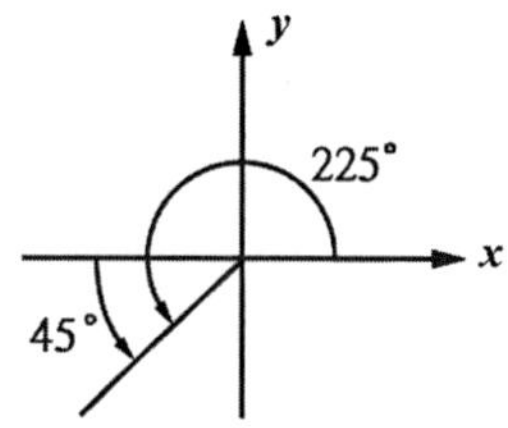

The reference angle for $225°$ is $45°$

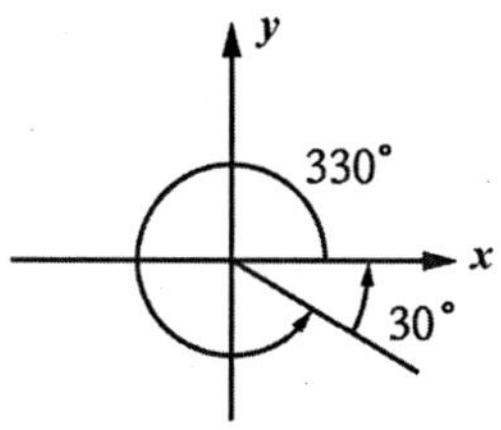

The reference angle for $330°$ is $30°$

Figure 1.18

Example 23

a. Find the reference angle associated with $-120°$.

b. Find the reference angle associated with $310°$.

Solution

a. The reference angle associated with $-120°$ is $60°$, since the terminal side of $\theta = -120°$ makes a $60°$ angle with the x-axis.

b. reference angle associated with $310°$ is $50°$, since the terminal side of $\theta = 310°$ makes a $50°$ angle with the x-axis.

1.24 Familiar Angles

Angles in standard position whose measures are multiples of $\frac{\pi}{6}$ radians or multiples of $\frac{\pi}{4}$ radians are called familiar angles. The sine and cosine values for angles with radian measures $0, \frac{\pi}{6}, \frac{\pi}{4}, \frac{\pi}{3}$, and $\frac{\pi}{2}$ are summarized in the following table.

θ	0	$\frac{\pi}{6}$ or 30°	$\frac{\pi}{4}$ or 45°	$\frac{\pi}{3}$ or 60°	$\frac{\pi}{2}$ or 90°
$\sin\theta$	$\frac{\sqrt{0}}{2}=0$	$\frac{\sqrt{1}}{2}=\frac{1}{2}$	$\frac{\sqrt{2}}{2}$	$\frac{\sqrt{3}}{2}$	$\frac{\sqrt{4}}{2}=1$
$\cos\theta$	$\frac{\sqrt{4}}{2}=1$	$\frac{\sqrt{3}}{2}$	$\frac{\sqrt{2}}{2}$	$\frac{\sqrt{1}}{2}=\frac{1}{2}$	$\frac{\sqrt{0}}{2}=0$

Example 24

Angles with a reference angle of 30° are 150°, 210°, and 330°

$(30^\circ=\frac{\pi}{6},\ 150^\circ=\frac{5\pi}{6},\ 210^\circ=\frac{7\pi}{6},\ 330^\circ=\frac{11\pi}{6})$

Solution

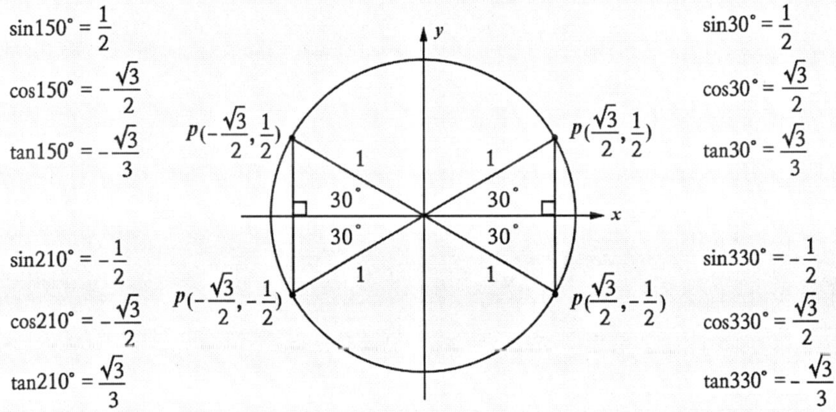

Figure 1.19

Angles with a reference angle of 60° are 120°, 240°, and 300°.

$(60^\circ=\frac{\pi}{3},\ 120^\circ=\frac{2\pi}{3},\ 240^\circ=\frac{4\pi}{3},\ 300^\circ=\frac{5\pi}{3})$

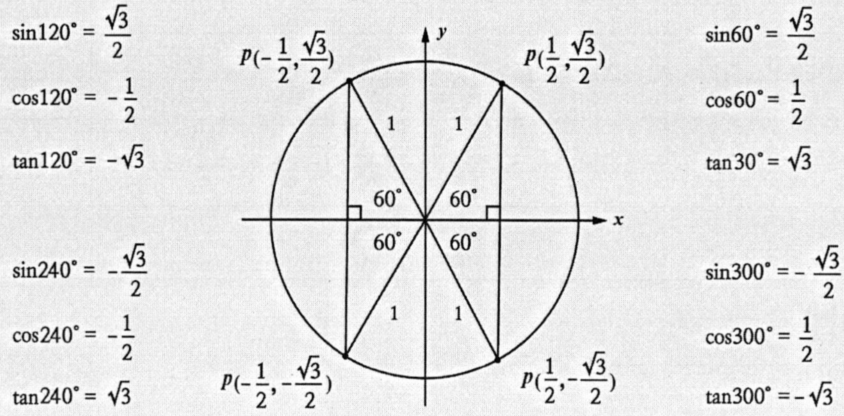

Figure 1.20

Angles with a reference angle of 45° are 135°, 225°, and 315°.

$(45° = \frac{\pi}{4},\ 135° = \frac{3\pi}{4},\ 225° = \frac{5\pi}{4},\ 315° = \frac{7\pi}{4})$

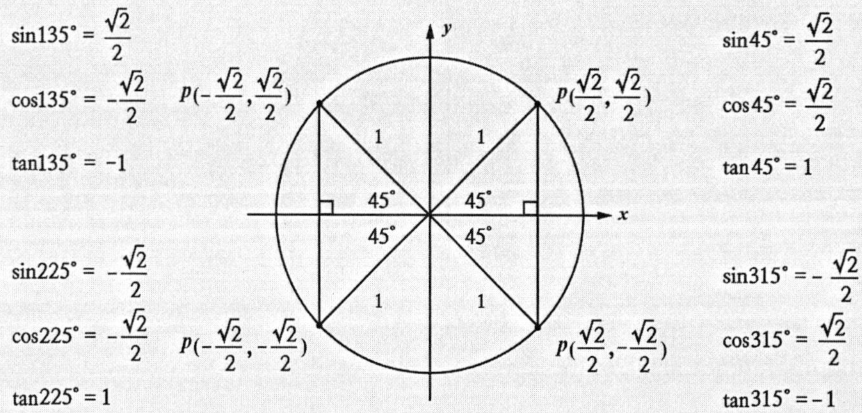

Figure 1.21

For the angles 0°, 90°, 180°, and 270°, $\cos\theta = x$ and $\sin\theta = y$

$\Rightarrow P(x,y) = P(\cos\theta, \sin\theta)$ $(90° = \frac{\pi}{2},\ 180° = \pi,\ 270° = \frac{3\pi}{2},\ 360° = 2\pi)$

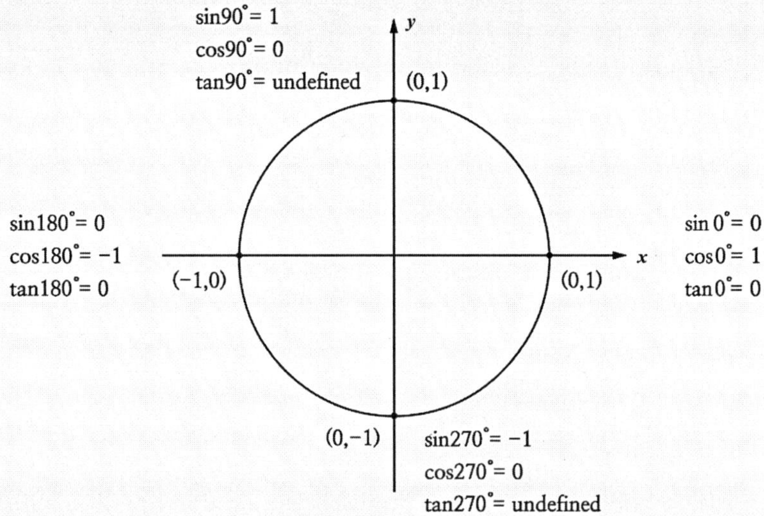

sin90°= 1
cos90°= 0
tan90°= undefined
y
(0,1)
sin180°= 0
cos180°= −1
tan180°= 0
(−1,0)
x
(0,1)
sin 0°= 0
cos0°= 1
tan0°= 0
(0,−1)
sin270°= −1
cos270°= 0
tan270°= undefined

Figure 1.22

1.25 Basic Trigonometric Identities

Identities from the Definitions of Trigonometric Functions

$\tan\theta = \dfrac{\sin\theta}{\cos\theta}$ $\sec\theta = \dfrac{1}{\cos\theta}$ $\csc\theta = \dfrac{1}{\sin\theta}$ $\cot\theta = \dfrac{1}{\tan\theta}$

Pythagorean Identities

$\sin^2\theta + \cos^2\theta = 1$ $\tan^2\theta + 1 = \sec^2\theta$ $\cot^2\theta + 1 = \csc^2\theta$

Opposite Angle Formulas

$\sin(-\theta) = -\sin\theta$ $\cos(-\theta) = \cos\theta$ $\tan(-\theta) = -\tan\theta$

Cofunction Identities

$\sin\theta = \cos(90^\circ - \theta)$ $\cos\theta = \sin(90^\circ - \theta)$ $\tan\theta = \cot(90^\circ - \theta)$

Example 25

$(\sin\theta + \cos\theta)^2 + (\sin\theta - \cos\theta)^2 =$

Solution

$(\sin\theta + \cos\theta)^2 + (\sin\theta - \cos\theta)^2$
$= (\sin^2\theta + 2\sin\theta\cos\theta + \cos^2\theta) + (\sin^2\theta - 2\sin\theta\cos\theta + \cos^2\theta)$
$= 2\sin^2\theta + 2\cos^2\theta = 2(\sin^2\theta + \cos^2\theta)$
$= 2$

1.26 Graphs of $\sin x$, $\cos x$ and $\tan x$

For the functions $y = a\sin bx$ and $y = a\cos bx$, the period is $\frac{2\pi}{|b|}$ and the amplitude is $|a|$.

For the graph of $y = \sin x$, the period $= 2\pi$ and the amplitude $= 1$.

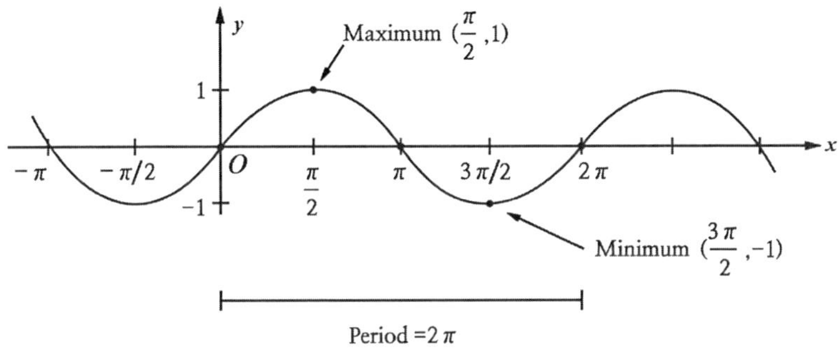

Figure 1.23

For the graph of $y = \cos x$, the period $= 2\pi$ and the amplitude $= 1$.

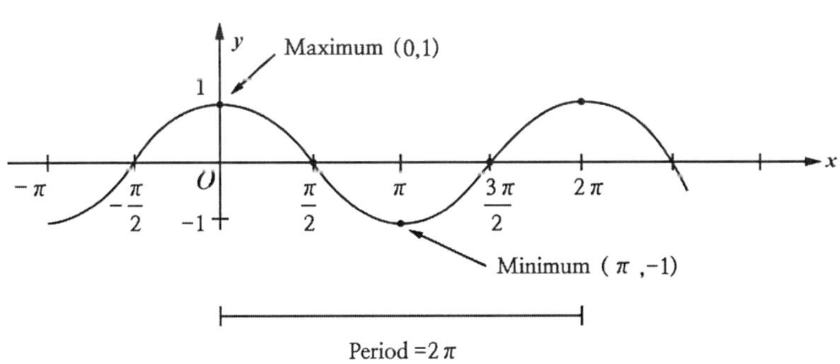

Figure 1.24

For the functions $y = \tan bx$ the period is $\frac{\pi}{|b|}$. Since the tangent function has no maximum or minimum value, it has no amplitude.

For the graph of $y = \tan x$, the period $= \pi$

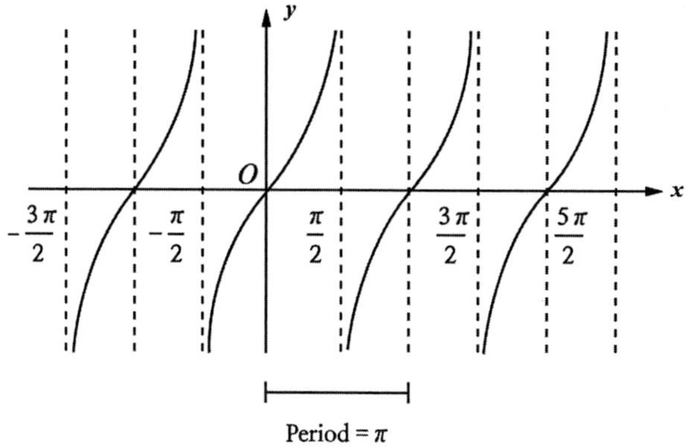

Figure 1.25

Example 26

Graph $y = 2\cos x$.

Solution

The period of the function $y = 2\cos x$ is 2π and the amplitude is 2.

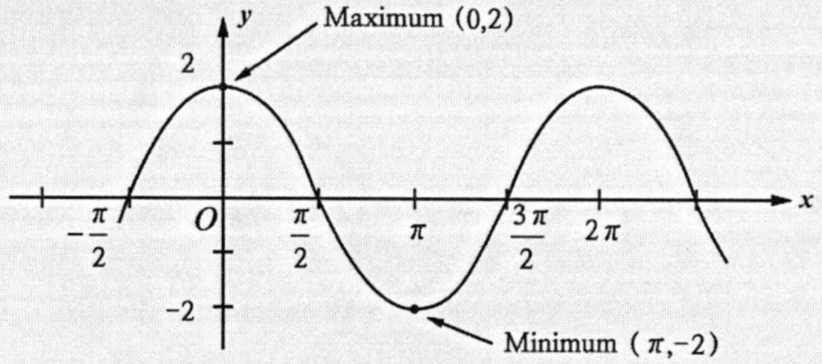

Figure 1.26

1.27 Law of Sines

In any triangle ABC, $\dfrac{\sin A}{a}=\dfrac{\sin B}{b}=\dfrac{\sin C}{c}$

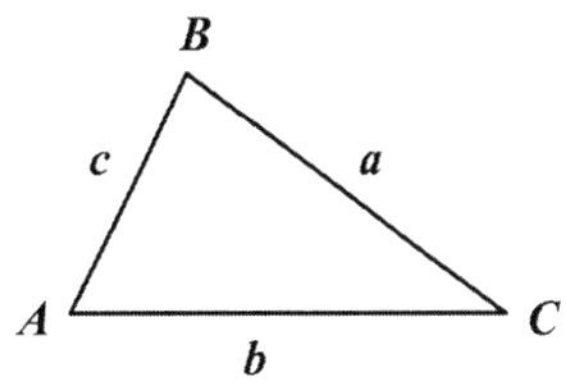

Figure 1.27

Example 27

What is the perimeter of $\triangle ABC$ shown below?

Solution

$m\angle A = m\angle C$ since $AB = BC$.

$m\angle A = m\angle C = \dfrac{180-84}{2} = 48$

$\dfrac{\sin 48^\circ}{14} = \dfrac{\sin 84^\circ}{AC} \Rightarrow AC = 18.74$

Perimeter of $\triangle ABC = 14+14+18.74 = 46.74$

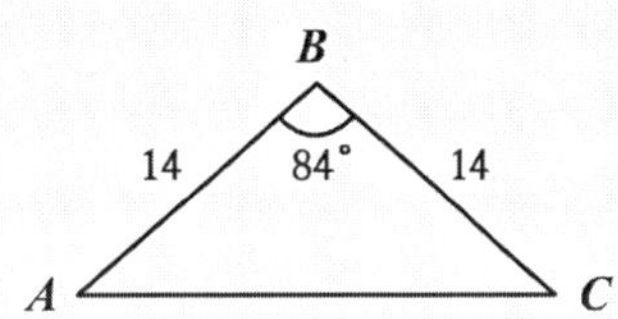

Figure 1.28

1.28 Law of Cosines

In any triangle ABC,

$c^2 = a^2 + b^2 - 2ab\cos C$

$b^2 = a^2 + c^2 - 2ac\cos B$

$a^2 = b^2 + c^2 - 2bc\cos A$

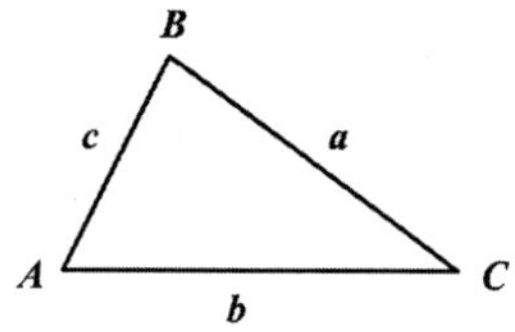

Figure 1.29

Example 28

What is the value of x in $\triangle ABC$ shown below?

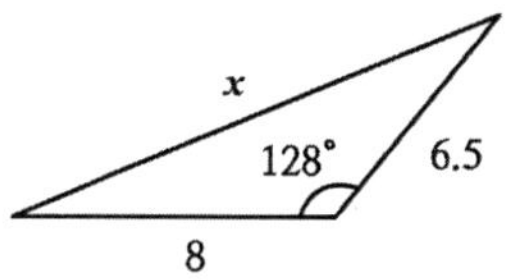

Figure 1.30

Solution

$x^2 = 8^2 + (6.5)^2 - 2(8)(6.5)\cos 128°$

$= 64 + 42.25 - (104)(-0.616) = 170.3$

$x = 13.05$

1.29 Trigonometric Identities

Sum and Difference Formulas

$\sin(A+B) = \sin A\cos B + \cos A\sin B$

$\cos(A+B) = \cos A\cos B - \sin A\sin B$

$\tan(A+B) = \dfrac{\tan A + \tan B}{1 - \tan A\tan B}$

$\sin(A-B) = \sin A\cos B - \cos A\sin B$

$\cos(A-B) = \cos A\cos B + \sin A\sin B$

$\tan(A-B) = \dfrac{\tan A - \tan B}{1 + \tan A\tan B}$

Double-Angle Formulas

$\sin 2\theta = 2\sin\theta\cos\theta$

$\cos 2\theta = \cos^2\theta - \sin^2\theta = 2\cos^2\theta - 1 = 1 - 2\sin^2\theta$

$\tan 2\theta = \dfrac{2\tan\theta}{1-\tan^2\theta}$

Example 29

a. Simplify $\cos(\theta + \frac{3\pi}{2})$.

b. If $0° \le \theta \le 90°$ and $\sin\theta = \frac{1}{3}$, then $\sin 2\theta =$

Solution

a. $\cos(\theta + \frac{3\pi}{2}) = \cos\theta\cos\frac{3\pi}{2} - \sin\theta\sin\frac{3\pi}{2} = \cos\theta(0) - \sin\theta(-1) = \sin\theta$

b. $\sin\theta = \frac{1}{3}$

By the Pythagorean theorem, the adjacent side $= \sqrt{3^2 - 1^2} = \sqrt{8} = 2\sqrt{2}$,

then $\cos\theta = \frac{2\sqrt{2}}{3}$.

$\sin 2\theta = 2\sin\theta\cos\theta = 2 \cdot \frac{1}{3} \cdot \frac{2\sqrt{2}}{3} = \frac{4\sqrt{2}}{9}$

1.30 Trigonometric Equations

A trigonometric equations is any statement involving the conditional equality of two trigonometric expressions.(Most trigonometric equations are true for some but not all values of the variable.) The values that satisfy the equation are called solutions of the equation.

Example 30

Solve each trigonometric equation over the interval $0 \leq x < 2\pi$.

a. $\sin x = \sqrt{3}\cos x$ b. $\cos 2x = -\sin x$

Solution

a. $\sin x = \sqrt{3}\cos x$

$\frac{\sin x}{\cos x} = \sqrt{3}$

$\tan x = \sqrt{3}$

$x = 60^\circ$ or 240°

b. $\cos 2x = -\sin x$

$1 - 2\sin^2 x = -\sin x$

$2\sin^2 x - \sin x - 1 = 0$

$(2\sin x + 1)(\sin x - 1) = 0$

$2\sin x + 1 = 0$ or $\sin x - 1 = 0$

$\sin x = -\frac{1}{2}$ or $\sin x = 1$

$x = 210^\circ$ or 330° or $x = 90^\circ$

1.31 Inverse Trigonometric Functions

The inverse sine function, denoted by $y=\sin^{-1}x$ or $y=\arcsin x$, is the function with a domain of $[-1,1]$ and a range of $[-\frac{\pi}{2},\frac{\pi}{2}]$ that satisfies the relation $\sin y=x$.

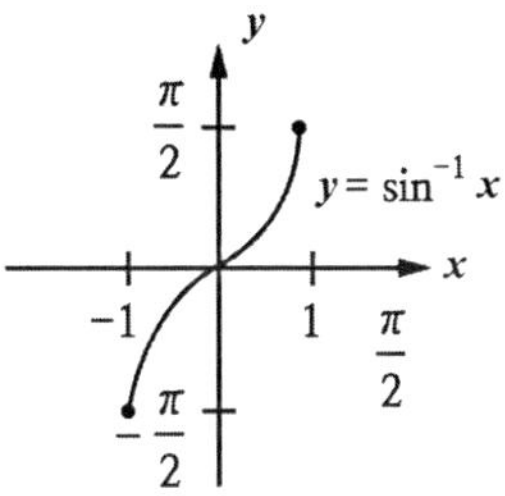

Figure 1.31

$\sin(\sin^{-1}x)=x$ for all x-values in the interval $[-1,1]$

$\sin^{-1}(\sin x)=x$ for all x-values in the interval $[-\frac{\pi}{2},\frac{\pi}{2}]$

The inverse cosine function. denoted by $y=\cos^{-1}x$ or $y=\arccos x$, is the function with a domain of [-1,1] and a range of $[0,\pi]$ that satisfies the relation $\cos y=x$.

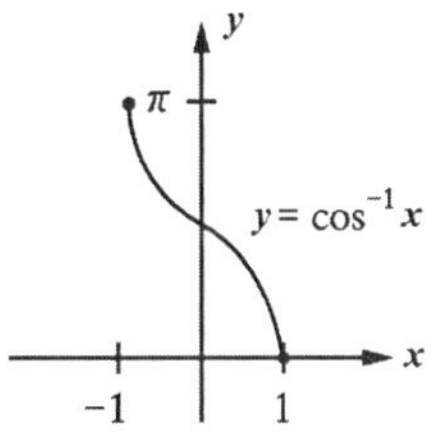

Figure 1.32

$\cos(\cos^{-1}x)=x$ for all x-values in the interval $[-1,1]$

$\cos^{-1}(\cos x)=x$ for all x-values in the interval $[0,\pi]$

The inverse tangent function, denoted by $y=\tan^{-1}x$ or $y=\arctan x$, is the function with a domain of $(-\infty,\infty)$ and a range of $(-\frac{\pi}{2},\frac{\pi}{2})$ that satisfies the relation $\tan y=x$.

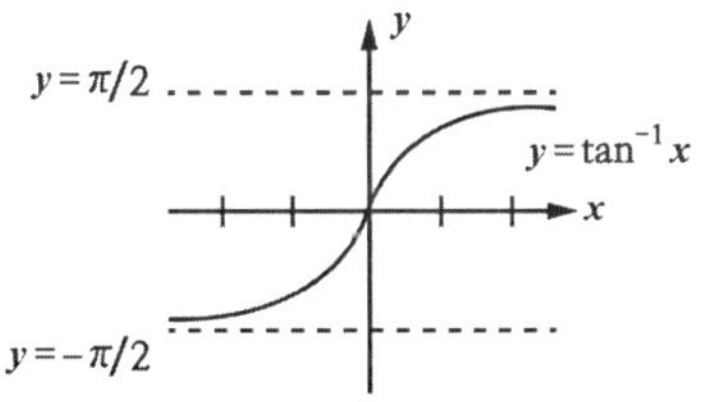

Figure 1.33

$\tan(\tan^{-1}x)=x$ for every real number x

$\tan^{-1}(\tan x)=x$ for all x-values in the interval $(-\frac{\pi}{2},\frac{\pi}{2})$

Example 31

Evaluate each of the following.

a. $\arcsin(-1)$ b. $\cos^{-1}(\cos\frac{2}{3})$ c. $\cos[\arctan(1.5)]$

Solution

a. $\arcsin(-1)$ is that number in the interval $[-\frac{\pi}{2}, \frac{\pi}{2}]$ whose sine is -1.

Since $\sin(-\frac{\pi}{2}) = -1$, we have $\arcsin(-1) = -\frac{\pi}{2}$

b. Since $\cos^{-1}(\cos x) = x$

$\cos^{-1}(\cos\frac{2}{3}) = \frac{2}{3}$

c. Use a calculator set in radian mode.

$\cos[\arctan(1.5)] = \cos(0.9828) = 0.5547$

1.32 Vectors

A vector is a directed line segment. The symbol $\overrightarrow{PQ}$ (read "vector PQ") denotes the vector extending from point $P(x_1, y_1)$, the initial point, to point $Q(x_2, y_2)$, the terminal point. $\overrightarrow{PQ} = \langle x_2 - x_1, y_2 - y_1 \rangle$.

Boldface letters such as **u** and **v** are often used to denote vectors, and the vector $\overrightarrow{PQ}$ can also be written as $\mathbf{u} = \langle a, b \rangle = \langle x_2 - x_1,\ y_2 - y_1 \rangle$.

The length or magnitude of the vector $\mathbf{u} = \langle a, b \rangle$ is. denoted by $|\mathbf{u}|$, and $|\mathbf{u}| = \sqrt{a^2 + b^2}$.
If two vectors **u** and **v** have the same length and the same direction, we say that they are equal and we write $\mathbf{u} = \mathbf{v}$

For each real number k and each vector $\mathbf{u} = \langle a, b \rangle$, the product $k\mathbf{u}$ is defined as $k\mathbf{u} = k\langle a, b \rangle = \langle ka, kb \rangle$.

Vector Addition and Subtraction, an Algebraic Approach

Let $\mathbf{u} = \langle a, b \rangle$ and $\mathbf{v} = \langle c, d \rangle$ be two vectors. Then

$\mathbf{u} + \mathbf{v} = \langle a, b \rangle + \langle c, d \rangle = \langle a + c, b + d \rangle$

$\mathbf{u} - \mathbf{v} = \langle a, b \rangle - \langle c, d \rangle = \langle a - c, b - d \rangle$

Example 32

a. Given the four points A=$(3,5)$, $B=(-4,1)$, $C=(-2,7)$, and $D=(6,6)$, computer $\overrightarrow{CD}-\overrightarrow{AB}$.

Let $\mathrm{u}=\langle 4,7\rangle$ and $\mathrm{v}=\langle -2,5\rangle$ for parts $b-d$. Compute each of the following.

b. $\mathrm{u}+\mathrm{v}$ c. $2\mathrm{u}-\mathrm{v}$ d. $|\mathrm{u}+\mathrm{v}|$

e. Given the points $\mathrm{u}=\langle 3,4\rangle$, compute magnitude of u

Solution

a. $\overrightarrow{AB}=\langle -4-3,1-5\rangle=\langle -7,-4\rangle$

$\overrightarrow{CD}=\langle 6+2,6-7\rangle=\langle 8,-1\rangle$

$\overrightarrow{CD}-\overrightarrow{AB}=\langle 8,-1\rangle-\langle -7,-4\rangle=\langle 15,3\rangle$

b. $\mathrm{u}+\mathrm{v}=\langle 4,7\rangle+\langle -2,5\rangle=\langle 4-2,7+5\rangle=\langle 2,12\rangle$

c. $2\mathrm{u}-\mathrm{v}=2\langle 4,7\rangle-\langle -2,5\rangle=\langle 8,14\rangle-\langle -2,5\rangle=\langle 10,9\rangle$

d. $|\mathrm{u}+\mathrm{v}|=|\langle 2,12\rangle|=\sqrt{2^2+12^2}=\sqrt{148}$ or $2\sqrt{37}$

e. $|\mathrm{u}|=|\langle 3,4\rangle|=\sqrt{3^2+4^2}=\sqrt{25}=5$

1.33 Polar Coordinates

A polar coordinate system consists of a point O called the pole, and a ray called the polar axis, whose endpoint is the fixed point O.

The polar coordinates of a point P are an ordered pair (r,θ), where r is the distance from the origin O to the point P, and θ is the measure of the angle formed by the polar axis and the segment $\overrightarrow{OP}$.

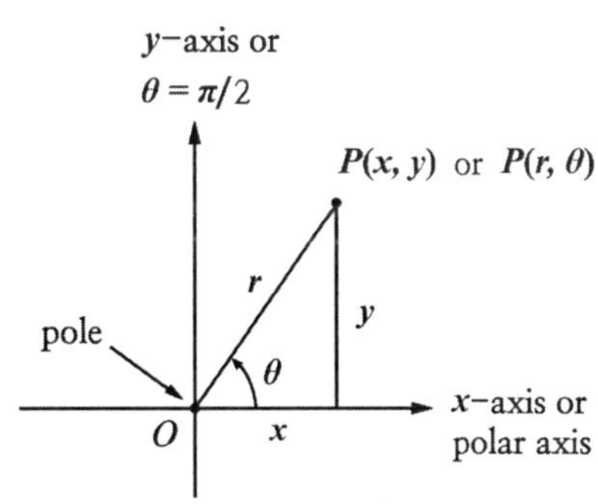

Figure 1.34

Comparing Rectangular and Polar Coordinate Systems

Rectangular Coordinates		Polar Coordinates
Origin	$\Leftrightarrow$	Pole
Positive x-axis	$\Leftrightarrow$	Polar axis or $\theta = 0$
$P(x,\ y)$	$\Leftrightarrow$	$P(r,\ \theta)$
Positive y-axis	$\Leftrightarrow$	$\theta = \pi/2$

Coordinate-System Conversion Formulas

$x = r\cos\theta \qquad y = r\sin\theta$

$r^2 = x^2 + y^2 \qquad \tan\theta = \dfrac{y}{x}(x \neq 0)$

Example 33

a. Convert $(3, \frac{\pi}{3})$ to rectangular coordinates.

b. Convert $(-1, -1)$ to polar coordinates.

c. Sketch a complete graph of $r = 3$.

Solution

a. $x = r\cos\theta = 3\cos\frac{\pi}{3} = 3 \cdot \frac{1}{2} = \frac{3}{2}$

$y = r\sin\theta = 3\sin\frac{\pi}{3} = 3 \cdot \frac{\sqrt{3}}{2} = \frac{3\sqrt{3}}{2}$

The rectangular coordinates are $(\frac{3}{2}, \frac{3\sqrt{3}}{2})$.

b. The point $(-1, -1)$ is in quadrant Ⅲ.

$r^2 = x^2 + y^2$

$r = \sqrt{x^2 + y^2} = \sqrt{(-1)^2 + (-1)^2} = \sqrt{2}$

$\tan\theta = \frac{y}{x} = \frac{-1}{-1} = 1 \Rightarrow \theta = \tan^{-1}(1) \Rightarrow \theta = \frac{\pi}{4}$

Since the point is in the third quadrant, $\theta = \pi + \frac{\pi}{4} = \frac{5\pi}{4}$.

The polar coordinates are $(\sqrt{2}, \frac{5\pi}{4})$.

c. $r = 3$

$r^2 = 9$

$x^2 + y^2 = 9$

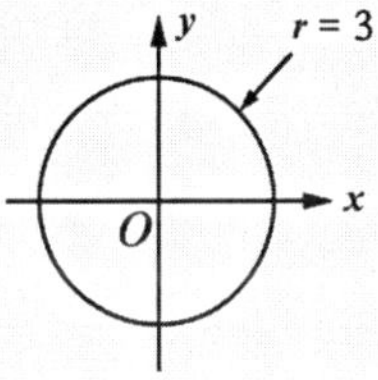

Figure 1.35

The graph of the equation is a circle whose center is $(0,0)$ and radius is 3.

CALCULUS

CHAPTER 02

Limits and Continuity

2.1 Limit Existence

The number L is the *limit of the function* $f(x)$ as x approaches c if, as the values of x get arbitrarily close (but not equal) to c the values of $f(x)$ approach (or equal) L. We write

$$\lim_{x\to c} f(x)=\text{L}$$

In order for $\lim_{x\to c} f(x)$ to exist, the values of f must tend to the same number L as x approaches c from either the left or the right. We write

$$\lim_{x\to c^-} f(x)$$

for the *left-hand limit of f* at c (as x approaches c through values less than c), and

$$\lim_{x\to c^+} f(x)$$

for the *right-hand limit of f* at c (as x approaches c through values greater than c)

Example 1

Given $f(x)=\dfrac{x^2-x-2}{x-2}$, find the value of $\lim_{x\to 2} f(x)$.

Solution

$$y=\frac{x^2-x-2}{x-2}$$

$$=\frac{(x-2)(x+1)}{x-2}=x+1,\ x\neq 2$$

$$\lim_{x\to 2^-} f(x)=3,\ \lim_{x\to 2^+} f(x)=3$$

$$\lim_{x\to 2^-} f(x)=\lim_{x\to 2^+} f(x)=3$$

$$\therefore \lim_{x\to 2} f(x)=3$$

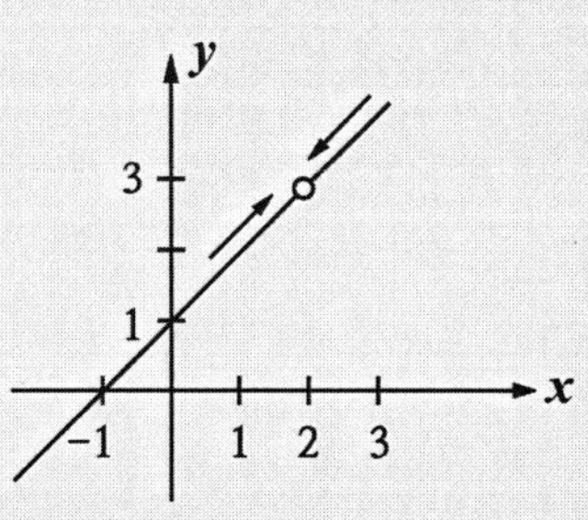

Figure 2.1

Example 2

Given $f(x)=\begin{cases} \dfrac{2(x-2)}{x^2-4}, & x \neq 2 \\ 2 & ,x=2 \end{cases}$, find the value of $\lim_{x\to 2} f(x)$.

Solution

$$y=\frac{2(x-2)}{x^2-4}=\frac{2(x-2)}{(x-2)(x+2)}$$

$$=\frac{2}{x+2}, x \neq 2$$

Vertical asymptote: $x=-2$

Horizontal asymptote: $y=0$

$$\lim_{x\to 2^-} f(x)=\frac{1}{2},\ \lim_{x\to 2^+} f(x)=\frac{1}{2}$$

$$\lim_{x\to 2^-} f(x)=\lim_{x\to 2^+} f(x)=\frac{1}{2}$$

$$\therefore \lim_{x\to 2} f(x)=\frac{1}{2}$$

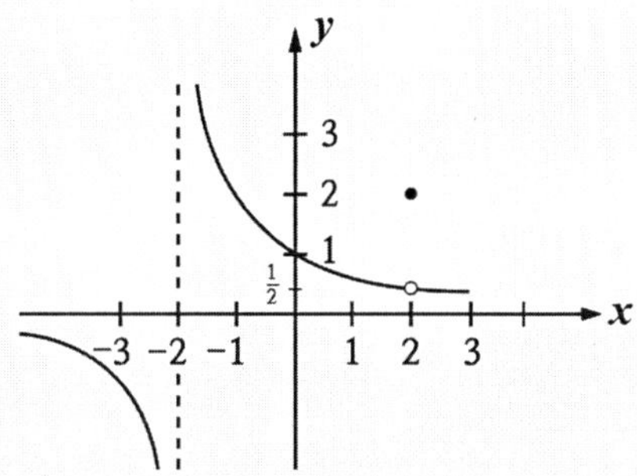

Figure 2.2

Example 3

Given $f(x)=\begin{cases} 3-x, & x \leq 2 \\ x & , x>2 \end{cases}$, find the value of $\lim_{x\to 2} f(x)$.

Solution

$$\lim_{x\to 2^-} f(x)=1$$

$$\lim_{x\to 2^+} f(x)=2$$

$$\lim_{x\to 2^-} f(x) \neq \lim_{x\to 2^+} f(x)$$

$\therefore \lim_{x\to 2} f(x)=DNE$ (Does Not Exist)

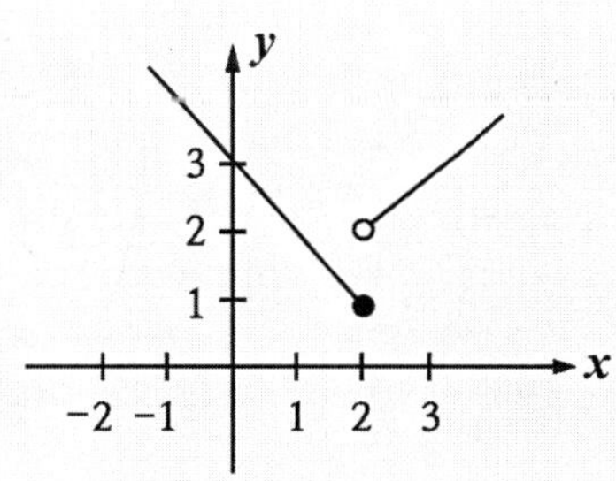

Figure 2.3

[EX 1] Use the graph of f to evaluate the following limits.

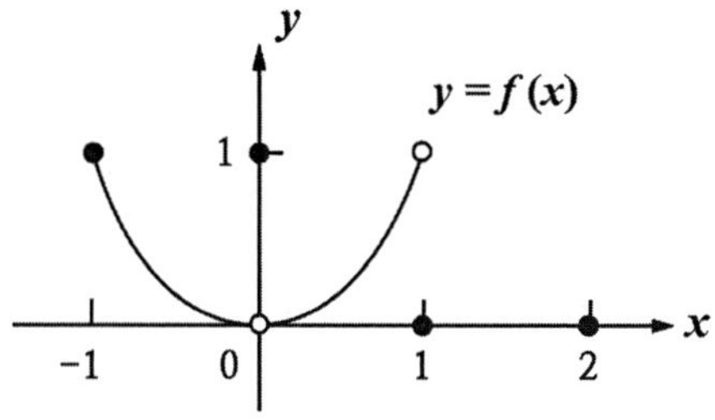

Figure 2.4

(1) $\lim_{x \to -1^+} f(x) =$

(2) $\lim_{x \to 0^+} f(x) =$

(3) $\lim_{x \to 0^-} f(x) =$

(4) $\lim_{x \to 0} f(x) =$

(5) $\lim_{x \to 1} f(x) =$

(6) $\lim_{x \to 1^+} f(x) =$

(7) $\lim_{x \to 2^-} f(x) =$

(8) $\lim_{x \to -1^-} f(x) =$

(9) $\lim_{x \to 2^+} f(x) =$

(10) $\lim_{x \to 1^-} f(x) =$

[EX 2] Use the graph of f to evaluate the following limits.

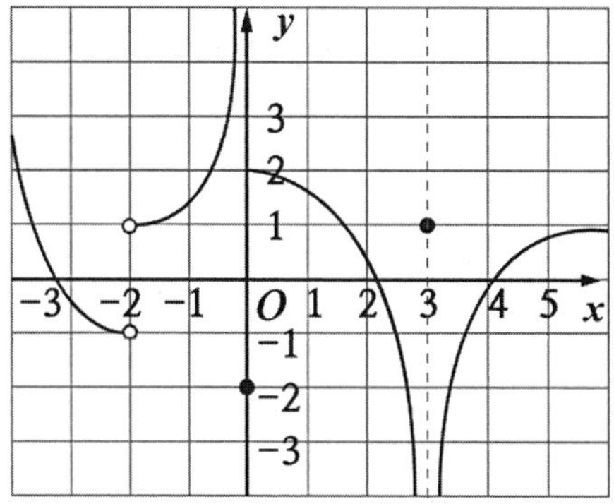

Figure 2.5

(1) (a) $\lim_{x \to -2^-} f(x) =$ (b) $\lim_{x \to -2^+} f(x) =$ (c) $\lim_{x \to -2} f(x) =$ (d) $f(-2) =$

(2) (a) $\lim_{x\to0^-} f(x)=$ (b) $\lim_{x\to0^+} f(x)=$ (c) $\lim_{x\to0} f(x)=$ (d) $f(0)=$

(3) (a) $\lim_{x\to3^-} f(x)=$ (b) $\lim_{x\to3^+} f(x)=$ (c) $\lim_{x\to3} f(x)=$ (d) $f(3)=$

(4) (a) $\lim_{x\to4^-} f(x)=$ (b) $\lim_{x\to4^+} f(x)=$ (c) $\lim_{x\to4} f(x)=$ (d) $f(4)=$

[EX 3] Sketch the graph of $y=f(x)$ and find the following limits.

(1) $f(x)=\begin{cases} 3-x, & x<2 \\ \dfrac{x}{2}+1, & x>2 \end{cases}$

(a) $\lim_{x\to2^+} f(x)=$ (b) $\lim_{x\to2^-} f(x)=$ (c) $\lim_{x\to2} f(x)=$

(d) $\lim_{x\to4^-} f(x)=$ (e) $\lim_{x\to4^+} f(x)=$ (f) $\lim_{x\to4} f(x)=$

(2) $f(x)=\begin{cases} 3-x, & x<2 \\ 2, & x=2 \\ \dfrac{x}{2}, & x>2 \end{cases}$

(a) $\lim_{x\to2^+} f(x)=$ (b) $\lim_{x\to2^-} f(x)=$ (c) $f(2)=$ (d) $\lim_{x\to2} f(x)=$

(e) $\lim_{x\to-1^-} f(x)=$ (f) $\lim_{x\to-1^+} f(x)=$ (g) $\lim_{x\to-1} f(x)=$

(3) $f(x)=\begin{cases} 1-x^2, & x\le0 \\ 1/x, & x>0 \end{cases}$

(a) $\lim_{x\to0^-} f(x)=$ (b) $\lim_{x\to0^+} f(x)=$ (c) $\lim_{x\to0} f(x)=$ (d) $f(0)=$

(4) $f(x)=\begin{cases} \dfrac{x^2-1}{x+1}, & x \neq -1 \\ 3, & x=-1 \end{cases}$

(a) $\lim_{x\to -1^-} f(x)=$ (b) $\lim_{x\to -1^+} f(x)=$ (c) $\lim_{x\to -1} f(x)=$ (d) $f(-1)=$

(5) $f(x)=\begin{cases} x^2+2x, & x \leq -1 \\ x, & -1 < x < 1 \\ 4-3x, & x \geq 1 \end{cases}$

(a) $\lim_{x\to -1^-} f(x)=$ (b) $\lim_{x\to -1^+} f(x)=$ (c) $\lim_{x\to -1} f(x)=$

(d) $\lim_{x\to 1^-} f(x)=$ (e) $\lim_{x\to 1^+} f(x)=$ (f) $\lim_{x\to 1} f(x)=$

(6) $f(x)=\begin{cases} \dfrac{x-1}{x^2-3x+2}, & x \neq 1 \\ 1, & x=1 \end{cases}$

(a) $\lim_{x\to 1^-} f(x)=$ (b) $\lim_{x\to 1^+} f(x)=$ (c) $\lim_{x\to 1} f(x)=$ (d) $f(1)=$

Limit Laws If L, M, c, and k are real numbers and $\lim_{x\to c} f(x)=L$ and $\lim_{x\to c} g(x)=M$, then

1. Sum Rule: $\lim_{x\to c}(f(x)+g(x))=L+M$
2. Difference Rule: $\lim_{x\to c}(f(x)-g(x))=L-M$
3. Constant Multiple Rule: $\lim_{x\to c}(k\cdot f(x))=k\cdot L$
4. Product Rule: $\lim_{x\to c}(f(x)\cdot g(x))=L\cdot M$
5. Quotient Rule: $\lim_{x\to c}\dfrac{f(x)}{g(x)}=\dfrac{L}{M},\ M\neq 0$
6. Power Rule: $\lim_{x\to c}[f(x)]^n=L^n$, n a positive integer
7. Root Rule: $\lim_{x\to c}\sqrt[n]{f(x)}=\sqrt[n]{L}=L^{1/n}$, n a positive integer

(If n is even, we assume that $\lim_{x\to c} f(x)=L>0$.)

[EX 4] Use the properties of limits to evaluate the following limits.

(1) $\lim_{x\to c}(x^3+4x^2-3)$ (2) $\lim_{x\to c}\frac{x^4+x^2-1}{x^2+5}$

(3) $\lim_{x\to 3}\frac{\sqrt{x+1}-1}{x-1}$ (4) $\lim_{x\to -2}\sqrt{4x^2-3}$

[EX 5] Suppose $\lim_{x\to c}f(x)=5$ and $\lim_{x\to c}g(x)=-2$. Find

(1) $\lim_{x\to c}f(x)g(x)$ (2) $\lim_{x\to c}2f(x)g(x)$

(3) $\lim_{x\to c}(f(x)+3g(x))$ (4) $\lim_{x\to c}\frac{f(x)}{f(x)-g(x)}$

Indeterminate form of type $\frac{0}{0}$

- Rational Expressions: Step 1. Factor the numerator and denominator.
  Step 2. Cancel the common factor.
- Radical Expressions: Step 1. Multiply the numerator and denominator by the conjugate.
  Step 2. Cancel the common factor.

Example 4

Evaluate the following limit.

Solution

$$\lim_{x\to 3}\frac{x^2-9}{x^3-27}=\lim_{x\to 3}\frac{(x-3)(x+3)}{(x-3)(x^2+3x+9)}=\lim_{x\to 3}\frac{x+3}{x^2+3x+9}=\frac{3+3}{3^2+3(3)+9}=\frac{6}{27}=\frac{2}{9}$$

Example 5

Evaluate the following limit.

Solution

$$\lim_{x\to 0}\frac{\sqrt{4+x}-2}{x}=\lim_{x\to 0}\frac{(\sqrt{4+x}-2)(\sqrt{4+x}+2)}{x(\sqrt{4+x}+2)}=\lim_{x\to 0}\frac{4+x-4}{x(\sqrt{4+x}+2)}=\lim_{x\to 0}\frac{x}{x(\sqrt{4+x}+2)}$$

$$=\lim_{x\to 0}\frac{1}{\sqrt{4+x}+2}=\frac{1}{\sqrt{4+0}+2}=\frac{1}{4}$$

[EX 6] Evaluate the following limit.

(1) $\lim_{t\to 1}\frac{t^2+t-2}{t^2-1}$

(2) $\lim_{x\to 4}\frac{x^2-4x}{x^2-5x+4}$

(3) $\lim_{x\to 2}\frac{x^3-8}{x^4-16}$

(4) $\lim_{x\to 2}\frac{x^3-8}{x^2-3x+2}$

(5) $\lim_{y\to 0}\frac{5y^3+8y^2}{3y^4-16y^2}$

(6) $\lim_{x\to 2}\frac{x^3-2x^2-4x+8}{x^3-2x^2}$

(7) $\lim_{h\to 0}\frac{(1+h)^4-1}{h}$

(8) $\lim_{x\to 1}\frac{(x+1)^3-8}{x-1}$

[EX 7] Evaluate the following limit.

(1) $\lim_{x\to 2}\frac{\sqrt{x^2+12}-4}{x-2}$

(2) $\lim_{x\to 4}\frac{x-4}{2-\sqrt{x}}$

(3) $\lim_{x\to 1}\frac{x-1}{\sqrt{x+3}-2}$

(4) $\lim_{x\to 3}\frac{\sqrt{x+1}-2}{\sqrt{x-2}-1}$

(5) $\lim_{x\to 2}\frac{\sqrt{x+2}-\sqrt{3x-2}}{\sqrt{5x-1}-\sqrt{4x+1}}$

[EX 8] Evaluate the following limit.

(1) $\lim_{x\to 0}\dfrac{\dfrac{1}{x-1}+\dfrac{1}{x+1}}{x}$

(2) $\lim_{h\to 0}\dfrac{\dfrac{1}{(h+3)^2}-\dfrac{1}{9}}{h}$

(3) $\lim_{x\to 9}\dfrac{\dfrac{1}{\sqrt{x}}-\dfrac{1}{3}}{x-9}$

(4) $\lim_{x\to 2}\dfrac{\dfrac{1}{\sqrt{x+2}}-\dfrac{1}{2}}{x^2-4}$

Definition of Absolute Value

$$|x|=\begin{cases} x & \text{if } x\geq 0 \\ -x & \text{if } x<0 \end{cases}$$

$[x]$ represents the greatest integer less than or equal to x

(eg) $[3.9]=3$, $[2.1]=2$, $[-1.9]=-2$, $[-2.1]=-3$

[EX 9] Evaluate the following limit.

(1) (a) $\lim_{x\to -2^+}(x+3)\dfrac{|x+2|}{x+2}$ (b) $\lim_{x\to -2^-}(x+3)\dfrac{|x+2|}{x+2}$

(2) (a) $\lim_{x\to 1^+}\dfrac{\sqrt{2}(x-1)}{|x-1|}$ (b) $\lim_{x\to 1^-}\dfrac{\sqrt{2}(x-1)}{|x-1|}$

(3) (a) $\lim_{\theta\to 3^+}\dfrac{[\theta]}{\theta}$ (b) $\lim_{\theta\to 3^-}\dfrac{[\theta]}{\theta}$

(4) (a) $\lim_{t\to 4^+}(t-[t])$ (b) $\lim_{t\to 4^-}(t-[t])$

$\lim_{x\to 0^+}\frac{1}{x}=+\infty$, $\lim_{x\to 0^-}\frac{1}{x}=-\infty$, $\lim_{x\to 0}\frac{1}{x}=DNE$ (Does Not Exist)

$\lim_{x\to 0^+}\frac{1}{x^2}=+\infty$, $\lim_{x\to 0^-}\frac{1}{x^2}=+\infty$, $\lim_{x\to 0}\frac{1}{x^2}=+\infty$

Example 6

Evaluate the following limit.

Solution

(1) $\lim_{x\to 2^+}\frac{4}{x-2}=\frac{4}{+0.000......1}=+\infty$

(2) $\lim_{x\to 2^-}\frac{4}{x-2}=\frac{4}{-0.000......1}=-\infty$

(3) $\lim_{x\to 2}\frac{4}{x-2}=DNE$

[EX 10] Evaluate the following limit.

(1) $\lim_{x\to 2^-}\frac{3}{x-2}$ (2) $\lim_{x\to 2^-}\frac{-1}{x-2}$ (3) $\lim_{x\to -8^+}\frac{2x}{x+8}$

(4) $\lim_{x\to 2}\frac{1}{x-2}$ (5) $\lim_{x\to 0}\frac{4}{x^{2/5}}$ (6) $\lim_{x\to 3^-}\frac{-1}{(x-3)^2}$

(7) $\lim_{x\to 7}\frac{4}{(x-7)^2}$ (8) $\lim_{x\to 3}\frac{1}{|x-3|}$ (9) $\lim_{x\to 0^+}\frac{2}{3x^{1/3}}$

(10) $\lim_{x\to 0^-}\frac{2}{3x^{(1/3)}}$ (11) $\lim_{x\to (\pi/2)^-}\tan x$ (12) $\lim_{\theta\to 0^-}(1+\csc\theta)$

[EX 11] Without sketching the graph of f. find the following limits. (Use the definition of the limit existence)

(1) Let $f(x)=\begin{cases} x+2 & , x<-1 \\ x^2+1 & , x\geq -1 \end{cases}$. Find $\lim_{x\to -1} f(x)$.

(2) Let $f(x)=\begin{cases} 4\cos(\pi\frac{x}{9}) & , x<3 \\ \sqrt{x^2-5} & , x\geq 3 \end{cases}$. Find $\lim_{x\to 3} f(x)$.

(3) Let $f(x)=\begin{cases} e^x+3 & , x\leq \ln 3 \\ e^{2x}-3 & , x> \ln 3 \end{cases}$. Find $\lim_{x\to \ln 3} f(x)$.

[EX 12] Find the values of a and b that satisfy the following equations.

(1) $\lim_{x\to 2}\frac{x^2-4x+a}{x-2}=b$

(2) $\lim_{x\to 1}\frac{x^2-a}{x-1}=b$

(3) $\lim_{x\to 1}\frac{a\sqrt{x+1}-b}{x-1}=\sqrt{2}$

(4) $\lim_{x\to 2}\frac{2-x}{ax^2+x-6}=b \ (b\neq 0)$

2.2. Squeezing Theorem (Sandwich Theorem)

The Sandwich Theorem

Suppose that $g(x) \le f(x) \le h(x)$ for all x in some open interval containing c, except possibly at $x = c$ itself.

Suppose also that $\lim_{x \to c} g(x) = \lim_{x \to c} h(x) = L$,

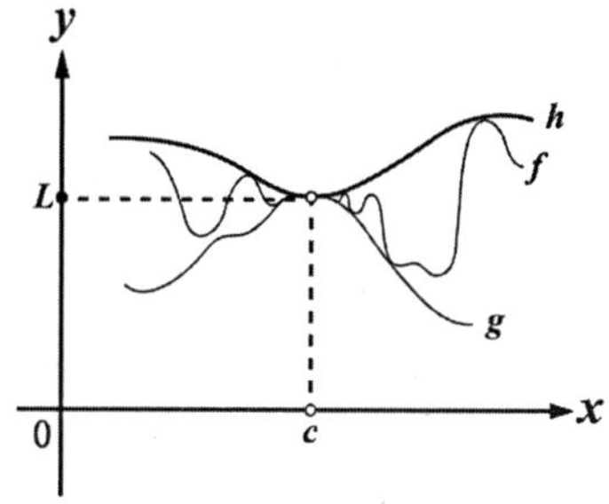

Figure 2.6

Then

$$\lim_{x \to c} f(x) = \mathrm{L}.$$

Example 1

Given that $1 - \dfrac{x^2}{4} \le u(x) \le 1 + \dfrac{x^2}{2}$ for all $x \neq 0$,

find $\lim_{x \to 0} u(x)$, no matter how complicated u is.

Solution

Since

$\lim_{x \to 0}(1-(x^2/4))=1$ and $\lim_{x \to 0}(1-(x^2/2))=1$,

the Sandwich Theorem implies that $\lim_{x \to 0} u(x)=1$

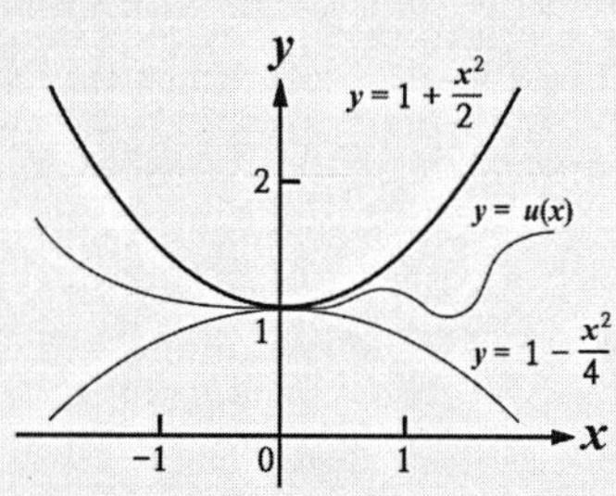

Figure 2.7

[EX 1] Use the squeezing Theorem to prove the following limits.

(1) $\lim_{x\to0} x^2 \cos\frac{1}{x} = 0$

(2) $\lim_{x\to0} x\sin(\frac{1}{x^2}) = 0$

(3) $\lim_{x\to\infty} \frac{\sin x}{x} = 0$

(4) $\lim_{x\to\infty} \frac{\cos 2x}{x^2} = 0$

[EX 2]

(1) Given $1-\frac{x^2}{6} < \frac{x\sin x}{2-2\cos x} < 1$ for all x, find $\lim_{x\to0}\frac{x\sin x}{2-2\cos x}$.

(2) Given $-6x-4 < f(x) < x^4+x^2$ for all x, find $\lim_{x\to-1} f(x)$.

(3) Given $\sqrt{5-2x} < f(x) < \sqrt{5}$ for all x, find $\lim_{x\to0} f(x)$.

2.3. Limit of Trigonometric Function

$\lim_{\theta \to 0} \frac{\sin\theta}{\theta} = 1$ (θ in radians)

proof The plan is to show that the right-hand and left-hand limits are both 1. Then we will know that the two-sided limit is 1 as well. (i) To show that the right-hand is 1, we begin with positive values of θ less than $\pi/2$,

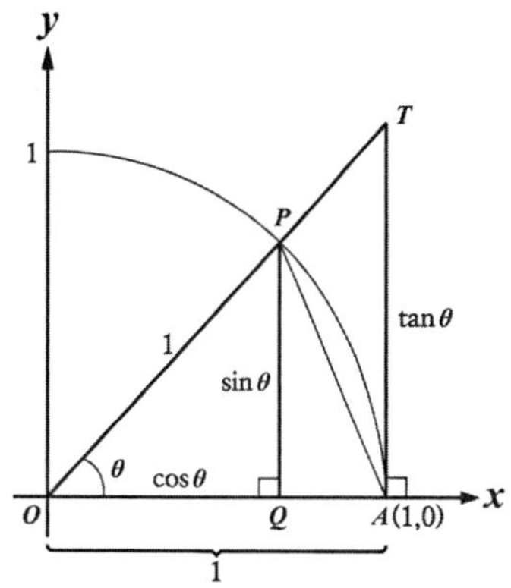

Figure 2.8

Area $\triangle OAP$ < area sector OAP < area $\triangle OAT$

We can express these areas in terms of θ as follows.

$$\text{Area } \triangle OAP = \frac{1}{2}\text{base} \times \text{height} = \frac{1}{2}(1)(\sin\theta) = \frac{1}{2}\sin\theta$$

$$\text{Area sector } OAP = \frac{1}{2}r^2\theta = \frac{1}{2}(1)^2\theta = \frac{\theta}{2}$$

$$\text{Area } \triangle OAT = \frac{1}{2}\text{base} \times \text{height} = \frac{1}{2}(1)(\tan\theta) = \frac{1}{2}\tan\theta$$

Thus,

$$\frac{1}{2}\sin\theta < \frac{1}{2}\theta < \frac{1}{2}\tan\theta.$$

This last inequality goes the same way if we divide all three terms by the number (1/2) $\sin\theta$, which is positive since $0 < \theta < \pi/2$:

$$1 < \frac{\theta}{\sin\theta} < \frac{1}{\cos\theta}$$

Taking reciprocals reverses the inequalities:

$$1 > \frac{\sin\theta}{\theta} > \frac{\cos\theta}{\theta}$$

Since $\lim_{\theta \to 0+} \cos\theta = 1$, the Sandwich Theorem gives

$$\lim_{\theta \to 0^+} \frac{\sin\theta}{\theta} = 1.$$

(ii) To show that the left-hand is 1, we begin with negative values of θ more than $-\pi/2$, let $\theta = -x$

$$\lim_{\theta \to 0^-} \frac{\sin\theta}{\theta} = \lim_{x \to 0^+} \frac{\sin(-x)}{(-x)} = \lim_{x \to 0^+} \frac{\sin x}{x} = 1.$$

by (i),(ii)

$$\lim_{\theta \to 0} \frac{\sin\theta}{\theta} = 1.$$

The formula $\lim_{\theta \to 0} \frac{\sin\theta}{\theta} = 1$ can be generalized. If $\lim_{x \to c} f(x) = 0$ and $f(x)$ is never zero in an open interval containing the point $x = c$, except possibly c itself, then $\lim_{x \to c} \frac{\sin f(x)}{f(x)} = 1$.

Example 1

Evaluato the following limits

(a) $\lim_{x \to 0} \frac{\sin x^2}{x^2}$ (b) $\lim_{x \to 0} \frac{\sin x^2}{x}$ (c) $\lim_{x \to -1} \frac{\sin(x^2 - x - 2)}{x+1}$ (d) $\lim_{x \to 1} \frac{\sin(1 - \sqrt{x})}{x-1}$

Solution

(a) $\lim_{x \to 0} \frac{\sin x^2}{x^2} = 1$

(b) $\lim_{x \to 0} \frac{\sin x^2}{x} = \lim_{x \to 0} \frac{\sin x^2}{x^2} \lim_{x \to 0} \frac{x^2}{x} = 1 \cdot 0 = 0$

(c) $\lim_{x \to -1} \frac{\sin(x^2 - x - 2)}{x+1} = \lim_{x \to -1} \frac{\sin(x^2 - x - 2)}{x^2 - x - 2} \cdot \lim_{x \to -1} \frac{(x^2 - x - 2)}{x+1}$

$= 1 \cdot \lim_{x \to -1} \frac{(x+1)(x-2)}{x+1} = -3$

(d) $\lim_{x \to 1} \frac{\sin(1 - \sqrt{x})}{x-1} = \lim_{x \to 1} \frac{\sin(1 - \sqrt{x})}{1 - \sqrt{x}} \frac{1 - \sqrt{x}}{x-1} = 1 \cdot \lim_{x \to 1} \frac{(1 - \sqrt{x})(1 + \sqrt{x})}{(x-1)(1 + \sqrt{x})}$

$= \lim_{x \to 1} \frac{1 - x}{(x-1)(1 + \sqrt{x})} = -\frac{1}{2}$

[EX 1] Evaluate the following limits.

(1) $\lim_{\theta \to 0} \frac{\sin \sqrt{2}\theta}{\sqrt{2}\theta}$

(2) $\lim_{x \to 0} \frac{\sin 2x}{3x}$

(3) $\lim_{y \to 0} \frac{\sin 3y}{4y}$

(4) $\lim_{x \to 0} \frac{\sin 5x}{\sin 2x}$

(5) $\lim_{x \to 0} \frac{\sin^3 2x}{x^3}$

(6) $\lim_{x \to 0} \frac{\sin^2 3x}{\sin^2 2x}$

(7) $\lim_{x \to 0} \frac{2x}{\tan x}$

(8) $\lim_{x \to 0} \frac{\tan 3x}{2x}$

(9) $\lim_{x \to 0} \frac{\tan 3x}{\sin 8x}$

(10) $\lim_{x \to 0} \frac{\tan^3 3x}{8x^3}$

(11) $\lim_{x \to 0} \frac{\sin 2x}{x^2 + 3x}$

(12) $\lim_{x \to 0} \frac{\sin^3 x}{x^2}$

(13) $\lim_{x \to 0} \frac{\sin(1 - \cos x)}{x}$

(14) $\lim_{x \to 0^+} \frac{\sin x}{\sin \sqrt{x}}$

(15) $\lim_{x \to 0} \frac{\sin(x^2 + x)}{x}$

(16) $\lim_{x \to 0} \frac{x - x\cos x}{\sin^2 3x}$

(17) $\lim_{x \to 0} \frac{\sin(1 - \cos x)}{x^2}$

2.4. Limit of Exponential and Logarithmic Functions.

Exponential e

$$\lim_{x \to \pm\infty}\left(1+\frac{1}{x}\right)^x = \lim_{x \to 0}(1+x)^{\frac{1}{x}} = 2.718281828459045235 = e$$

x	$(1+x)^{\frac{1}{x}}$	x	$(1+x)^{\frac{1}{x}}$
0.1	2.59374···	−0.1	2.86797···
0.01	2.70481···	−0.01	2.73199···
0.001	2.71692···	−0.001	2.71964···
0.0001	2.71814···	−0.0001	2.71841···
0.00001	2.71827···	−0.00001	2.71829···
······	······	······	······

Example 1

$$\lim_{x \to \infty}\left(1+\frac{3}{x}\right)^{\frac{x}{3}} =$$

Solution

$$\lim_{x \to \infty}\left(1+\frac{3}{x}\right)^{\frac{x}{3}} = \lim_{x \to 0}(1+5x)^{\frac{1}{5x}} = e$$

Example 2

$$\lim_{x \to \infty}(1+\frac{3}{x})^x =$$

Solution

$$\lim_{x \to \infty}(1+\frac{3}{x})^x = \lim_{x \to \infty}[(1+\frac{3}{x})^{\frac{x}{3}}]^3 = e^3$$

[EX 1] Evaluate the following limits.

(1) $\lim_{x\to 0}(1+3x)^{\frac{1}{x}}$ (2) $\lim_{x\to\infty}\left(1+\frac{3}{x}\right)^{x}$ (3) $\lim_{x\to 0}(1+x)^{\frac{3}{x}}$

(4) $\lim_{x\to -\infty}\left(1+\frac{2}{3x}\right)^{x}$ (5) $\lim_{x\to 0}(1+3x)^{\frac{1}{2x}}$ (6) $\lim_{n\to\infty}P\left(1+\frac{r}{n}\right)^{nt}$

(7) $\lim_{x\to\infty}\left(1+\frac{3}{x}\right)^{x}$ (8) $\lim_{x\to 0}\left(1+\frac{x}{2}\right)^{\frac{1}{2x}}$ (9) $\lim_{x\to\infty}\left(1+\frac{1}{x}\right)^{3x}$

(10) $\lim_{x\to 0}\left(1-\frac{x}{2}\right)^{\frac{1}{x}}$ (11) $\lim_{x\to\infty}\left(1+\frac{3}{x}\right)^{2x}$ (12) $\lim_{x\to\infty}\left(\frac{2x+1}{x}\right)^{3x}$

[EX 2] Evaluate the following limits.

(1) $\lim_{x\to 0}\frac{\ln(x+1)}{x}$

(2) $\lim_{x\to 0}\frac{e^{x}-1}{x}$ (Hint: Let $e^{x}-1=t$)

(3) $\lim_{x\to 0}\frac{\log_a(x+1)}{x}$

(4) $\lim_{x\to 0}\frac{a^{x}-1}{x}$

2.5. Limits Involving Infinity

Limit of Rational Function in the form $\frac{\infty}{\infty}$.

Divide both the numerator and the denominator by the highest power of x in the denominator.

Example 1

Evaluato the following limits

(1) $\lim_{x\to\infty}\frac{5x^2+8x-3}{3x^2+2}$ (2) $\lim_{x\to-\infty}\frac{11x+2}{2x^3-1}$

Solution

(1) $\lim_{x\to\infty}\frac{5x^2+8x-3}{3x^2+2}=\lim_{x\to\infty}\frac{5+(8/x)-(3/x^2)}{3+(2/x^2)}=\frac{5+0-0}{3+0}=\frac{5}{3}$

(2) $\lim_{x\to-\infty}\frac{11x+2}{2x^3-1}=\lim_{x\to-\infty}\frac{(11/x^2)+(2/x^3)}{2-(1/x^3)}=\frac{0+0}{2-0}=0$

[EX 1] Evaluate the following limits.

(1) $\lim_{x\to-\infty}\left(\frac{1-x^3}{x^2+7x}\right)^5$ (2) $\lim_{x\to\infty}\frac{3x^3+4}{2x^3-3x+5}$ (3) $\lim_{x\to\infty}\frac{2\sqrt{x}+x^{-1}}{3x-7}$

(4) $\lim_{x\to\infty}\frac{3x^2+1}{5x^2-2}$ (5) $\lim_{x\to-\infty}\frac{\sqrt[3]{x}-\sqrt[5]{x}}{\sqrt[3]{x}+\sqrt[5]{x}}$ (6) $\lim_{x\to-\infty}\frac{2x^4-4x}{x^5+9x-5}$

(7) $\lim_{x\to\infty}\frac{2x^{5/3}-x^{1/3}+7}{x^{8/5}+3x+\sqrt{x}}$ (8) $\lim_{x\to\infty}\frac{\sqrt{x^2+1}}{x+1}$ (9) $\lim_{x\to-\infty}\frac{\sqrt{x^2+1}}{x+1}$

(10) $\lim_{x\to\infty}\frac{x-3}{\sqrt{4x^2+25}}$ (11) $\lim_{x\to-\infty}\frac{4-3x^3}{\sqrt{x^6+9}}$

Horizontal Asymptote

A line $y=b$ is a horizontal asymptote of the graph of a function $y=f(x)$ if either

$$\lim_{x\to\infty} f(x)=b \text{ or } \lim_{x\to-\infty} f(x)=b.$$

Example 2

Find the equation of the horizontal asymptote for thd following function

Solution

The graph of the function

$$f(x)=\frac{5x^2+8x-3}{3x^2+2}$$

sketched in Figure has the line $y=5/3$ as a horizontal asymptote on both the right and the left because

$$\lim_{x\to\infty} f(x)=\frac{5}{3} \text{ and } \lim_{x\to-\infty} f(x)=\frac{5}{3}.$$

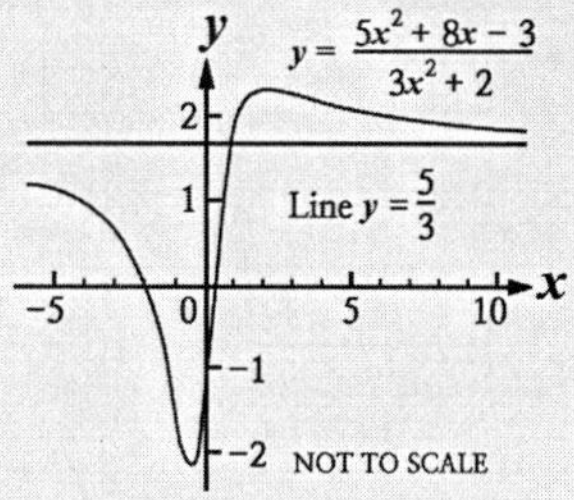

Figure 2.9

Horizontal Asymptote of Rational Functions

$$y=\frac{ax^m+\cdots\cdots}{bx^n+\cdots\cdots}$$

(1) $m=n$: $y=\frac{a}{b}$ is the horizontal asymptote

(2) $m<n$: $y=0$ is the horizontal asymptote

(3) $m>n$: there is no horizontal asymptote

[EX 2] Find the equation of horizontal asymptote for the following functions.

(1) $y=\dfrac{2x+3}{5x+7}$ (2) $y=\dfrac{1-x^2}{2x^2+1}$ (3) $y=\dfrac{2x^4+7}{x^3-x^2+x+7}$

(4) $y=\dfrac{x+1}{1-x^4}$ (5) $y=\dfrac{x+1}{x^2+3}$ (6) $y=\dfrac{(1-x)(2x-1)(x+2)}{(x-2)(x^2+x+2)}$

(7) $y=\dfrac{7}{x+4}+5$ (8) $y=\dfrac{2x}{x^2+4}+2$ (9) $y=-\dfrac{4x}{2x-3}+5$

Vertical Asymptote

A line $x=a$ is a vertical asymptote of the graph of a function. $y=f(x)$ if either

$$\lim_{x\to a^+} f(x)=\pm\infty \text{ or } \lim_{x\to a^-} f(x)=\pm\infty.$$

Example 3

Find the horizontal and vertical asymptotes of graph of $f(x)=-\dfrac{8}{x^2-4}$

Solution

The behavior as $x\to\pm 2$. Since

$\lim_{x\to 2^+} f(x)=-\infty$ and $\lim_{x\to 2^-} f(x)=\infty$,

the line $x=2$ is a vertical asymptote both from the right and from the left. By symmetry, the line $x=-2$ is also a vertical asymptote.

There are no other asymptotes because f has a finite limit at every other point.

[EX 3] (a) Find the equation(s) of the vertical asymptote(s). (b) Describe the behaviour of the function to the left and right of each horizontal asymptote.

(1) $y=\dfrac{2+x}{x-3}$ (2) $y=\dfrac{x^2}{x+2}$ (3) $y=\dfrac{1}{x^2}$

(4) $y=\dfrac{x^2}{1-x^2}$ (5) $y=\dfrac{2x}{x^2-4}$ (6) $y=\dfrac{x}{\sqrt{2-x}}+3$

[EX 4] Find the equation of the (a) horizontal and (b) vertical asymptotes for the following functions.

(1) $y=\dfrac{2x}{|x|-1}$ (2) $y=\dfrac{x-1}{2|x|+1}$

(3) $y=\dfrac{5(4)^x+2}{4^x-1}$ (4) $y=\dfrac{3(2)^x+1}{2^x-3}$

[EX 5] Evaluate the following limits.

(1) $\lim\limits_{x\to\left(\frac{\pi}{2}\right)^-}\tan x$ (2) $\lim\limits_{x\to\left(\frac{\pi}{2}\right)^+}\sec x$ (3) $\lim\limits_{x\to 0^-}\csc x$

(4) $\lim\limits_{x\to 0^+}\cot x$ (5) $\lim\limits_{x\to 0^+}\ln x$ (6) $\lim\limits_{x\to 2^-}\log_3(2-x)$

[EX 6] Evaluate the following limits.

(1) $\lim_{x \to \infty} e^{-x}$

(2) $\lim_{x \to -\infty} (0.9)^{x}$

(3) $\lim_{x \to 0^-} e^{\frac{1}{x}}$

(4) $\lim_{x \to 3^+} (0.9)^{\frac{1}{x-3}}$

(5) $\lim_{x \to 0^+} e^{\csc x}$

(6) $\lim_{x \to 1^-} \frac{1}{\ln(1-x)}$

(7) $\lim_{x \to -\infty} \frac{3^x + 1}{3^x - 1}$

(8) $\lim_{x \to -\infty} \tan^{-1} x$

[EX 7] Determine the following limits.

(1) $\lim_{x \to \infty} \sin x$

(2) $\lim_{x \to \infty} \frac{\sin x}{x}$

(3) $\lim_{x \to 0} \sin\left(\frac{1}{x}\right)$

(4) $\lim_{x \to \infty} x \sin\frac{1}{x}$

[EX 8] Determine the following limits.

(1) $\lim_{x \to -\infty} \frac{-3x+1}{\sqrt{2x^2+1}}$

(2) $\lim_{x \to \infty} \frac{x+3}{\sqrt{x^2+1}+2}$

(3) $\lim_{x \to -\infty} \frac{x+3}{\sqrt{x^2+1}+2x}$

(4) $\lim_{x \to \infty} \frac{\sqrt{x^2-1}+3}{2x}$

[EX 9] Determine the following limits.

(1) $\lim_{x\to\infty}(\sqrt{x^2+3x}-\sqrt{x^2-2x})$

(2) $\lim_{x\to-\infty}(\sqrt{x^2+x}+x)$

(3) $\lim_{x\to-\infty}(\sqrt{4x^2+3x-2}+2x)$

(4) $\lim_{x\to\infty}(2x-\sqrt{4x^2+x})$

[EX 10] For the function f whose graph is given, determine the following limits.

(1) $\lim_{x\to 4} f(x)$

(2) $\lim_{x\to 2^+} f(x)$

(3) $\lim_{x\to 2^-} f(x)$

(4) $\lim_{x\to 2} f(x)$

(5) $\lim_{x\to -3^+} f(x)$

(6) $\lim_{x\to -3^-} f(x)$

(7) $\lim_{x\to -3} f(x)$

(8) $\lim_{x\to 0^+} f(x)$

(9) $\lim_{x\to 0^-} f(x)$

(10) $\lim_{x\to 0} f(x)$

(11) $\lim_{x\to\infty} f(x)$

(12) $\lim_{x\to-\infty} f(x)$

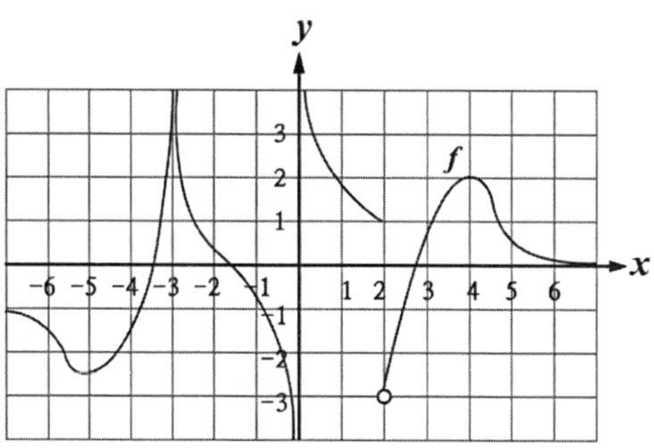

Figure 2.10

[EX 11] Sketch a graph of function $f(x)$ that satisfies all of the following conditions.

(1) - The domain of f is $(-\infty, 3)$.
- The range of f is $(-\infty, 1)$.
- $\lim_{x\to 3^-} f(x) = -\infty$.
- $\lim_{x\to -\infty} f(x) = 1$.
- $f(-2) = f(-1) = f(1) = 0$.
- $\lim_{x\to -1} f(x) = -1$.

(2) - The domain of f is all real numbers.
- The range of f is $(-3, \infty)$.
- $\lim_{x\to \infty} f(x) = 1$ and $\lim_{x\to -\infty} f(x) = -3$.
- $\lim_{x\to 0} f(x) = +\infty$.
- $\lim_{x\to -2^-} f(x) = -1$ and $\lim_{x\to -2^+} f(x) = 0$.
- $f(-2) = f(0) = -2$.

2.6. Continuity

Continuity Test

A function $f(x)$ is continuous at an interior point $x=c$ of its domain if and only if it meets the following three conditions.

1. $f(c)$ exists (*c* lies in the domain of *f*).
2. $\lim_{x\to c} f(x)$ exists (*f* has a limit as $x\to c$).
3. $\lim_{x\to c} f(x) = f(c)$ (the limit equals the function value).

Example 1

Explain why the function f below is discontinuous at $x = -2, 0, 3$ and 4.

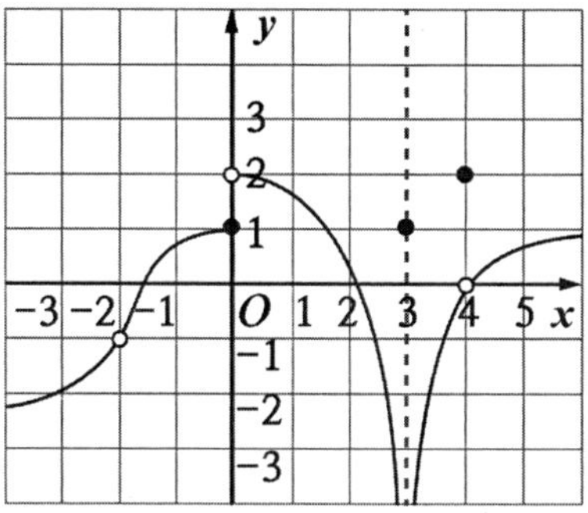

Figure 2.11

Solution

At $x=-2$, $f(-2)$ does not exist. (Removable Discontinuity)

At $x=0$, $\lim_{x\to 0} f(x)$ does not exist. (Jump Discontinuity)

At $x=3$, $\lim_{x\to 3} f(x) = -\infty$. (Infinite Discontinuity)

At $x=4$, $0=\lim_{x\to 4} f(x) \neq f(4)=2$. (Removable Discontinuity)

[EX 1] (a) Explain why the function f below is discontinuous at x = - 3, -1 , 1, 3 and 5
(b) State the type of discontinuity at each point.

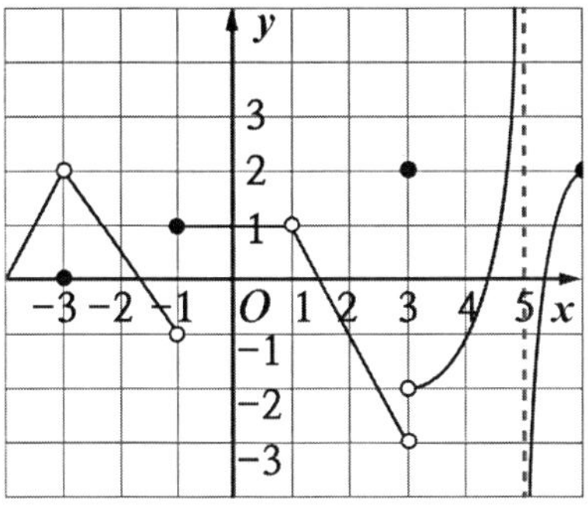

Figure 2.12

(1) At $x = -3$ (2) At $x = -1$ (3) At $x = 1$

(4) At $x = 3$ (5) At $x = 5$

[EX 2] Sketch the graph of the following function and determine any values of x at which f is discontinuous.

(1) $f(x)=\begin{cases} x+3 & , x \leq -2 \\ x^2 & , -2 < x < 1 \\ 1/x & , x \geq 1 \end{cases}$

(2) $f(x)=\begin{cases} |x+3|-1 & , x < -2 \\ \sqrt{4-x^2} & , |x| \leq 2 \\ \sqrt{x-2} & , x > 2 \end{cases}$

(3) $f(x)=\begin{cases} x+4, & x \leq -1 \\ \dfrac{|x|}{x}, & |x| < 1 \\ 1, & x = 1 \\ (x-2)^2, & x > 1 \end{cases}$

(4) $f(x)=\begin{cases} \sqrt{1-x^2}, & |x| < 1 \\ \dfrac{x+1}{x-1}, & |x| > 1 \\ 0, & x = -1 \\ 1, & x = 1 \end{cases}$

[EX 3] Determine any values of x of which f is continuous

(1) Define $f(3)$ in a way that extends $f(x)=(x^2-9)/(x-3)$ to be continuous at $x=3$.

(2) Define $f(2)$ in a way that extends $f(t)=(t^2+3t-10)/(t-2)$ to be continuous at $t=2$.

(3) Define $f(1)$ in a way that extends $f(s)=(s^3-1)/(s^2-1)$ to be continuous at $s=1$.

(4) Define $f(4)$ in a way that extends $f(x)=(x^2-16)/(x^2-3x-4)$ to be continuous at $x=4$.

[EX 4] Use the three-part definition of continuity to prove f in continuous

(1) $f(x)=\begin{cases} x+3 & , x \le -1 \\ \sqrt{x+5} & , x > -1 \end{cases}$

(2) $f(x)=\begin{cases} 3-\ln x & , 0 < x < 1 \\ 2e^{\sin(\pi x)}+1 & , x \ge 1 \end{cases}$

[EX 5] Find the constant k that makes $f(x)$ continuous everywhere.

(1) $f(x)=\begin{cases} \dfrac{x^2-4}{x-2} & , x \ne 2 \\ k & , x=2 \end{cases}$

(2) $f(x)=\begin{cases} \dfrac{x^2-2x}{x} & , x \ne 0 \\ k & , x=0 \end{cases}$

(3) $f(x)=\begin{cases} \dfrac{x^3+1}{x+1} & , x \ne -1 \\ k+3 & , x=-1 \end{cases}$

(4) $f(x)=\begin{cases} \dfrac{\sin x}{x} & , x \ne 0 \\ k & , x=0 \end{cases}$

[EX 6] Find the constants a, b that makes $f(x)$ continuous everywhere.

(1) For what values of a is $f(x)=\begin{cases} a^2x-2a, & x \geq 2 \\ 12, & x < 2 \end{cases}$ continuous at every x?

(2) For what values of b is $g(x)=\begin{cases} \dfrac{x-b}{b+1}, & x < 0 \\ x^2+b, & x > 0 \end{cases}$ continuous at every x?

(3) For what values of a, b is $f(x)=\begin{cases} -2, & x \leq -1 \\ ax-b, & -1 < x < 1 \\ 3, & x \geq 1 \end{cases}$ continuous at every x?

(4) For what values of a, b is $g(x)=\begin{cases} ax+2b, & x \leq 0 \\ x^2+3a-b, & 0 < x \leq 2 \\ 3x-5, & x \geq 2 \end{cases}$ continuous at every x?

2.7. Average Rate of Change and Instantaneous Rate of Change

The average rate of change of $y = f(x)$ with respect to x over the interval $[x_1, x_2]$ is

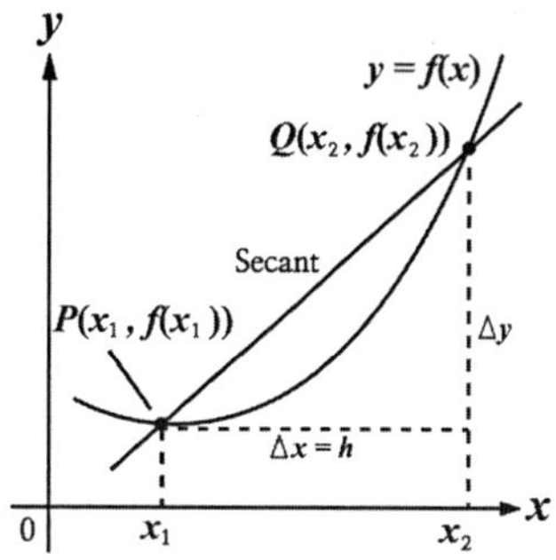

Figure 2.13

$$\frac{\Delta y}{\Delta x} = \frac{f(x_2) - f(x_1)}{x_2 - x_1} = \frac{f(x_1 + h) - f(x_1)}{h}, \; h \neq 0.$$

A secant to the graph $y = f(x)$. Its slope is $\Delta y / \Delta x$, the average rate of change of f over the interval $[x_1, x_2]$.

Example 1

Find the average rate of change for $f(x) = x^2 + 1$, with respect to x over the interval $[1, 3]$.

Solution

$$\frac{f(3) - f(1)}{3 - 1} = \frac{10 - 2}{3 - 1} = 4$$

[Ex 1] Find the average rate of change for the function over the indicated interval.

(1) $f(x) = x^3$ $\quad [a, b]$

(2) $f(x) = x^2 + 2x$ $\quad [2, 4]$

(3) $f(x) = \cot x$ $\quad \left[\frac{\pi}{4}, \frac{3\pi}{4}\right]$

(4) $f(x) = \sqrt{2x - 1}$ $\quad [1, 5]$

(5) $f(x)=\dfrac{1}{x}$ $[a,\ a+h]$ (6) $f(x)=\sin x+\cos x$ $\left[-\dfrac{\pi}{6},\ \dfrac{\pi}{3}\right]$

The Instantaneous Rate of Change of f with respect to x at $x=a$: Tangent Slope

$$f'(a)=\frac{dy}{dx}\Big|_{x=a}=\lim_{x\to a}\frac{f(x)-f(a)}{x-a} \text{ or } \lim_{h\to 0}\frac{f(a+h)-f(a)}{h}$$

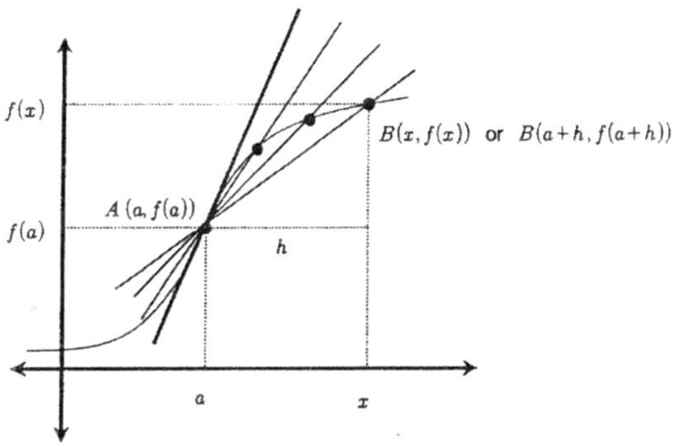

Figure 2.14

The tangent slope is the slope of the line tangent
to the curve at a point $x=a$.
As point B approaches point A,
the secant slope of
$\overline{AB}$ becomes the slope of the tangent at point A.

Example 2

Find the slope of the tangent for $f(x)=x^2+1$ at $x=2$.

Solution

(Method 1) $f'(2)=\lim\limits_{x\to 2}\dfrac{f(x)-f(2)}{x-2}=\lim\limits_{x\to 2}\dfrac{x^2+1-5}{x-2}=\lim\limits_{x\to 2}\dfrac{x^2-4}{x-2}=\lim\limits_{x\to 2}(x+2)=4$

(Method 2) $f'(2)=\lim\limits_{h\to 0}\dfrac{f(2+h)-f(2)}{h}=\lim\limits_{h\to 0}\dfrac{(2+h)^2+1-5}{h}=\lim\limits_{h\to 0}\dfrac{h^2+4h}{h}=\lim\limits_{h\to 0}(h+4)=4$

[EX 2]

(a) Find the slope of the tangent to each curve at the point indicated by using two methods.

(Method 1) $f'(a)=\lim_{x\to a}\dfrac{f(x)-f(a)}{x-a}$ (Method 2) $f'(a)=\lim_{h\to 0}\dfrac{f(a+h)-f(a)}{h}$

(b) Write the equation of the tangent line in point-slope form $y-y_1=m(x-x_1)$ $(m=f'(x_1))$

(1) $f(x)=x^2+2x$ *at* $(1,\ 3)$

(Method 1) (Method 2)

(2) $f(x)=x^2-6x$ *at* $(2,\ -8)$

(Method 1) (Method 2)

(3) $f(x)=x^3$ *at* $(-1,\ -1)$

(Method 1) (Method 2)

(4) $f(x)=\dfrac{1}{x+1}$ *at* $(1,\ \dfrac{1}{2})$

(Method 1) (Method 2)

(5) $f(x)=\sqrt{x-2}$ *at* $x=6$

(Method 1) (Method 2)

(6) $f(x)=\dfrac{1}{\sqrt{x+1}}$ *at* $x=3$

(Method 1) (Method 2)

2.8. Intermediate Value Theorem (I.V.T.)

If f is a continuous function on a closed interval $[a, b]$, and if y_0 is any value between $f(a)$ and $f(b)$, then $f(c) = y_0$ for some c in $[a, b]$.

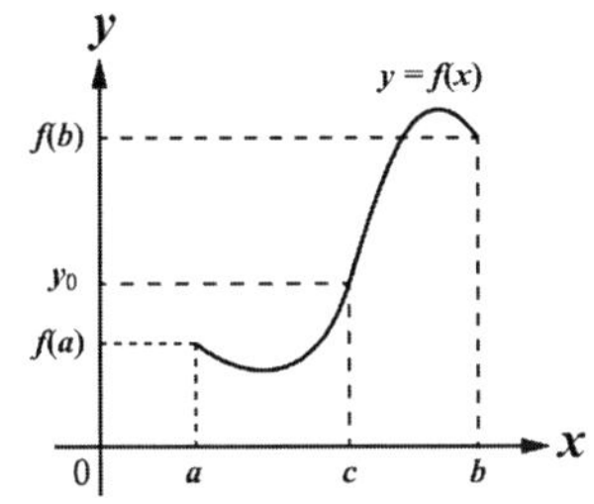

Figure 2.15

Example 1

Show that the equation $x^3 - 2x - 2 = 0$ has a solution in the interval $[1, 2]$.

Solution

Let $f(x) = x^3 - 2x - 2$.
Since f is a polynomial function, it is continuous on the interval $[1, 2]$.
$f(1) = 1 - 2 - 2 = -3$ and $f(2) = 8 - 4 - 2 = 2$, so $f(1) < 0 < f(2)$.
By the Intermediate Value Theorem, there must be at least one solution c, $1 < c < 2$, such that $f(c) = 0$.

[EX 1]

(1) Use the Intermediate Value Theorem to show that the equation $x^3 - 3x = 8$ has at least one solution in the interval $[2, 3]$.

(2) Use the Intermediate Value Theorem to show that the equation $x^3 + x^2 - 2x - 6 = 0$ has at least one solution in the interval $[1, 2]$.

[EX 2] A particle is moving along a horizontal line. The particle's velocity v and acceleration a are continuous functions of time t. The values are shown below.

t (seconds)	0	5	10	15	20
$v(t)$ (meters/second)	−3	−1	5	10	13
$a(t)$ (meters/second^2)	1	3	2	6	5

(1) Explain why there must be a value t for $5 < t < 10$, such that $v(t) = 0$.

(2) Explain why there must be a value t for $10 < t < 20$, such that $a(t) = 4$.

[EX 3]

(1) A function f is continuous on the closed interval $[1, 3]$ and has values that are shown in the table. The equation $f(x) = 0$ has at least two roots on the interval $[1, 3]$. Find the values of M.

x	1	2	3
$f(x)$	-2	M	-3

(2) A function f is continuous on the closed interval $[-2, 2]$ and has values that are shown in the table. The equation $f(x) = 0$ has at least one positive root in the interval $[-2, 2]$. Find the values of M.

x	-2	0	2
$f(x)$	-2	M	3

memo

CALCULUS

CHAPTER

03

Derivatives

3.1 Definition of Derivative

The derivative of the function $f(x)$ with respect to the variable x is the function f' whose value at x is

$$f'(x) = \lim_{h \to 0} \frac{f(x+h) - f(x)}{h}$$

provided the limit exists

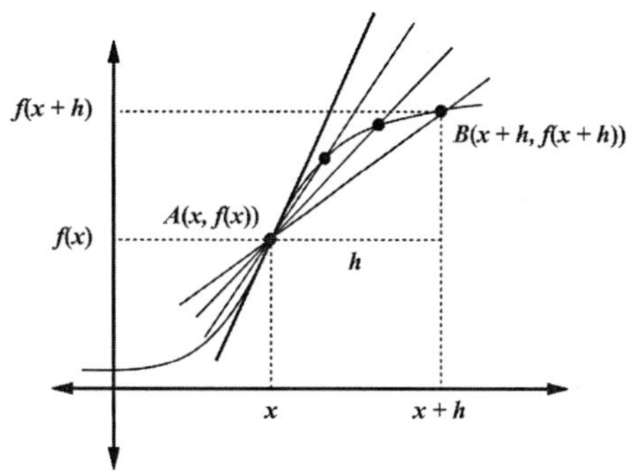

f' is the tangent slope function f.
As point B approaches point A,
the secant slope of $\overline{AB}$ becomes slope of the tangent at point A

Figure 3.1

Derivative Notation

$$f'(x) = y' = \frac{dy}{dx} = \frac{df}{dx} = \frac{d}{dx} f(x) = D(f)(x) = D_x f(x)$$

The symbols d/dx and D indicate the operation of differentiation

---

Example 1

Use the formula $f'(a) = \lim_{h \to 0} \frac{f(a+h) - f(a)}{h}$ to find the slope of the tangent to the curve $f(x) = x^2 + x$ at $x = 1$, 2, and 3.

Solution

$f'(1)=\lim_{h\to0}\frac{f(1+h)-f(1)}{h}=\lim_{h\to0}\frac{(1+h)^2+f(1+h)-2}{h}=\lim_{h\to0}\frac{h^2+3h}{h}=\lim_{h\to0}(h+3)=3$

$f'(2)=\lim_{h\to0}\frac{f(2+h)-f(2)}{h}=\lim_{h\to0}\frac{(2+h)^2+(2+h)-6}{h}=\lim_{h\to0}\frac{h^2+5h}{h}=\lim_{h\to0}(h+5)=5$

$f'(3)=\lim_{h\to0}\frac{f(3+h)-f(3)}{h}=\lim_{h\to0}\frac{(3+h)^2+(3+h)-12}{h}=\lim_{h\to0}\frac{h^2+7h}{h}=\lim_{h\to0}(h+7)=7$

Example 2

Use the of the definition to find the slope of the tangent to the curve $f(x)=(x-1)^2+1$ at $x=$-1, 0, and 2.

Solution

$f(x)=(x-1)^2+1$ and $f(x+h)=(x+h-1)^2+1\Rightarrow f'(x)=\lim_{h\to0}\frac{[(x+h-1)^2+1]-[(x-1)^2+1]}{h}$

$=\lim_{h\to0}\frac{(x^2+2xh+h^2-2x-2h+1+1)-(x^2-2x+1+1)}{h}=\lim_{h\to0}\frac{2xh+h^2-2h}{h}=\lim_{h\to0}(2x+h-2)$

$=2(x-1)$; $f'(-1)=-4$, $f'(0)=-2$, $f'(2)=2$

[EX 1] Use the definition of the derivative to find the derivative of each function.

(1) $f(x)=\frac{1}{\sqrt{3x-2}}$

(2) $f(x)=x^3-2x$

(3) $f(x)=x-\frac{1}{x}$

(4) $f(x)=\frac{1}{x^2}$

(5) $y=\frac{3x}{x+1}$

(6) $f(x)=\sqrt{x^2+1}$

3.2 Differentiability

If f is continuous at $x=a$, the left-hand derivative at $x=a$ is equal to the right-hand derivative at $x=a$,

$$\left(\lim_{h\to 0^-}\frac{f(a+h)-f(a)}{h}=\lim_{h\to 0^+}\frac{f(a+h)-f(a)}{h}\right)$$

then $f(x)$ is differentiable at $x=a$.

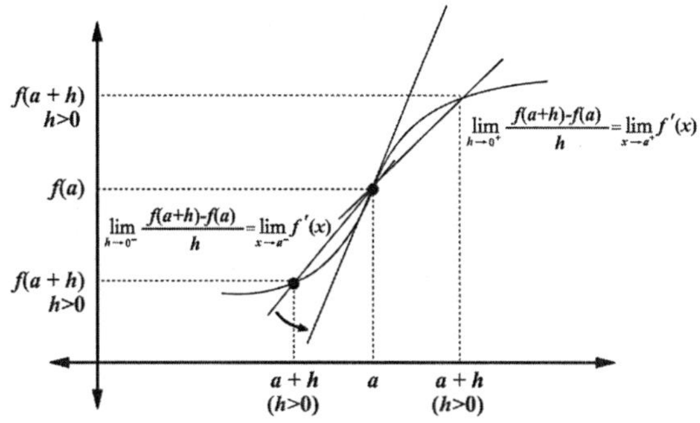

Figure 3.2

For the tangent slope to exist at x=a, the function must be continuous, and the left-hand slope must equal to the right-hand slope

Functions that are not differentiable

① Discontinuity $y=\dfrac{|x|}{x}$ ② Corner $y=|x|$ ③ Cusp $y=\sqrt[3]{x^2}$ ④ Vertical tangent $y=\sqrt[3]{x}$

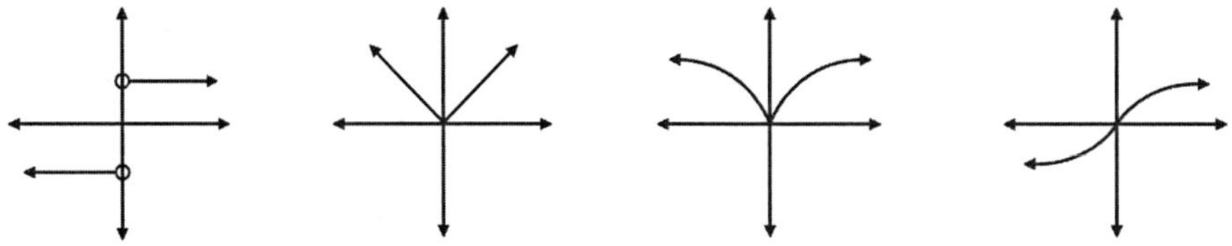

① Discontinuity	② Corner	③ Cusp	④ Vertical tangent
discontin uous at $x=0$	$\lim_{x\to 0^-} f'(x)=-1$	$\lim_{x\to 0^-} f'(x)=-\infty$	$\lim_{x\to 0^-} f'(x)=+\infty$
	$\lim_{x\to 0^+} f'(x)=1$	$\lim_{x\to 0^+} f'(x)=\infty$	$\lim_{x\to 0^+} f'(x)=+\infty$

Figure 3.3

[EX 1] Each figure in (1)~(6) shows the graph of a function over a closed interval D.

t what domain points does the function appear to be

a. differentiable?

b. continuous but not differentiable?

c. neither continuous nor differentiable?

Give reasons for your answers.

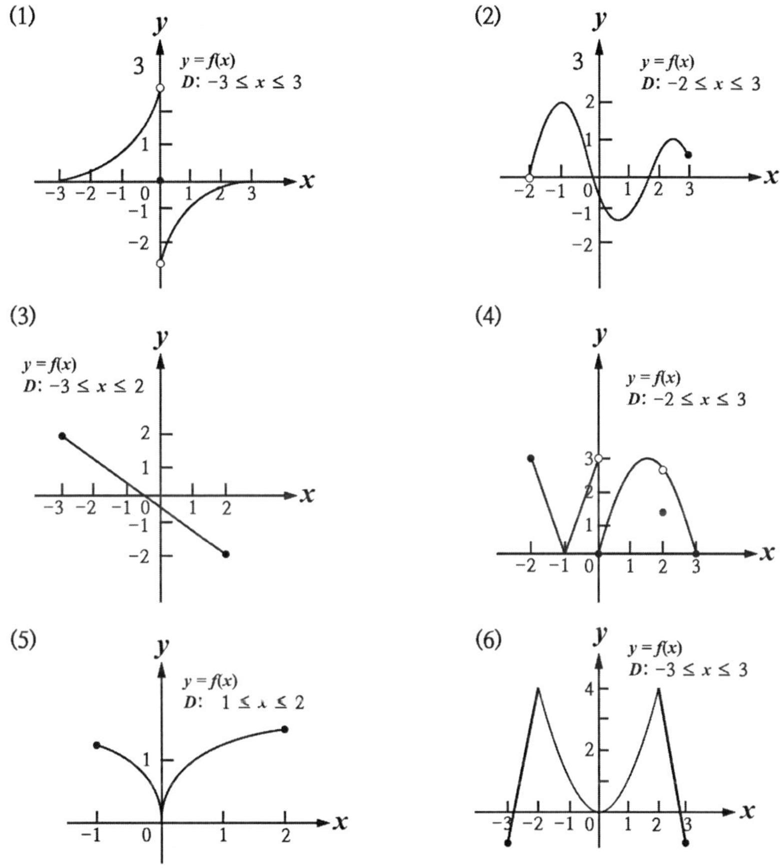

Figure 3.4

[EX 2] The graph of f is shown in the figure below. Determine whether each of the following statements is true or false.

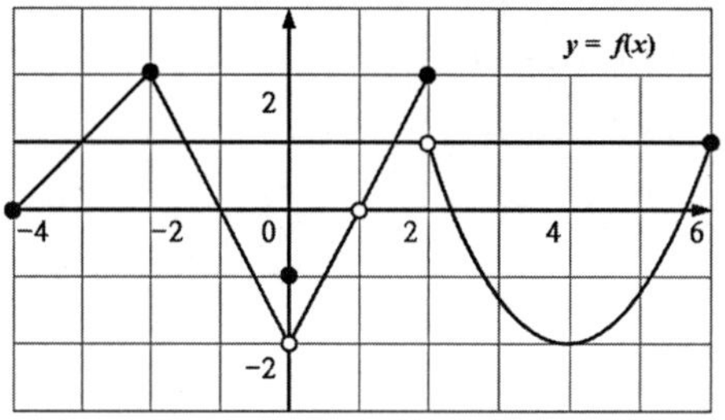

Figure 3.5

(1) $\lim_{x\to-2^-} f(x) = \lim_{x\to-2^+} f(x)$

(2) $\lim_{x\to-2} f(x) = f(-2)$

(3) $\lim_{x\to-2^-} f'(x) = \lim_{x\to-2^+} f'(x)$

(4) $\lim_{x\to0} f(x)$

(5) $\lim_{x\to0} f(x) = f(0)$

(6) $\lim_{x\to0^-} f'(x) = \lim_{x\to0^+} f'(x)$

(7) $f(1)$ is not defined

(8) $\lim_{h\to0} \dfrac{f(1+h)-f(1)}{h}$ exists

(9) $\lim_{h\to0^+} \dfrac{f(h)-f(0)}{h} = 2$

(10) $\lim_{x\to1} \dfrac{f(x)-f(1)}{x-1}$ does not exist

(11) $\lim_{x\to1^-} f'(x) = \lim_{x\to1^+} f'(x)$

(12) $\lim_{x\to2} f(x)$ does not exist

(13) $\lim_{x\to2^-} f(x) \neq f(2)$

(14) $\lim_{h\to0} \dfrac{f(3+h)-f(3)}{h}$ exists

(15) $\lim_{x\to4} f(x) = f(4)$

(16) $\lim_{x\to4} \dfrac{f(x)-f(4)}{x-4} = -2$

(17) $\lim_{x\to c} f(x)$ exists at every c in the interval (2,6)

(18) $\lim_{h\to 0^+} \frac{f(c+h)-f(c)}{h} = \lim_{h\to 0^-} \frac{f(c+h)-f(c)}{h}$ at every c in the interval (2, 6)

(19) $f'(-3)=1$

(20) $f'(-1)=0$

(21) $f'(3) < f'(4) < f'(5) < f'(5.5)$

Example 1

Sketch the graph of $f'(x)$ using the graph of $f(x)$ given below.

(1)

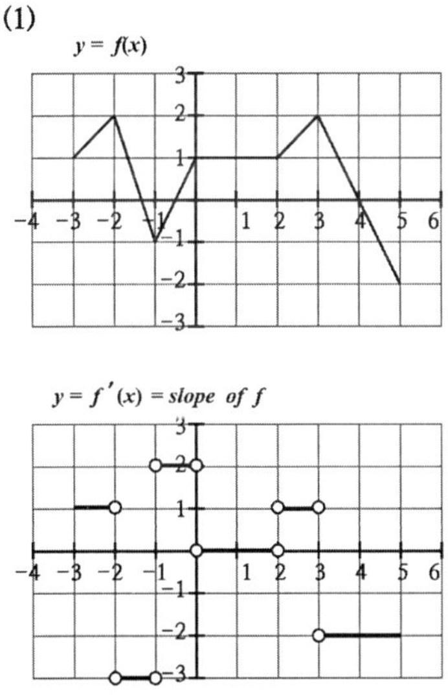

(2)

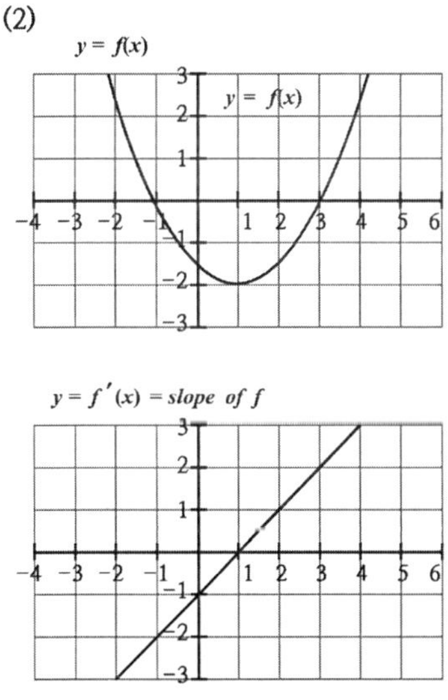

Figure 3.6

Solution

[EX 3] Sketch the graph of $f'(x)$ using the graph of $f(x)$ given below.

(1)

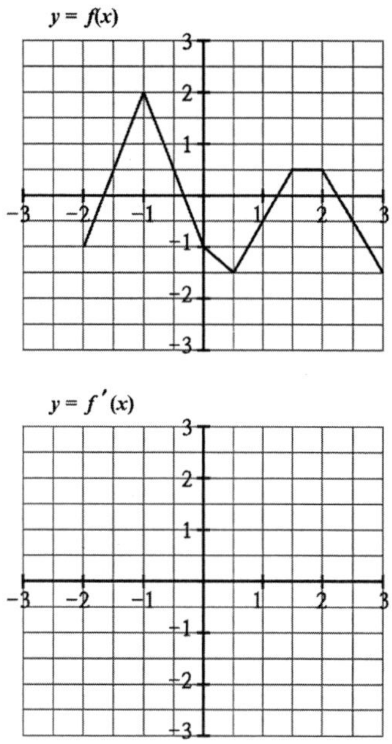

(2)

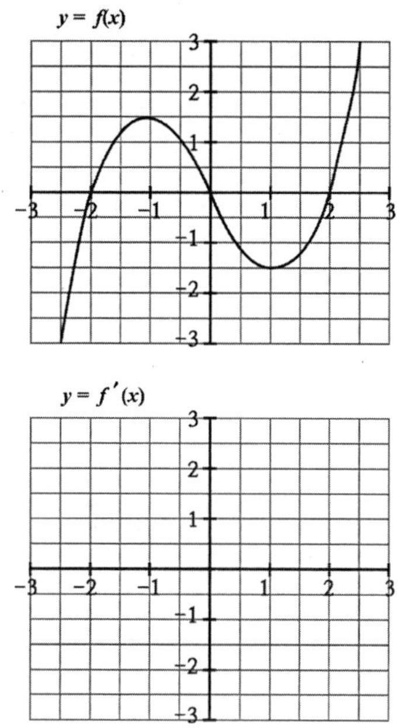

(3)

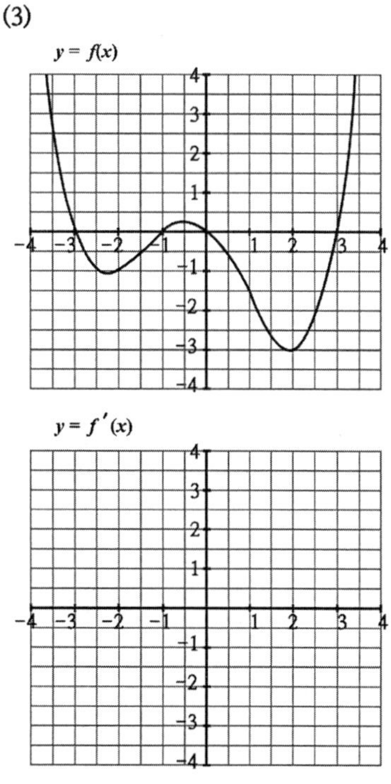

(4)

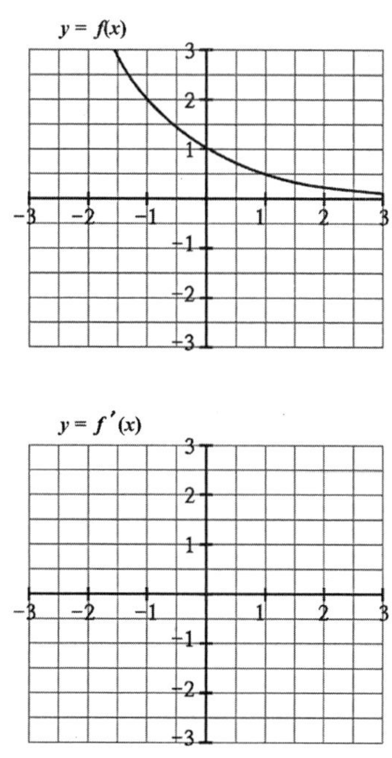

Figure 3.7

3.3 Power Rule

If n is any real number, $\frac{d}{dx}(x^n) = nx^{n-1}$

proof Let $f(x) = x^n$

$$f'(x) = \frac{d}{dx}(x^n) = \lim_{h \to 0} \frac{f(x+h) - f(x)}{h}$$

$$= \lim_{h \to 0} \frac{(x+h)^n - x^n}{h}$$

$$= \lim_{h \to 0} \left[{}_nC_0 x^n + {}_nC_1 x^{n-1}h + {}_nC_2 x^{n-2}h^2 + {}_nC_3 x^{n-3}h^3 + \cdots\cdots + {}_nC_{n-1} x h^{n-1} + {}_nC_n h^n \right] - x^n$$

$$= \lim_{h \to 0} \left[x^n + nx^{n-1}h + \frac{n(n-1)}{2!}x^{n-2}h^2 + \frac{n(n-1)(n-2)}{3!}x^{n-3}h^3 + \cdots\cdots + nxh^{n-1} + h^n \right] - x^n$$

$$= \lim_{h \to 0} \frac{nx^{n-1}h + \frac{n(n-1)}{2!}x^{n-2}h^2 + \frac{n(n-1)(n-2)}{3!}x^{n-3}h^3 + \cdots\cdots + nxh^{n-1} + h^n}{h}$$

$$= \lim_{h \to 0} \left[nx^{n-1} + \frac{n(n-1)}{2!}x^{n-2}h + \frac{n(n-1)(n-2)}{3!}x^{n-3}h^2 + \cdots\cdots + nxh^{n-2} + h^{n-1} \right]$$

$$= nx^{n-1} + 0 + 0 + 0 + \cdots\cdots + 0$$

$$= nx^{n-1}$$

Example 1

Find the derivative of the following functions.

(1) $y = x^3$ (2) $y = \frac{1}{x^4}$ (3) $y = x^{\frac{2}{3}}$

Solution

(1) $y = x^3$

$y' = 3x^2$

(2) $y = \frac{1}{x^4}$

$y' = -4x^{-5}$

(3) $y = x^{\frac{2}{3}}$

$y' = \frac{2}{3}x^{-\frac{1}{3}}$

Meaning of the notation $\frac{d}{dx}(x^3)$: Given $y=x^3$, find $\frac{dy}{dx}$ or y'.

Example 2

(1) $\frac{d}{dx}(x^{\sqrt{2}})=$ (2) $\frac{d}{dx}(x^{10})=$ (3) $\frac{d}{dx}(x^{-\frac{4}{3}})=$

Solution

(1) $\frac{d}{dx}(x^{\sqrt{2}})=\sqrt{2}x^{\sqrt{2}-1}$ (2) $\frac{d}{dx}(x^{10})=10x^9$ (3) $\frac{d}{dx}(x^{-\frac{4}{3}})=-\frac{4}{3}x^{-\frac{7}{3}}$

$\frac{d}{dx}(c)=0$

proof $y'=\lim_{h\to 0}\frac{c-c}{h}=0$

Example 3

$\frac{d}{dx}(7)=$

Solution

$\frac{d}{dx}(7)=0$

$\frac{d}{dx}(cf(x))=cf'(x)$

proof

$$y'=\lim_{h\to 0}\frac{cf(x+h)-cf(x)}{h}$$
$$=c\lim_{h\to 0}\frac{f(x+h)-f(x)}{h}=cf'(x)$$

Example 4

(1) $\frac{d}{dx}(3x^2)=$ (2) $\frac{d}{dx}(-4x^{-\frac{1}{2}})=$

Solution

(1) $\frac{d}{dx}(3x^2)=6x$ (2) $\frac{d}{dx}(-4x^{-\frac{1}{2}})=2x^{-\frac{3}{2}}$

$$\frac{d}{dx}(f(x) \pm g(x)) = \frac{d}{dx}f(x) \pm \frac{d}{dx}g(x) = f'(x) \pm g'(x)$$

proof If $y = f(x) + g(x)$,

$$y' = \lim_{h \to 0} \frac{[f(x+h)+g(x+h)]-[f(x)+g(x)]}{h} = \lim_{h \to 0} \frac{[f(x+h)-f(x)]+[g(x+h)-g(x)]}{h}$$
$$= \lim_{h \to 0} \frac{f(x+h)-f(x)}{h} + \lim_{h \to 0} \frac{g(x+h)-g(x)}{h} = f'(x)+g'(x)$$

Example 5

(1) $\frac{d}{dx}(x^3 - \frac{4}{3}x^2 - 5x) =$

(2) $\frac{d}{dx}(3x^3 - 6x^{\frac{1}{2}} + 4) =$

Solution

(1) $\frac{d}{dx}(x^3 - \frac{4}{3}x^2 - 5x) = 3x^2 - \frac{8}{3}x - 5$

(2) $\frac{d}{dx}(3x^3 - 6x^{\frac{1}{2}} + 4) = 9x^2 - 3x^{-\frac{1}{2}} = 9x^2 - \frac{3}{\sqrt{x}}$

[EX 1] Find the derivative of the following functions.

(1) $y=8$

(2) $y=k$ (k : constant)

(3) $y=-5x^4$

(4) $y=\sqrt[5]{x^4}$

(5) $y=-\dfrac{2}{x^3}$

(6) $y=\dfrac{2}{\sqrt{x}}$

[EX 2] Find the derivative of the following functions.

(1) $y=3x^2-5x+2$

(2) $y=5x^3-2\sqrt{x}$

(3) $y=1+\dfrac{1}{x}+\dfrac{2}{x^2}$

(4) $y=2(\sqrt{x}+\dfrac{1}{x})$

(5) $y=-2x^{-1}+\dfrac{4}{x^2}$

(6) $y=\dfrac{12}{x}-\dfrac{4}{x^3}+\dfrac{1}{x^4}$

[EX 3] Find the derivative of the following functions.

(1) $y=\dfrac{x^2+5x-1}{x^2}$

(2) $y=\dfrac{2x^3-3x\sqrt{x}-1}{\sqrt{x}}$

(3) $y=\dfrac{3x^{-3}+2x^{-1}-1}{x^{-2}}$

(4) $f(x)=6x^{\frac{2}{3}}(2x-1)$

(5) $f(x)=\sqrt{x^3}(x-3x^2)$

The point–slope form of the equation of a line : $y - y_1 = m(x - x_1)$ ($m = f'(x_1)$).

[EX 4]

(1) Find the equation of the tangent line to the curve $y = f(x)$ at $x = -2$, if $f(-2) = 3$ and $f'(-2) = 2$.

(2) Find the equation of the tangent line to the curve $y = g(x)$ at $x = 1$, if $g(1) = -2$ and $g'(1) = 5$.

[EX 5] Find the equation of the tangent line to the curves at given point.

(1) $y = 4 - x^2$ at $(-1, 3)$

(2) $f(x) = 4\sqrt{x} - \dfrac{1}{x}$ at $x = 1$

(3) $g(x) = \dfrac{8}{x^2} + \dfrac{2}{\sqrt{x}}$ at $x = 4$

[EX 6]

(1) Find equations for the horizontal tangents to the curve $y = x^3 - 3x - 2$. Also find equations for the lines that are perpendicular to these tangents at the points of tangency.

(2) Find all points (x, y) on the graph of $g(x) = \dfrac{1}{3}x^3 - \dfrac{3}{2}x^2 + 1$ with tangent lines parallel to the line $8x - 2y = 1$.

[EX 7]

(1) The curve $y=ax^2+bx+c$ passes through the point (1,2) and is tangent to the line $y=x$ at the origin. Find a, b, and c.

(2) The curves $y=x^2+ax+b$ and $y=cx-x^2$ have a common tangent line at the point (1,0). Find a, b, and c.

[EX 8] Find the values of a and b that make $f(x)$ differentiable everywhere.

(1) $f(x)=\begin{cases} ax, & x<0 \\ x^2-3x, & x\geq 0 \end{cases}$

(2) $f(x)=\begin{cases} ax+b, & x>-1 \\ bx^2-3, & x\leq -1 \end{cases}$

(3) $f(x)=\begin{cases} 4\sqrt{x}, & x\leq 4 \\ ax^2-bx, & x>4 \end{cases}$

(4) $f(x)=\begin{cases} ax+b, & x\leq -1 \\ ax^3+x+2b, & x>-1 \end{cases}$

3.4 Product Rule

If f and g are differentiable at x, then $f \cdot g$ is differentiable and

$$\frac{d}{dx}(f(x) \cdot g(x)) = f'(x)g(x) + f(x)g'(x)$$

proof

$$\begin{aligned}
\frac{d}{dx}(f(x) \cdot g(x)) &= \lim_{h\to 0}\frac{f(x+h)g(x+h)-f(x)g(x)}{h} \\
&= \lim_{h\to 0}\frac{[f(x+h)g(x+h)-f(x)g(x+h)]+[f(x)g(x+h)-f(x)g(x)]}{h} \\
&= \lim_{h\to 0}\frac{[f(x+h)-f(x)]g(x+h)+f(x)[g(x+h)-g(x)]}{h} \\
&= \lim_{h\to 0}\frac{[f(x+h)-f(x)]g(x+h)}{h}+\lim_{h\to 0}\frac{f(x)[g(x+h)-g(x)]}{h} \\
&= \lim_{h\to 0}\frac{[f(x+h)-f(x)]}{h} \cdot \lim_{h\to 0}g(x+h)+\lim_{h\to 0}f(x) \cdot \lim_{h\to 0}\frac{[g(x+h)-g(x)]}{h} \\
&= f'(x)g(x)+f(x)g'(x)
\end{aligned}$$

Example 1

Use the product rule to find the derivative.

$$y = (x^2 - 2)(3x^2 - 2x)$$

Solution

$$\begin{aligned}
y' &= 2x(3x^2-2x)+(x^2-2)(6x-2) \\
&= 6x^3-4x^2+6x^3-2x^2-12x+4 \\
&= 12x^3-6x^2-12x+4
\end{aligned}$$

[EX 1] Use the product rule to find the derivative.

(1) $y=(3-x^2)(x^3-x+1)$

(2) $y=(1+x^2)(x^{\frac{3}{4}}-x^{-3})$

(3) $y=(3-2x)\sqrt{x}$

[EX 2]

(1) Find the equation of the tangent line the curve $y=(x^2-2)(x^2-3x-1)$ at $x=-1$.

(2) Find the equation of the tangent line the curve $y=(x-2\sqrt{x})(x^2-6x+2)$ at $x=4$.

[EX 3] Suppose u and v are function of x that are differentiable at $x=0$ and that $u(0)=5$, $u'(0)=-3$, $v(0)=-1$, $v'(0)=2$.
Find the values of the following derivatives at $x=0$.

(1) $\frac{d}{dx}(uv)$ (2) $\frac{d}{dx}(\frac{u}{v})$ (3) $\frac{d}{dx}(\frac{v}{u})$ (4) $\frac{d}{dx}(7v-2u)$

[EX 4] Show that $\frac{d}{dx}(f \cdot g \cdot h)=f'gh+fg'h+fgh'$.

3.5 Quotient Rule

If f and g are differentiable at x, then $\frac{f}{g}$ is differentiable and

$$\frac{d}{dx}\left(\frac{f(x)}{g(x)}\right)=\frac{f'(x)g(x)-f(x)g'(x)}{g^2(x)}$$

proof

$$\frac{d}{dx}\left(\frac{f(x)}{g(x)}\right)$$

$$=\lim_{h\to 0}\frac{\frac{f(x+h)}{g(x+h)}-\frac{f(x)}{g(x)}}{h}=\lim_{h\to 0}\frac{\left[\frac{f(x+h)}{g(x+h)}-\frac{f(x)}{g(x)}\right]\cdot g(x+h)g(x)}{h\cdot g(x+h)g(x)}$$

$$=\lim_{h\to 0}\frac{f(x+h)g(x)-f(x)g(x+h)}{hg(x+h)g(x)}$$

$$=\lim_{h\to 0}\frac{[f(x+h)g(x)-f(x)g(x)]+[f(x)g(x)-f(x)g(x+h)]}{hg(x+h)g(x)}$$

$$=\lim_{h\to 0}\frac{[f(x+h)g(x)-f(x)g(x)]-[f(x)g(x+h)-f(x)g(x)]}{hg(x+h)g(x)}$$

$$=\lim_{h\to 0}\frac{[f(x+h)-f(x)]g(x)-f(x)[g(x+h)-g(x)]}{hg(x+h)g(x)}$$

$$=\lim_{h\to 0}\frac{[f(x+h)-f(x)]g(x)}{hg(x+h)g(x)}-\lim_{h\to 0}\frac{f(x)[g(x+h)-g(x)]}{hg(x+h)g(x)}$$

$$=\lim_{h\to 0}\frac{[f(x+h)-f(x)]}{h}\cdot\lim_{h\to 0}\frac{g(x)}{g(x+h)g(x)}-\lim_{h\to 0}\frac{f(x)}{g(x+h)g(x)}\cdot\lim_{h\to 0}\frac{[g(x+h)-g(x)]}{h}$$

$$=\frac{f'(x)g(x)}{g(x)g(x)}-\frac{f(x)g'(x)}{g(x)g(x)}=\frac{f'(x)g(x)-f(x)g'(x)}{g^2(x)}$$

[EX 1] Find derivative of the function $y=\dfrac{2x}{x+1}$

[EX 2] Differentiable the following functions.

(1) $y=\dfrac{2x+5}{3x-2}$

(2) $y=\dfrac{4-3x}{3x^2+x}$

(3) $y=\dfrac{(x+1)(x+2)}{(x-1)(x-2)}$

(4) $y=\dfrac{\sqrt{x}-1}{\sqrt{x}+1}$

[EX 3]

(1) Find the equation of the tangent line to the curve $y=\dfrac{1+x}{2x-1}$ at $(1,2)$

(2) Find the equation of the normal line to the curve $y=\dfrac{x}{\sqrt{x}-3}$ at $x=4$

[EX 4] At what point does the function $y=\dfrac{4x}{x^2+1}$ have a horizontal tangent line?

[EX 5] Suppose that functions f and g and their derivatives with respect to x have the following values at $x=2$ and$x=3$.

x	$f(x)$	$g(x)$	$f'(x)$	$g'(x)$
2	8	2	1/3	-3
3	3	-4	2π	5

Find the derivatives with respect to of the following combinations at the given value of x.

(1) $2f(x),\ x=2$

(2) $f(x)+g(x),\ x=3$

(3) $f(x)\cdot g(x),\ x=3$

(4) $f(x)/g(x),\ x=2$

[EX 6] If $f(3)=2, g(3)=-2, f'(3)=4$ and $g'(3)=-5$, find $\left(\frac{f}{g}\right)'(3)$.

[EX 7]

(1) Find the tangents to newton's serpentine (graphed here) at the origin and the point (1,2).

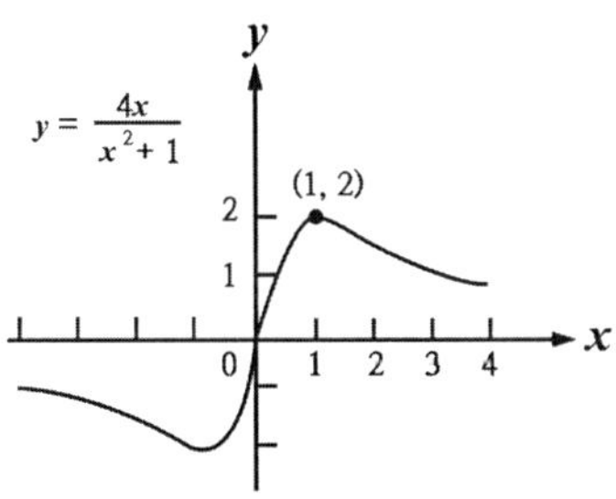

Figure 3.8

(2) Find the tangent to the Witch of Agnesi (graphed here)at the point (2,1).

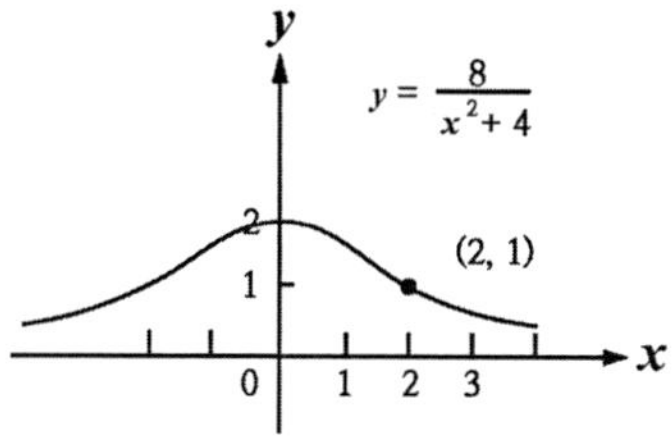

Figure 3.9

3.6 Chain Rule

If $f(u)$ is differentiable at the point $u = g(x)$ and $g(x)$ is differentiable at x, then the composite function $(f \circ g)(x) = f(g(x))$ is differentiable at x, and

$$(f \circ g)'(x) = f'(g(x)) \cdot g'(x).$$

In Leibniz's notation, if $y = f(u)$ and $u = g(x)$, then

$$\frac{dy}{dx} = \frac{dy}{du} \cdot \frac{du}{dx}$$

where dy/du is evaluated $u = g(x)$.

Example 1

Given $y = (x^3 - 4x + 1)^{10}$, find $\frac{dy}{dx}$.

Solution

$y' = 10(x^3 - 4x + 1)^9(3x^2 - 4)$

[EX 1] Find the derivative of the following functions.

(1) $y=(4-3x)^9$

(2) $y=(3x^2-\sqrt{x})^4$

(3) $y=(\frac{x^2}{8}+x-\frac{1}{x})^4$

(4) $y=\sqrt[3]{2x-x^2}$

[EX 2] Find the derivative functions.

(1) $y=\frac{1}{(2-3x)^4}$

(2) $y=(2x+1)^3(3x-4)^5$

(3) $y=(2-5x)^4(3x^2+1)^3$

(4) $y=\frac{(x+1)^2}{(2x+1)^3}$

Square Root Rule $\frac{d}{dx}(\sqrt{f(x)})=\frac{1}{2\sqrt{f(x)}}\cdot f'(x)$

proof $\frac{d}{dx}(\sqrt{f(x)})=\frac{d}{dx}[f(x)]^{\frac{1}{2}}=\frac{1}{2}[f(x)]^{-\frac{1}{2}}\cdot f'(x)$

Example 2

$\frac{d}{dx}(\sqrt{3x^2-1})=$

Solution

$$\frac{d}{dx}(\sqrt{3x^2-1})=\frac{1}{2\sqrt{3x^2-1}}\cdot 6x=\frac{3x}{\sqrt{3x^2-1}}$$

[EX 3] Find the derivative of the following functions.

(1) $y = \sqrt{x^2 - 4x}$

(2) $y = 3x^2\sqrt{x^2 + 1}$

(3) $y = \dfrac{1 - 2x}{\sqrt{4x + 1}}$

(4) $y = \dfrac{\sqrt{x^2 + 1}}{x - 1}$

(5) $y = \left(\dfrac{x + 2}{2x - 1}\right)^5$

(6) $y = \left(\dfrac{\sqrt{x} - 1}{\sqrt{x} + 1}\right)^4$

[EX 4]

(1) Find the tangent to $y = \sqrt{x^2 - x + 7}$ at $x = 2$.

(2) Find the tangent to $y = ((x - 1)/(x + 1))^2$ at $x = 0$.

[EX 5] Differentiate the following functions.

(1) $y = [f(x)]^3$

(2) $y = f(x^3)$

(3) $y = [f(x^3)]^3$

(4) $y = \sqrt{f(x) - 3x}$

(5) $y = \dfrac{1}{\sqrt{f(x)}}$

(6) $y = 3x^2 f(g(\sqrt{x}))$

(7) $y = 3[f(x^2) + 2x]^4$

[EX 6] Suppose that functions f and g and their derivatives with respect to x have the following values at $x = 0$ and $x = 1$.

x	$f(x)$	$g(x)$	$f'(x)$	$g'(x)$
0	1	1	5	1/3
1	3	-4	-1/3	-8/3

Find the derivatives with respect to of the following combinations at the given value of x.

(1) $5f(x) - g(x),\ x = 1$

(2) $f(x)g^3(x),\ x = 0$

(3) $\dfrac{f(x)}{g(x)+1},\ x = 1$

(4) $f(g(x)),\ x = 0$

(5) $g(f(x)),\ x = 0$

(6) $(x^{11} + f(x))^{-2},\ x = 1$

(7) $f(x + g(x)),\ x = 0$

3.7 Implicit Differentiation

Most of the function we have dealt with so far have beem described by an equation of the form $y = f(x)$ that expresses y explicitly in terms of the variable x. We have learned rules for differentiating functions defines in this way. Another situation occurs when we encounter equations like

$$x^3 + y^3 - 9xy = 0, \quad y^2 - x = 0, \quad \text{or} \quad x^2 + y^2 - 25 = 0$$

These equations define an *implicit* relation between the variables x and y, In some cases we may be able to solve such an equation for y as an explicit function (or even several functions) of x. When we cannot put an equation $F(x,y) = 0$ in the form $y = f(x)$ to differentiable it in the usual way, we may still be able to find dy/dx by *implicit differentiation*

1. Differentiation both sides of the equation with repect to x treating y as a differentiation function of x
2. Collect the terms with dy/dx on one side of the equation and solve for dy/dx.

Example 1

Find $\dfrac{dy}{dx}$ for the equation $xy = 2y^3 + x$

Solution

Step 1. $\dfrac{d}{dx}(xy) = \dfrac{d}{dx}(2y^3 + x)$

$1 \cdot y + x \cdot y' = 6y^2y' + 1$

$y + xy' = 6y^2y' + 1$

Step 2. $xy' - 6y^2y' = 1 - y$

$(x - 6y^2)y' = 1 - y$

$\therefore \dfrac{dy}{dx} = \dfrac{1-y}{x-6y^2}$

[EX 1] If y is a function of x, find

(1) $\frac{d}{dx}(3y^3)$ (2) $\frac{d}{dx}(-\sqrt{y})$ (3) $\frac{d}{dx}(\frac{4}{y})$

(4) $\frac{d}{dx}(xy^2)$ (5) $\frac{d}{dx}(\frac{7x}{y})$

[EX 2] Use implicit differentiation to find $\frac{dy}{dx}$.

(1) $x^3+y^3=16$ (2) $x^{\frac{2}{3}}+y^{\frac{2}{3}}=1$ (astroid)

(3) $x^2-xy+y^2=6$ (ellipse) (4) $x^3+y^3=18xy$

(5) $2\sqrt{y}=x-y$ (6) $x^2y^3-x^3y^2=-5$

[EX 3] Find the slope of the tangent to the curves as the given point.

(1) $x^2+xy-y^2=1$, $(2,3)$ (2) $(x^2+y^2)^2=(x-y)^2$, $(1,-1)$

(3) $\sqrt{4x-y^2}-4\sqrt{xy}=-6$, $(2,2)$ (4) $x=y^2\sqrt{y-1}$, $(4,2)$

[EX 4]

(1) Find the equation of the tangent line to the curve $x^3+y^3-9xy=0$ at the point $(4,2)$

(2) The line that is normal to the curve $x^2+2xy-3y^2=0$ at $(1,1)$ intersect the curve at what other point?

[EX 5]

(1) Consider the curve defined by $xy^2-x^2y=16$. Find the coordinates of the point on the curve where the tangent line is horizontal.

(2) Consider the curve defined by $x^2y-xy^3=6$. Find the coordinates of the point on the curve where the tangent line is vertical.

3.8 Higher Order Derivatives

First derivative : $y'=\dfrac{dy}{dx}$

Second derivative : $y''=\dfrac{dy'}{dx}=\dfrac{d}{dx}\left(\dfrac{dy}{dx}\right)=\dfrac{d^2y}{dx^2}$

Third derivative : $y'''=\dfrac{dy''}{dx}=\dfrac{d}{dx}\left(\dfrac{d^2y}{dx^2}\right)=\dfrac{d^3y}{dx^3}$

Fourth derivative : $y^{(4)}=\dfrac{dy'''}{dx}=\dfrac{d}{dx}\left(\dfrac{d^3y}{dx^3}\right)=\dfrac{d^4y}{dx^4}$

Fifth derivative : $y^{(5)} = \dfrac{dy^{(4)}}{dx} = \dfrac{d}{dx}\left(\dfrac{d^4y}{dx^4}\right) = \dfrac{d^5y}{dx^5}$

⋮ ⋮

nth derivative : $y^{(n)} = \dfrac{dy^{(n-1)}}{dx} = \dfrac{d}{dx}\left(\dfrac{d^{n-1}y}{dx^{n-1}}\right) = \dfrac{d^ny}{dx^n}$

Example 1

Find the first six derivatives of $y = \dfrac{1}{4}x^4 - \dfrac{2}{3}x^3 + x^2 - 6x + 2$

Solution

$$y' = x^3 - 2x^2 - 2x - 6$$
$$y'' = 3x^2 - 4x - 2$$
$$y''' = 6x - 4$$
$$y^{(4)} = 6$$
$$y^{(5)} = 0$$
$$y^{(6)} = 0$$

[EX 1] Find the second derivative of following functions.

(1) $y = \dfrac{1}{2x-1}$

(2) $y = (x^2 - 2x)^6$

(3) $y = \sqrt{x^2 + 2}$

(4) $x^2 - y^2 = 4$

[EX 2]

(1) If $f(x)=\dfrac{3x-2}{2x+3}$, find $f''(-2)$

(2) If $y=\dfrac{1}{\sqrt{1-2x}}$, find $\left.\dfrac{d^2y}{dx^2}\right|_{x=0}$

(3) If $g(x)=(3-2x^2)^5$, find $g''(1)$

(4) Find $\left.\dfrac{d^3}{dx^3}(x^n-1)\right|_{x=1}$

[EX 3]

(1) Given $G(x)=xf(x^2)$, $\dfrac{d^2G(x)}{dx^2}$

(2) Let $f(2)=-1, f'(2)=-2$ and $f''(2)=3$. For the function $F(x)=[f(x)]^2$, find $F''(2)$.

(3) Let $f(1)=2, f'(1)=3$ and $f''(1)=-2$. Find the value of $\dfrac{d^2}{dx^2}[f(x^3)]_{x=1}$.

[EX 4]

(1) Consider the curve defined by $y^2+xy=2$. Find the value of $\dfrac{d^2y}{dx^2}|_{(1,1)}$.

(2) Consider the curve defined by $x^2 - xy + y^2 = 3$. Find the value of $\frac{d^2y}{dx^2}|_{(1,-1)}$.

[EX 5] Let f be twice-differentiable function defined on interval$-1.2 < x < 3.2$ with $f(1)=2$. The graph of f' crosses the derivative of f, is shown above. The graph of f' crosses the x-axis at $x=-1$ and $x=3$ and has a horizontal tangent at $x=2$. Let g be the function given by $g(x)=e^{f(x)}$.

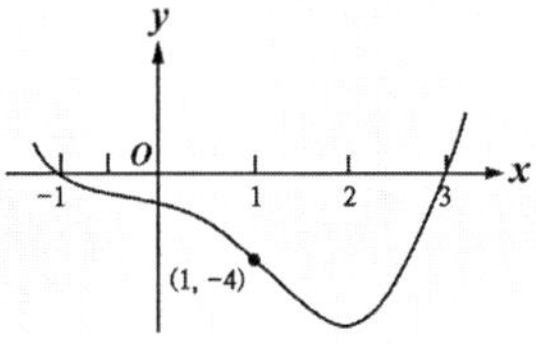

Graph of f'

Figure 3.10

(1) Write an equation for the line tangent to the graph of g at $x=1$

(2) For $-1.2 < x < 3.2$, find all values of x at which g has a local maximum. justify your answer.

(3) The second derivative of g is $g''(x) = e^{f(x)}[(f'(x))^2 + f''(x)]$. Is $g''(-1)$ positive, negative, or zero? justify your answer.

(4) Find the average rate of change of change of g', the derivative of g, over the interval [1,3]

3.9 Derivatives of Trigonometric Functions

$\frac{d}{dx}(\sin x) = \cos x$ $\frac{d}{dx}(\cos x) = -\sin x$

$\frac{d}{dx}(\tan x) = \sec^2 x$ $\frac{d}{dx}(\cot x) = -\csc^2 x$

$\frac{d}{dx}(\sec x) = \sec x \tan x$ $\frac{d}{dx}(\csc x) = -\csc x \cot x$

proof

① Let $f(x) = \sin x$.

$$f'(x) = \lim_{h\to 0}\frac{f(x+h)-f(x)}{h} = \lim_{h\to 0}\frac{\sin(x+h)-\sin x}{h} = \lim_{h\to 0}\frac{[\sin x\cos h + \cos x\sin h]-\sin x}{h}$$
$$= \lim_{h\to 0}\frac{\sin x(\cos h-1)}{h} + \lim_{h\to 0}\frac{\cos x\sin h}{h}$$
$$= \lim_{h\to 0}\sin x \cdot \lim_{h\to 0}\frac{\cos h-1}{h} + \lim_{h\to 0}\cos x \cdot \lim_{h\to 0}\frac{\sin h}{h}$$
$$= \sin x \cdot 0 + \cos x \cdot 1$$
$$= \cos x$$

② Let $f(x) = \cos x$.

$$f'(x) = \lim_{h\to 0}\frac{f(x+h)-f(x)}{h} = \lim_{h\to 0}\frac{\cos(x+h)-\cos x}{h} = \lim_{h\to 0}\frac{[\cos x\cos h + \sin x\sin h]-\cos x}{h}$$
$$= \lim_{h\to 0}\frac{\cos x(\cos h-1)}{h} + \lim_{h\to 0}\frac{\sin x\sin h}{h}$$
$$= \lim_{h\to 0}\cos x \cdot \lim_{h\to 0}\frac{\cos h-1}{h} - \lim_{h\to 0}\sin x \cdot \lim_{h\to 0}\frac{\sin h}{h}$$
$$= \cos x \cdot 0 - \sin x \cdot 1$$
$$= -\sin x$$

③ Let $f(x) = \tan x$.

$$f'(x) = \frac{d}{dx}(\tan x) = \frac{d}{dx}\left(\frac{\sin x}{\cos x}\right) = \frac{(\sin x)'\cos x - \sin x(\cos x)'}{(\cos x)^2} = \frac{\cos x\cos x - \sin x(-\sin x)}{(\cos x)^2}$$
$$= \frac{\cos^2 x + \sin^2 x}{\cos^2 x} = \frac{1}{\cos^2 x} = \sec^2 x$$

④ Let $f(x) = \cot x$.

$$f'(x) = \frac{d}{dx}(\cot x) = \frac{d}{dx}\left(\frac{\cos x}{\sin x}\right) = \frac{(\cos x)'\sin x - \cos x(\sin x)'}{(\sin x)^2} = \frac{(-\sin x)\sin x - \cos x(\cos x)}{(\sin x)^2}$$

$$= \frac{-\sin^2 x - \cos^2 x}{\sin^2 x} = \frac{-1}{\sin^2 x} = -\csc^2 x$$

⑤ Let $f(x) = \sec x$.

$$f'(x) = \frac{d}{dx}(\sec x) = \frac{d}{dx}\left(\frac{1}{\cos x}\right) = \frac{(1)'\cos x - 1(\cos x)'}{(\cos x)^2} = \frac{0-(-\sin x)}{(\cos x)^2}$$

$$= \frac{\sin x}{\cos^2 x} = \frac{1}{\cos x}\frac{\sin x}{\cos x} = \sec x \tan x$$

⑥ Let $f(x) = \csc x$.

$$f'(x) = \frac{d}{dx}(\csc x) = \frac{d}{dx}\left(\frac{1}{\sin x}\right) = \frac{(1)'\sin x - 1(\sin x)'}{(\sin x)^2} = \frac{0-(\cos x)}{(\sin x)^2}$$

$$= \frac{-\cos x}{\sin^2 x} = \frac{-1}{\sin x}\frac{\cos x}{\sin x} = -\csc x \cot x$$

Example 2

(1) $\frac{d}{dx}(\cos(x^2+1)) =$

(2) $\frac{d}{dx}(\sin\sqrt{x^2+1}) =$

Solution

(1) $\frac{d}{dx}(\cos(x^2+1)) = -\sin(x^2+1) \cdot 2x = -2x\sin(x^2+1)$

(2) $\frac{d}{dx}(\sin\sqrt{x^2+1}) = \cos(\sqrt{x^2+1}) \cdot \frac{1}{2\sqrt{x^2+1}} \cdot 2x = \frac{x\cos(\sqrt{x^2+1})}{\sqrt{x^2+1}}$

[EX 1] Find the derivative of each of the following trigonometric functions.

(1) $y = \sin(3x+1)$

(2) $y = \cos(-3x+2)$

(3) $y = 2\sin(x^3 - \frac{1}{x})$

(4) $y = \cos(\sin x)$

(5) $y = \tan(10x-5)$

(6) $y = \cot(3x^2+2x)$

(7) $y = \sec(\frac{1}{2}x)$

(8) $y = -\frac{1}{5}\csc(5x)$

(9) $y = \sec(\tan x)$

(10) $y = \sqrt{1+2\cot x}$

(11) $y = \cot(\pi - \frac{1}{x})$

(12) $y = \sin(x^3-1)^4$

[EX 2] Differentiable each of the following trigonometric functions.

(1) $y = (1+\cos 2t)^{-4}$

(2) $y = \sin^3 x$

(3) $y = 3\cos^2(x^2)$

(4) $y = \tan x^3 + \tan^3 x$

(5) $y = 2\sec^3(\sqrt[3]{x})$

(6) $y = \sin^3(2x) + \cos^3(2x)$

(7) $y = \tan^2(\sin^3 t)$

[EX 3]Differentiable each of the following trigonometric functions.

(1) $y = x\sin x$

(2) $y = x^2\cos\frac{1}{x}$

(3) $y = 3x\sqrt{\csc x}$

(4) $y = \frac{\sin x}{1+\cos x}$

(5) $y = \sin 4x \cos^2 x$

(6) $y = \sec\sqrt{x}\tan(\frac{1}{x})$

(7) $y = \frac{\cos^2 x}{\sin x}$

[EX 4] Use implicit differentiation to find $\frac{dy}{dx}$.

(1) $x\cos(2x+3y)=y\sin x$

(2) $\cos(x+y)+\sin(x-y)=1$

(3) $x+\tan(xy)=0$

(4) $x^2+y^2=\sec(xy)$

[EX 5] Find the equation of the tangent line to the curves at the given point

(1) $2xy+\pi\sin y=2\pi$ at $(1,\frac{\pi}{2})$

(2) $y=x^2\cos x$ at $x=\pi$

(3) $y=\sin x\cos 2x$ at $x=\frac{\pi}{6}$

(4) $y=2\cot^2 x-\cos^2 x$ at $x=\frac{\pi}{4}$

(5) $x\sin 2y=y\cos 2x$ at $(\frac{\pi}{4},\frac{\pi}{2})$

[EX 6] Find the second derivative of the following functions.

(1) $y=\sin^3 2x$

(2) $y=\cos^3 3x$

3.10 Derivatives of Exponential Functions

$$\frac{d}{dx}(e^x)=e^x \qquad \frac{d}{dx}(e^{f(x)})=e^{f(x)}\cdot f'(x)$$

$$\frac{d}{dx}(a^x)=a^x\cdot \ln a\ (a\neq 1, a>0) \qquad \frac{d}{dx}(a^{f(x)})=a^{f(x)}\cdot f'(x)\cdot \ln a\ (a\neq 1, a>0)$$

proof

① $f(x)=e^x$

$$f'(x)=\lim_{h\to 0}\frac{f(x+h)-f(x)}{h}=\lim_{h\to 0}\frac{e^{x+h}-e^x}{h}=\lim_{h\to 0}e^x\cdot\frac{e^{h-1}}{h}=e^x\cdot\lim_{h\to 0}\frac{e^h-1}{h}$$

Let $e^h-1=t \Rightarrow e^h=1+t$

$h=\ln(1+t)$

if $h\to 0$, then $t\to 0$.

$$f'(x)=e^x\cdot\lim_{h\to 0}\frac{e^h-1}{h}=e^x\cdot\lim_{h\to 0}\frac{t}{\ln(1+t)}=e^x\cdot\lim_{h\to 0}\frac{1}{\frac{1}{t}\ln(1+t)}$$

$$=e^x\cdot\lim_{t\to 0}\frac{1}{\ln(1+t)^{1/t}}=e^x\cdot\frac{1}{\ln e}=e^x\cdot 1=e^x$$

② $y=e^{f(x)}$

Let $y=e^u$ and $u=f(x)$

$\frac{dy}{du}=e^u$ and $\frac{du}{dx}=f'(x)$

$$\frac{dy}{dx}=\frac{dy}{du}\cdot\frac{du}{dx}=e^u\cdot f'(x)=e^{f(x)}\cdot f'(x)$$

③ $y=a^x=(e^{\ln a})^x=e^{x\ln a}$ (where $\ln a$ is a constant)

$y'=e^{x\ln a}\cdot\ln a$

$=a^x\cdot\ln a$

④ $y=a^{f(x)}=(e^{\ln a})^{f(x)}=e^{f(x)\ln a}$ (where $\ln a$ is a constant)

$y'=e^{f(x)\ln a}\cdot f'(x)\ln a$

$=a^{f(x)}\cdot f'(x)\ln a$

Example 1

(1) $\frac{d}{dx}(e^{4x}) =$ (2) $\frac{d}{dx}(e^{-3x}) =$ (3) $\frac{d}{dx}(2^{x^2}) =$

Solution

(1) $\frac{d}{dx}(e^{4x}) = e^{4x} \cdot 4$
$= 4e^{4x}$

(2) $\frac{d}{dx}(e^{-3x}) = e^{-3x} \cdot (-3)$
$= -3e^{-3x}$

(3) $\frac{d}{dx}(2^{x^2}) = 2^{x^2} \cdot 2x \cdot \ln 2$
$= 2x(2)^{x^2} \ln 2$

[EX 1] Find the derivative of each of the following functions.

(1) $y = e^{5-7x}$ (2) $y = e^{\sec x}$ (3) $y = \cos(e^{-x^2})$

(4) $y = 3^{-x^2}$ (5) $y = xe^{-x} + e^{3x}$ (6) $y = a^{\cos x}$

(7) $y = x^3 e^{2x}$ (8) $y = (9x^2 - 6x + 2)e^{x^3}$ (9) $y = \frac{e^{-2x}}{x}$

(10) $y = (e^{\sin\frac{x}{2}})^3$ (11) $y = \pi^{x\cos x}$ (12) $y = \frac{e^{-x}+1}{e^x - 1}$

(13) $e^{x^2 y} = 2x + 2y$ (14) $y = \sin(x^2 e^x)$ (15) $y = \frac{1}{\sqrt{4e^{-x}+2}}$

(16) $e^{x+y} = 3x^2 y$ (17) $e^{2x} = \sin(x + 3y)$

[EX 2] Find the equation of the tangent line to the curves at the given point

(1) $y=\dfrac{e^{-x}}{\sqrt{x}}$ at $x=1$

(2) $y=x^2e^{\frac{1}{x}}$ at $x=1$

[EX 3]

(1) Find the equation of the tangent line to the curves $y=e^{3x}$ from the external point $(0,0)$.

(2) Find the equation of the tangent line to the curves $y=xe^{2x}$ from the external point $(-2,0)$.

[EX 4] Given $f(x)=e^x\cos x$, the equation of $f''(x)-pf'(x)+qf(x)=0$ is satisfied for all real number x. Find the values of p and q.

3.11 Derivatives of Logarithmic Functions

$$\frac{d}{dx}(\ln x)=\frac{1}{x} \qquad \frac{d}{dx}(\log_a x)=\frac{1}{x}\cdot\frac{1}{\ln a}$$

$$\frac{d}{dx}(\ln f(x))=\frac{1}{f(x)}\cdot f'(x) \qquad \frac{d}{dx}(\log_a f(x))=\frac{1}{f(x)}\cdot f'(x)\cdot\frac{1}{\ln a}$$

proof

① $f(x)=\ln x$

$$f'(x)=\lim_{h\to0}\frac{f(x+h)-f(x)}{h}=\lim_{h\to0}\frac{\ln(x+h)-\ln x}{h}=\lim_{h\to0}\frac{\ln\left(\frac{x+h}{x}\right)}{h}$$

$$=\lim_{h\to0}\frac{1}{h}\ln\left(\frac{x+h}{x}\right)=\lim_{h\to0}\ln\left(1+\frac{h}{x}\right)^{\frac{1}{h}}$$

$$=\lim_{h\to0}\frac{1}{h}\ln\left[\left(\frac{x+h}{x}\right)^{\frac{x}{h}}\right]^{\frac{1}{x}}=\ln e^{\frac{1}{x}}=\frac{1}{x}\ln e=\frac{1}{x}$$

② $f(x)=\log_a x$.

$$f(x)=\log_a x=\frac{\ln x}{\ln a} \quad \text{(where } \ln a \text{ is a constant)}$$

$$f'(x)=\frac{(\ln x)'}{\ln a}=\frac{\frac{1}{x}}{\ln a}=\frac{1}{x}\cdot\frac{1}{\ln a}$$

③ $f(x)=\ln f(x)$

Let $y=\ln u$ and $u=f(x)$

$$\frac{dy}{du}=\frac{1}{u},\ \frac{du}{dx}=f'(x)$$

$$\frac{dy}{dx}=\frac{dy}{du}\cdot\frac{du}{dx}=\frac{1}{u}\cdot f'(x)=\frac{1}{f(x)}\cdot f'(x)$$

④ $f(x)=\log_a f(x)=\dfrac{\ln f(x)}{\ln a}$ (where $\ln a$ is a constant)

$$\frac{dy}{dx}=\frac{1}{f(x)}\cdot f'(x)\cdot\frac{1}{\ln a}$$

$y = \ln|f(x)|$

If $f(x) > 0$, $y = \ln|f(x)| = \ln f(x)$

$$y = \frac{1}{f(x)} \cdot f'(x)$$

If $f(x) < 0$, $y = \ln|f(x)| = \ln(-f(x))$

$$y' = \frac{1}{-f(x)} \cdot -f'(x) = \frac{1}{f(x)} \cdot f'(x)$$

$$\therefore \frac{d}{dx}(\ln|f(x)|) = \frac{1}{f(x)} \cdot f'(x)$$

Example 1

(1) $\frac{d}{dx}(\ln(x^2+1)) =$

(2) $\frac{d}{dx}(\ln(\ln(x^2+\sin x))) =$

Solution

(1) $\frac{d}{dx}(\ln(x^2+1)) = \frac{1}{x^2+1} \cdot 2x = \frac{2x}{x^2+1}$

(2) $\frac{d}{dx}(\ln(\ln(x^2+\sin x))) = \frac{1}{\ln(x^2+\sin x)} \cdot \frac{1}{x^2+\sin x} \cdot (2x+\cos x)$

$$= \frac{2x+\cos x}{(x^2+\sin x)\ln(x^2+\sin x)}$$

[EX 1] Find the derivative of each of the following functions.

(1) $y = \ln 3x$ (2) $y = \ln(\ln x)$ (3) $y = \ln(\sin x)$

(4) $y = \ln(\tan x)$ (5) $y = (\ln x)^5$ (6) $y = \log_3(\sin x)$

(7) $y = \log_2 5x$ (8) $y = \ln|xcosx|$ (9) $y = 3^{\log_2 x}$

(10) $y = x^2 \ln x$ (11) $y = x\sqrt{\ln x}$ (12) $y = \ln(\sec x + \tan x)$

(13) $y = \tan[\log(x^2+4)]$ (14) $y = (x^2 \ln x)^4$ (15) $y = \dfrac{\ln x}{x^2}$

(16) $\ln(xy) = x - y$ (17) $y + 1 = xln^3 y$ (18) $y = \ln(\sec(\ln x))$

[EX 2] Find the derivative of each of the following functions.

(1) $y = \ln((2x+1)^3(x^2+1)^4)$ (2) $y = \ln(3xe^{-x})$

(3) $y = \ln\left(\dfrac{(x^2+1)^5}{\sqrt{1-x}}\right)$ (4) $y = \ln\sqrt{\dfrac{(x+1)^5}{(x+2)^{20}}}$

(5) $y = \ln\left(\dfrac{\sqrt{\sin x \cos x}}{1+2\ln x}\right)$ (6) $y = \log_7\left(\dfrac{\sin\theta \cos\theta}{e^\theta 2^\theta}\right)$

Logarithmic Differentiation :

Step 1. Take the natural logarithm of both sides.
Step 2. Use the logarithmic properties to simplify the right side of the equation
Step 3. Differentiate both sides implicitly
Step 4. Solve for y'

Example 2

Use logarithmic differentiation to find the derivative of $y = x^x$.

Solution

Step 1. $\ln y = \ln x^x$
Step 2. $\ln y = x \ln x$
Step 3. $\frac{1}{y} \cdot y' = 1 \cdot \ln x + x \cdot \frac{1}{x}$
step 4. $y' = y(\ln x + 1)$ or $x^x(\ln x + 1)$

[EX 3] Use logarithmic differentiation to find the derivative of each of the following functions.

(1) $y = x^{\ln x}$

(2) $y = (\sin x)^x$

(3) $y = (x+1)^x$

(4) $y = (\sin x)^{\sin x}$

(5) $y = \sqrt{(x^2+1)(x-1)^2}$

(6) $y = (\ln x)^{\ln x}$

(7) $y = \dfrac{x\sqrt{x^2+1}}{(x+1)^{\frac{2}{3}}}$

(8) $y = \dfrac{\theta \sin\theta}{\sqrt{\sec\theta}}$

(9) $y = \dfrac{e^{-x}\cos x}{\sqrt[3]{x^2 - x}}$

[EX 4]

(1) Find the slope of the tangent line to the curve $y=\dfrac{(x+1)^2}{x^3(x-2)}$ at $x=1$.

(2) Find the slope of the tangent line to the curve $y=\dfrac{\sqrt[4]{2x-1}\,\sqrt{x^2+1}}{\sqrt{x+1}\,\sqrt{5-x}}$ at $x=1$.

[EX 5] Given $f(2)=1$ and $f'(2)=4$, find value of $\dfrac{d}{dx}[\sqrt{2x+\ln(f(x))}]$ at $x=2$

3.12 Inverse Trigonometric Functions

[1] $y=\sin^{-1}x=\arcsin x$

$y=\sin x$

Domain : $-\dfrac{\pi}{2}\le x\le\dfrac{\pi}{2}$

Range : $-1\le y\le 1$

$y=\sin^{-1}x$

Domain : $-1\le x\le 1$

Range : $-\dfrac{\pi}{2}\le y\le\dfrac{\pi}{2}$

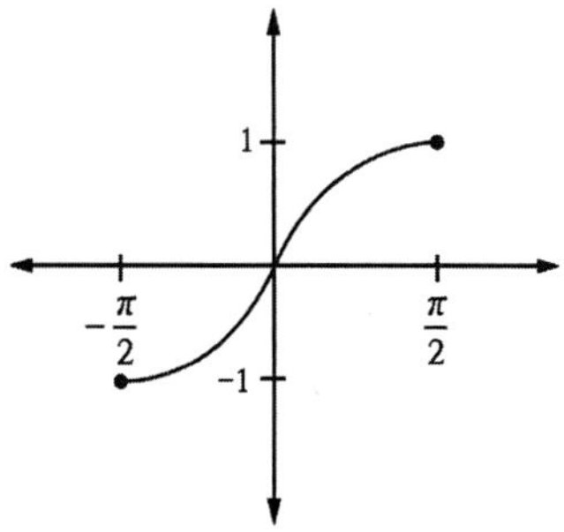

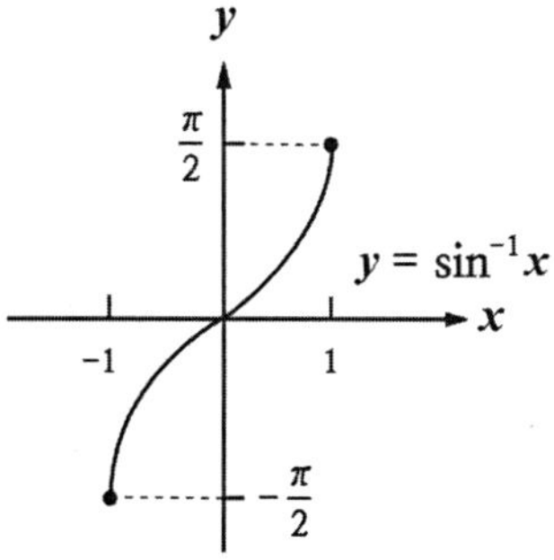

Figure 3.11

[EX 1] Evaluate

(1) $\sin^{-1}(-\frac{1}{2})$

(2) $\sin^{-1}(-\frac{\sqrt{3}}{2})$

(3) $\sin^{-1}(\frac{\sqrt{3}}{2})$

(4) $\sin^{-1}(1)$

[EX 2] Evaluate

(1) $\sin^{-1}\left(\sin\left(\frac{7\pi}{6}\right)\right)$

(2) $\sin\left(\sin^{-1}\left(-\frac{1}{2}\right)\right)$

(3) $\sin^{-1}\left(\sin\left(\frac{2\pi}{3}\right)\right)$

(4) $\cos(\sin^{-1}(-1))$

[2] $y=\cos^{-1}x=\arccos x$

$y=\cos x$

Domain : $0 \le x \le \pi$

Range : $-1 \le y \le 1$

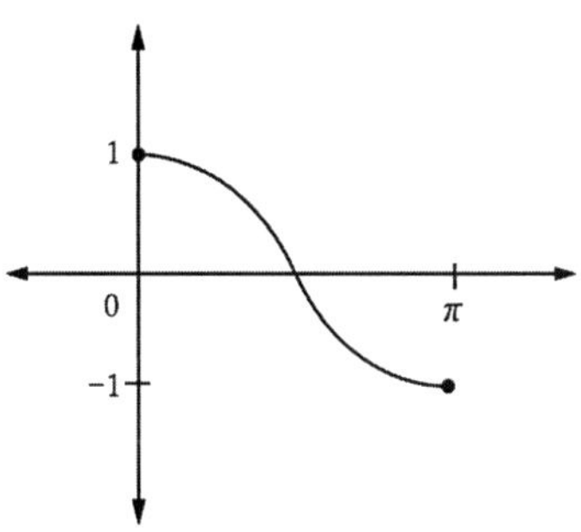

$y=\cos^{-1}x$

Domain : $-1 \le x \le 1$

Range : $0 \le y \le \pi$

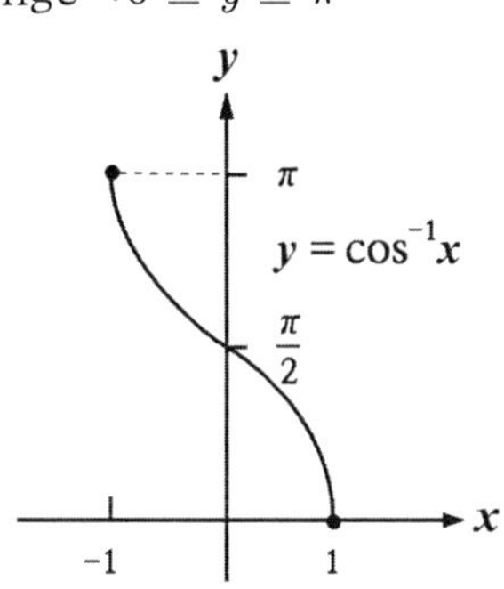

Figure 3.12

[EX 3] Evaluate

(1) $\cos^{-1}(-\frac{1}{2})$

(2) $\cos^{-1}(-\frac{\sqrt{3}}{2})$

(3) $\cos^{-1}(\frac{\sqrt{3}}{2})$

(4) $\cos^{-1}(0)$

[EX 4] Evaluate

(1) $\cos^{-1}\left(\cos\left(\frac{5\pi}{6}\right)\right)$

(2) $\cos\left(\cos^{-1}\left(-\frac{1}{2}\right)\right)$

(3) $\cos^{-1}\left(\cos\left(-\frac{\pi}{6}\right)\right)$

(4) $\tan\left(\cos^{-1}(1)\right)$

[3] $y = \tan^{-1}x = \arctan x$

$y = \tan x$

Domain : $-\frac{\pi}{2} < x < \frac{\pi}{2}$

Range : $y \in R$

$y = \tan^{-1}x$

Domain : $\infty < x < \infty$

Range : $-\frac{\pi}{2} < y < \frac{\pi}{2}$

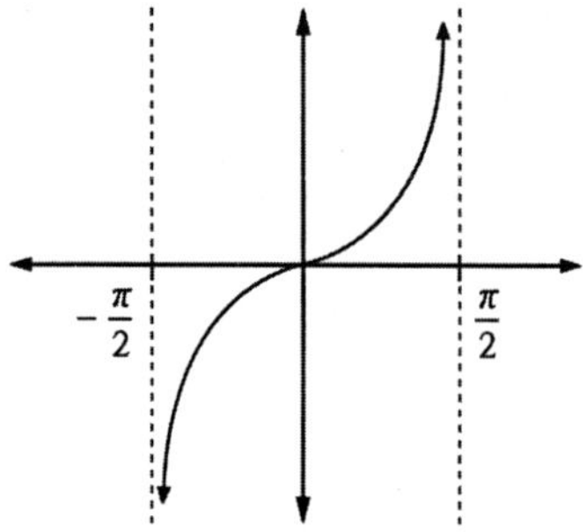

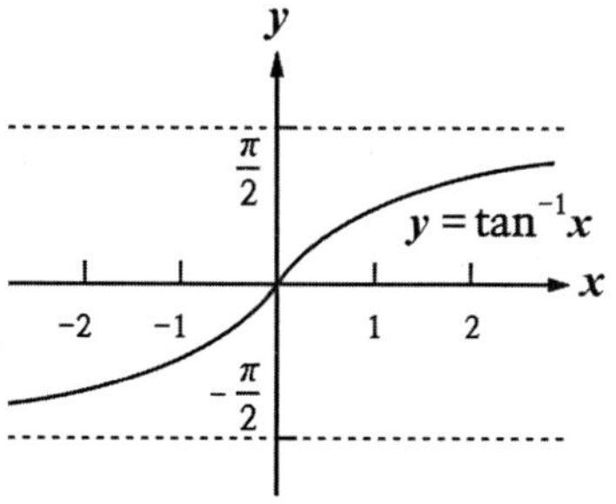

Figure 3.13

[EX 5] Evaluate

(1) $\tan^{-1}(-1)$

(2) $\tan^{-1}(\frac{1}{\sqrt{3}})$

(3) $\tan^{-1}(\sqrt{3})$

(4) $\tan^{-1}(0)$

[EX 6] Evaluate

(1) $\tan^{-1}\left(\tan\left(\frac{3\pi}{4}\right)\right)$

(2) $\tan\left(\tan^{-1}\left(-\frac{1}{\sqrt{3}}\right)\right)$

(3) $\tan^{-1}\left(\tan\left(-\frac{7\pi}{6}\right)\right)$

(4) $\cos(\tan^{-1}(-\sqrt{3}))$

[4] $y=\csc^{-1}x$, $y=\sec^{-1}x$, and $y=\cot^{-1}x$

Domain: $x \le -1$ or $x \ge 1$
Range: $0 \le y \le \pi,\ y \ne \frac{\pi}{2}$

Domain: $x \le -1$ or $x \ge 1$
Range: $-\frac{\pi}{2} \le y \le \frac{\pi}{2},\ y \ne 0$

Domain: $-\infty < x < \infty$
Range: $0 < y < \pi$

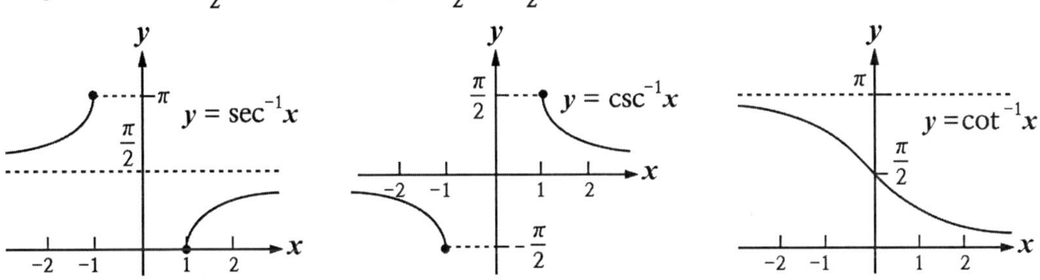

Figure 3.14

[EX 7] Evaluate

(1) $\csc^{-1}(2)$

(2) $\csc^{-1}\left(-\dfrac{2}{\sqrt{3}}\right)$

(3) $\sec^{-1}(-\sqrt{2})$

(4) $\csc^{-1}(undefined)$

(5) $\sec^{-1}(2)$

(6) $\sec^{-1}(undefined)$

(7) $\cot^{-1}\left(\dfrac{1}{\sqrt{3}}\right)$

(8) $\cot^{-1}(-\sqrt{3})$

[EX 8] Evaluate

(1) $\sin\left(\cos^{-1}\left(\dfrac{\sqrt{2}}{2}\right)\right)$

(2) $\sec\left(\cos^{-1}\left(\dfrac{1}{2}\right)\right)$

(3) $\tan\left(\sin^{-1}\left(-\dfrac{1}{2}\right)\right)$

(4) $\cot\left(\sin^{-1}\left(-\dfrac{\sqrt{3}}{2}\right)\right)$

3.13 Derivatives of Inverse Trigonometric Functions

$$\frac{d}{dx}(\sin^{-1}x)=\frac{1}{\sqrt{1-x^2}} \qquad \frac{d}{dx}(\cos^{-1}x)=-\frac{1}{\sqrt{1-x^2}}$$

$$\frac{d}{dx}(\tan^{-1}x)=\frac{1}{1+x^2} \qquad \frac{d}{dx}(\cot^{-1}x)=-\frac{1}{1+x^2}$$

$$\frac{d}{dx}(\sec^{-1}x)=\frac{1}{|x|\sqrt{x^2-1}} \qquad \frac{d}{dx}(\csc^{-1}x)=-\frac{1}{|x|\sqrt{x^2-1}}$$

proof

① $y=\sin^{-1}x$ where $-\frac{\pi}{2}\le y\le\frac{\pi}{2}$

$$\sin y=x$$

$$\frac{d}{dx}(\sin y)=\frac{d}{dx}(x)$$

$$\cos y \cdot y'=1$$

$$y'=\frac{1}{\cos y}$$

$$=\frac{1}{\sqrt{1-x^2}}$$

$$\sin^2 y+\cos^2 y=1$$

$$\cos y=\pm\sqrt{1-\sin^2 y}=\pm\sqrt{1-x^2}$$

For $-\frac{\pi}{2}\le y\le\frac{\pi}{2}$, $\cos y>0$

$$\therefore \cos y=\sqrt{1-x^2}$$

② $y=\cos^{-1}x$ where $0\le y\le\pi$

$$\cos y=x$$

$$\frac{d}{dx}(\cos y)=\frac{d}{dx}(x)$$

$$-\sin y \cdot y'=1$$

$$y'=-\frac{1}{\sin y}$$

$$=-\frac{1}{\sqrt{1-x^2}}$$

$$\sin^2 y+\cos^2 y=1$$

$$\sin y=\pm\sqrt{1-\cos^2 y}=\pm\sqrt{1-x^2}$$

For $0\le y\le\pi$, $\sin y>0$

$$\therefore \sin y=\sqrt{1-x^2}$$

③ $y = \tan^{-1}x$ $\qquad \tan y = x \qquad$ where $-\frac{\pi}{2} < y < \frac{\pi}{2}$

$$\frac{d}{dx}(\tan y) = \frac{d}{dx}(x)$$

$\sec^2 y \cdot y' = 1 \qquad\qquad \tan^2 y + 1 = \sec^2 y$

$$y' = \frac{1}{\sec^2 y}$$

$$= \frac{1}{1+x^2}$$

④ $y = \cot^{-1}x$ $\qquad \cot y = x \qquad$ where $0 < y < \pi$

$$\frac{d}{dx}(\cot y) = \frac{d}{dx}(x)$$

$-\csc^2 y \cdot y' = 1 \qquad\qquad \cot^2 y + 1 = \csc^2 y$

$$y' = -\frac{1}{\csc^2 y}$$

$$= -\frac{1}{1+x^2}$$

⑤ $y = \sec^{-1}x$ $\qquad$ where $0 \le y \le \pi,\ y =\neq \frac{\pi}{2}$ $\qquad \sec y = x$

$$\frac{d}{dx}(\sec y) = \frac{d}{dx}(x)$$

$\sec y \tan y \cdot y' = 1$

$$y' = \frac{1}{\sec y \tan y} \qquad\qquad \tan^2 y + 1 = \sec^2 y$$

$$= \pm\frac{1}{x\sqrt{x^2-1}} \qquad\qquad \tan y = \pm\sqrt{\sec^2 y - 1}$$

$$\tan y = \pm\sqrt{x^2-1} \text{ and } \sec y = x$$

$$= \frac{1}{|x|\sqrt{x^2-1}}$$

⑥ $y = \csc^{-1}x$ $\qquad$ where $0 < y < \pi,\ y = \frac{\pi}{2}$

$\csc y = x$

$$\frac{d}{dx}(\csc y) = \frac{d}{dx}(x) \qquad \cot^2 y + 1 = \csc^2 y$$

$$-\csc y \cot y \cdot y' = 1 \qquad \cot y = \pm\sqrt{\csc^2 y - 1}$$

$$y' = -\frac{1}{\csc y \cot y} \qquad \longleftarrow \qquad \cot y = \pm\sqrt{x^2-1}\ ,\ \csc y = x$$

$$= \pm\frac{-1}{x\sqrt{x^2-1}}$$

$$= \frac{-1}{|x|\sqrt{x^2-1}}$$

Example 2

(1) $\frac{d}{dx}(\sin^{-1}x^2) =$ (2) $\frac{d}{dx}(\tan^{-1}(\sin x)) =$

Solution

(1) $\frac{d}{dx}(\sin^{-1}x^2) = \frac{1}{\sqrt{1-(x^2)^2}} \cdot 2x = \frac{2x}{\sqrt{1-x^4}}$

(2) $\frac{d}{dx}(\tan^{-1}(\sin x)) = \frac{1}{1+(\sin x)^2} \cdot \cos x = \frac{\cos x}{1+\sin^2 x}$

[EX 1] Find the derivative of each of the following inverse trigonometric functions.

(1) $y = \sin^{-1}(\frac{3}{x^2})$ (2) $y = \cos^{-1}(-3x+2)$

(3) $y = \sin^{-1}(1-x)$ (4) $y = 2\cos^{-1}(\sqrt{x-1})$

(5) $y = \tan^{-1}(\ln x)$

(6) $y = \cot^{-1}\left(\frac{1}{x}\right)$

(7) $y = \ln(x^2+4) - x\tan^{-1}\left(\frac{x}{2}\right)$

(8) $y = \cot^{-1}(\csc x^2)$

(9) $y = \sec^{-1}(2x+1)$

(10) $y = \sin^{-1}((1-x^2)^2)$

(11) $y = \tan^{-1}\sqrt{x^2-1} + \csc^{-1}x, \ \ x > 1$

(12) $y = \sqrt{x^2-1} + \csc^{-1}x, \ \ x > 1$

[EX 2] Find the slope of the tangent line to the curves at the given point

(1) $y = (\cos^{-1}x)^4 \ \ at \ x = 0$

(2) $y = x\tan^{-1}(x^2-1) \ \ at \ x = \sqrt{2}$

[EX 3] Find the equation of the tangent line to the curves at the given point

(1) $y = \sec^{-1}(\sqrt{x}) \ at \ x = 4$

(2) $f(x) = (6\tan^{-1}x)^3 \ \ at \ \ x = \sqrt{3}$

3.14 Derivatives of Inverse Functions

Let f be differentiable and a one-to-one function on a domain.

$$\frac{d}{dx}[f^{-1}(x)] = (f^{-1})'(x) = \frac{1}{f'(f^{-1}(x))}$$

proof if f has an inverse function, then $f(f^{-1}(x)) = x$ for every x in the domain.

$$\begin{aligned} f(f^{-1}(x)) &= x \\ \frac{d}{dx} f(f^{-1}(x)) &= 1 \\ f'(f^{-1}(x)) \cdot \frac{d}{dx} f^{-1}(x) &= 1 \\ \frac{d}{dx} f^{-1}(x) &= \frac{1}{f'(f^{-1}(x))} \end{aligned}$$

If f has an interval I as domain and $f'(x)$ exist and is never zero on I. then f^{-1} is differentiable at every point in its domain (the range of f). The value of the $(f^{-1})'$ at a point b in the domain of f^{-1} is the reciprocal of the value of f' at the point $a = f^{-1}(b)$:

$$(f^{-1})'(b) = \frac{1}{f'(f^{-1}(b))} \quad \text{or} \quad \left.\frac{df^{-1}}{dx}\right|_{x=b} = \frac{1}{\left.\frac{df}{dx}\right|_{x=f^{-1}(b)}}$$

Example 1

Let $y = f(x) = x^3 + x - 2$, and let g be the inverse function. Evaluate $g'(0)$.

Solution

Since $f'(x) = 3x^2 + 1$, $g'(y) = \dfrac{1}{3x^2 + 1}$. To find x

when $y = 0$, we must solve the equation $x^3 + x - 2 = 0$. Note by inspection that $x = 1$,

so $g'(0) = \dfrac{1}{3(1)^2 + 1} = \dfrac{1}{4}$.

[EX 1]

(1) A function f and its derivative take on the values shown in the table. If g is the inverse of f, find $g'(6)$.

x	$f(x)$	$f'(x)$
2	6	$\frac{1}{3}$
6	8	$\frac{3}{2}$

(2) If $f(3)=8$ and $f'(3)=5$, what do we know about f^{-1}?

[EX 2]

(1) Suppose $y=f(x)=2x^3-3x$. If $h(x)$ is the inverse function of f, then $h'(-1)=$

(2) Let $f(x)=x^3-3x^2-1, x\geq 2$. Find the value of df^{-1}/dx at the point $x=-1=f(3)$.

(3) Let $f(x)=x^2-4x-5, x>2$. Find the value of df^{-1}/dx at the point $x=0=f(5)$.

(4) Suppose that the differentiable function $y=f(x)$ has an inverse and that the graph of f passes through the point (2,4)and has a slope of 1/3 there. Find the value of df^{-1}/dx *at* $x=4$.

(5) Suppose that the differentiable function $y = g(x)$ has an inverse and that the graph of g passes through the origin with slope 2. Find the slope of the graph of g^{-1} at the origin.

(6) Let $f(x) = e^x + x + 1$ and $f(g(x)) = x$. Find the value of $g'(2)$.

Basic Derivative Formulas

$$\frac{d}{dx}(x^n) = nx^{n-1}, \qquad \frac{d}{dx}(\sqrt{x}) = \frac{1}{2\sqrt{x}}, \qquad \frac{d}{dx}\left(\frac{1}{x}\right) = -\frac{1}{x^2}$$

$$\frac{d}{dx}(f(x) \cdot g(x)) = f'(x)g(x) + f(x)g'(x), \qquad \frac{d}{dx}\left(\frac{f(x)}{g(x)}\right) = \frac{f'(x)g(x) - f(x)g'(x)}{g^2(x)}$$

$$\frac{d}{dx}f(g(x)) = f'(g(x)) \cdot g'(x), \qquad \frac{d}{dx}f(g(h(x))) = f'(g(h(x))) \cdot g'(h(x)) \cdot h'(x)$$

$$\frac{d}{dx}(\sqrt{f(x)}) = \frac{1}{2\sqrt{f(x)}} \cdot f'(x)$$

Derivatives of Trigonometric Formulas

$$\frac{d}{dx}(\sin x) = \cos x \qquad \frac{d}{dx}(\cos x) = -\sin x$$

$$\frac{d}{dx}(\tan x) = \sec^2 x \qquad \frac{d}{dx}(\cot x) = -\csc^2 x$$

$$\frac{d}{dx}(\sec x) = \sec x \tan x \qquad \frac{d}{dx}(\csc x) = -\csc x \cot x$$

Derivatives of Inverse Trigonometric Formulas

$$\frac{d}{dx}(\sin^{-1}x) = \frac{1}{\sqrt{1-x^2}} \qquad \frac{d}{dx}(\cos^{-1}x) = -\frac{1}{\sqrt{1-x^2}}$$

$$\frac{d}{dx}(\tan^{-1}x) = \frac{1}{1+x^2} \qquad \frac{d}{dx}(\cot^{-1}x) = -\frac{1}{1+x^2}$$

$$\frac{d}{dx}(\sec^{-1}x) = \frac{1}{|x|\sqrt{x^2-1}} \qquad \frac{d}{dx}(\csc^{-1}x) = -\frac{1}{|x|\sqrt{x^2-1}}$$

Derivatives of Exponential and Logarithmic Fuctions Formulas

$$\frac{d}{dx}(e^x) = e^x \qquad \frac{d}{dx}(a^x) = a^x \ln a \quad (a \neq 1,\ a > 0)$$

$$\frac{d}{dx}(e^{f(x)}) = e^{f(x)} \cdot f'(x) \qquad \frac{d}{dx}(a^{f(x)}) = a^{f(x)} \cdot f'(x) \cdot \ln a$$

$$\frac{d}{dx}(\ln x) = \frac{1}{x} \qquad \frac{d}{dx}(\log_a x) = \frac{1}{x} \cdot \frac{1}{\ln a}$$

$$\frac{d}{dx}(\ln f(x)) = \frac{1}{f(x)} \cdot f'(x) \qquad \frac{d}{dx}(\log_a f(x)) = \frac{1}{f(x)} \cdot f'(x) \cdot \frac{1}{\ln a}$$

Derivatives of Inverse Formulas

$$(f^{-1})'(x) = \frac{1}{f'(f^{-1}(x))}$$

memo

CALCULUS

CHAPTER 04

Application of Derivatives

4.1 Critical Points

An interior point of the domain of a function f where f' is zero or undefined is a **critical point** of f.

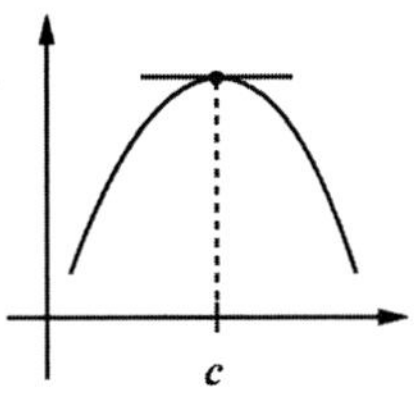

$f'(c) = 0$

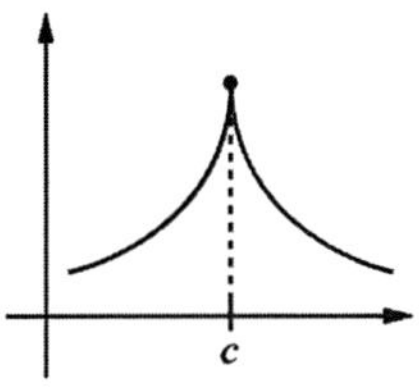

$f'(c) =$ undefined

Figure 4.1

Example 1

Determine the critical points of the function $y = x^2 - 32\sqrt{x}$

Solution

$y = x^2 - 32\sqrt{x} \Rightarrow y' = 2x - \dfrac{16}{\sqrt{x}} = \dfrac{2x^{3/2} - 16}{\sqrt{x}} = 0 \Rightarrow 2x^{3/2} - 16 = 0 \Rightarrow x = 4; \dfrac{2x^{3/2} - 16}{\sqrt{x}} =$ undefined $\Rightarrow \sqrt{x} = 0 \Rightarrow x = 0$. The critical points are $x = 4$ and $x = 0$

[EX 1] Determine the critical points of the following functions.

(1) $y = 6x^2 - x^3$

(2) $y = x^2 + \dfrac{2}{x}$

(3) $y = \sqrt{2x - x^2}$

(4) $y = \sqrt[3]{x}(4 + x)$

(5) $y = x^2 \ln x$

(6) $y = x + 2\sin x, \ [0, 2\pi]$

4.2 Extrema : Maximum & Minimum value

Definition Let f be a function with domain D. Then f has an **absolute maximum** value on D at a point c if

$$f(x) \le f(c) \qquad \text{for all } x \text{ in } D$$

and an **absolute minimum** value on D at c if

$$f(x) \ge f(c) \qquad \text{for all } x \text{ in } D.$$

Definition A function f has a local maximum value at a point c within its domain D if $f(x) \le f(c)$ for all $x \in D$ lying in some open interval containing c.

A function f has a local minimum value at a point c within its domain D if $f(x) \ge f(c)$ for all $x \in D$ lying in some open interval containing c.

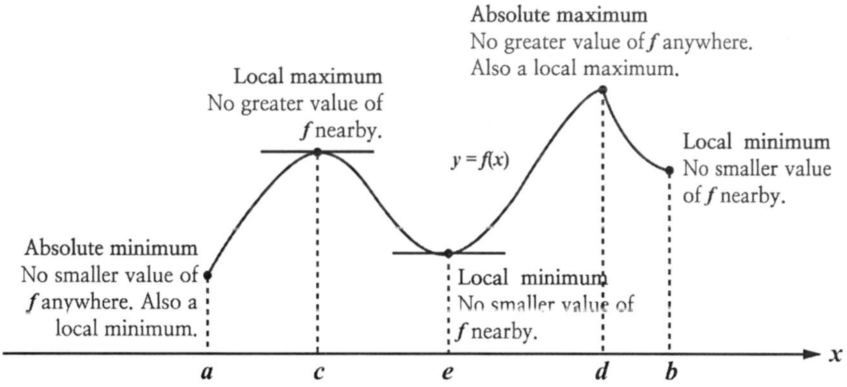

Figure 4.2

[EX 1] Determine whether the function has an absolute maximum or minimum, a relative maximum or minimum, or neither a maximum nor a minimum for each point.

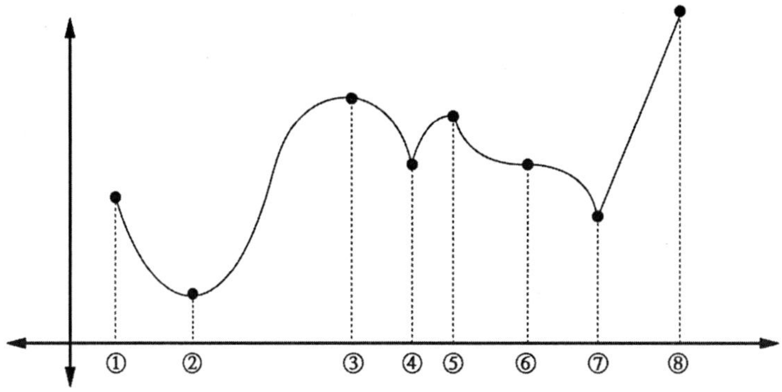

Figure 4.3

Increasing & Decreasing Functions

Suppose that f is continuous on [a, b] and differentiable on (a, b)

If $f'(x) > 0$ at each point $x \in (a, b)$, then f is increasing on [a, b].
If $f'(x) < 0$ at each point $x \in (a, b)$, then f is decreasing on [a, b].

$f'(x) > 0$ $\Leftrightarrow$ The graph of $f(x)$ is

Figure 4.4

$f'(x) < 0$ $\Leftrightarrow$ The graph of $f(x)$ is

Figure 4.5

Example 1

Find the intervals on which the function $h(x)=-x^3+2x^2$ is increasing/decreasing.

Solution

$h(x)=-x^3+2x^2 \Rightarrow h'(x)=-3x^2+4x=x(4-3x) \Rightarrow$ critical points at $x=0, \frac{4}{3}$

$\Rightarrow h' = \underset{0}{---|}\underset{4/3}{+++|}---$, increasing on $(0,\frac{4}{3})$, decreasing on $(-\infty,0)$ and $(\frac{4}{3},\infty)$

[EX 2] Find the intervals on which the fnction is increasing/decreasing.

(1) $y=x^4-4x^3+4x^2$

(2) $h(x)=\ln x+\frac{1}{x}$

(3) $y=x+2\cos x$, $[0, \pi]$

(4) $y=x^3e^{-x}$

(5) $y=\frac{2x}{x^2+4}$

(6) $h(x)=x^2\ln x$

First Derivative Test for Local Extrema

Suppose that c is a critical point of a continuous function f, and that f is differentiable at every point in some interval containing c except possibly at c itself. Moving across this interval from left to right,

1. if f' changes from negative to positive at c, then f has a local minimum at c ;
2. if f' changes form positive to negative at c, then f has a local maximum at c ;
3. if f' does not change sign at c (that is, f' is positive on both sides of c or negative on both sides), then f has no local extremum at c.

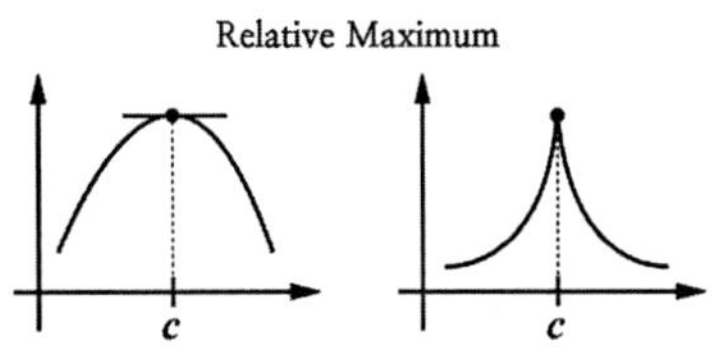

Figure 4.6

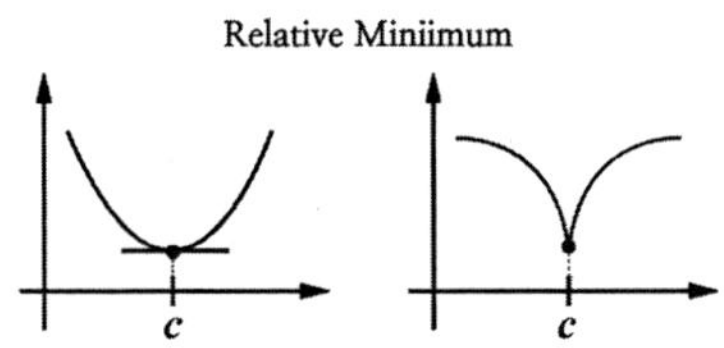

Figure 4.7

Example 2

Find the relative extreme values of the function $f(\theta)=3\theta^2-4\theta^3$

Solution

$f(\theta)=3\theta^2-4\theta^3 \Rightarrow f'(\theta)=6\theta-12\theta^2=6\theta(1-2\theta) \Rightarrow$ critical points at $\theta=0, \frac{1}{2}$

$\Rightarrow f' = ---|+++|---$, increasing on $(0,\frac{1}{2})$, decreasing on $(-\infty,0)$ and $(\frac{1}{2},\infty)$
(sign changes at 0 and $1/2$)

a local maximum is $f(\frac{1}{2})=\frac{1}{4}$ at $\theta=\frac{1}{2}$, a local minimum is $f(0)=0$ at $\theta=0$, no absolute extrema

[EX 3] Use the first derivative test to find the relative extreme values of the function and where they occur.

(1) $k(x)=15t^3-t^5$

(2) $y=\dfrac{e^x}{x}$

(3) $y=x^2e^{-x}$

(4) $f(x)=x^{\frac{1}{3}}(x+8)$

(5) $g(x)=x^{\frac{2}{3}}(x+5)$

The Extreme Value Theorem

If f is continuous on a closed interval [a, b], then f attains both an absolute maximum value M and an absolute minimum value m in [a, b]. That is, there are numbers x_1 and x_2 in [a, b] with $f(x_1)=m$, $f(x_2)=M$, and $m \le f(x) \le M$ for every other x in [a, b].

Example 3

Find the absolute extreme values of the function given by $g(x)=e^{-x^2}$, $-2 \le x \le 1$

Solution

$g(x)=e^{-x^2} \Rightarrow g'(x)=-2xe^{-x^2}$

$\Rightarrow$ a critical point at

$x=0$; $g(-2)=e^{-4}$, $g(0)=1$, and $g(1)=e^{-1}$

$\Rightarrow$ the absolute maximum is 1 at $x=0$ and the absolute minimum is e^{-4} at $x=-2$

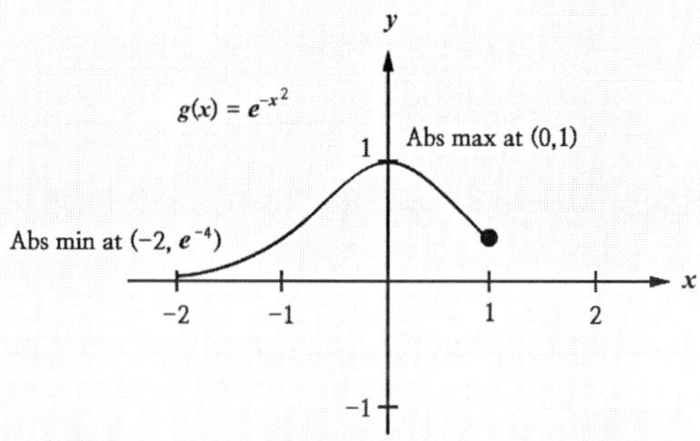

Figure 4.8

[EX 4] Find the absolute extreme values of the function and where they occur.

(1) $h(x)=-3x^{\frac{2}{3}}$, $[-1,1]$

(2) $f(x)=12x-x^3$, $[-3,4]$

(3) $y=\dfrac{\ln x}{x}$, $[1,e]$

(4) $y=xe^{-x}$, $[-1,1]$

(5) $y=x\sqrt{4-x^2}$, $[0,2]$

(6) $g(x)=4x-6\sqrt[3]{x^2}$, $[-1,8]$

(7) $f(x)=2\cos x-\cos 2x$, $[0,2\pi]$

4.3 Concavity

The graph of a differentiable function $y = f(x)$ is

(a) **concave up** on an open interval I if f' is increasing on I;
(b) **concave down** on an open interval I if f' is decreasing on I.

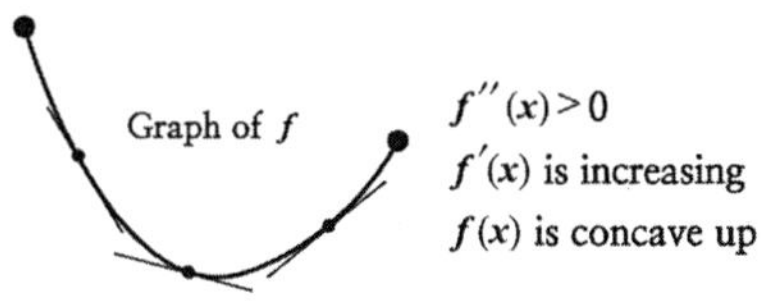

Figure 4.9

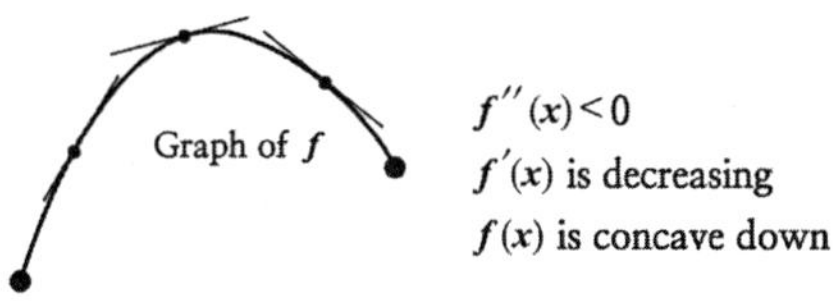

Figure 4.10

Point of Inflection

A point where the graph of a function has a tangent line and where the concavity changes is a point of inflection.

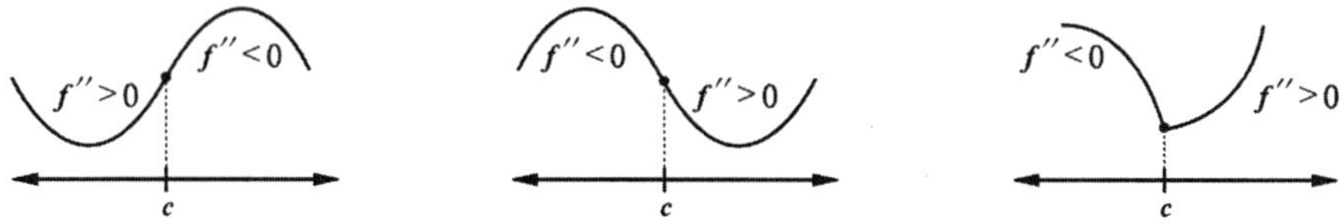

Figure 4.11

Example 1

Find the coordinated of the inflection point of the function $f(x) = x^3 - 6x^2 + 12$.

Solution

$f'(x) = 3x^2 - 12x$
$f''(x) = 6x - 12 = 6(x-2)$
$f''(x) = 0$ at $x = 2$
$\therefore$ The point of inflection is (2, -4)

x		2	
y''	−	0	+
y	∩	-4	∪

[EX 1] Find (a) all intervals on which the function is concave up/down,
(b) all values of x at which the function has a point of inflection.

(1) $y = 4x^3 - x^4$

(2) $y = x^5 - 5x^4$

(3) $y = x^2 \ln x$

(4) $y = \dfrac{e^x}{x}$

[EX 2]

(1) The cubic function $y = x^3 + ax^2 + b$ has a point of inflection at (2, 8). Find the values of a and b.

(2) The cubic function $y = x^3 + ax^2 + bx + c$ has a relative maximum at point (-1, 3) and a point of inflection at $x = -3$. Find the values of a, b, and c.

(3) The quartic function $f(x) = x^4 + ax^2 - b\ln x$ has a horizontal tangent and a point of inflection at $x = 1$. Find the values of a and b.

(4) The cubic function $y = ax^3 + bx^2 + c$ has a tangent slope of -9 at $x = 1$ and has a point of inflection at $(2, -12)$. Find the values of a, b, and c.

[EX 3] The graph of f is shown below. Determine whether f' and f'' are positive, negative or zero and describe the behaviour of the graph of f at each point given in the table below.

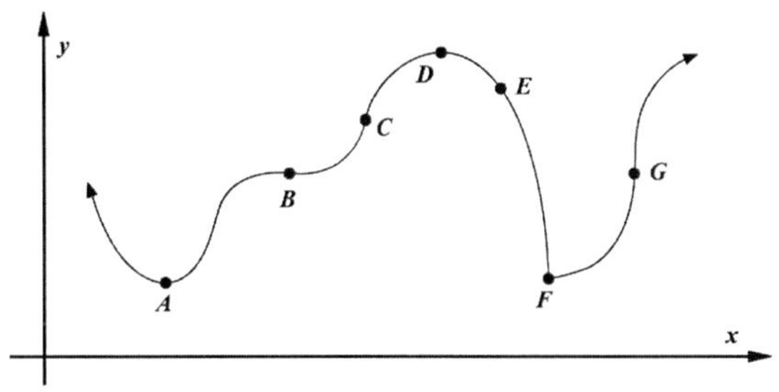

Figure 4.12

	$\frac{dy}{dx}$	$\frac{d^2y}{dx^2}$	Behaviour of the graph of f
A			
B			
C			
D			
E			
F			
G			

Second Derivative Test for Local Extrema

Suppose f'' is continuous on an open interval that contains $x = c$.

1. If $f'(c) = 0$ and $f''(c) < 0$, then f has a local maximum at $x = c$.
2. If $f'(c) = 0$ and $f''(c) > 0$, then f has a local minimum at $x = c$.
3. If $f'(c) = 0$ and $f''(c) = 0$, then the test fails. The function f may have a local maximum, a local minimum, or neither.

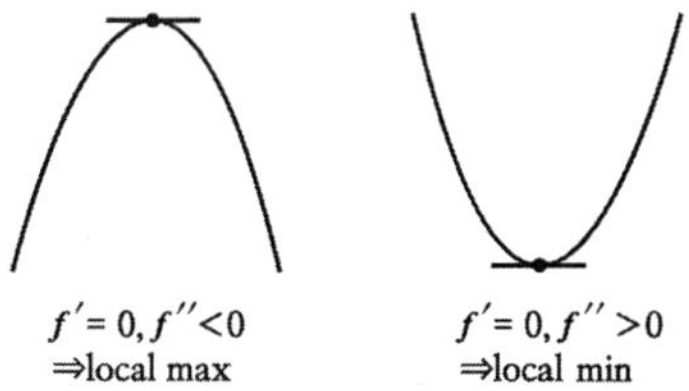

Figure 4.13

Example 2

Find the relative extreme values of the function $y = x^3 - 3x + 2$.

Solution

$f'(x) = 3x^2 - 3$
$= 3(x^2 - 1)$
$= 3(x+1)(x-1)$
$f'(x) = 0$ at $x = -1, 1$
$f''(x) = 6x$

$f''(-1) = -6 < 0$
f has a relative maximum at $x = -1$.
$\therefore$ Relative Maximum $= f(-1)$
$= -1 + 3 + 2 = 4$

$f''(1) = 6 > 0$
f has a relative minimum at $x = 1$.
$\therefore$ Relative Minimum $= f(1)$
$= 1 - 3 + 2 = 0$

[EX 4] Use the second derivative test to find the relative extreme values of the function.

(1) $f(x) = x^3 - 3x^2 - 9x$

(2) $f(x) = 2x\,e^x$

[EX 5] Consider the curve defined by $3x^2 - 2y^3 + 12x + 28 = 0$.

(1) Show that $\dfrac{dy}{dx} = \dfrac{x+2}{y^2}$.

(2) Find the coordinates of point P on the curve where $\dfrac{dy}{dx} = 0$.

(3) Find the value of $\dfrac{d^2y}{dx^2}$ at point P.

(4) Determine whether f has a relative maximum, relative minimum, or neither at point P. Explain.

[EX 6] Use the second derivative test to find the relative extrema of the curve given by $x^2 + y^2 - 2y = 0$.

4.4 Curve Sketching

Curve Sketching of Polynomial Functions

Step 1. Find y'
Step 2. Find y''
Step 3. Make a sign diagram.
(Identify the critical points, extreme values, concavity, points of inflection and y-intercept).
Step 4. Sketch the graph.

Example 1

Sketch the graph of the function $y=-2x^3+6x^2-3$. Label the relative extrema, point of inflection, and y-intercept.

Solution

When $y=-2x^3+6x^2-3$,
then $y'=-6x^2+12x=-6x(x-2)$
and $y''=-12x+12=-12(x-1)$.
The curve rises on $(0,-2)$
and falls on $(-\infty,0)$ and $(2,\infty)$.
At $x=0$ there is a local minimum a
nd at $x=2$ a local maximum.
The curve is concave up on $(-\infty,1)$
and concave down on $(1,\infty)$.
At $x=1$ there is a point of inflection.

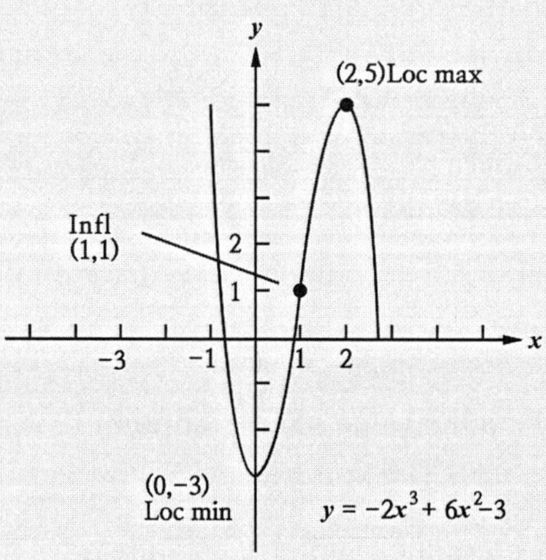

Figure 4.14

Example 2

Sketch the graph of the function $y = \cos x + \sqrt{3}\sin x$. Label the relative extrema, points of inflection, and y-intercept.

Solution

When $y = \cos x + \sqrt{3}\sin x$,

then $y' = -\sin x + \sqrt{3}\cos x$

and $y'' = -\cos x - \sqrt{3}\sin x$.

The curve is increasing on $(0, \frac{\pi}{3})$ and $(\frac{4\pi}{3}, 2\pi)$, and decreasing on $(\frac{\pi}{3}, \frac{4\pi}{3})$.

At $x = 0$ there is a local minimum, at $x = \frac{\pi}{3}$ there is a local and absolute maximum, and at $x = \frac{4\pi}{3}$ there is a local and absolute minimum, and at $x = 2\pi$ there is a local maximum.

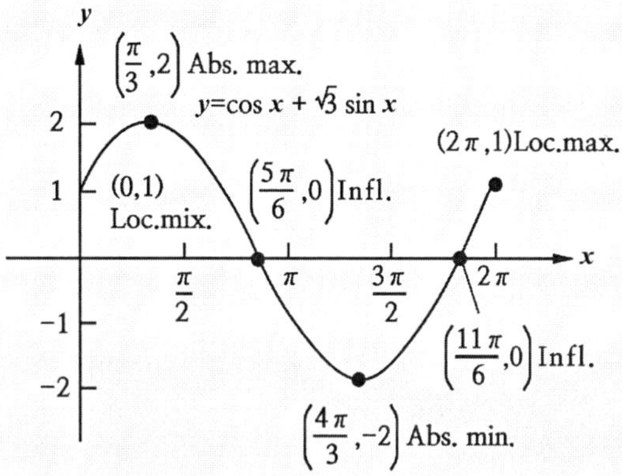

Figure 4.15

The curve is concave down on $(0, \frac{5\pi}{6})$ and $(\frac{11\pi}{6}, 2\pi)$, and is concave up on $(\frac{5\pi}{6}, \frac{11\pi}{6})$. At $x = \frac{5\pi}{6}$ and $x = \frac{11\pi}{6}$ there are point of inflection.

[EX 1] Sketch the graph on a separate sheet of paper and label the relative extrema, points of inflection, and y-intercept of the following functions.

(1) $f(x) = x^3 + 3x^2 - 9x + 5$

(2) $y = -2x^3 - 3x^2 + 12x + 4$

(3) $y = x^4 - 4x^3 + 5$

(4) $f(x) = -x^4 + 6x^2 + 1$

(5) $y = x(x+2)^3$

[EX 2] Sketch the graph on a separate sheet of paper and label the relative extrema, points of inflection, y-intercept and asymptotes for the following functions.

(1) $y = 2x - 3x^{2/3}$

(2) $y = xe^{\frac{1}{x}}$

(3) $y = xe^{-x}$

(4) $y = \dfrac{\ln x}{\sqrt{x}}$

(5) $y = \ln(3 - x^2)$

(6) $y = \dfrac{4}{3}x - \tan x, \quad (-\dfrac{\pi}{2}, \dfrac{\pi}{2})$

[EX 3] Sketch the graph on a separate sheet of paper and label the relative extrema, point of inflection, y-intercept and asymptotes for the following functions.

(1) $y = \dfrac{8x}{x^2 + 4}$

(2) $f(x) = \dfrac{x^2 - 3}{x - 2}$

(3) $y = \dfrac{e^x}{1 + e^x}$

(4) $y = \dfrac{x}{\sqrt{x^2 + 1}}$

4.5 Relation of $f'(x)$, $f''(x)$ to the Graph of $f(x)$

- $f'(x) > 0 \Leftrightarrow f(x)$ is increasing.
- $f'(x)$ is increasing $\Leftrightarrow f''(x) > 0 \Leftrightarrow f(x)$ is concave up.

f'	+	+	−	−	+	−	0
f''	+	−	+	−	0	0	0
graph of f	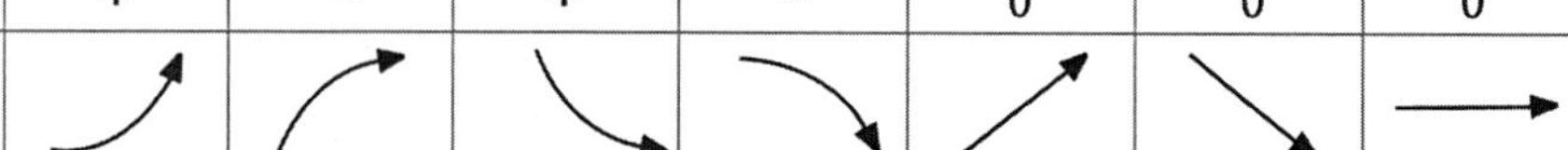						

Figure 4.16

[EX 1] Each of Exercises (1)-(4) shows the graphs of the first and second derivatives of a function $y=f(x)$. Copy the picture and add to it a sketch of the approximate graph of f, given that the graph passes through the point P.

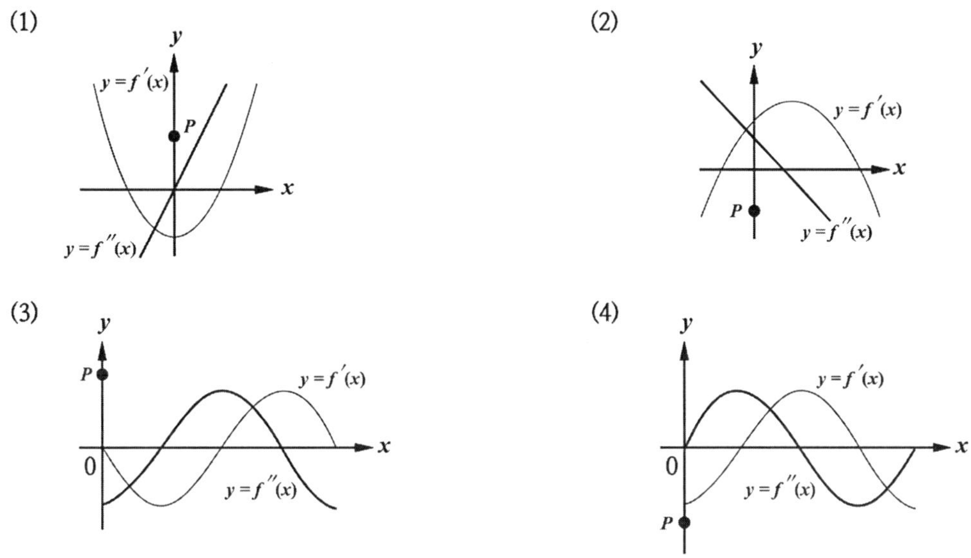

Figure 4.17

[EX 2] The accompanying figure shows a portion of the graph of a twice-differentiable function $y=f(x)$. At each of the five labeled points, classify y' and y'' as positive, negative, or zero.

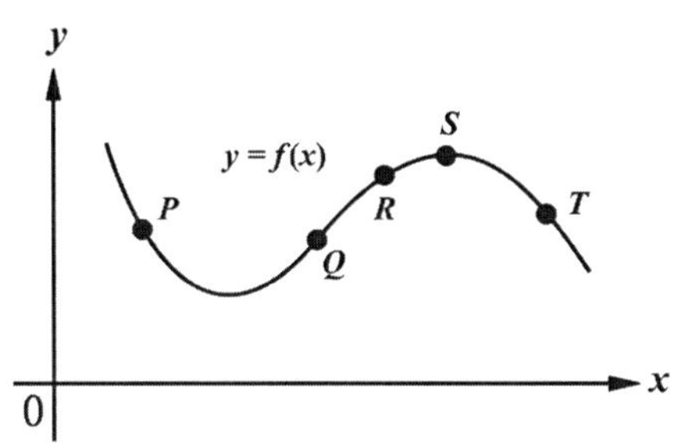

Figure 4.18

[EX 3]

(1) Sketch a smooth connected curve $y = f(x)$ with

$f(-2) = 8$	$f'(2) = f'(-2) = 0$,
$f(0) = 4$,	$f'(x) < 0$ for $\vert x \vert < 2$,
$f(2) = 0$,	$f''(x) < 0$ for $x < 0$,
$f'(x) > 0$ for $\vert x \vert > 2$,	$f''(x) > 0$ for $x > 0$.

(2) Sketch the graph of a twice-differentiable function $y = f(x)$ with the following properties.
Label coordinates where possible.

x	y	Derivatives
$x < 2$		$y' < 0,\ y'' > 0$
2	1	$y' = 0,\ y'' > 0$
$2 < x < 4$		$y' > 0,\ y'' > 0$
4	4	$y' > 0,\ y'' = 0$
$4 < x < 6$		$y' > 0,\ y'' < 0$
6	7	$y' = 0,\ y'' < 0$
$x > 6$		$y' < 0,\ y'' < 0$

(3) Sketch the graph of a twice-differentiable function $y = f(x)$ that passes through the points $(-2, 2), (-1, 1), (0, 0), (1, 1)$, and $(2, 2)$ and whose first two derivatives have the following sign patterns.

y': $+$ | -2 | $-$ | 0 | $+$ | 2 | $-$

y'': $-$ | -1 | $+$ | 1 | $-$

[EX 4] The graph of f' is shown below.

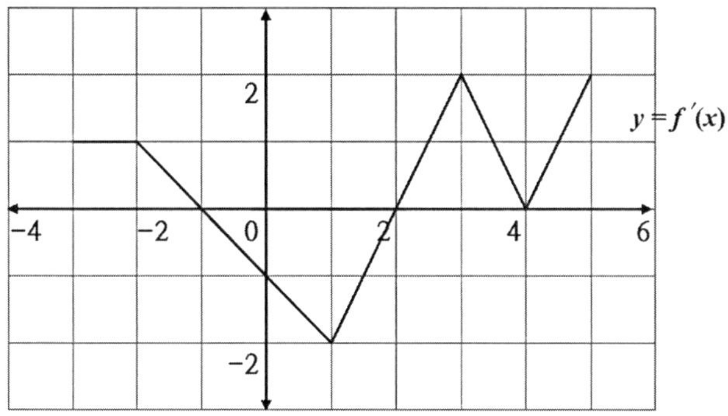

Figure 4.19

(1) Find the values of $f'(-1), f'(1), f'(5), f''(-1), f''(1), f''(2)$ and $f''(3)$.

(2) Find the average rate of change of f' over the interval $-1 \le x \le 3$.

(3) Find all intervals on which the graph of f is increasing. Justify your answer.

(4) Find all intervals on which the graph of f is concave down. Justify your answer.

(5) Find all values of x at which the graph of f has a relative minimum/maximum. Explain.

(6) Find all values of x at which the graph of f has a point of inflection. Explain.

[EX 5] The graph of f is shown below.

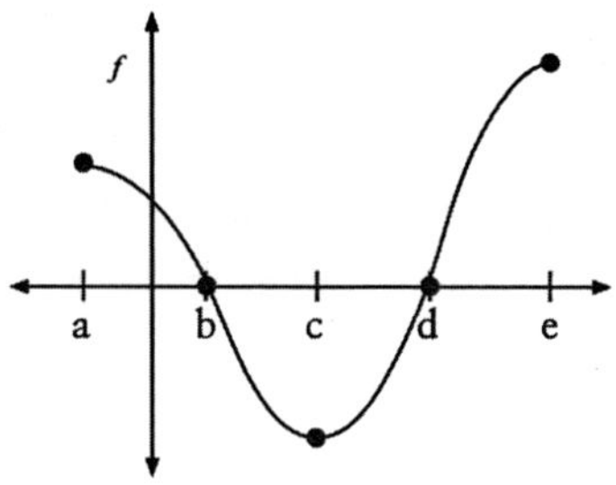

Figure 4.20

(1) Find all intervals on which the graph of f is increasing.

(2) Find all intervals on which the graph of f' is increasing.

(3) Find all values of x at which the graph of f has a relative minimum.

(4) Find all values of x at which the graph of f has a point of inflection.

[EX 6] The figure above shows the graph of f', the derivative of a twice-differentiable function f, on the closed interval $0 \leq x \leq 8$. The graph of f' has horizontal tangent lines at $x = 1$, $x = 3$, and $x = 5$. The areas of the regions between the graph of f' and the $x - axis$ are labeled in the figure. The function f is defined for all real numbers and satisfies $f(8) = 4$.

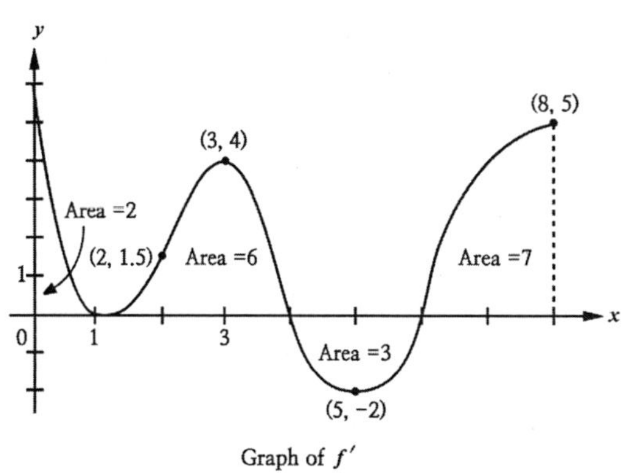

Figure 4.21

(1) Find all values of x on the open interval $0 < x < 8$ for which the function f has a local minimum. Justify your answer.

(2) Determine the absolute minimum value of f on the closed interval $0 \le x \le 8$. Justify your answer.

(3) On what open intervals contained in $0 < x < 8$ is graph of f both concave down and increasing? Explain your reasoning.

(4) The function g is defined by $g(x) = (f(x))^3$. If $f(x) =-\frac{5}{2}$, find the slope of the line tangent to the graph of g at $x = 3$.

[EX 7] The graph of f'' is shown below.

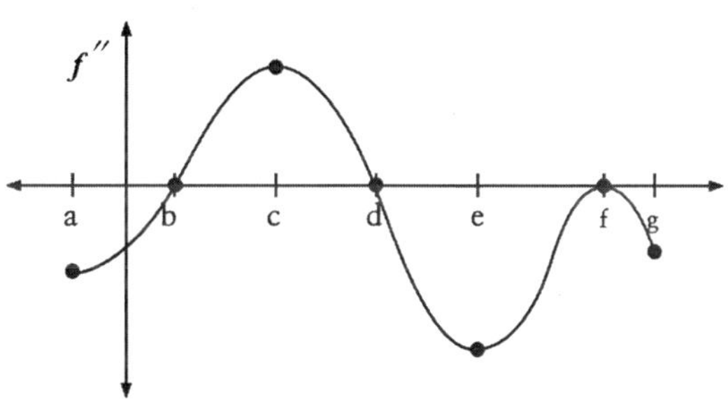

Figure 4.22

(1) Find all values of x at which the graph of f has a point of inflection. Explain briefly.

(2) Find all intervals on which the graph of f is concave up. Explain briefly.

(3) Find all intervals on which the graph of f' is increasing. Explain briefly.

(4) Find all values of x at which the graph of f' has a relative maximum. Justify your answer.

(5) Find all intervals on which the graph of f' is concave down. Justify your answer.

4.6 Motion Along a Line (1-Dimension)

If a particle moves along a line according to the law $s = f(t)$, where s represents the position of the particle P on the line at time t, then the velocity v of P at time t is given by $\frac{ds}{dt}$ and its acceleration a by $\frac{dv}{dt}$ or by $\frac{d^2s}{dt^2}$. The speed of the particle is $|v|$, the magnitude of v. If the line of motion is directed positively to the right, then the motion of the particle P is subject to the following: at any instant,

(1) if $v > 0$, then P is moving to the right and its distance s is increasing; if $v < 0$, then P is moving to the left and its distance s is decreasing;
(2) if $a > 0$, then v is increasing; if $a < 0$, then v is decreasing;
(3) if a and v are both positive or both negative, then (1) and (2) imply that the speed of P is increasing or that P is accelerating; if a and v have opposite signs, then the speed of P is decreasing or P is decelerating;
(4) if s is a continuous function of t, then P reverses direction whenever v is zero and a is different from zero; note that zero velocity does not necessarily imply a reversal in direction.

Example 1

A particle moves along a line according to the law $s = 2t^3 - 9t^2 + 12t - 4$, where $t \geq 0$.

(1) Find all t for which the distance s is increasing.

(2) Find all t for which the velocity is increasing.

(3) Find all t for which the speed of the particle is increasing.

(4) Find the speed when $t = \frac{3}{2}$.

(5) Find the total distance traveled between $t = 0$ and $t = 4$.

Solution

We have

$v = \frac{ds}{dt} = 6t^2 - 18t + 12 = 6(t^2 - 3t + 2) = 6(t-2)(t-1)$ and $a = \frac{dv}{dt} = \frac{d^2s}{dt^2} = 12t - 18 = 12\left(t - \frac{3}{2}\right)$.

The sign of v behaves as follows:

if	$t < 1$,	then	$v > 0$,
	$1 < t < 2$,		$v < 0$,
	$t > 2$,		$v > 0$.

For a, We have:

if	$t < \frac{3}{2}$,	then	$a < 0$,
	$t > \frac{3}{2}$,	then	$a > 0$.

These signs of v and a immediately yield the answers, as follows:

(1) s increases when $t < 1$ or $t > 2$.

(2) v increases when $t > \frac{3}{2}$.

(3) The speed $|v|$ is increasing when v and a are both positive, that is, for $t > 2$, and when v and a are both negative, that is for $1 < t < \frac{3}{2}$.

(4) The speed when $t = \frac{3}{2}$ equals

$$|v| = \left|-\frac{3}{2}\right| = \frac{3}{2}.$$

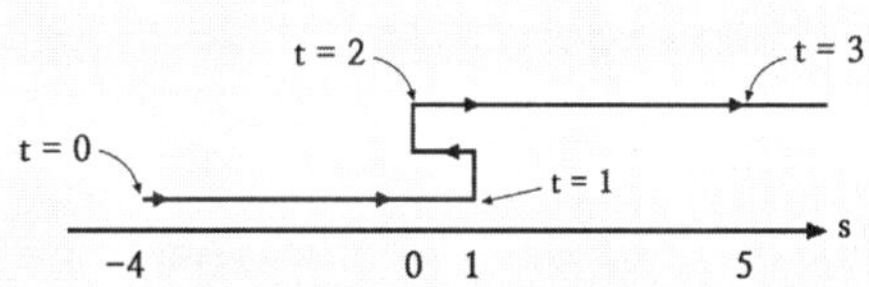

Figure 4.23

(5) P's motion can be indicated as shown in Figure N4-15. P moves to the right if $t < 1$, reverses its direction at $t = 1$, moves to the left when $1 < t < 2$, reversed again at $t = 2$, and continues to the right for all $t > 2$. The position of P at certain times t are shown in the following table:

t:	0	1	2	4
s:	-4	1	0	28

Thus P travels a total of 34 units between times $t = 0$ and $t = 4$.

[EX 1] A particle is moving along a horizontal line. Its position as a function of time is given by $s(x) = t^3 - 12t + 1$, $t \geq 0$ where s is in meters and t is in seconds.

(1) Find the velocity and acceleration functions of the particle.

(2) At what time does the particle change direction?

(3) For what time intervals is the velocity of the particle increasing?

(4) For what time intervals is the speed of the particle decreasing? Justify your answer.

[EX 2] A particle is moving along a horizontal line. Its position as a function of time is given by $s(t) = t^3 - 6t^2 + 9t - 3$, $t \geq 0$ where s is in meters and t is in seconds.

(1) Find the displacement of the particle from $t = 1$ to $t = 2$.

(2) Find the average velocity of the particle form $t = 1$ to $t = 2$.

(3) Find the velocity and acceleration functions of the particle.

(4) At what time(s) is the particle at rest?

(5) Find the position, velocity, speed, and acceleration of the particle at time $t = 4$.

(6) Is the particle moving away from the origin or toward the origin at time $t = 4$? Justify your answer.

(7) Find the average acceleration of the particle from $t = 0$ to $t = 2$.

(8) For what time intervals is the particle moving in the positive direction?

(9) For what time intervals is the velocity of the particle decreasing?

(10) For what time intervals is the speed of the particle increasing? Justify your answer.

(11) Find the total distance travelled by the particle during the first 4 seconds. Justify your answer.

[EX 3] The velocity, in m/sec, of a particle moving along the x-axis is shown in the graph below.

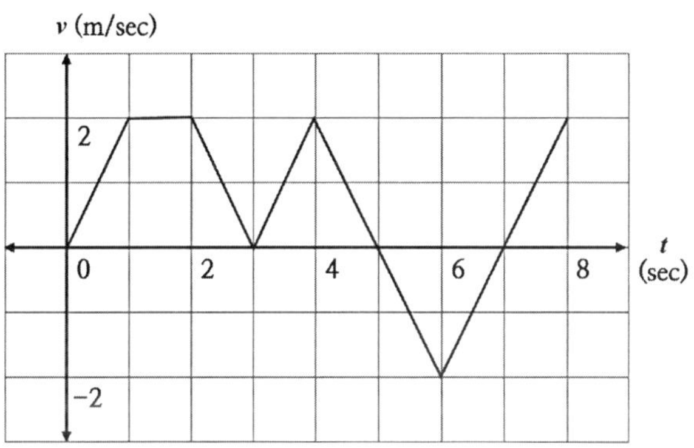

Figure 4.24

(1) At what time(s) is the particle at rest?

(2) At what time(s) does the particle change direction?

(3) Find the speed of the particle at time $t=6$.

(4) Find the acceleration of the particle at time $t=5$.

(5) Find the average acceleration of the particle from $t=1$ to $t=7$.

(6) For what time intervals is the particle moving in the negative direction?

(7) For what time intervals is the speed of particle decreasing? Justify your answer.

(8) Find the total distance travelled by the particle from $t=0$ to $t=8$.

(9) How many seconds will it take the particle to reach the farthest point from the origin? Explain.

[EX 4] A particle moves along the x-axis so that its velocity at time t, for $0 \le t \le 6$, is given by a differentiable function v whose graph is shown above. The velocity is 0 at $t=0$, $t=3$, and $t=5$, and the graph has horizontal tangents at $t=1$ and $t=4$. The areas of the regions bounded by the t-axis and the graph of v on the intervals $[0,3]$, $[3,5]$, and $[5,6]$ are 8, 3, and 2, respectively. At time $t=0$, the particle is at $x=-2$.

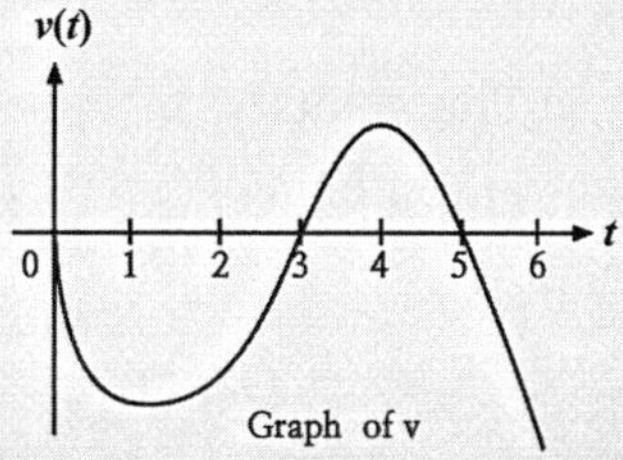

Figure 4.25

(1) For $0 \le t \le 6$, find both the time and the position of the particle when the particle is farthest to the left. Justify your answer.

(2) For how many values of t, where $0 \le t \le 6$, is the particle at $x=-8$? Explain your reasoning.

(3) On the interval $2 < t < 3$, is the speed of the particle increasing or decreasing)? Give a reason for your answer.

(4) During what time intervals, if any, is the acceleration of the particle negative? Justify your answer.

4.7 Motion Along a Curve: Velocity and Acceleration Vectors

If a point P moves along a curve defined parametrically by $P(t)=(x(t),y(t))$, where t represents time, then the vector from the origin to P is called the *position vector*, with x as its *horizontal component* and y as its vertical *component*. The set of position vectors for all values of t in the domain common to $x(t)$ and $y(t)$ is called *vector function*.

The position vector, then, may be written as $\vec{R}(t)=(x,y)$ or as $R=xi+yj$, where i is the horizontal unit vector from $(0,0)$ to $(1,0)$ and j is the vertical unit vector from $(0,0)$ to $(0,1)$.

The *velocity vector* is the derivative of the vector function (the position vector):

$$v=\frac{dR}{dt}=\frac{dx}{dt}i+\frac{dy}{dt}j \text{ or } \vec{v}(t)=(\frac{dx}{dt},\frac{dy}{dt}).$$

Alternative notations for $\frac{dx}{dt}$ and $\frac{dy}{dt}$ are v_x and v_y respectively; these are the components of v in the horizontal and vertical directions, respectively.

The slope of v is $\frac{\frac{dy}{dt}}{\frac{dx}{dt}}=\frac{dy}{dx}$, (speed) the *magnitude* of v is

$$|v|=\sqrt{(\frac{dx}{dt})^2+(\frac{dy}{dt})^2}=\sqrt{v_x^2+v_y^2}$$

The *acceleration vector* a is $\frac{dv}{dt}$ or $\frac{d^2R}{dt^2}$, and can be obtained by a second differentiation of the components of R. Thus

$$a=\frac{d^2x}{dt^2}i+\frac{d^2y}{dt^2}j;$$

the direction of **a** is

$$\tan^{-1}\frac{\frac{d^2y}{dt^2}}{\frac{d^2x}{dt^2}};$$

and its magnitude is

$$|\mathbf{a}| = \sqrt{\left(\frac{d^2x}{dt^2}\right)^2 + \left(\frac{d^2y}{dt^2}\right)^2} = \sqrt{{a_x}^2 + {a_y}^2},$$

where we have used a_x and a_y for $\frac{d^2x}{dt^2}$ and $\frac{d^2y}{dt^2}$, respectively.

Example 1

A particle moves according to the equations $x = 3\cos t$, $y = 2\sin t$.

(1) Find a single equation in x and y for the path of the particle and sketch the curve.

(2) Find the velocity and acceleration vectors at any time t, and show that a=-R at all time.

(3) Find R, v and a when $t_1 = \frac{\pi}{6}$, $t_2 = \pi$,

(4) Find the speed of the particle and the magnitude of its acceleration at each instant in (3).

(5) When is the speed a maximum? A minimum?

Solution

(1) Since $\frac{x^2}{9} = \cos^2 t$ and $\frac{y^2}{4} = \sin^2 t$, therefore $\frac{x^2}{9} + \frac{y^2}{4} = 1$ and the particle moves in a counterclockwise direction along an ellipse, starting, when $t=0$, at (3,0) and at (3,0) and returning to this point when $t=2\pi$.

(2) We have

$R = 3\cos ti + 2\sin tj,$

$v = -3\sin ti + 2\cos tj$

$a = -3\cos ti - 2\sin tj = -R.$

The acceleration, then, is always directed toward the center of the ellipse.

(3) At $t_1 = \frac{\pi}{6}$, $R_1 = \frac{3\sqrt{3}}{2}i + j$
$v_1 = -\frac{3}{2}i + \sqrt{3}j$
$a_1 = -\frac{3\sqrt{3}}{2}i - j.$

At $t_2 = \pi$, $R_2 = -3i$
$v_2 = -2j$
$a_2 = 3i.$

(4) At $t_1 = \frac{\pi}{6}$, $|v_1| = \sqrt{\frac{9}{4} + 3} = \frac{\sqrt{21}}{2}$,
$|a_1| = \sqrt{\frac{27}{4} + 1} = \frac{\sqrt{31}}{2}.$

At $t_2 = \pi$ $|v_2| = \sqrt{0+4} = 2,$
$|a_2| = \sqrt{9+0} = 3$

(5) For the speed $|v|$ at any time t we have
$$v = \sqrt{9\sin^2 t + 4\cos^2 t} = \sqrt{4\sin^2 t + 4\cos^2 t + 5\sin^2 t} = \sqrt{4 + 5\sin^2 t}.$$

We see immediately that the speed is a maximum when $t = \frac{\pi}{2}$ or $\frac{3\pi}{2}$, and a minimum when $t = 0$ or π.

[EX 1] In Question (1)-(4) $R = 3\cos\frac{\pi}{3}ti + 2\sin\frac{\pi}{3}tj$ is the (position) vector $xi + yj$ from the origin to a moving point $P(x, y)$ at time t,

(1) A single equation in x and y for the path of the point is

(A) $x^2 + y^2 = 13$ (B) $9x^2 + 4y^2 = 36$ (C) $2x^2 + 3y^2 = 13$

(D) $4x^2 + 9y^2 = 1$ (E) $4x^2 + 9y^2 = 36$

(2) When $t=3$, the speed of the particle is

(A) $\frac{2\pi}{3}$ (B) 2 (C) 3 (D) π (E) $\frac{\sqrt{13}}{3}\pi$

(3) The magnitude of the acceleration when $t=3$ is

(A) 2 (B) $\frac{\pi^2}{3}$ (C) 3 (D) $\frac{2\pi^2}{9}$ (E) π

(4) At the point where $t=\frac{1}{2}$, the slope of the curve along which the particle moves is

(A) $-\frac{2\sqrt{3}}{9}$ (B) $-\frac{\sqrt{3}}{2}$ (C) $\frac{2}{\sqrt{3}}$

(D) $-\frac{2\sqrt{3}}{3}$ (E) none of these

[EX 2] In Question (1)~(4) the motion of a particle in a plane is given by the pair of equations $x=2t$ and $y=4t-t^2$.

(1) The particle moves along

(A) an ellipse (B) a circle (C) a hyperbola
(D) a line (E) a parabola

(2) The speed of the particle at any time t is

(A) $\sqrt{6-2t}$ (B) $2\sqrt{t^2-4t+5}$ (C) $2\sqrt{t^2-2t+5}$
(D) $\sqrt{8}(|t-2|)$ (E) $2(|3-t|)$

(3) The minimum speed of the particle is

(A) 2 (B) $2\sqrt{2}$ (C) 0 (D) 1 (E) 4

(4) The acceleration of the particle

(A) depends on t
(B) is always directed upward
(C) is constant both in magnitude and in direction
(D) never exceeds 1 in magnitude
(E) is non of these

[EX 3] For $t \geq 0$, a particle is moving along a curve so that its position at time t is $(x(t), y(t))$. At time $t=2$, the particle is at position (1,5). It is known that $\frac{dx}{dt}=\frac{\sqrt{t+2}}{e^t}$ and $\frac{dy}{dt}=\sin^2 t$.

(1) Is the horizontal movement of the particle to the left or to the right at time $t=2$? Expain your answer. Find the slope of the path of the particle at time $t=2$.

(2) Find the x-coordinate of the particle's postion at time $t=4$.

(3) Find the speed of the particle at time $t=4$. Find the acceleration vector of the particle at time $t=4$.

[EX 4] An object moving along a curve in the xy-plane is at position $(x(t), y(t))$ at time t with

$$\frac{dx}{dt} = \arctan\left(\frac{t}{1+t}\right) \text{ and } \frac{dy}{dt} = \ln(t^2+1)$$

for $t \geq 0$. At time $t = 0$, the object is at position (-3, -4). (Note: $\tan^{-1}x = \arctan x$)

(1) Find the speed of the object at time $t = 4$.

(2) For $t > 0$, there is a point on the curve where the line tangent to the curve has slope 2. At what time t is the object at this point? Find the acceleration vector at this points.

4.8 Modeling and Optimization

Step 1. Draw a large, clear diagram with the variable and the constants, Identify the quantity to be maximized or minimized.

Step 2. Write an equation connecting the variables and the constants.
Express the quantity to be maximized or minimized as a function of a single variable.

Step 3. Differentiate and simplify the function. Consider the domain to the function. find the critical Pont.

Step 4. Make a sign diagram and find the answers.

Example 1

What is the smallest perimeter possible for a rectangle whose area is 16in^2, and what are its dimensions?

Solution

Let ℓ and w represent the length and width of the rectangle, respectively. With an area of 16 in^2, we have that $(\ell)(w)=16 \Rightarrow w=16\ell^{-1} \Rightarrow$ the perimeter is $P=2\ell+2w = 2\ell+32\ell^{-1}$ and $p'(\ell)=2-\dfrac{32}{\ell^2}=\dfrac{2(\ell^2-16)}{\ell^2}$.

Solving $p'(\ell)=0 \Rightarrow \dfrac{2(\ell+4)(\ell-4)}{\ell^2}=0 \Rightarrow \ell=-4$ or 4. Since $\ell>0$ for the length of a rectangle ℓ must be 4 and $w=4 \Rightarrow$ The perimeter is 16 in., a minimum since $p''(\ell)=\dfrac{64}{\ell^3}>0$.

[EX 1] A rectangular plot of farmland will be bounded on one side by a river and on the other three sides by a single-strand electric fence. With 800 m of wire at your disposal, what is the largest area you can enclose, and what are its dimensions?

[EX 2]

(1) Two number have a difference of 10 and their product is a minimum. Determine the numbers.

(2) The sum of two positive numbers is 18. The product of one and the square of the other is a maximum. Determine the numbers.

[EX 3]

(1) You are planning to make an open rectangular box from an 8-in by-15-in. piece of cardboard by cutting congruent squares from the corners and folding up the sides. What are the dimensions of the box of largest volume you can make this way, and what is its volume?

(2) An <u>open-top</u>, rectangular box is to be made from a 10 in by 16 in sheet of tin. It is made by cutting out equal squares from the four corners measuring x cm, and folding up the sides to create the box. What dimensions will yield a box of maximum volume?

[EX 4]

(1) A container company is designing an <u>open-top</u>, square based, rectangular box that will have a volume of 500 ml. What is the minimum surface area the of the box?

(2) What are the dimensions of the lightest open-top right circular cylindrical can that will hold a volume of 1000cm^3?

(3) A container company is designing a square based, rectangular box <u>with top</u> that will have a volume of 256 ft^3. The cost of the top is 154 $\$/\text{ft}^2$, the bottom is 54 $\$/\text{ft}^2$ and the sides is 2.54 $\$/\text{ft}^2$. Find the dimensions of the box that will minimize the total cost.

(4) An oil can shaped like a right circular cylinder has a volume of 1000 ml. Find the dimensions of the can that will minimize the material required to build it.

[EX 5]

(1) A piece of string measuring 24 cm is cut into two pieces. One is bent into a circle and the other is bent into a square. The sum of the areas of the circle and square is a minimum. what is the length of the piece bent into the circle?

(2) A piece of string measuring 20 ㎝ is cut into two pieces. One is bent into an equilateral triangle and the other is bent into a square. The sum of the areas of the triangle and square is a minimum. What is the length of the piece bent into the equilateral triangle?

[EX 6]

(1) Find the dimensions of the largest rectangle that can be inscribed in a circle with a radius of 10 ㎝.

(2) Determine the dimensions of the rectangle of largest area that can be inscribed in a semicircle of radius 3.

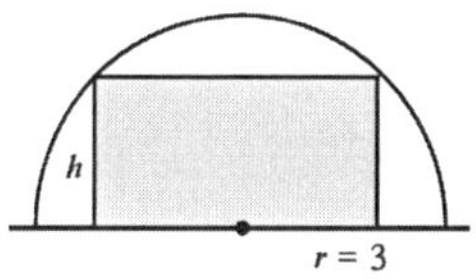

Figure 4.26

(3) A right circular cylinder is inscribed in a sphere. Find the volume of the largest cylinder that can be inscribed inside a sphere with a radius of $10\sqrt{3}$ ㎝.

(4) A right circular cone is inscribed in a sphere. Find the dimensions of the largest cone that can be inscribed inside a sphere with a radius of 3㎝.

[EX 7]

(1) A utilities company is designing a power line from point A to point C, which is located directly across a straight river 1 ㎞ from point B. Point B is 4 ㎞ down shore from point A. It costs $50 per meter to lay the line under the water but only $30 per meter to lay the line underground. How far from B should P be located to minimize the cost of the line?

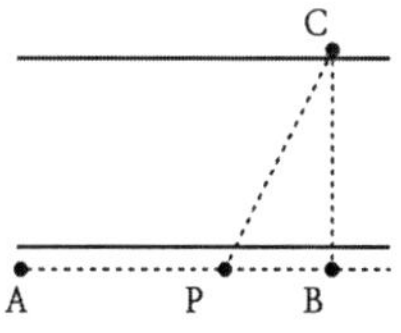

Figure 4.27

(2) An oil company is construction a pipeline from their refinery to storage tanks. The storage tanks are located 10 ㎞ away from the point A, which is situated directly across the 3 ㎞ wide river from the refinery. The cost of lying pipe is \$20 per miter underground and \$25 per meter under the river. how far from A should P be located to minimize the cost of the pipeline?

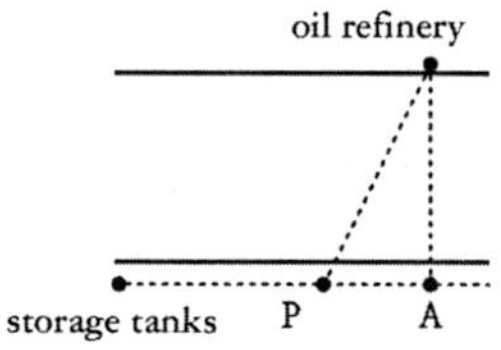

Figure 4.28

(3) Jane is 2 mi offshore in a boat and wishes to reach coastal village 6 mi down a straight shoreline from the point nearest the boat. She can row 2 mph and can walk 5 mph. Where should she land her boat to reach toe village in the least amount of time.

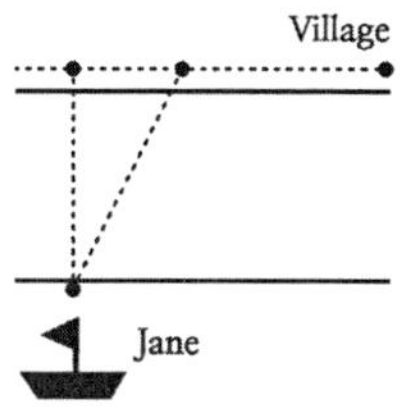

Figure 4.29

[EX 8]

(1) Find the coordinates of the point (x,y), on the curve $y=2\sqrt{x}$, which is closest to the pont $A(6,0)$.

(2) Find the coordinates of the point (x,y), on the curve $y=\frac{1}{4}x^2-1$, which is closest to $(0,3)$.

(3) Find the shortest distance from the origin to a point on the curve $xy^2=16$.

(4) Find the shortest distance from the pont $(4,0)$ to a point on the curve $x^2-y^2=-1$.

(5) Find the coordinates of the point (x,y), on the curve $y=x^2-3$, which is closest to $(6,0)$.

[EX 9]

(1) An isosceles triangle is inscribed in the parabola of $y = 9 - x^2$ above the x-axis. One vertex is at the origin and the other two vertices are on the curve. Find the maximum area of the isosceles triangle.

(2) A rectangle is inscribed in the parabola of $y = x^2 - 12$ below the x-axis. Two vertices are on the x-axis and the other two vertices are on the curve. Find the maximum area of the rectangle.

(3) a rectangle with two vertices on the x-axis is inscribed under the curve $y = e^{-2x^2}$ as shown in the figure. Find the maximum area of the rectangle.

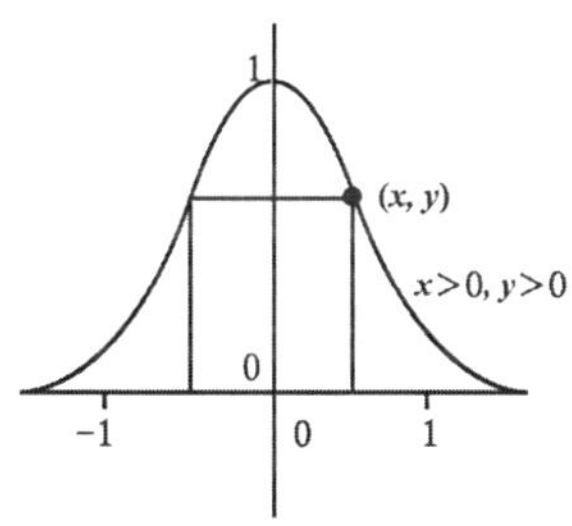

Figure 4.30

4.9 Related Rates

1. Draw a picture and name the variables and constants. Use t for time.
2. Write down the numerical information
3. Write down what you are asked to find
4. Write and equation that relates the variables.
5. Differentiate with respect to t.
6. Evaluate. Use known values to find the unknown rate.

Example 1

If the original 24 m edge length x of a cube decreases at the rate of 5 m/min, when $x = 3$m at what rate does the cube's.

(1) surface area change?

(2) volume change?

Solution

(1) $S=6x^2, \frac{dx}{dt}=-5\frac{m}{min} \Rightarrow \frac{ds}{dt}=12x\frac{dx}{dt}$;when $x=3 \Rightarrow \frac{dS}{dt}=12(3)(-5)=-180\frac{m^2}{min}$

(2) $V=x^3, \frac{dx}{dt}=-5\frac{m}{min} \Rightarrow \frac{dV}{dt}=3x^2\frac{dx}{dt}$;when $x=3 \Rightarrow \frac{dV}{dt}=3(3)^2(-5)=-135\frac{m^3}{min}$

[EX 1] The voltage V (volts), current I (amperes), and resistance R (ohms) of an electric circuit like the one shown here are related by the equation $V=IR$. Suppose that V is increasing at the rate of 1 volt/sec while I is decreasing at the rate of 1/3 amp/sec. Let t denote time in seconds.

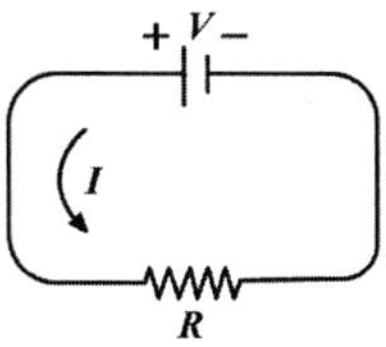

Figure 4.31

(1) What is the value of dV/dt?

(2) What is the value of dI/dt?

(3) What equation relates dR/dt to dV/dt and dI/dt?

(4) Find the rate at which R is changing when $V=12$ volts and $I=2$ amp. Is R increasing, or decreasing?

[EX 2]

(1) A conical tank with a radius of 3 m and a height of 4 m is being filled with water at a rate of $18\pi\ m^3/min$. How fast is <u>the water lever rising</u> when the water is 2 m deep?

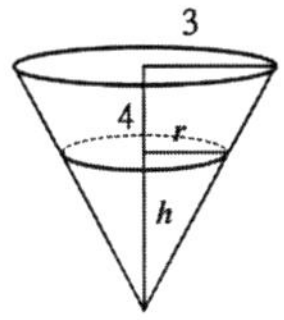

Figure 4.32

(2) A conical tank with a radius of 45 m and a height of 6 m is filled with water. The water is leaking out at a constant rate of $16\pi\ m^3/min$. How fast is the radius of the water's surface decreasing when it has a 2 m deep.

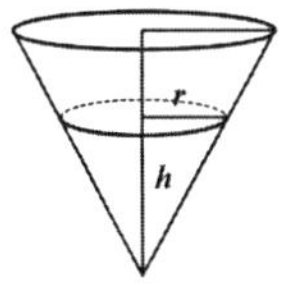

Figure 4.33

(3) A spherical balloon is inflated with helium at the rate of $100\pi\ ft^3/min$. How fast is the balloon's radius increasing at the instant the radius is 5 ft? How fast is the surface area increasing?

[EX 3] A 13-ft ladder is leaning against a house when its base starts to slide away (see accompanying figure). By the time the base is 12 ft from the house, the base is moving at the rate of 5ft/sec.

(1) How fast is the top of the ladder sliding down the wall then?

(2) At what rate is the area of the triangle formed by the ladder, wall, and ground changing then?

(3) At what rate is the angle θ between the ladder and the ground changing then?

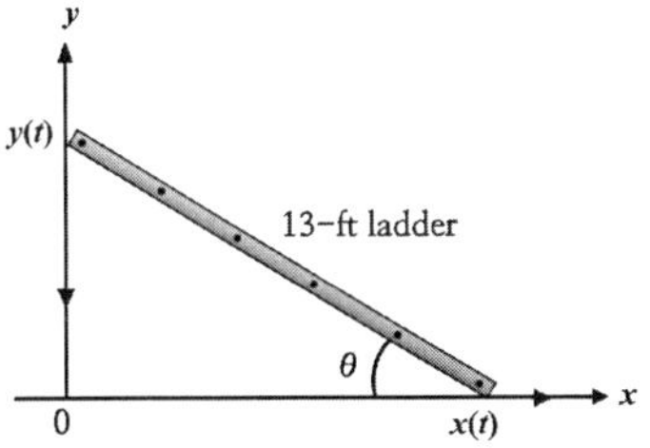

Figure 4.34

[EX 4]

(1) A dinghy is pulled toward a dock by a rope from the bow through a ring on the dock 6 ft above the bow. The rope is hauled in at the rate of 2 ft/sec.

(a) How fast is the boat approaching the dock when 10 ft of rope are out?

(b) At what rate is the angle θ changing at this instant (see the figure)?

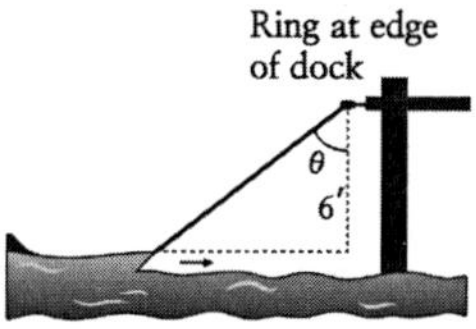

Figure 4.35

(2) You are videotaping a race from a stand 132 ft from the track, following a car that is moving at 180 mi/h (264 ft/sec), as shown in the accompanying figure. How fast will your camera angle θ be changing when the car is right in front of you? A half second later?

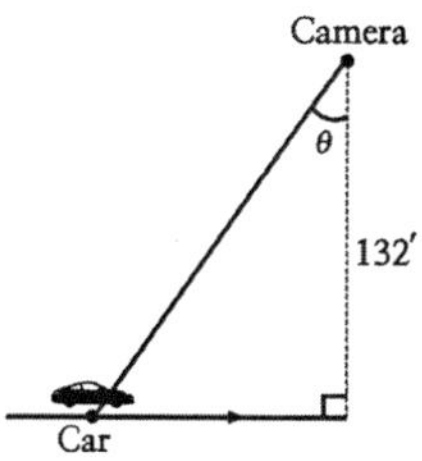

Figure 4.36

[EX 5]

(1) A 6 ft tall man is walking man toward a 20 ft lamp post at a rate of 3 ft/sec. (a) How fast is the <u>length of his shadow changing?</u> (b) How fast is <u>the tip of the shadow moving?</u>

(2) A balloon is rising vertically above a level, straight road at a constant rate of 1 ft/sec. Just when the balloon is 65ft above the ground, a bicycle moving at a constant rate of 17 ft/sec passes under it. How fast is the distance $s(t)$ between the bicycle and

balloon increasing 3 sec later?

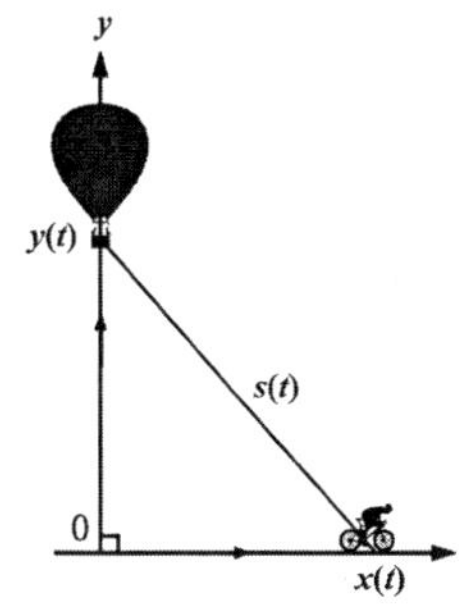

Figure 4.37

(3) Car A travels north at a rate 30 mph and car B travels east at 50 mph, as shown in the figure. When car A is 80 miles south from the intersection, car B is 70 miles east from the intersection. How fast is the distance between the cars changing 1 hour later?

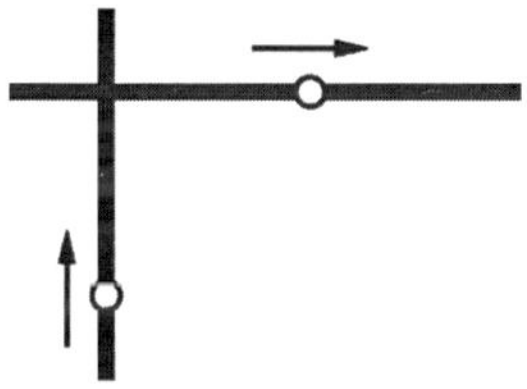

Figure 4.38

[EX 6]

(1) One side of a triangle is 7 ㎝ long and another side is 8 ㎝ long. The angle between the two sides is decreasing at a rate of $\frac{\pi}{14}$ rad/sec. How fast is the area of the triangle changing when the angle is $\frac{\pi}{3}$?

(2) The height of a triangle is increasing at a rate of 2 ㎝/sec and the base of the triangle is decreaing in length at a rate of 3 ㎝/sec. How fast is the area of the triangle changing when the base is 8 ㎝ and the area is 40 ㎠ ?

(3) The height of a cylindrically shaped rubber tube, with a volume of $240\pi\,\text{cm}^3$, is stretched at a constant rate of 6 cm/min. The volume of the rubber tube does not change as it stretches. Find the rate of change in radius when the height is 15 cm.

[EX 7]

(1) A point is moving along the curve $y = \sqrt{x}$ and x is increasing at a rate of 4 units/sec.

(2) How fast is y changing when x is 4?

(3) How fast is the distance from the origin changing when x is 4 ?

(4) Find the rate of change in the angle of elevation of the point from the origin when x is 4.

[EX 8] Two ships are steaming straight away from a point O along routes that make a 120° angle. Ship A moves at 14 knots (nautical miles per hour; a nautical mile is 2000 yd). Ship B moves at 21 knots. How fast are the ships moving apart when $OA = 5$ and $OB = 3$ nautical miles?

4.10 Mean Value Theorem (M.V.T)

The Mean Value Theorem

Suppose $y = f(x)$ is continuous on a closed interval $[a,b]$ and differentiable on the interval's interior (a,b). Then there is at least on point c in (a,b) at which

$$\frac{f(b)-f(a)}{b-a} = f'(c).$$

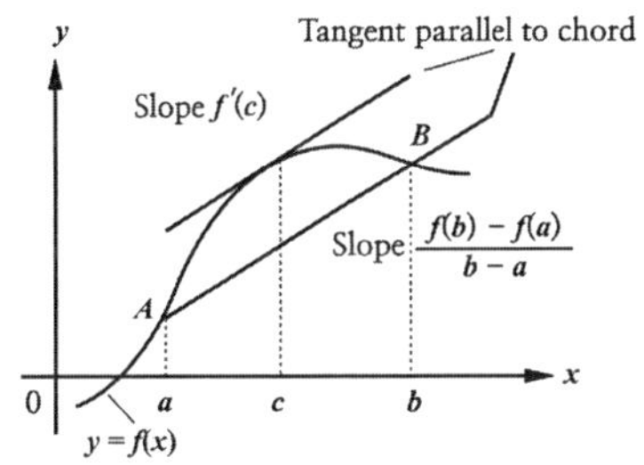

Figure 4.39

Rolle's Theorem

Suppose that $y = f(x)$ is continuous at every point of the closed interval $[a,b]$ and differentiable at every point of its interior (a,b). It $f(a) = f(b)$, then there is at least one number c in (a,b) at which $f'(c) = 0$.

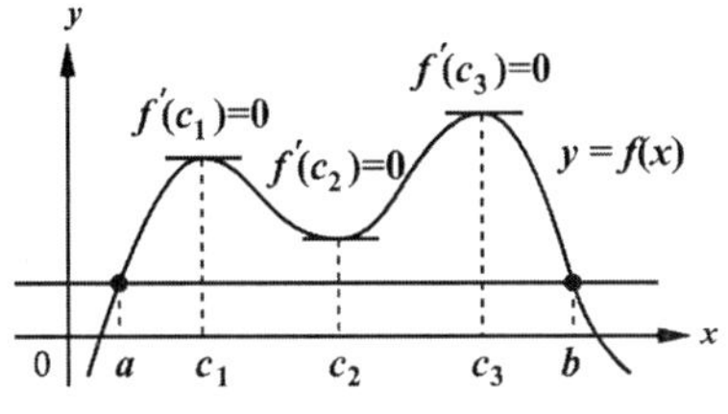

Figure 4.40

Example 1

Let $f(x)=x^2-2x$. Find all numbers c, $1<c<3$, such that $f'(c)=\dfrac{f(3)-f(1)}{3-1}$.

Solution

$f'(x)=2x-2$, $\dfrac{f(3)-f(1)}{3-1}=\dfrac{3-(-1)}{3-1}=2 \Rightarrow 2c-2=2$

$\therefore\ c=2$

[EX 1] Find the value or values of c that satisfy the equation

$$\frac{f(b)-f(a)}{b-a}=f'(c)$$

in the conclusion of the Mean Value Theorem for the functions and intervals in Exercises 1-8.

(1) $f(x)=x^2+3x-1,\ [0,1]$

(2) $f(x)=x^{2/3},\ [0,1]$

(3) $f(x)=x+\dfrac{1}{x},\ [\dfrac{1}{2},2]$

(4) $f(x)=\sqrt{x-1},\ [1,3]$

(5) $f(x)=\sin^{-1}x,\ [-1,1]$

(6) $f(x)=\ln(x-1),\ [2,4]$

(7) $f(x)=x^3-x^2,\ [-1,2]$

(8) $g(x)=\begin{cases} x^3, & -2\le x\le 0 \\ x^2, & 0\le x\le 2 \end{cases}$

[EX 2] Explain why the Mean Value Theorem cannot be applied to the following functions on the given interval.

(1) $y=\dfrac{1}{x},\ [-2,4]$

(2) $y=\dfrac{x}{|x|},\ [-1,1]$

(3) $y = |x-1| + 2,\ [0,2]$

(4) $y = \sqrt[3]{x^2},\ [-1,8]$

[EX 3] Use a graphing calculator to find all numbers c in the interval $[a,b]$ that satisfy the Mean Value Theorem for the following functions.

(1) $f(x) = 2^x - \sin x,\ [0,\pi]$

(2) $f(x) = 3x\cos^{-1}x,\ [0,\frac{1}{2}]$

(3) If $f(x) = 2x^3 - 6x$, at what point on the interval $0 \le x \le \sqrt{3}$, is any, is the tangent to the curve parallel to the secant line?

(4) Let $f(x) = 3^x - x^3$. Find all numbers $c, 0 < c < 3$, such that the tangent to the curve is parallel to the secant through $(0,1)$ and $(3,0)$ for x equal.

4.11 Local Linear Approximation

If f is differentiable at $x = a$, then the approximating function

$$L(x) = f(a) + f'(a)(x - a)$$

is the linearization of f at a. The approximation

$$f(x) \approx L(x)$$

of f by L is the standard linear approximation of f at a. The point $x = a$ is the center of the approximation.

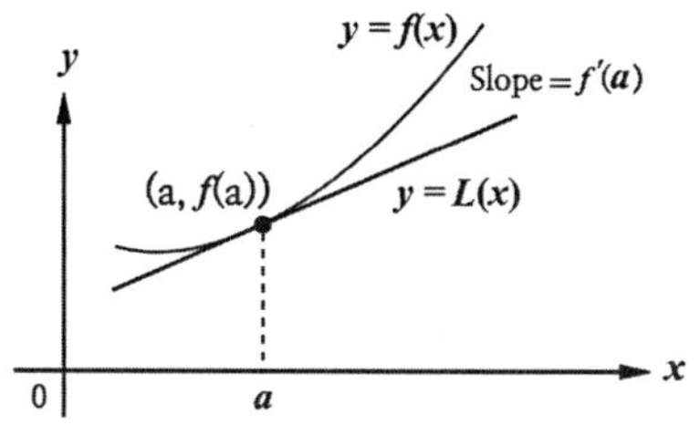

Figure 4.41

Example 1

Find the linearization of $f(x) = \sqrt{1+x}$ at $x = 0$ since

Approximation	True value
$\sqrt{1.2} \approx 1 + \frac{0.2}{2} = 1.10$	1.095445
$\sqrt{1.05} \approx 1 + \frac{0.05}{2} = 1.025$	1.024695
$\sqrt{1.005} \approx 1 + \frac{0.005}{2} = 1.00250$	1.002497

Solution

$f'(x) = \frac{1}{2}(1+x)^{-1/2}$, we have $f(0) = 1$ and $f'(0) = 1/2$, giving the linearization

$L(x) = f(a) + f'(a)(x-a) = 1 + \frac{1}{2}(x-0) = 1 + \frac{x}{2}.$

[EX 1] Use local linear approximation to estimate the values of the following.

(1) $\sqrt{16.08}$

(2) $\sqrt[3]{7.4}$

(3) $\sqrt{63}$

(4) $(1.01)^{10}$

(5) $2^{3.99}$

(6) $\tan^{-1}(1.01)$

[EX 2] Find the linearization $L(x)$ of $f(x)$ at $x = a$.

(1) $f(x) = x^3 - 2x + 3,\ a = 2$

(2) $f(x) = \sqrt{x^2 + 9},\ a = -4$

(3) $f(x) = x + \dfrac{1}{x},\ a = 1$

(4) $f(x) = \sqrt[3]{x},\ a = -8$

(5) $f(x) = \tan x,\ a = \pi$

[EX 3]

t (minutes)	0	2	5	7	11	12
$r'(t)$ (feet per minute)	5.7	4.0	2.0	1.2	0.6	0.5

The volume of a spherical hot air balloon expands as the air inside the balloon is heated. The radius of the balloon, in feet, is modeled by a twice-differentiable function r of time t, where t is measured in minutes.

For $0 < t < 12$, the graph of r is concave down. The table above gives selected values of the rate of change, $r'(t)$, of the radius of the balloon over the time interval $0 \le t \le 12$. The radius of the balloon is 30 feet when $t = 5$.

(Note: The volume of a sphere of radius r is given by $V = \frac{4}{3}\pi r^3$.)

(1) Estimate the radius of the balloon when $t = 5.4$ using the tangent line approximation at $t = 5$. Is your estimate grater than or less than the true value? Give a reason for your answer.

(2) Find the rate of change of the volume of the balloon with respect to time when $t = 5$. Indicate units of measure.

[EX 4] A table of values of a differentiable function f and its derivative f' is shown below.

x	0	1	2	3
f	1	2	4	7
f'	0.5	1	3	4

The second derivative of f has the property that $f'' > 0$ on the interval $0 < x < 3$.

(1) Use local linear approximation to estimate $f(2.1)$.

(2) Is this approximate value greater than or less than the actual value of $f(2.1)$? Give a reason to support your answer.

4.12 Newton's Method

1. Guess a first approximation to a solution of the equation $f(x)=0$. A graph of $y=f(x)$ may help.
2. Use the first approximation to get a second, the second to tet a third, and so on, using the formula

$$x_{n+1}=x_n-\frac{f(x_n)}{f'(x_n)}, \qquad \text{if } f'(x_n)\neq 0,$$

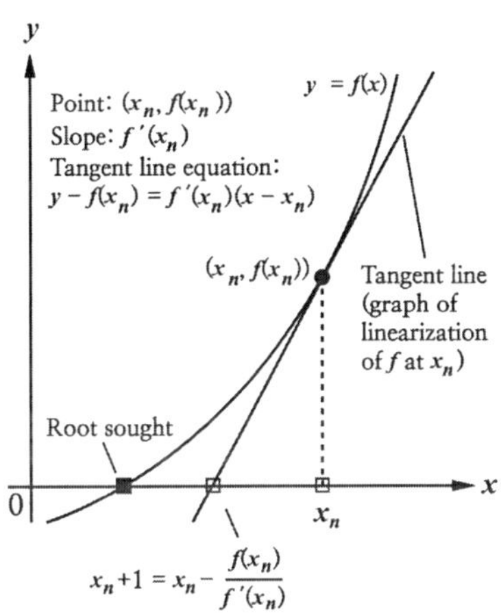

Figure 4.42

Given the approximation x_n, the point

Slope equation for the tangent to the curve at $(x_n, f(x_n))$ is

$$y=f(x_n)+f'(x_n)(x-x_n).$$

We can find where it crosses the x-axis by setting $y=0$(Figure):

$$0=f(x_n)+f'(x_n)(x-x_n)-\frac{f(x_n)}{f'(x_n)}=x-x_n$$

$$x=x_n-\frac{f(x_n)}{f'(x_n)}$$

If $f(x_n)\neq 0$. The value of x is the next approximation x_{n+1}.

Here is a summary of Newton's method.

Example 1

Use Newton's method to approximate the solution of the equation $x^3-5=0$. Give the first four iterations starting with $x_1=2$.

Solution

Let $f(x)=x^3-5$. $f'(x)=3x^2$

$x_1 = 2$

$$x_2 = x_1 - \frac{f(x_1)}{f'(x_1)} = 2 - \frac{f(2)}{f'(2)} = 2 - \frac{(2)^3 - 5}{3(2)^2} = \frac{7}{4} = 1.75$$

$$x_3 = x_2 - \frac{f(x_2)}{f'(x_2)} = 1.75 - \frac{f(1.75)}{f'(1.75)} = 1.75 - \frac{(1.75)^3 - 5}{3(1.75)^2} = 1.7108844$$

$$x_4 = x_3 - \frac{f(x_3)}{f'(x_3)} = 1.7108844 - \frac{f(1.7108844)}{f'(1.7108844)} = 1.7108844 - \frac{(1.7108844)^3 - 5}{3(1.7108844)^2} = 1.7099764$$

$$x_5 = x_4 - \frac{f(x_4)}{f'(x_4)} = 1.7099764 - \frac{f(1.7099764)}{f'(1.7099764)} = 1.7099764 - \frac{(1.7099764)^3 - 5}{3(1.7099764)^2} = 1.7099759$$

[EX 1] Use Newton's method to approximate the solution of the following equation correct to six decimal places.

(1) $x^3 - x - 2 = 0 \quad (x_1 = 2)$

$x_1 = 2$

$x_2 =$

$x_3 =$

$x_4 =$

$x_5 =$

(2) $x^5 + 2x - 5 = 0 \quad (x_1 = 2)$

$x_1 = 2$

$x_2 =$

$x_3 =$

$x_4 =$

$x_5 =$

(3) $x - 2\cos x = 0 \quad (x_1 = 2)$

$x_1 = 2$

$x_2 =$

$x_3 =$

$x_4 =$

$x_5 =$

(4) $\sin x = 1 - x \quad (x_1 = 0)$

$x_1 = 0$

$x_2 =$

$x_3 =$

$x_4 =$

$x_5 =$

[EX 2] Use Newton's method to estimate the solutionof the equation $x^2+x-1=0$. Start with $x_0=-1$ for the left-hand solution and with $x_0=1$ for the solution on the right. Then, in each case, find x_2.

[EX 3] Use Newton's method to estimate the one real solution of $x^3+3x+1=0$. Start with $x_0=0$ and then find x_2.

[EX 4] Use Newton's method to estimate the two zeros of the function $f(x)=x^4+x-3$. Start with $x_0=-1$ for the left-hand zero and with $x_0=1$ for the zero on the right. Then, in each case, find x_2.

4.13 Approximating Derivatives Numerically

Let f be differentiable on an open interval containing a.

- Forward Difference Quotient : $f'(a) \approx \dfrac{f(a+h)-f(a)}{h}$

- Backward Difference Quotient : $f'(a) \approx \dfrac{f(a)-f(a-h)}{h}$

- Symmetric Difference Quotient :
  - the average of the forward and backward difference quotient.
  - a better approximation of the derivative than the forward/backward difference quotient.

$$f'(a) \approx \frac{f(a+h)-f(a-h)}{2h}$$

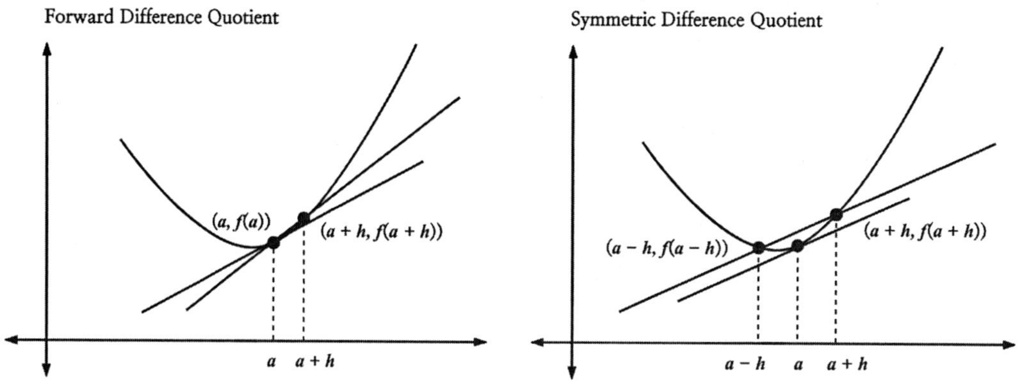

$$f'(a) = \frac{f(a+h) - f(a)}{h}$$

$$f'(a) = \frac{f(a+h) - f(a-h)}{(a+h) - (a-h)} = \frac{f(a+h) - f(a-h)}{2h}$$

Figure 4.43

Example 1

Figure is a radioactive substance. The amount of mass remaining is given by a differentiable function f of time t. The table below shows the amount of the substance remaining as measured for a 40 minute period.

t(minutes)	0	10	20	30	40
$f(t)$(mg)	100	51	24	13	6

Approximate $f'(0), f'(20)$, and $f'(40)$.

Solution

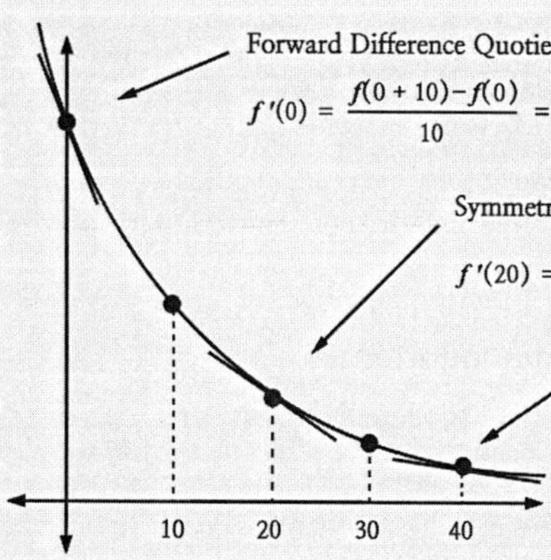

Forward Difference Quotient :

$$f'(0) = \frac{f(0+10)-f(0)}{10} = \frac{f(10)-f(0)}{10} = \frac{51-100}{10} = -4.9 \text{ mg/min}$$

Symmetric Difference Quotient :

$$f'(20) = \frac{f(20+10)-f(20-10)}{2(10)} = \frac{f(30)-f(10)}{20} = \frac{13-51}{20} = -1.9\text{mg/min}$$

Backward Difference Quotient :

$$f'(40) = \frac{f(40)-f(40-10)}{10} = \frac{f(40)-f(30)}{10} = \frac{6-13}{10} = -0.7\text{mg/min}$$

Figure 4.44

[EX 1] The table gives the values of a function f that is differentiable on the interval $[0,1]$:

x	0.10	0.20	0.30	0.40	0.50	0.60
$f(x)$	0.171	0.288	0.357	0.384	0.375	0.336

According to this table, the best approximation of $f'(0.10)$ is

[EX 2] A differentiable function f has the values shown. Estimate $f'(1.5)$.

x	1.0	1.2	1.4	1.6
$f(x)$	8	10	14	22

[EX 3] For $f(x)=5^x$, what is the estimate of $f'(2)$ obtained by using the symmetric difference quotient with $h=0.03$?

[EX 4] From the values of f shown, estimate $f'(2)$.

x	1.92	1.94	1.96	1.98	2.00
$f(x)$	6.00	5.00	4.40	4.10	4.00

[EX 5] Using the values shown in the table for Question [EX 4] estimate $(f^{-1})'(4)$.

[EX 6] The values of differentiable function f are shown below.

x	0	2	4	6	8	10	12
$f(x)$	9	13	19	28	37	43	45

(1) Find the average rate of change of f over the interval [0, 12].

(2) Approximate $f'(0)$, $f'(4)$, $f'(9)$ and $f'(12)$.

(3) Approximate $(f^{-1})'(19)$.

[EX 7]

t(minutes)	0	1	2	3	4	5	6
$C(t)$(ounces)	0	5.3	8.8	11.2	12.8	13.8	14.5

How water is dripping through a coffee maker, filling a large cup with coffee. The amount of coffee in the cup at time $t, 0 \le t \le 6$, is by a differentiable function C, where t is measured in minutes. Selected values of $C(t)$, measured in ounces, are given in the table above.

(1) Use he date in the table to approximate $C'(3.5)$, Show the computations that lead to your answer, and indicate units of measure.

(2) Is there a time $t, 2 \le t \le 4$ at which $C'(t) = 2$? Justify your answer.

(3) The amount of coffee in the cup in ounces is modeled b $B(t) = 16 - 16e^{-0.4t}$. Using this model, find the rate at which the amount of coffee in the cup is changing when $t = 5$.

4.14 Indeterminate Forms and L'Hospital's Rule

To find the limit of an indeterminate form of the type $\frac{0}{0}$ or, $\frac{\infty}{\infty}$, we apply L' Hospital's rule, which involves taking derivatives. In the following, a is a finite number. The rule has several parts:

(a) If $\lim_{x \to a} f(x) = \lim_{x \to a} g(x) = 0$ and if $\lim_{x \to a} \frac{f'(x)}{g'(x)}$ exists, then

$$\lim_{x \to a} \frac{f(x)}{g(x)} = \text{Lim}_{x \to a} \frac{f'(x)}{g'(x)};$$

if $\lim_{x \to a} \frac{f'(x)}{g'(x)}$ does not exist, then L'Hopital's rule cannot be applied.

(b) If $\lim_{x \to a} f(x) = \lim_{x \to a} g(x) = \infty$, the same consequences follow as in case (a). The rules in (a) and (b) both hold for one-sides limits.

(c) $\lim_{x \to \infty} f(x) = \lim_{x \to \infty} g(x) = 0$ and if $\lim_{x \to \infty} \frac{f'(x)}{g'(x)}$ exists, then

$$\lim_{x \to \infty} \frac{f(x)}{g(x)} = \text{Lim}_{x \to \infty} \frac{f'(x)}{g'(x)};$$

if $\lim_{x \to \infty} \frac{f'(x)}{g'(x)}$ does not exist, then L'Hopital's rule cannot be applied.(Here the notation "$x \to \infty$" represents either "$x \to +\infty$" or "$x \to +\infty$.")

(d) If $\lim_{x \to \infty} f(x) = \lim_{x \to \infty} g(x) = \infty$, the same consequences follow as in case (c).

In applying any of the above rules, if we obtain $\frac{0}{0}$ or $\frac{\infty}{\infty}$ again, we can apply the rule once more,
repeating the process until the form we obtain is no longer indeterminate.

Example 1

$\lim_{x \to -2} \frac{x^3+8}{x^2-4}$ is of type $\frac{0}{0}$ and thus equals $\lim_{x \to -2} \frac{3x^2}{2x} = -3$

Solution

Example 2

$\lim_{x \to \infty} \frac{x^3-4x^2+7}{3-6x-2x^3}$ is of type $\frac{\infty}{\infty}$, so that it equals $\lim_{x \to \infty} \frac{3x^2-8x}{-6-x^2}$, which is again of type $\frac{\infty}{\infty}$, Apply L'Hopital's rule twice more:

Solution

$$\lim_{x \to \infty} \frac{6x-8}{-12x} = \lim_{x \to \infty} \frac{6}{-12} = -\frac{1}{2}$$

L'Hopital's rule can be applied also to indeterminate forms of the types $0 \cdot \infty$ and $\infty - \infty$, if the forms can be transformed to either $\frac{0}{0}$ or $\frac{\infty}{\infty}$.

Example 3

$\lim_{x \to \infty} x \sin\frac{1}{x}$ is of the type $\infty \cdot 0$. Since $x \sin\frac{1}{x} = \frac{\sin 1/x}{1/x}$ and, as $x \to \infty$, the latter is the indeterminate form $\frac{0}{0}$, we see that

Solution

$$\lim_{x \to \infty} x \sin\frac{1}{x} = \lim_{x \to \infty} \frac{-\frac{1}{x^2}\cos\frac{1}{x}}{-\frac{1}{x^2}} = \lim_{x \to \infty} \cos\frac{1}{x} = 1.$$

Example 4

Find in $\lim_{x \to 0^+}(\ln x - \ln sinx)$

Solution

$$\lim_{x \to 0^+}(\ln x - \ln sinx) = \lim_{x \to 0^+}\ln(\frac{x}{\sin x}) = \ln(\lim_{x \to 0^+}\frac{x}{\sin x}) = \ln(\lim_{x \to 0^+}\frac{1}{\cos x}) = \ln 1 = 0.$$

Other indeterminate forms, such as $0^0, 1^\infty$ and ∞^0, may be resolved by taking the natural logarithm and then applying L'Hopital's rule.

Example 5

$\lim_{x \to 0}(1+x)^{1/x}$ is of type 1^∞. Let $y = (1+x)^{1/x}$, so that $\ln y = \frac{1}{x}\ln(1+x)$. Then

$\lim_{x \to 0}\ln y = \lim_{x \to 0}\frac{\ln(1+x)}{x}$, which is of type $\frac{0}{0}$. Thus,

Solution

$\lim_{x \to 0}\ln y = \lim_{x \to 0}\frac{\frac{1}{1+x}}{1} = \frac{1}{1} = 1$, and since $\lim_{x \to 0}\ln y = 1$, $\lim_{x \to 0} y = e^1 = e$.

Example 6

Find in $\lim_{x \to \infty}(\ln x)^{1/x}$

Solution

The limit leads to the indeterminate form ∞^0.

Let $f(x) = (\ln x)^{1/x} \Rightarrow \ln f(x) = \ln(\ln x)^{1/x} = \frac{\ln(\ln x)}{x}$.

Now $\lim_{x \to \infty}\ln f(x) = \lim_{x \to \infty}\frac{\ln(\ln x)}{x} = \lim_{x \to \infty}\frac{(\frac{1}{x \ln x})}{1} = 0$.

Therefore $\lim_{x \to \infty}(\ln x)^{1/x} = \lim_{x \to \infty} f(x) = \lim_{x \to \infty} e^{\ln f(x)} = e^0 = 1$

[EX 1] Use L'Hopital's rule to find the limits in Exercises (1) - (19)

(1) $\lim_{x\to 2}\frac{x-2}{x^2-4}$

(2) $\lim_{x\to\infty}\frac{5x^3-2x}{7x^3+3}$

(3) $\lim_{x\to 0}\frac{8x^2}{\cos x-1}$

(4) $\lim_{\theta\to\pi/2}\frac{1-\sin\theta}{1+\cos 2\theta}$

(5) $\lim_{t\to 0}\frac{t(1-\cos t)}{t-\sin t}$

(6) $\lim_{\theta\to 0}\frac{3^{\sin\theta}-1}{\theta}$

(7) $\lim_{x\to\infty}\frac{\ln(x+1)}{\log_2 x}$

(8) $\lim_{y\to 0}\frac{\sqrt{5y+25}-5}{y}$

(9) $\lim_{x\to\infty}(\ln 2x-\ln(x+1))$

(10) $\lim_{x\to 1^+}(\frac{1}{x-1}-\frac{1}{\ln x})$

(11) $\lim_{x\to 0}\frac{x-\sin x}{x\tan x}$

(12) $\lim_{\theta\to 0}\frac{\theta-\sin\theta\cos\theta}{\tan\theta-\theta}$

(13) $\lim_{x\to 1^+}x^{1/(x-1)}$

(14) $\lim_{x\to e^+}(\ln x)^{1/(x-e)}$

(15) $\lim_{x\to\infty}x^{1/\ln x}$

(16) $\lim_{x\to 0}(e^x+x)^{1/x}$

(17) $\lim_{x\to 0^+}(1+\frac{1}{x})^x$

(18) $\lim_{x\to 0^+}x(\ln x)^2$

(19) $\lim_{x\to 0^+}\sin x\cdot\ln x$

memo

CALCULUS

CHAPTER 05

Integration

5.1 Indefinite Integrals

The antiderivative or indefinite integral of a function $f(x)$ is a function $F(x)$ whose derivative is $f(x)$. Since the derivative of a constant equals zero, the antiderivative of $f(x)$ is not unique; that is, if $F(x)$ is an integral of $f(x)$, then so is $F(x)$ + C, where C is any constant. The arbitrary constant C is called the constant of integration. the indefinite integral of $f(x)$ is written as $\int f(x)\,dx$; thus

$$\int f(x)\,dx = F(x) + C \quad \text{if} \quad \frac{dF(x)}{dx} = f(x)$$

The function $F(x)$ is called the integrand.

Basic Integration Formulas

$$\int kf(x)\,dx = k\int f(x)\,dx \quad (k \neq 0)$$

$$\int [f(x) + g(x)]\,dx = \int f(x)\,dx + \int g(x)dx$$

$$\int x^n\,dx = \frac{x^{n+1}}{n+1} + C \qquad (n \neq -1)$$

$$\int \frac{dx}{x} = \ln|x| + C$$

Example 1

Evaluate the following indefinite integrals.

(1) $\int 7\,dx$ (2) $\int 4x^3\,dx$ (3) $\int \left(x^6 + \frac{1}{x}\right)dx$

Solution

(1) $\int 7\,dx = 7x + C$

(2) $\int 4x^3\,dx = x^4 + C$

(3) $\int \left(x^6 + \frac{1}{x}\right)dx = \frac{1}{7}x^7 + \ln|x| + C$

[EX 1] Evaluate the following indefinite integrals.

(1) $\int (3-4x+2x^3)\,dx$

(2) $\int 6x^2 \sqrt[3]{x}\,dx$

(3) $\int x^4 + \sqrt[3]{x^2} - \frac{2}{x^2} - \frac{1}{3\sqrt[3]{x}}\,dx$

(4) $\int \left(\frac{2}{x} - \frac{1}{x^2} + \frac{1}{2\sqrt{x}}\right)dx$

(5) $\int \frac{1}{\ln 2}\frac{dx}{x}$

(6) $\int (x-3)(x^2+3x+9)dx$

(7) $\int \frac{x^4}{x^2+1}dx - \int \frac{1}{x^2+1}dx$

(8) $\int (2\sin x + \cos x)^2\,dx + \int (\sin x - 2\cos x)^2 dx$

(9) $\int (t+\sqrt{t})(t^2+3)dt$

[EX 2] Evaluate the following indefinite integrals.

(1) $\int \frac{(x-2)^3}{x^2}dx$

(2) $\int \sqrt{x}\,(x-2)dx$

(3) $\int \frac{x^3-x-1}{x^2}dx$

(4) $\int \left(x-\frac{1}{2x}\right)^2 dx$

(5) $\int (\sqrt{x}+x)\sqrt[3]{x^2}\,dx$

(6) $\int \frac{(2-y)^2}{4\sqrt{y}}dy$

Integration of Exponential Functions

$$\int e^x\,dx = e^x + C$$

$$\int a^x\,dx = \frac{a^x}{\ln a} + C \qquad (a > 0, a \neq 1)$$

[EX 3] Evaluate the following indefinite integrals.

(1) $\int (e^x + 4)dx$

(2) $\int (5^x - 7^x)dx$

(3) $\int \left(10^x + \frac{10}{x} - \frac{7}{x^3}\right)dx$

Integration of Trigonometric Functions

$$\int \cos x\,dx = \sin x + C$$

$$\int \tan x\,dx = \ln|\sec x| + C$$

$$\text{or } -\ln|\cos x| + C$$

$$\int \sec^2 x\,dx = \tan x + C$$

$$\int \sec x \tan x\,dx = \sec x + C$$

$$\int \sec x\,dx = \ln|\sec x + \tan x| + C$$

$$\int \sin x\,dx = -\cos x + C$$

$$\int \cot x\,dx = \ln|\sin x| + C$$

$$\text{or } -\ln|\csc x| + C$$

$$\int \csc^2 x\,dx = -\cot x + C$$

$$\int \csc x \cot x\,dx = -\csc x + C$$

$$\int \csc x\,dx = \ln|\csc x - \cot x| + C$$

[EX 4] Evaluate the following indefinite integrals.

(1) $\int (5\cos x + \sin x)dx$

(2) $\int (2\tan x + 3\cot x)dx$

(3) $\int \frac{1}{\sin x} dx$

(4) $\int (\tan x + 3\sin x) dx$

(5) $\int (\sec x \tan x + 5\cos x) dx$

(6) $\int (\sec^2 x - 4\csc x) dx$

(7) $\int \frac{1}{1-\cos^2 x} dx$

(8) $\int \frac{\sin^2 x}{1-\cos x} dx$

(9) $\int \frac{\sin x}{\cos^2 x} dx$

(10) $\int \frac{\cos x}{\sin^2 x} dx$

(11) $\int \tan^2 x \, dx$

(12) $\int \cot^2 x dx$

(13) $\int \frac{e^x \sin^2 x + 1}{\sin^2 x} dx$

(14) $\int \frac{10^x \cot x - 1}{\cot x} dx$

Integration of Inverse Trigonometric Functions

$$\int \frac{dx}{\sqrt{1-x^2}} = \sin^{-1} x + C$$

or $\arcsin x + C$

$$\int \frac{dx}{1+x^2} = \tan^{-1} x + C$$

or $\arctan x + C$

$$\int \frac{dx}{x\sqrt{x^2-1}} = \sec^{-1}|x| + C$$

or $\mathrm{arc}sec|x| + C$

[EX 5] Evaluate the following indefinite integrals.

(1) $\int \frac{5}{\sqrt{1-x^2}} dx$

(2) $\int \frac{3}{x^2+1} dx$

(3) $\displaystyle\int \frac{-7}{x\sqrt{x^2-1}}dx$

(4) $\displaystyle\int \frac{4}{1+x^2}dx$

Integration Using Reverse Chain Rule (1)

$$\int (ax+b)^n dx = \frac{1}{a(n+1)}(ax+b)^{n+1} + C \quad \text{(a and b are constants)}$$

[EX 6] Evaluate the following indefinite integrals.

(1) $\displaystyle\int (2x+1)^3 dx$

(2) $\displaystyle\int (3x-2)^8 dx$

(3) $\displaystyle\int \left(\frac{1}{2}x+3\right)^5 dx$

(4) $\displaystyle\int \sqrt{1-3x}\,dx$

(5) $\displaystyle\int \frac{7}{\sqrt{2x+1}}dx$

(6) $\displaystyle\int \frac{1}{(1-x)^2}dx$

Integration Using Reverse Chain Rule (2)

$$\int e^{kx}dx = \frac{1}{k}e^{kx} + C \qquad \int a^{kx}dx = \frac{1}{k}\frac{a^{kx}}{\ln a} + C \quad \text{(k and a are constants)}$$

$$\int \cos kx\,dx = \frac{1}{k}\sin kx + C \qquad \int \sin kx\,dx = -\frac{1}{k}\cos kx + C$$

[EX 7] Evaluate the following indefinite integrals.

(1) $\int (e^{7x}+e^{3x})dx$

(2) $\int (e^{-2x}+\cos 2x)dx$

(3) $\int \left(e^{-2x}+e^{-\frac{1}{3}x}\right)dx$

(4) $\int (4e^{-x}+3^{x-1})dx$

(5) $\int (10^{5x}+\sin 2x)dx$

(6) $\int 4^{x}(4^{2x}+1)dx$

(7) $\int \sqrt{3^{x+1}}\,dx$

(8) $\int \left[\cos 3x+\sin\left(\frac{x}{2}\right)\right]dx$

(9) $\int \cos(3x+5)dx$

(10) $\int \sec 5x\tan 5x\,dx$

(11) $\int (e^{x}+e^{-x})^2 dx$

(12) $\int (\sec^2 2x+\tan 2x)dx$

Integration Using Double Angle Identities

$\sin 2x = 2\sin x\cos x \Rightarrow \sin x\cos x = \frac{1}{2}\sin 2x$

$\cos 2x = 2\cos^2 x - 1 \Rightarrow \cos^2 x = \frac{1+\cos 2x}{2} = \frac{1}{2}+\frac{1}{2}\cos 2x$

$\cos 2x = 1-2\sin^2 x \Rightarrow \sin^2 x = \frac{1-\cos 2x}{2} = \frac{1}{2}-\frac{1}{2}\cos 2x$

[EX 8] Evaluate the following indefinite integrals.

(1) $\int 3\sin x\cos x\,dx$

(2) $\int \sin\frac{\theta}{2}\cos\frac{\theta}{2}\,d\theta$

(3) $\int 8\sin 2x\cos 2x\,dx$

(4) $\int 12\sin 3x\cos 3x\,dx$

(5) $\int \sin^2\frac{1}{2}x\,dx$

(6) $\int \sin^2 3x\,dx$

(7) $\int \cos^2 x\,dx$

(8) $\int \cos^2 4x\,dx$

Integration of Special Fractions

$$\int \frac{f'(x)}{f(x)}dx = \ln|f(x)| + C$$

[EX 9] Evaluate the following indefinite integrals.

(1) $\int \frac{5\cos x}{\sin x}dx$

(2) $\int \frac{3x^2+1}{x^3+x}dx$

(3) $\int \frac{7}{x+1}dx$

(4) $\int \frac{x}{x^2+1}dx$

(5) $\int \frac{6x}{x^2-4}dx$

(6) $\int \frac{2}{1-3x}dx$

(7) $\int \frac{5\sin x}{1+\cos x}dx$

(8) $\int \frac{x}{2x^2+1}dx$

(9) $\int \frac{e^x-e^{-x}}{e^x+e^{-x}}dx$

[EX 10] Determine $f(x)$.

(1) $f'(x) = 1 + \dfrac{1}{x}$, $f(1) = 3$

(2) $\dfrac{dy}{dx} = 2^x - x^2\sqrt{x} + \sin x$, $y(0) = \dfrac{1}{\ln 2}$

(3) $f'(x) = \dfrac{8}{1+x^2} + \sec^2 x$, $f(0) = 1$

(4) $y' = \dfrac{4}{2x-1} + 8(2x+1)^3$, $y(0) = 3$

(5) $f'(x) = \dfrac{2x}{x^2+4} + \cos^2 x$, $f(0) = 0$

(6) $f''(x) = e^{\frac{1}{2}x} - 12x + 2$, $f(0) = 1$, $f'(0) = -1$

(7) $\dfrac{d^2y}{dx^2} = \dfrac{2}{x^3}$, $y(1) = 1$, $y'(1) = 1$

(8) $\dfrac{d^2y}{dx^2} = e^{-x} + 6x$, $y(0) = 2$, $y(1) = \dfrac{1}{e} - 1$

Integration Formulas

Basic Integration Formulas

$$\int k\,dx = kx + c \quad (k : a\,constant)$$

$$\int x^n\,dx = \frac{1}{n+1}x^{n+1} + C,\ n \neq -1$$

$$\int \frac{1}{x}dx = \ln|x| + C$$

Integration of Exponential Functions

$$\int e^x\,dx = e^x + C$$

$$\int a^x\,dx = \frac{a^x}{\ln a} + C \quad (a > 0,\ a \neq 1)$$

Integration of Trigonometric Functions

$$\int \cos x\,dx = \sin x + C$$

$$\int \sin x\,dx = -\cos x + C$$

$$\int \sec^2 x\,dx = \tan x + C$$

$$\int \csc^2 x\,dx = -\cot x + C$$

$$\int \sec x \tan x\,dx = \sec x + C$$

$$\int \csc x \cot x\,dx = -\csc x + C$$

$$\int \tan x\,dx = \ln|\sec x| + C = -\ln|\cos x| + c$$

$$\int \cot x\,dx = \ln|\sin x| + C = -\ln|\csc x| + c$$

$$\int \sec x\,dx = \ln|\sec x + \tan x| + C$$

$$\int \csc x\,dx = \ln|\csc x - \cot x| + C$$

$$\int \sin^2 x\,dx = \int\left(\frac{1}{2} - \frac{1}{2}\cos 2x\right)dx = \frac{1}{2}x - \frac{1}{4}\sin 2x + C$$

$$\int \cos^2 x\,dx = \int\left(\frac{1}{2} + \frac{1}{2}\cos 2x\right)dx = \frac{1}{2}x + \frac{1}{4}\sin 2x + C$$

Integration Using Reverse Chain Rule

$$\int e^{kx} = \frac{1}{k}e^{kx} + C$$

$$\int \sin kx\,dx = -\frac{1}{k}\cos kx + C$$

$$\int \cos kx\,dx = \frac{1}{k}\sin kx + C$$

$$\int a^{kx}\,dx = \frac{1}{k}\,\frac{a^{kx}}{\ln a} + C$$

$$\int (kx+b)^n dx = \frac{1}{n+1} \cdot \frac{1}{k}(kx+b)^{n+1} + C$$

(k, a and b are constants)

Integration of Special Fractions

$$\int \frac{f'(x)}{f(x)}dx = \ln|f(x)| + C$$

Integration of Inverse Trigonometric Functions

$$\int \frac{dx}{\sqrt{1-x^2}} = \sin^{-1}x + C$$

$$\int \frac{dx}{\sqrt{a^2-x^2}} = \sin^{-1}\left(\frac{x}{a}\right) + C$$

$$\int \frac{dx}{x^2+1} = \tan^{-1}x + C$$

$$\int \frac{dx}{x^2+a^2} = \frac{1}{a}\tan^{-1}\left(\frac{x}{a}\right) + C$$

$$\int \frac{dx}{x\sqrt{x^2-1}} = \sec^{-1}|x| + C$$

5.2 Integration by Substitution

Substitute $u = g(x)$ and $du = g'(x)dx$, integrate with respect to u, and replace u with $g(x)$.

$$\int f(g(x))g'(x)dx = \int f(u)du = F(u) + C = F(g(x)) + C$$

Example 1

(1) Find $\int \sec^2(5t+1) \cdot 5\,dt$

(2) Find $\int \cos(7\theta+3)d\theta$.

Solution

(1) Find $\int \sec^2(5t+1) \cdot 5\,dt$

We substitute $u = 5t+1$ and $du = 5dt$.

Then, $\int \sec^2(5t+1) \cdot 5\,dt = \int \sec^2 u\,du = \tan u + C = \tan(5t+1) + C.$

(2) Find $\int \cos(7\theta+3)d\theta$.

$$\int \cos(7\theta+3)d\theta = \frac{1}{7}\int \cos(7\theta+3) \cdot 7\,d\theta = \frac{1}{7}\int \cos u\,du = \frac{1}{7}\sin u + C = \frac{1}{7}\sin(7\theta+3) + C.$$

[EX 1] Evaluate the following indefinite integrals.

(1) $\int x\sqrt[4]{1-x^2}\,dx$

(2) $\int (x^2-2x+3)^{10}(x-1)dx$

(3) $\int \frac{4x^3}{(x^4+1)^2}dx$

(4) $\int \frac{x-3}{\sqrt[4]{x^2-6x}}dx$

(5) $\int \frac{9x^2}{\sqrt{1-x^3}}dx$

(6) $\int \frac{x-2}{\sqrt{4x-x^2}}dx$

(7) $\int 3x\sqrt{7-3x^2}\,dx$

(8) $\int \frac{(1+\sqrt{x})^5}{\sqrt{x}}dx$

(9) $\int \frac{2^{\sqrt{x}}}{\sqrt{x}}dx$

(10) $\int \frac{1}{\sqrt{x}(\sqrt{x}-1)}dx$

[EX 2] Evaluate the following indefinite integrals.

(1) $\int \frac{(\ln x)^3}{2x}dx$

(2) $\int \frac{1}{xlnx}dx$

(3) $\int \frac{\ln x}{x\sqrt{\ln^2 x+1}}dx$

(4) $\int \frac{1}{xlnx \cdot \ln(\ln x)}dx$

(5) $\int \sin^5\frac{x}{3}\cos\frac{x}{3}dx$

(6) $\int \cos^3 3x\sin 3x\,dx$

(7) $\int e^{\sec\pi x}\text{scc}\,\pi x\tan\pi x\,dx$

(8) $\int \sec^2\theta\tan^2\theta\,d\theta$

(9) $\int \frac{e^{\sin^{-1}x}}{\sqrt{1-x^2}}dx$

(10) $\int \sec^4\theta\,d\theta$

(11) $\int \frac{1}{(\tan^{-1}x)(1+x^2)}dx$

(12) $\int \frac{\sec^2 2\theta}{1+\tan 2\theta}d\theta$

(13) $\int \frac{\sin(2x+1)}{\cos^2(2x+1)}dx$

(14) $\int \frac{\cos 2x}{\sqrt{1+\sin 2x}}dx$

(15) $\int \sin 2x \, e^{\sin^2 x} dx$

(16) $\int \frac{\sin x}{1+\cos^2 x} dx$

[EX 3] Evaluate the following indefinite integrals.

(1) $\int \frac{e^x}{1+e^x} dx$

(2) $\int \frac{e^{2x}}{e^{2x}+1} dx$

(3) $\int 2x \, e^{-x^2} dx$

(4) $\int \frac{x}{e^{x^2+2}} dx$

(5) $\int \frac{\sqrt{\tan^{-1} x}}{1+x^2} dx$

(6) $\int \frac{\sin^{-1}(2x)}{\sqrt{1-4x^2}} dx$

(7) $\int \frac{e^x}{1+e^{2x}} dx$

(8) $\int \frac{e^{2x}}{1+e^{4x}} dx$

(9) $\int \frac{x^2+3}{x+1} dx$

(10) $\int \frac{e^{2x}}{e^x+1} dx$

Integration of Inverse Trigonometric Functions

$$\int \frac{dx}{\sqrt{a^2-x^2}} = \sin^{-1}\left(\frac{x}{a}\right) + C \qquad \int \frac{dx}{a^2+x^2} = \frac{1}{a}\tan^{-1}\left(\frac{x}{a}\right) + C$$

proof

(1) $\int \frac{dx}{\sqrt{a^2-x^2}} = \int \frac{\frac{1}{a}dx}{\frac{1}{a}\sqrt{a^2-x^2}} = \int \frac{\frac{1}{a}dx}{\sqrt{1-\frac{x^2}{a^2}}} = \sin^{-1}\left(\frac{x}{a}\right) + C$

(2) $\displaystyle\int \frac{dx}{a^2+x^2} = \int \frac{\frac{1}{a^2}dx}{\frac{1}{a^2}(a^2+x^2)} = \int \frac{\frac{1}{a}\cdot\frac{1}{a}dx}{1+\frac{x^2}{a^2}} = \frac{1}{a}\tan^{-1}(\frac{x}{a}) + C$

[EX 4] Evaluate the following indefinite integrals.

(1) $\displaystyle\int \frac{1}{1+4x^2}dx$

(2) $\displaystyle\int \frac{1}{\sqrt{16-x^2}}dx$

(3) $\displaystyle\int \frac{1}{x^2+2x+2}dx$

(4) $\displaystyle\int \frac{3}{9+x^2}dx$

(5) $\displaystyle\int \frac{1}{\sqrt{9-x^2}}dx$

(6) $\displaystyle\int \frac{1}{9+4x^2}dx$

(7) $\displaystyle\int \frac{2x-1}{\sqrt{4x-4x^2}}dx$

(8) $\displaystyle\int \frac{e^x}{9+e^{2x}}dx$

(9) $\displaystyle\int \frac{2x+1}{4+x^2}dx$

(10) $\displaystyle\int \frac{x+1}{\sqrt{4-x^2}}dx$

(11) $\displaystyle\int \frac{1}{x^2-2x+5}dx$

[EX 5] Given that $f'(x) = \dfrac{1}{\sqrt{9-4x^2}}$ and $f(-\dfrac{3}{2}) = \dfrac{\pi}{12}$, find $f(x)$

5.3 Integration by Parts

$$\int uv'\,dx = uv - \int u'v\,dx$$

proof $\frac{d}{dx}(uv) = u'v + uv'$ $\quad \int \frac{d}{dx}(uv)dx = \int (u'v + uv')dx$

$$\int \frac{d}{dx}(uv)dx = \int (u'v)dx + \int (uv')dx$$

$$uv = \int u'v dx + \int uv' dx$$

$$\therefore \int uv' dx = uv - \int u'v dx$$

Example 1

Evaluate the integral.

(1) $\int x \cdot e^x dx = x \cdot e^x - \int 1 \cdot e^x dx = xe^x - e^x + C$

(2) $\int x \cdot e^{-2x} dx = x \cdot (-\frac{1}{2}e^{-2x}) - \int 1 \cdot (-\frac{1}{2}e^{-2x})dx = -\frac{1}{2}xe^{-2x} - \frac{1}{4}e^{-2x} + C$

Solution

(1) $\int x \cdot e^x dx = x \cdot e^x - \int 1 \cdot e^x dx = xe^x - e^x + C$

(2) $\int x \cdot e^{-2x} dx = x \cdot (-\frac{1}{2}e^{-2x}) - \int 1 \cdot (-\frac{1}{2}e^{-2x})dx = -\frac{1}{2}xe^{-2x} - \frac{1}{4}e^{-2x} + C$

[EX 1]Evaluate the following indefinite integrals.

(1) $\int x\cos x\, dx$ (2) $\int xe^{-\frac{1}{2}x} dx$

(3) $\int x\cos 3x\,dx$

(4) $\int x\sin 2x\,dx$

(5) $\int x^2 e^x\,dx$

(6) $\int \ln x\,dx$

(7) $\int x^3 \ln x\,dx$

(8) $\int \frac{\ln x}{x^2}dx$

(9) $\int \frac{\ln x}{\sqrt{x}}dx$

(10) $\int \tan^{-1}x\,dx$

(11) $\int x\sec^{-1}x\,dx,\ x>0$

[EX 2] Evaluate the following indefinite integrals.

(1) $\int x(\ln x)^2 dx$

(2) $\int \frac{(\ln x)^2}{x^2}dx$

Circular Integration by Parts : $\int$ (Exponential Function) × (Trigonometric Function) dx

Example 2

Evaluate $\int e^x \cos x\, dx$.

$= e^x \sin x - \int e^x \sin x\, dx$

$= e^x \sin x - \left[- e^x \cos x - \int - e^x \cos x\, dx \right]$

$= e^x \sin x + e^x \cos x - \int e^x \cos x\, dx$

Solution

Let $\int e^x \cos x dx = I$

$I = e^x \sin x + e^x \cos x - I \quad 2I = e^x (\sin x + \cos x)$

$\therefore I = \dfrac{e^x (\sin x + \cos x)}{2} + C$

[EX 3] Evaluate the following indefinite integrals.

(1) $\int e^{2x} \sin x\, dx$

(2) $\int e^x \cos 2x\, dx$

(3) $\int e^{2x} \cos 3x\, dx$

(4) $\int e^{-2x} \sin 2x\, dx$

[EX 4] Evaluate the following indefinite integrals.

(1) $\int x^2 e^x dx$

(2) $\int x^3 e^{-x} dx$

(3) $\int x^3 e^{-\frac{x}{2}} dx$

(4) $\int x^2 \cos 3x\, dx$

(5) $\int x^3 \sin \frac{1}{2} x\, dx$

[EX 5] Evaluate the following indefinite integrals.

(1) $\int e^{\sqrt{x}}dx$ (2) $\int \cos\sqrt{x}\,dx$

(3) $\int x^3e^{-x^2}dx$ (4) $\int x^3\sin(x^2)dx$

5.4 Rectangular Approximation Method (RAM) :

We may approximate $\int_a^b f(x)dx$ by any of the following sums:

(1) Left sum: $f(x_0)\triangle x+f(x_1)\triangle x+\cdots+f(x_{n-1})\triangle x$, using the value of f at the left endpoint of each subinterval.

(2) Right sum: $f(x_1)\triangle x+f(x_2)\triangle x+\cdots+f(x_n)\triangle x$, using the value of f at the right end of each subinterval.

(3) Midpoint sum: $f(\frac{x_0+x_1}{2})\triangle x+f(\frac{x_1+x_2}{2})\triangle x+\cdots+f(\frac{x_{n-1}+x_n}{2})\triangle x$, using the value of f at the midpoint of each subinterval.

Example 1

Approximate $\int_0^2 x^3dx$ by using four subintervals and calculating (a) the left sum, (b) the right sum, and (c) the midpoint sum. Evaluate the Integral exactly.

Here $\triangle x=\frac{2-0}{4}=\frac{1}{2}$.

(a) LRAM (b) RRAM (c) MRAM

Solution

(1) For a left sum we use the left-hand altitudes at $x=0, \frac{1}{2}, 1$, and $\frac{3}{2}$.

The approximating sum is $(0)^3 \cdot \frac{1}{2}+\left(\frac{1}{2}\right)^3 \cdot \frac{1}{2}+(1)^3 \cdot \frac{1}{2}+\left(\frac{3}{2}\right)^3 \cdot \frac{1}{2}=\frac{9}{4}$.

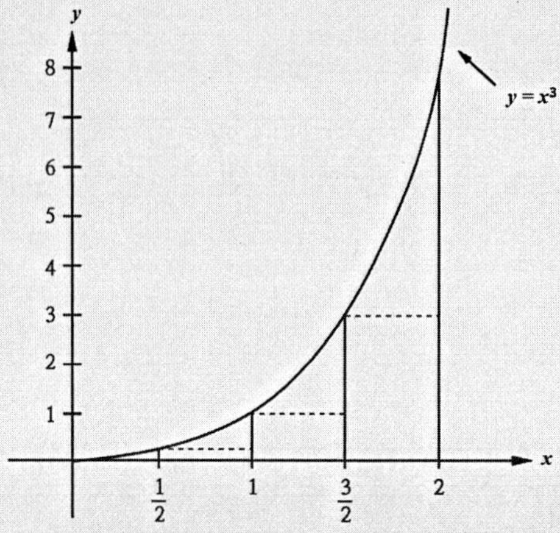

Figure 5.1

(2) Fore the right sum we use right-hand altitudes at $x=\frac{1}{2}, 1, \frac{3}{2}$, and 2.

The approximating sum is $\left(\frac{1}{2}\right)^3 \cdot \frac{1}{2}+(1)^3 \cdot \frac{1}{2}+\left(\frac{3}{2}\right)^3 \cdot \frac{1}{2}+(2)^3 \cdot \frac{1}{2}=\frac{25}{4}$.

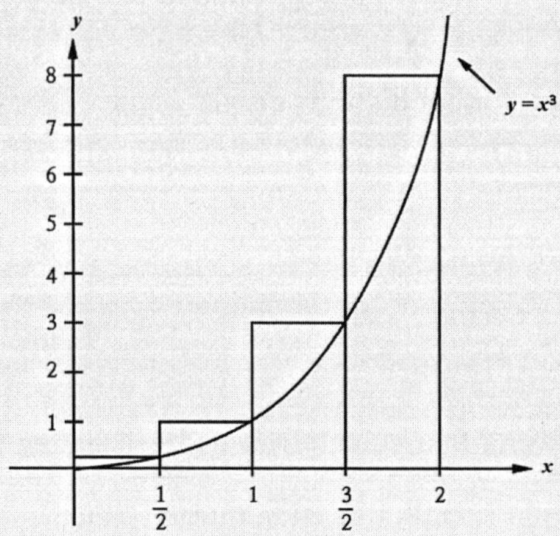

Figure 5.2

(3) The midpoint sum uses the heights at the midpoints of the subintervals.

The approximating sum is $\left(\frac{1}{4}\right)^3 \cdot \frac{1}{2}+\left(\frac{3}{4}\right)^3 \cdot \frac{1}{2}+\left(\frac{5}{4}\right)^3 \cdot \frac{1}{2}+\left(\frac{7}{4}\right)^3 \cdot \frac{1}{2}=\frac{31}{8}$ or $\frac{15.5}{4}$.

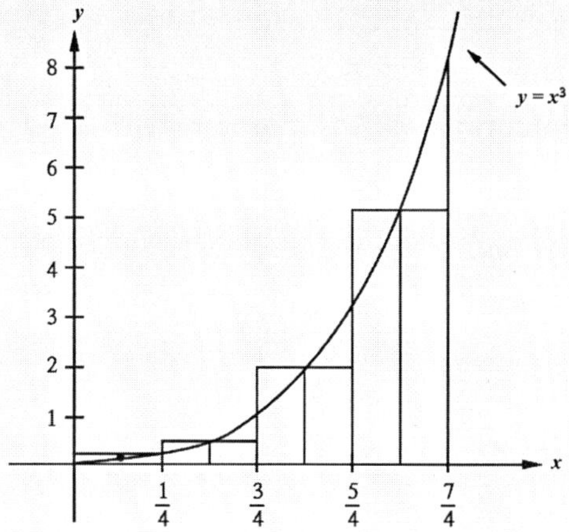

Figure 5.3

[EX 1] Approximate the area under the curve $y = 4 - x^2$ from $x = 0$ to $x = 2$ using (a) LRAM (b) RRAM and (c) MRAM respectively with 4 equal subintervals.

(a) LRAM

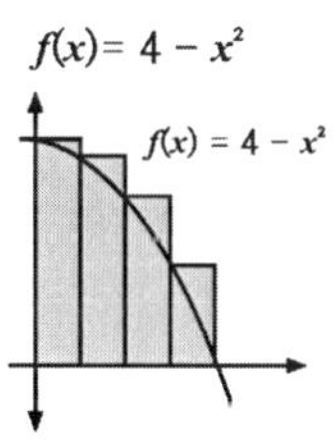

Figure 5.4

(b) RRAM

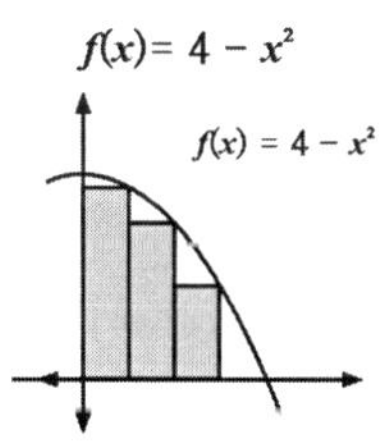

Figure 5.5

(c) MRAM

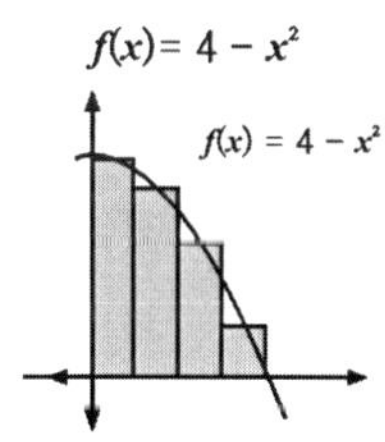

Figure 5.5

5.5 Trapezoidal Approximation

To approximate the area under a curve on the interval $[a,b]$,

$$T(n)=\frac{h}{2}(y_0+2y_1+2y_2 \ldots\ldots+2y_{n-1}+y_n)$$

$$=\frac{LRAM+RRAM}{2}, \qquad where\ \ h=\frac{b-a}{n}$$

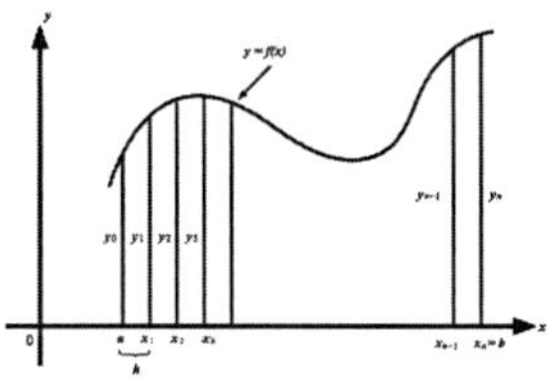

Figure 5.6

$T(n)$

$$=\frac{(y_0+y_1)h}{2}+\frac{(y_1+y_2)h}{2}+\frac{(y_2+y_3)h}{2}\cdots+\frac{(y_{n-1}+y_n)h}{2}$$

$$=\frac{h[(y_0+y_1)+(y_1+y_2)+(y_2+y_3)+(y_{n-1}+y_n)]}{2}$$

$$=\frac{h}{2}(y_0+2y_1+2y_2+\cdots\cdots+2y_{n-1}+y_n)$$

Example 1

Approximate the area under the curve $y=x^3$ from $x=0$ to $x=2$ using the trapezoidal rule with 4 equal subintervals.

Solution

$$T(4)=\frac{h}{2}[y_0+2y_1+2y_2+2y_3+y_4]=\frac{1/2}{2}\left(0^3+2\left(\frac{1}{2}\right)^3+2(1)^3+2\left(\frac{3}{2}\right)^3+2^3\right)=\frac{17}{4}.$$

[EX 1] Find $L(5)$, $R(5)$, and $T(5)$ for $\int_1^6 \frac{120}{x}dx$.

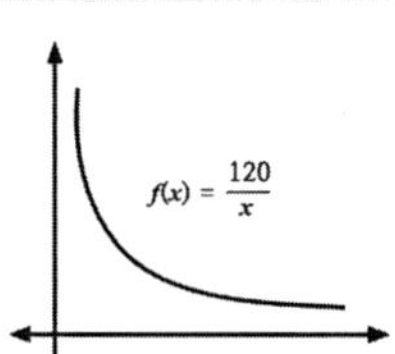

Figure 5.7

[EX 2] Approximate the area under the curve $y = \sin x$ from $x = 0$ to $x = \pi$ using the trapezoidal rule with 6 equal subintervals.

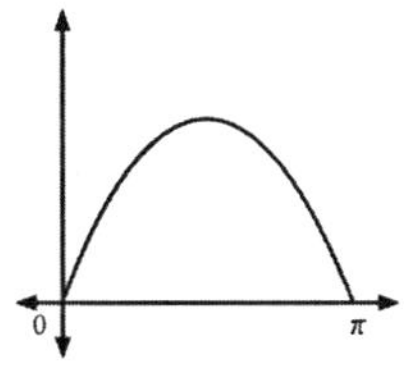

Figure 5.8

[EX 3] Approximate the area under the curve $y = x^3 + 1$ from $x = -1$ to $x = 3$ using (a) LRAM (b) RRAM (c) the trapezoidal rule, with 4 equal subintervals.

(a) LRAM

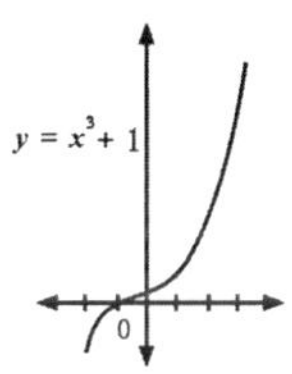

Figure 5.9

(b) RRAM

(c) Trapezoidal rule

[EX 4] Readings from a car's speedometer at 10-minute intervals during a 1-hour period are given in the table; t=time in minutes, v=speed in miles per hour:

t	0	10	20	30	40	50	60
v	26	40	55	10	60	32	45

(1) Draw a graph that could represent the car's speed during the hour.

(2) Approximate the distance traveled, using L(6), R(6), and T(6).

[EX 5] The rate at which oil flows out of a tank, in iters per minute, is shown in the table below. Approximate the total amount of oil flowing out of the tank during 10 minutes using (a) LRAM (b) RRAM and (c) the trapezoidal rule, with 5 subintervals indicated by the data.

t (min)	R(t) (L/min)
0	4
1	6
3	8
6	9
9	7
10	5

5.6 Definition of Area as a Limit

Let f be continuous on the interval $[a, b]$ and the interval be divided into n subintervals of equal length. The area under the curve $y = f(x)$ over the interval $[a, b]$ is defined by

$$Area = \lim_{n \to \infty} \sum_{k=1}^{n} f(x_k) \triangle x \ , \quad \text{where } \triangle x = \frac{b-a}{n} \text{ and } x_k = a + k\triangle x$$

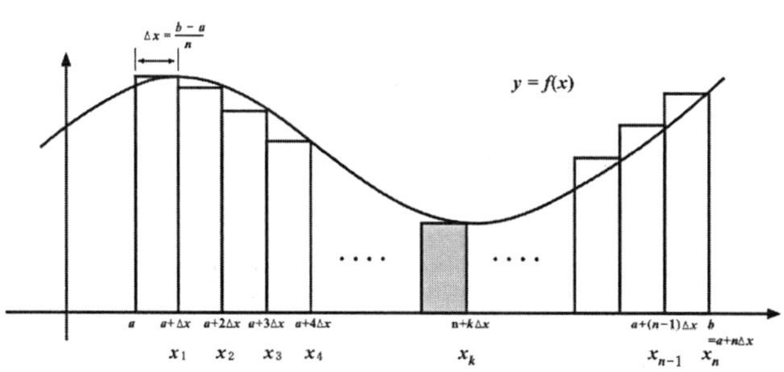

Figure 5.10

$$Area \approx f(x_1)\triangle x + f(x_2)\triangle x + f(x_3)\triangle x + \cdots\cdots + f(x_k)\triangle x + \cdots\cdots f(x_n)\triangle x \ : \text{RRAM}$$

$$= \sum_{k=1}^{n} f(x_k)\triangle x$$

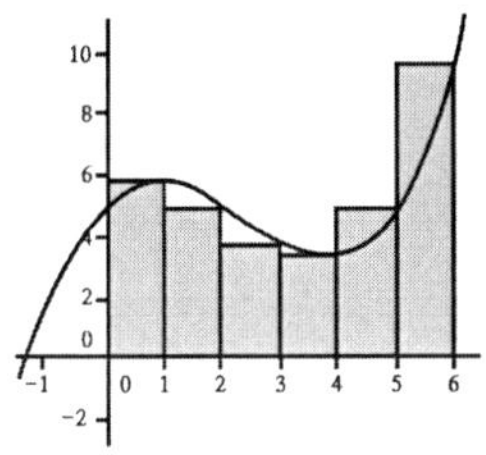

$n = 6$

Figure 5.11

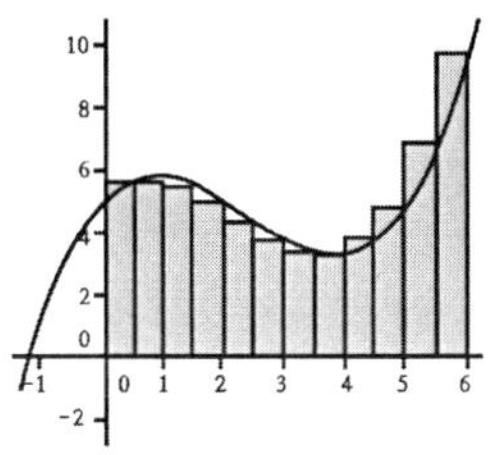

$n = 12$

Figure 5.12

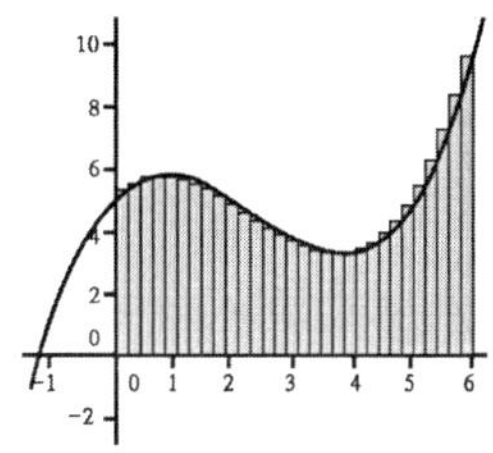

$n = 30$

Figure 5.13

The approximation becomes more accurate as the number of strips increases.

Therefore, $Area = \lim_{n\to\infty}\sum_{k=1}^{n} f(x_k)\triangle x$

This method can also be applied to LRAM and MRAM.

$Area = \lim_{n\to\infty}\sum_{k=1}^{n} f(x_k{}^*)\triangle x$ for RRAM : $x_k^* = a + k\triangle x$,

LRAM : $x_k^* = a + (k-1)\triangle x$,

MRAM : $x_k^* = a + (k - \frac{1}{2})\triangle x$,

Example 1

Find the area under the cureve $y = 16 - x^2$ from $x = 1$ to $x = 3$ by summing the areas of the rectangles.

Solution

$$y = 16 - x^2,\ [1,3]\left(\Delta x = \frac{2}{n}\right)$$

$$s(n) = \sum_{i=1}^{n} f\left(1 + \frac{2i}{n}\right)\left(\frac{2}{n}\right) = \sum_{i-1}^{n}\left[16 - \left(1 + \frac{2i}{n}\right)^2\right]\left(\frac{2}{n}\right) = \frac{2}{n}\sum_{i=1}^{n}\left[15 - \frac{4i^2}{n^2} - \frac{4i}{n}\right]$$

$$= \frac{2}{n}\left[15n - \frac{4}{n^2}\frac{n(n+1)(2n+1)}{6} - \frac{4}{n}\frac{n(n+1)}{2}\right]$$

$$= 30 - \frac{8}{6n^2}(n+1)(2n+1) - \frac{4}{n}(n+1)$$

$$Area = \lim_{n\to\infty} s(n) = 30 - \frac{8}{3} - 4 = \frac{70}{3} = 23\frac{1}{3}.$$

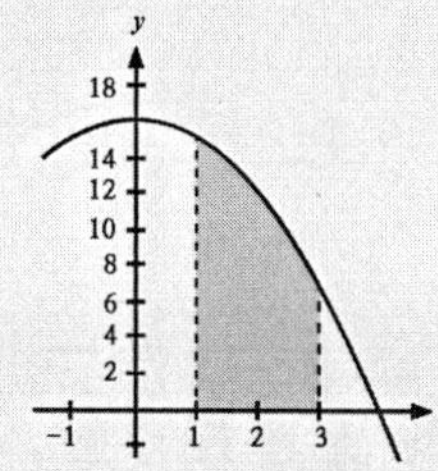

Figure 5.14

[EX 1] Find the area under the curve $y = -2x + 3$ from $x = 0$ to $x = 1$ by summing the areas of the rectangles.

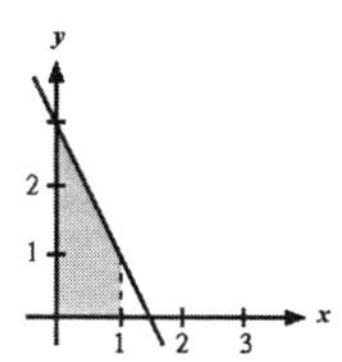

Figure 5.15

[EX 2] Find the area under the curve $y = x^2 + 1$ from $x = 0$ to $x = 3$ by summing the areas of the rectangles.

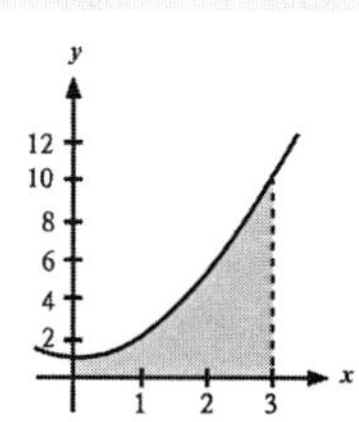

Figure 5.16

[EX 3] Find the area under the curve $y = x^3$ from $x = 0$ to $x = 2$ by summing the areas of the rectangles.

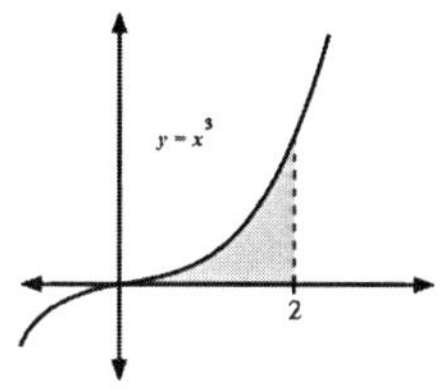

Figure 5.17

[EX 4] Find the area under the curve $y = x^2$ from $x = 1$ to $x = 3$ by summing the areas of the rectangles.

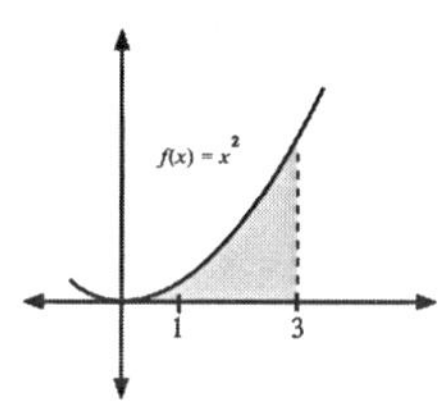

Figure 5.18

[EX 5] Find the area under the curve $y = 9 - x^2$ from $x = 2$ to $x = 3$ by summing the areas of the rectangles.

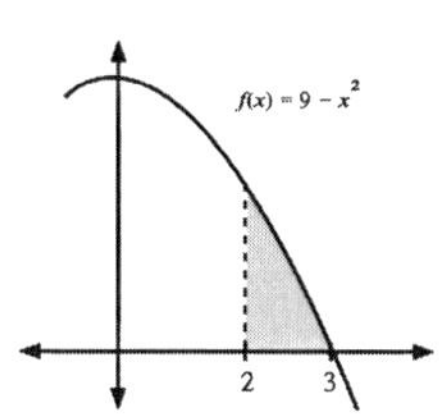

Figure 5.19

5.7 Definite Integrals

Let f be continuous on the interval $[a, b]$. The definite integral of f from a to b is

$$\int_a^b f(x)dx = \lim_{n\to\infty}\sum_{k=1}^{n} f(x_k^*)\Delta x$$

$f(x)$ is called the integrand and a and b are called the lower limit and the upper limit, respectively.

Rules satisfied by definite integrals

1. Order of Integration : $\int_b^a f(x)dx = -\int_a^b f(x)dx$
2. Zero Width Interval : $\int_a^a f(x)dx = 0$
3. Constant Multiple : $\int_a^b kf(x)dx = k\int_a^b f(x)dx$
4. Sum and Difference : $\int_a^b (f(x)\pm g(x))dx = \int_a^b f(x)dx \pm \int_a^b g(x)dx$
5. Additivity : $\int_a^b f(x)dx + \int_b^c f(x)dx = \int_a^c f(x)dx$
6. Max-Min Inequality : If f has maximum alue max f and minimum value min f on [a, b], then

$$\min f \cdot (b-a) \le \int_a^b f(x)dx \le \max f \cdot (b-a).$$

7. Domination : $f(x) \ge g(x)\, on\, [a,b] \Rightarrow \int_a^b f(x)dx \ge \int_a^b g(x)dx$

$$f(x) \ge 0\ on\ [a,b] \Rightarrow \int_a^b f(x)dx \ge 0$$

[EX 1] Suppose that f and h are integrable and that find each of the following.

$$\int_1^9 f(x)dx = -1,\ \int_7^9 f(x)dx = 5,\ \int_7^9 h(x)dx = 4.$$

(1) $\int_1^9 -2f(x)dx$

(2) $\int_7^9 [f(x)+h(x)]dx$

(3) $\int_7^9 [2f(x)-3h(x)]dx$

(4) $\int_9^1 f(x)dx$

(5) $\int_1^7 f(x)dx$

(6) $\int_9^7 [h(x)-f(x)]dx$

[EX 2] Given that $\int_{-1}^2 f(x)dx = 2$, $\int_2^4 f(z)dz = 3$, $\int_{-1}^2 g(s)ds = -1$, and $\int_4^2 g(x)dx = 4$, find each of the following.

(1) $\int_{-1}^2 f(t)dt$

(2) $\int_4^2 f(x)dx$

(3) $\int_4^{-1} f(y)dy$

(4) $\int_3^3 [3f(x)+2g(x)]dx$

(5) $\int_{-1}^2 [3f(x)+2g(x)]dx$

(6) $\int_{-1}^4 [2f(x)-3g(x)]dx$

[EX 3] Given that $\int_a^b e^x \cos x\,dx = r$, $\int_b^c e^x \cos x\,dx = s$, and $\int_c^a x^2 \ln(x+1)dx = p$, find each of the following.

(1) $\int_b^b e^x \cos x\,dx$

(2) $\int_c^a e^y \cos y\,dy$

(3) $\int_a^c t^2 \ln(t+1)dt$

(4) $\int_a^c [2e^x \cos x - 3x^2 \ln(x+1)]\,dx$

[EX 4]

(1) Given that $\int_{-2}^4 f(x)dx = 3$, $\int_2^4 f(x)dx = 2$, and $\int_2^5 f(x)dx = 6$, evaluate $\int_{-2}^5 f(x)dx$.

(2) Given that

$\int_6^{-3} f(x)dx = 4$, $\int_6^{-1} f(x)dx = -1$, and $\int_3^{-3} f(x)dx = -2$, evaluate $\int_{-1}^3 f(x)dx$.

(3) Given that

$\int_3^{-2} f(x)dx = 2$, $\int_5^{-2} f(x)dx = -1$, and $\int_3^{-3} f(x)dx = -4$, evaluate $\int_0^2 f(x+3)dx$.

(4) Given that $\int_1^2 f(x)dx = 3$, $\int_0^4 f(x)dx = 7$, and $\int_4^1 f(x)dx = -5$, evaluate $\int_1^2 f(2x)dx$.

5.8 Fundamental Theorem of Calculus, Part Ⅰ

If $f(x)$ is continuous on the closed interval $[a.b]$ and $F(x)$ is an antiderivative of $f(x)$, then

$$\int_a^b f(x)dx = F(b) - F(a)$$

proof Let $A(x)$ be the area under the curve $y = f(x)$ from a to x.

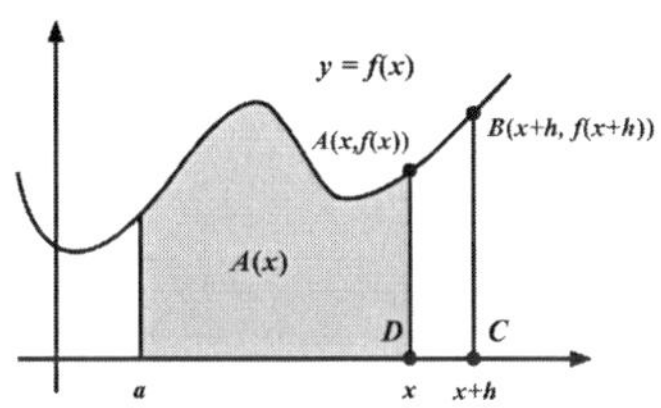

Figure 5.20

$$A(x) = \int_a^x f(x)dx \Rightarrow A(a) = 0 \text{ and } A(b) = \int_a^b f(x)dx \quad \cdots\cdots \text{①}$$

$$A'(x) = \lim_{h\to 0}\frac{A(x+h) - A(x)}{h} = \lim_{h\to 0}\frac{\triangle A}{h} = \lim_{h\to 0}\frac{Area\ of\ \square ABCD}{h}$$

$$= \lim_{h\to 0}\frac{\frac{1}{2}[f(x)+f(x+h)]h}{h} = \lim_{h\to 0}\frac{1}{2}[f(x)+f(x+h)] = \frac{1}{2}[f(x)+f(x)] = f(x)$$

$\therefore A'(x) = f(x)$

$A(x) = F(x) + C$

When $x = a,\ A(a) = F(a) + C = 0$

$\therefore C = -F(a)$

$\therefore A(x) = F(x) - F(a)$

When $x = b,\ A(b) = F(b) - F(a)$ ⋯⋯ ②

By ① and ② $\int_a^b f(x)dx = F(b) - F(a)$

Example 1

(1) $\int_{-1}^{2}(3x^2-2x)dx$ (2) $\int_{1}^{2}\frac{x^2+x-2}{2x^2}dx$

Solution

(1) $\int_{-1}^{2}(3x^2-2x)dx$

$=\left[x^3-x^2\right]_{-1}^{2}$

$=(8-4)-(-1-1)$

$=6$

(2) $\int_{1}^{2}\frac{x^2+x-2}{2x^2}dx$

$=\frac{1}{2}\int_{1}^{2}\left(1+\frac{1}{x}-\frac{2}{x^2}\right)dx$

$=\frac{1}{2}\left[x+\ln x+\frac{2}{x}\right]_{1}^{2}$

$=\frac{1}{2}\ln 2$

[EX 1] Evaluate the following definite integrals.

(1) $\int_{-1}^{1}(x^2-x-1)dx$ (2) $\int_{0}^{2}(4+8x-6x^2)dx$

(3) $\int_{4}^{9}\frac{2+x}{2\sqrt{x}}dx$ (4) $\int_{-1}^{8}\frac{1}{\sqrt[3]{x}}dx$

(5) $\int_{1}^{2}\frac{3x-1}{3x}dx$ (6) $\int_{2}^{1}\left(3x+\frac{2}{x}\right)^2dx$

(7) $\int_{-\frac{\pi}{4}}^{\frac{\pi}{4}}\cos 2x\,dx$ (8) $\int_{0}^{\frac{\pi}{3}}[\tan\theta-\sec\theta\tan\theta]\,d\theta$

[EX 2] Evaluate the following definite integrals.

(1) $\int_{0}^{\ln 2}(e^{2x}+e^{-x})dx$

(2) $\int_{0}^{1}\frac{4x}{x^2+1}dx$

(3) $\int_{-\frac{1}{2}}^{1}\frac{4\,dx}{\sqrt{1-x^2}}$

(4) $\int_{-1}^{\sqrt{3}}\frac{3\,dx}{x^2+1}$

(5) $\int_{1}^{\sqrt{3}}\frac{dx}{\sqrt{4-x^2}}$

(6) $\int_{0}^{\sqrt{3}}\frac{dx}{x^2+9}$

(7) $\int_{0}^{\frac{\pi}{4}}\sin^2 x\,dx$

(8) $\int_{\frac{\pi}{12}}^{\frac{\pi}{6}}\cos^2 2\theta\,d\theta$

[EX 3] Use a graphing calculator to evaluate the following definite integrals.

(1) $\int_{-2}^{2}(x^2+2x)dx$

(2) $\int_{1}^{4}\sqrt{x^2-1}\,dx$

(3) $\int_{0}^{3}\frac{10x}{x^2+2}dx$

(4) $\int_{-0.5}^{5}6e^{-x^2}dx$

5.9 Fundamental Theorem of Calculus, Part II

If f is continuous on $[a, b]$, then $F(x) = \int_a^x f(t)dt$ is continuous on $[a, b]$ and differentiable on (a, b) and its derivative is $f(x)$:

$$F'(x) = \frac{d}{dx}\int_a^x f(t)dt = f(x)$$

proof Let

$$\int f(x)\,dx = F(x) + C$$

$$\begin{aligned}\frac{d}{dx}\int_a^x f(t)dt &= \frac{d}{dx}[F(t)]_a^x \\ &= \frac{d}{dx}[F(x) - F(a)] \quad (\textit{Fundamental Theorem of Calculus, Part I}) \\ &= F'(x) - 0 \quad (\because F(a) \textit{ is a constant}) \\ &= f(x)\end{aligned}$$

Example 1

Find dy/dx if

(a) $y = \int_a^x (t^3 + 1)dt$ (b) $y = \int_x^5 3t\sin t\,dt$

Solution

(a) $\frac{dy}{dx} = \frac{d}{dx}\int_a^x (t^3 + 1)dt = x^3 + 1$

(b) $$\begin{aligned}\frac{dy}{dx} = \frac{d}{dx}\int_x^5 3t\sin t\,dt &= \frac{d}{dx}\left(-\int_5^x 3t\sin t\,dt\right) \\ &= -\frac{d}{dx}\int_5^x 3t\sin t\,dt \\ &= -3x\sin x\end{aligned}$$

[EX 1] Evaluate.

(1) $\frac{d}{dx}\int_0^x e^{-t}\sec^2 t\,dt$

(2) $\frac{d}{dx}\int_2^x \sqrt{1+t^2}\,dt$

(3) $\frac{d}{dx}\int_3^x \left(y^4+\frac{3}{\sqrt{1-y^2}}\right)dy$

(4) $\frac{d}{dx}\int_1^x (u-3)\ln|u+1|\,du$

[EX 2] Differentiate the following integrals with respect to x.

(1) $\int_2^x \sqrt{2+\cos^3 t}\,dt$

(2) $\int_e^x \frac{\ln(u)}{u}du$

(3) $\int_\pi^x \frac{6}{3+t^4}dt$

(4) $\int_0^x u\cos u\,du$

$$\frac{d}{dx}\int_{p(x)}^{q(x)} f(t)\,dt = f(q(x))q'(x) - f(p(x))p'(x)$$

proof Let $\int f(x)dx = F(x)+C$

$$\frac{d}{dx}\int_{p(x)}^{q(x)} f(t)dt = \frac{d}{dx}\left[F(t)\right]_{p(x)}^{q(x)}$$

$$= \frac{d}{dx}\left[F(q(x)) - F(p(x))\right]$$

$$= F'(q(x))q'(x) - F'(p(x))p'(x) = f(q(x))q'(x) - f(p(x))p'(x)$$

Example 2

Given $y=\int_{2x}^{x^2}\frac{t^3}{t^2+1}dt$, find $\frac{dy}{dx}$

Solution

$$\frac{dy}{dx}=\frac{d}{dx}\int_{2x}^{x^2}\frac{t^3}{t^2+1}dt=\frac{(x^2)^3}{(x^2)^2+1}\cdot 2x-\frac{(2x)^3}{(2x)^2+1}\cdot 2=\frac{2x^7}{x^4+1}-\frac{16x^3}{4x^2+1}$$

[EX 3] Differentiate the following integrals with respect to x

(1) $\int_{\tan x}^{0}\frac{dt}{1+t^2}$

(2) $\int_{2x}^{x^2}f(t)dt$

(3) $\int_{0}^{\sqrt{x}}\cos t\,dt$

(4) $\int_{-x}^{x^2}\sqrt{t^2+1}\,dt$

(5) $\int_{\cos x}^{\sin x}\frac{1}{1-t^2}dt$

(6) $\int_{\frac{1}{x}}^{3x}\cos 2t\,dt$

[EX 4] Find $\frac{dy}{dx}$

(1) $y=\int_{\sqrt{x}}^{x^2}\frac{e^t}{t}dt$

(2) $y=\int_{x+1}^{\ln x}\ln t\,dt$

(3) $y=\int_{3x}^{\tan x}\frac{1}{1-t^2}dt$

(4) $y=\int_{\cos^2 x}^{\sin^2 2x}\cos^2 t\,dt$

[EX 5]

(1) Let $\int_2^x f(t)\,dt = x^2\sin 3x$. Find $f(\pi)$.

(2) Let $A(x)=\int_4^{x^2} \frac{1}{1+t^2}dt$. Find $A(2)$ and $A'(2)$.

[EX 6] Find the intervals on which the function $f(x)=\int_1^x \frac{1}{t^4+1}dt$ is concave up.

5.10 Area between a curve and the x-axis.

The definite integral can be used to find the area enclosed by a curve and the x-axis in the interval $[a, b]$.

If a curve is above the x−axis,

$$Area = \int_a^b f(x)dx, \quad \text{where } a < b$$

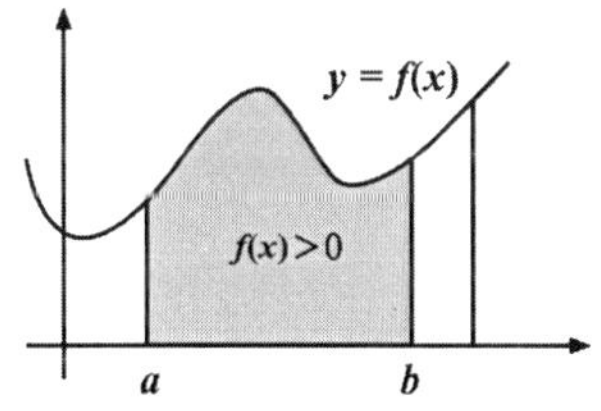

Figure 5.21

If a curve is below the x−axis,

$$Area = -\int_a^b f(x)dx, \quad \text{where } a < b$$

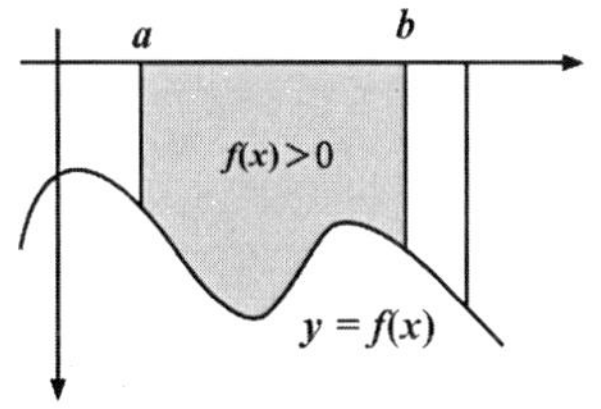

Figure 5.22

Example 1

Find the area enclosed by the curve $y = x^2 - 1$ and the x-axis in the interval [0, 2].

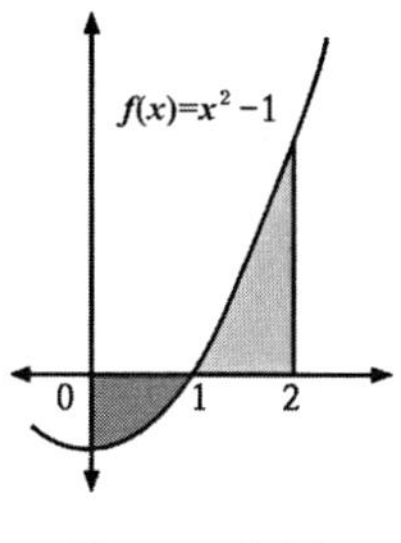

Figure 5.23

Solution

$$Area = -\int_0^1 (x^2-1)dx + \int_1^2 (x^2-1)dx$$

$$= -\left[\frac{1}{3}x^3 - x\right]_0^1 + \left[\frac{1}{3}x^3 - x\right] = -\left[\left(\frac{1}{3}-1\right)-0\right] + \left[\left(\frac{8}{3}-2\right)-\left(\frac{1}{3}-1\right)\right]$$

$$= -\left(-\frac{2}{3}\right) + \left[\frac{2}{3} - \left(-\frac{2}{3}\right)\right] = \frac{2}{3} + \frac{4}{3} = 2$$

[EX 1] Sketch the curve and find the area enclosed by the x-axis in the give interval.

(1) $y = x^2,\ [-1, 2]$

(2) $y = \cos x,\ \left[\frac{\pi}{2}, \pi\right]$

(3) $y = \cot x,\ \left[\frac{\pi}{4}, \frac{\pi}{2}\right]$

(4) $y = 4 - x^2,\ [1, 3]$

(5) $y = \tan x,\ \left[-\frac{\pi}{4}, \frac{\pi}{3}\right]$

(6) $y = \frac{4}{x^2+4},\ [-2, 2]$

(7) $y = 2\sin x,\ \left[\frac{\pi}{3}, \frac{3\pi}{2}\right]$

(8) $y = x^3 - 2x^2 - 3x,\ x - axis$

[EX 2] Given $f(x)=\begin{cases}2x+1, & x\le 1\\ x^2+2, & x>1\end{cases}$, find the value of $\int_{-1}^{3} f(x)dx$.

[EX 3] Evaluate each of the following.

(1) $\int_0^3 |x^2-4|\,dx$

(2) $\int_{-1}^{3} |x^2-4x|\,dx$

(3) $\int_0^{2\pi} |\cos x|\,dx$

(4) $\int_0^2 |e^x-2|\,dx$

[EX 4] The graph of f is shown below. The areas enclosed by the curve and the x-axis in the intervals $[a, b]$, $[b, c]$ and $[c, d]$ are 10, 10, and 5, respectively. Evaluate the following definite integrals.

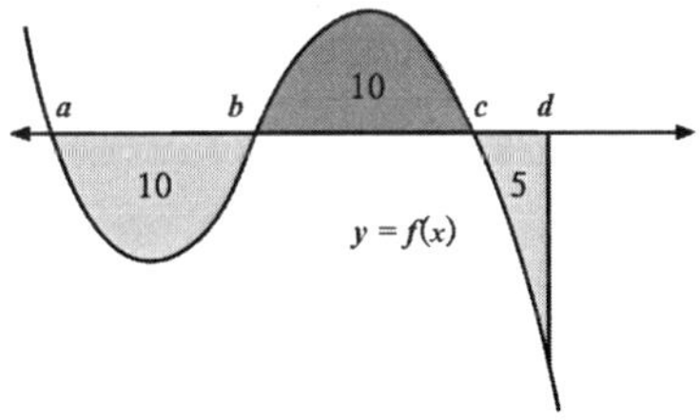

Figure 5.24

(1) $\int_a^b f(x)dx$

(2) $\int_b^c f(x)dx$

(3) $\int_a^d f(x)dx$

(4) $\int_a^d |f(x)|\,dx$

(5) $\int_b^a f(x)dx$

(6) $\int_d^b f(x)dx$

[EX 5] Sketch a graph to evaluate the following definite integrals.

(1) $\int_{0}^{2} \sqrt{4-x^2}\,dx$

(2) $\int_{-2}^{2} \left(-\sqrt{4-x^2}\right)dx$

(3) $\int_{-3}^{3} \left(5-\sqrt{9-x^2}\right)dx$

(4) $\int_{0}^{3} |x-2|\,dx$

(5) $\int_{-2}^{4} (|x+1|-1)dx$

(6) $\int_{0}^{2} (|x-1|+3x^2)dx$

[EX 6] The graph of f is shown below. Evaluate each of the following.

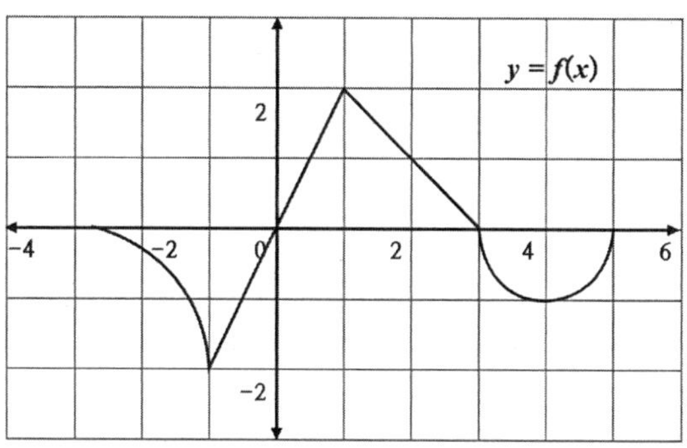

Figure 5.25

(1) $\int_{-1}^{1} f(x)dx$

(2) $\int_{1}^{5} f(x)dx$

(3) $\int_{5}^{0} f(x)dx$

(4) $\int_{-3}^{0} f(x)dx$

(5) $\int_{1}^{5} |f(x)|\,dx$

(6) $\int_{0}^{8} f(x-3)dx$

(7) $\int_0^1 [3f(x)+2x]\,dx$

(8) $\int_1^2 f(2x)\,dx$

[EX 7] The graph of f is shown below. Let g be the function defined by $g(x)=\int_1^x f(t)dt$.

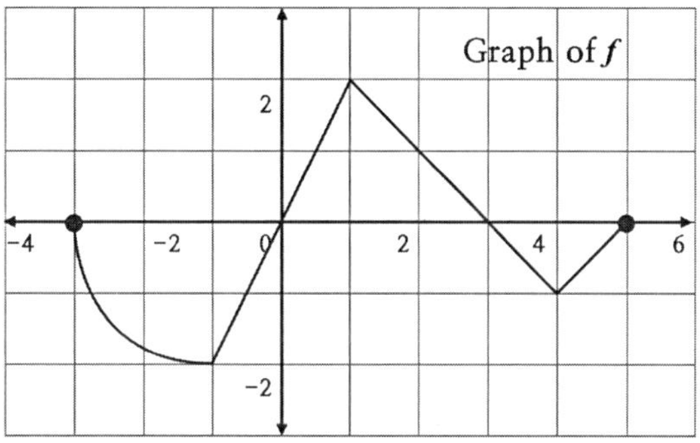

Figure 5.26

(1) Find $g(3)$, $g(-3)$, $g'(3)$ and $g''(3)$.

(2) Find the average rate of change of g over the interval $-3 \le x \le 3$.

(3) Find the average rate of change of g' over the interval $-1 \le x \le 3$.

(4) Find the equation of the line tangent to the graph of g at $x=-1$.

(5) Use local linearization to approximate the value of $g(-1.1)$.

(6) Find all intervals on shich the graph of g is decreasing. Justify your answer.

(7) Find all intervals on which the graph of g is concave up. Justify your answer.

(8) Find values of x at which the graph of g has a relative maximum. Explain.

(9) Find all values of x at which the graph of g has a point of inflection. Explain.

(10) Find the absolute minimum value of g on the closed interval $[-3, 5]$. Show an analysis to support your answer.

5.11 Substitution for Definite Integrals

If g' is continuous on the interval $[a, b]$ and f is continuous on the range of $g(x)=u$, then

$$\int_a^b f(g(x)) \cdot g'(x)dx = \int_{g(a)}^{g(b)} f(u)du.$$

Example 2

Evaluate the following definite integral. $\int_0^2 x(x^2-1)^2 dx$

Solution

Let $u=x^2-1$ $\frac{du}{dx}=2x$ $\frac{1}{2}du=x\,dx$

If $x=0, u=-1$

If $x=2, u=3$

$$\int_0^2 x(x^2-1)^2 dx = \int_{-1}^3 u^2\frac{1}{2}du = \frac{1}{6}[u^3]_{-1}^3 = \frac{1}{6}(27-(-1)) = \frac{28}{6} = \frac{14}{3}$$

[EX 1] Evaluate the following definite integrals.

(1) $\displaystyle\int_0^{\sqrt{3}} \frac{4x}{\sqrt{x^2+1}}dx$

(2) $\displaystyle\int_0^1 t^3(1+t^4)^3 dt$

(3) $\displaystyle\int_1^{e^{\pi/4}} \frac{4\,dt}{t(1+\ln^2 t)}$

(4) $\displaystyle\int_{-1}^1 \frac{5r}{(4+r^2)^2}dr$

(5) $\displaystyle\int_0^{\sqrt[3]{\pi^2}} \sqrt{\theta}\cos^2(\theta^{3/2})d\theta$

(6) $\displaystyle\int_2^4 \frac{dx}{x(\ln x)^2}$

(7) $\displaystyle\int_0^{\pi} 5(5-4\cos t)^{1/4}\sin t\,dt$

(8) $\displaystyle\int_0^{\pi/4} (1+e^{\tan\theta})\sec^2\theta\,d\theta$

(9) $\displaystyle\int_0^{2\pi} \frac{\cos z}{\sqrt{4+3\sin z}}dz$

(10) $\displaystyle\int_0^{\pi/6} \cos^{-3}2\theta\sin 2\theta\,d\theta$

(11) $\displaystyle\int_{-\pi/2}^{\pi/2} \frac{2\cos\theta\,d\theta}{1+(\sin\theta)^2}$

(12) $\displaystyle\int_{\sqrt{2}}^2 \frac{\sec^2(\sec^{-1})}{x\sqrt{x^2-1}}dx$

(13) $\displaystyle\int_2^{16} \frac{dx}{2x\sqrt{\ln x}}$

(14) $\displaystyle\int_0^{\pi/4} \tan x\sec^2 x\,dx$

[EX 2] Let $g(x)$ be a continuous function on the interval $[-4, 4]$ and $f'(x)= g(x)$. A table of values of differentiable function f is shown below. Evaluate the following definite integrals.

x	−4	−3	−2	−1	0	1	2	3	4
f	−3	1	−1	0	−2	5	4	6	2

(1) $\int_{-1}^{1} [g(x)+3x^2]dx$

(2) $\int_{-2}^{2} g(2x)dx$

(3) $\int_{-2}^{2} 3g\left(\frac{x}{2}\right)dx$

(4) $\int_{-2}^{0} xg(x^2)dx$

5.12 Integration by Parts for Definite Integrals

$$\int_a^b uv'\,dx = [uv]_a^b - \int_a^b u'v\,dx$$

Example 1

$\int_0^4 xe^{-x}dx =$

Solution

$$\int_0^4 xe^{-x}dx = [-xe^{-x}]_0^4 - \int_0^4 (-e^{-x})dx = [-4e^{-4}-0] + \int_0^4 e^{-x}dx$$

$$= -4e^{-4} - [e^{-x}]_0^4 = -4e^{-4} - e^{-4} + 1$$

$$= 1 - 5e^{-4}$$

[EX 1] Evaluate each of the following definite integrals.

(1) $\int_{-1}^{1} xe^{x}\,dx$

(2) $\int_{\frac{\pi}{3}}^{\frac{\pi}{2}} x\sin x\,dx$

(3) $\int_{0}^{\frac{\pi}{2}} x^2\sin 2x\,dx$

(4) $\int_{1}^{e} x\ln x\,dx$

(5) $\int_{0}^{e} \ln x\,dx$

(6) $\int_{0}^{1} \tan^{-1}x\,dx$

(7) $\int_{\frac{1}{2}}^{1} \sin^{-1}x\,dx$

(8) $\int_{0}^{4} e^{\sqrt{x}}\,dx$

(9) $\int_{0}^{\pi} e^{x}\sin x\,dx$

[EX 2] Let f be the function defined for $x > 0$, with $f(e)=2$ and f', the first derivative of f, given by $f'(x)=x^2\ln x$.

(1) Write an equation for the line tangent to the graph of f at the point $(e,2)$.

(2) Is the graph of f concave up of concave down on the interval $1 < x < 3$? Give a reason for your answer.

(3) Use antidifferentiation to find $f(x)$.

[EX 3] The derivative of a function f is given by $f'(x)=(x-3)e^x$ for $x > 0$, and $f(1)=7$.

(1) The function f has a critical point at $x=3$. At this point, does f have a relative minimum, a relative maximum, or neither? Justify your answer.

(2) On what intervals, if any, is the graph of f both decreasing and concave up? Explain your reasoning.

(3) Find the value of $f(3)$.

5.13 Even and Odd Functios

Let f be continuous on the symmetric interval $[-a, a]$.

(a) If f is even, then $\int_{-a}^{a} f(x)dx = 2\int_{0}^{a} f(x)dx$. (e.g.) $y = x^4, y = x^6, y = \cos x$ etc

(b) If f is odd, then $\int_{-a}^{a} f(x)dx = 0$. (e.g.) $y = x, y = x^3, y = \sin x, y = \tan x$ etc

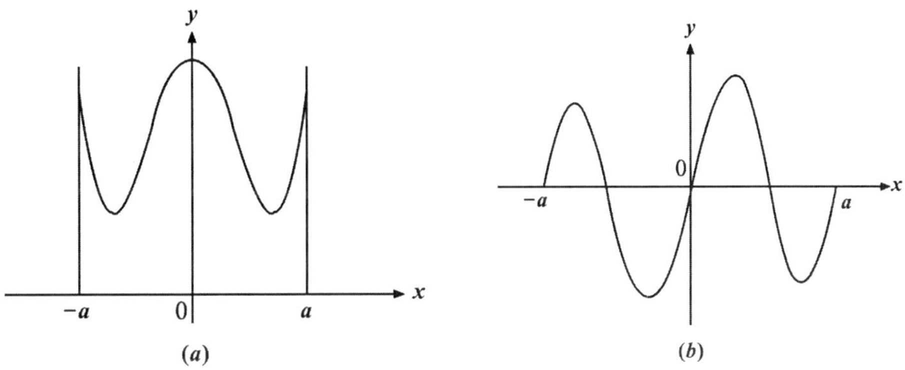

Figure 5.27

Example 1

$$\int_{-2}^{2} (x^4 - 4x^2 + 6)dx =$$

Solution

$$\int_{-2}^{2} (x^4 - 4x^2 + 6)dx = 2\int_{0}^{2} (x^4 - 4x^2 + 6)dx = 2\left[\frac{x^5}{5} - \frac{4}{3}x^3 + 6x\right]_0^2$$

$$= 2\left(\frac{32}{5} - \frac{32}{3} + 12\right) = \frac{232}{15}$$

[EX 1] Evaluate the following definite integrals.

(1) $\int_{-1}^{1} (x^4 + 2x^3 + 3x^2 + 4x + 5)dx$

(2) $\int_{-3}^{3} (x^5 - 3x^3 + x - 2\sin x)dx$

(3) $\int_{-2}^{2} (x^5 - 3x^3 + 3x^2 + \tan x + 2)dx$

(4) $\int_{-\pi}^{\pi} (x^4\sin x + x^3\cos x + 3x^2\tan x + x\cos x + 1)dx$

5.14 Integration by partial fraction

The method of partial fractions makes it possible to express a rational function $\frac{f(x)}{g(x)}$ as a sum of simpler fractions. Here $f(x)$ and $g(x)$ are real polynomials in x and it is assumed that $\frac{f(x)}{g(x)}$ is a proper fraction; that is of lower degree than $g(x)$. If not, we divide $f(x)$ by $g(x)$ to express the given rational function as the sum of a polynomial and a proper rational function. Thus,

$$\frac{x^3 - x^2 - 2}{x(x-1)} = x - \frac{2}{x(x-1)},$$

where the fraction on the right is proper.

Theoretically, every real polynomial can be expressed as a product of (powers of) real linear factors and (powers of) real quadratic factors.

In the following, the capital letters denote constants to be determined. We consider only nonrepeationg linear factors. For each distinct linear factor $(x-a)$ of $g(x)$ we set up one partial fraction of the type $\frac{A}{x-a}$. The techniques for determining the unknown constants are illustrated in the following examples.

Example 1

Find $\int \frac{x-1}{x(x-2)} dx$

Solution

Use partial fractions; find A and B such that $\frac{x-1}{x(x-2)} = \frac{A}{x} + \frac{B}{x-2}$.

Then $x-1 = A(x-2) + Bx$.

Set $x=0 : -1 = -2A$ and $A = \frac{1}{2}$.

Set $x=2 : 1 = 2B$ and $B = \frac{1}{2}$.

So the given integral equals $\int \left(\frac{1}{2x} + \frac{1}{2(x-2)}\right) dx = \frac{1}{2}\ln|x| + \frac{1}{2}\ln|x-2| + C$

$$= \frac{1}{2}\ln|x(x-2)| + C.$$

[EX 1] Evaluate the following indefinite integrals.

(1) $\int_{\frac{1}{2}}^{1} \frac{y+4}{y^2+y} dy$

(2) $\int \frac{dx}{x^2+2x}$

(3) $\int \frac{y-1}{y+1} dy$

(4) $\int \frac{x+4}{x^2+5x-6} dx$

(5) $\int \frac{dt}{t^3+t^2-2t}$

(6) $\int_{0}^{1} \frac{dx}{(x+1)(x^2+1)}$

(7) $\int \frac{4x dx}{(x-1)^2(x+1)}$

(8) $\int \frac{2x^2-x+4}{x(x^2+4)} dx$

memo

CALCULUS

CHAPTER

06

Applications of Definite Integrals

6.1 Average Value of a Function

If f is integrable on $[a, b]$, then its average value on $[a, b]$, also called its mean, is

$$f_{avg} = \frac{1}{b-a}\int_a^b f(x)dx.$$

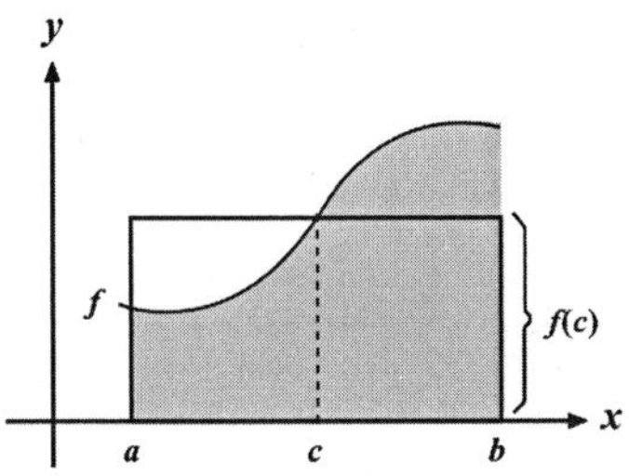

Figure 6.1

If $f(x)$ is continuous on the interval $[a, b]$, then there must be a number c such that the area of the rectangle, with a height of f_{avg} and a base of length $b-a$, is equal to the area under the graph f over the interval $[a, b]$.

$$(b-a) \cdot f_{avg} = \int_a^b f(x)dx$$

$$\therefore f_{avg} = \frac{1}{b-a}\int_a^b f(x)dx$$

[EX 1] Determine the average value of each of the following functions on the given interval.

(1) $y = x^2 - 1,\ [0,\ \sqrt{3}\,]$

(2) $y = (x-3)^2,\ [-1,\ 2]$

(3) $y = |x| - 1,\ [-1,\ 3]$

(4) $y = \dfrac{20}{x^2+1},\ [-1,\ 1]$

(5) $y = e^x - 1,\ [0,\ \ln 3]$

(6) $y = \sin x,\ [0,\ \pi]$

[EX 2] The rate at which of flows out of a tank, in liters per minute, is shown in the table below. Approximate the average flow rate of oil over the time interval $0 \le t \le 18$ by using the trapezoidal rule with 3 subintervals of equal length.

t (min)	$R(t)$ (L/min)
0	8
3	11
6	13
9	16
12	12
15	13
18	8

[EX 3]

(1) The velocity of a particle moving along a along a line is $v(t) = 5t - t^2 + 100$ at time t. Find the average velocity of the paticle over the time interval $0 \le t \le 4$

(2) The velocity of a particle moving along a line is $v(t) = 12t^2 + 1$ at time t.

(a) Fine the average velocity of the particle over the time interval $1 \le t \le 3$.

(b) Fine the average acceleration of the particle over the time interval $1 \le t \le 3$

[EX 4] The average area, in square inches, of all circles with radii between 2 and 5 in. is

(A) 7π (B) 11π (C) 13π (D) $\frac{29\pi}{2}$ (E) 17π

[EX 5] Suppose the amount of a drug in a patient's bloodstream t hr after intravenous administraion is $30/(t+1)^2$ mg. The average amount in the bloodstream during the first 4 hr is

(A) 6.0 mg (B) 11.0 mg (C) 16.6 mg

(D) 24.0 mg (E) none of these

[EX 6] Suppose the current world population is 6 billion and the population t yr from now is estimated to be $P(t)=6e^{0.024t}$. On The basis of this supposition, the averager popu-lation of the world, in billions, over the next 25 yr will be approximately

(A) 6.75 (B) 7.2 (C) 7.8 (D) 8.2 (E) 9.0

[EX 7] Use a graphing calculator to solve the following.

(1) Suppose that the value of a car in dollars after t years is given by the function $V(t)=35{,}000e^{-0.223t}$. Determine the average value of the car over its first 10 years.

(2) Suppose that the number of bacteria in a culture is given by the function $P(t)=\dfrac{20000}{1+299e^{-0.2t}}$ where t is measured in hours. Determine the average number of bacteria in the culture during the time interval $10 \le t \le 30$.

6.2 Area Between Curves

To find the area between curves (Figure), we first find where they intersect and then write the area of a typical element for each region between the points of intersec-tion. For the total area bounded by the curves $y=f(x)$ and $y=g(x)$ between $x=a$ and $x=e$, we see that, if they intersect at $[c, d]$, the total area is given exactly by

$$\int_a^c [f(x)-g(x)]\,dx + \int_c^e [g(x)-f(x)]\,dx.$$

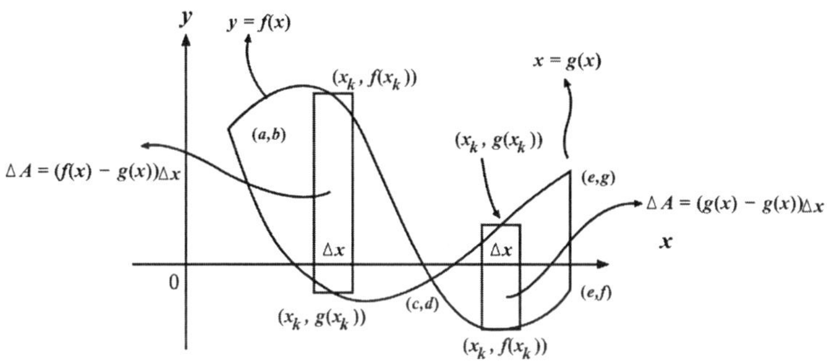

Figure 6.2

Example 1

Find the area of the region Bounded above by the curve $y=\sin x$ and below by $y=\cos x$ from $x=\dfrac{\pi}{4}$ to $x=\dfrac{5\pi}{4}$

Solution

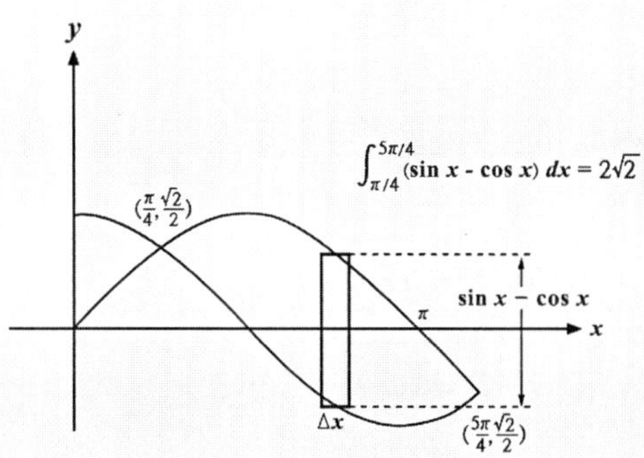

Figure 6.3

Example 2

Find the area of the region enclosed by $y = -x^2 + 3x$ and $y = x$ on the interval $[0, 3]$.

Solution

Step 1. Sketch the graphs on the same grid and find the x-coordinate(s) of the intersection(s).

Step 2. Determine the parts on which the graphs lie above one another

Step 3. Calculate the separate integrals.

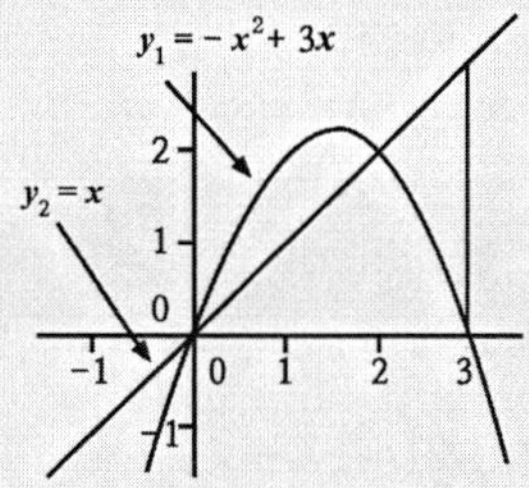

Figure 6.4

$x = -x^2 + 3x$

$x^2 - 2x = 0$

$x(x-2)$

$\therefore x = 0, 2$

$$Area = \int_0^2 (y_1 - y_2)dx + \int_2^3 (y_2 - y_1)\,dx$$

$$= \int_0^2 (-x^2 + 3x - x)\,dx + \int_2^3 (x + x^2 - 3x)dx$$

$$= \int_0^2 (-x^2 + 2x)\,dx + \int_2^3 (x^2 - 2x)dx$$

$$= \left[-\frac{1}{3}x^3 + x^2\right]_0^2 + \left[\frac{1}{3}x^3 - x^2\right]_2^3 = \frac{4}{3} + \frac{4}{3} = \frac{8}{3}$$

[EX 1] Find the areas of the regions enclosed by the lines and curves is area.

(1) $y = x^4$, $y = 8x$

(2) $y = -x^2 + 3x$, $y = -x$, $x = 0$, and $x = 3$

(3) $y = 2\sin x$, $y = \sin 2x$, $x = 0$, and $x = \pi$

(4) $y = e^x$, $y = \sin x$, $x = 0$, and $x = \pi$

(5) $y = \sec^2 x$, $y = \tan^2 x$, $x = -\frac{\pi}{4}$, and $x = \frac{\pi}{4}$

(6) $y = \sin x$ and $y = \sin 2x$, $x = 0$, and $x = \pi$

[EX 2] Find the area of the regions enclosed by the lines and curves

(1) $y = x^2 - 2x$ and $y = x$

(2) $y = \dfrac{10}{x^2+1}$ and $y = 5$

(3) $y = \dfrac{2}{x}$ and $x + y = 3$

(4) $y = x^3 - 2x$ and $y = x^2$

(5) $y = x^3 - 4x$ and $y = 3x - 6$

(6) $y = |x^2 - 4|$ and $y = (x^2/2) + 4$

[EX 3] Use a graphing calculator to find the area of the region enclosed by the curves.

(1) $y = x - 1$ and $y = 2\ln(x+2)$

(2) $y = 3\sin x$ and $y = x^2 - 2$

(3) $y = e^x$ and $y = -x^2 + 20$

(4) $y = -x^2$ and $y = 0.3x^3 - 5x$

If $f(y)$ and $g(y)$ are continuous and $f(y) \geq g(y)$ for all y in $[c, d]$, then the area of the region bounded by the graphs of $x = f(y), x = g(y)$, and the horizontal lines $y = c$, $y = d$ is

$$Area = \int_c^d (x_1 - x_2)\,dy = \int_e^d [f(y) - g(y)]\,dy \qquad \text{where } c < d$$

proof

When $y = y_k$, the horizontal strip is a rectangle with a height of $f(y_k) - g(y_k)$ and a width of $\triangle y$. By summing the areas of these rectangles,

$$Area = \lim_{n\to\infty}\sum_{k=1}^{n}[f(y_k) - g(y_k)]\triangle y = \int_c^d [f(y) - g(y)]\,dy \qquad \text{where } c < d$$

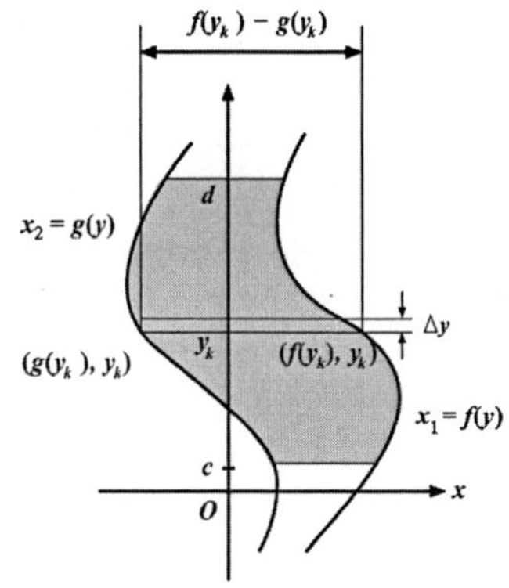

Figure 6.5

Example 3

Find the area of the region enclosed by $x=y^2$ and $x=y+2$

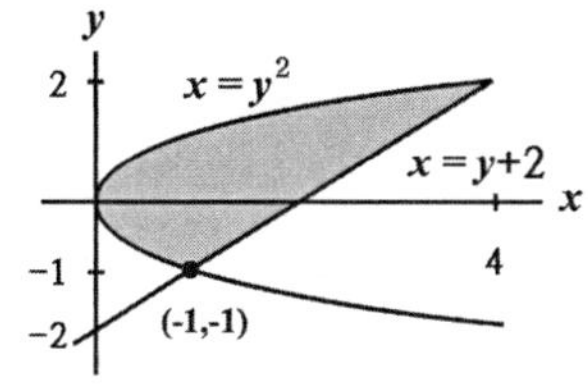

Figure 6.6

Solution

Limits of integration : $y^2=y+2 \Rightarrow (y+1)(y-2)=0$

$\Rightarrow c=-1$ and $d=2; f(y)-g(y)=(y+2)-y^2$

$$\Rightarrow A=\int_{-1}^{2}(y+2-y^2)\,dy=\left[\frac{y^2}{2}+2y-\frac{y^3}{3}\right]_{-1}^{2}=\left(\frac{4}{2}+4-\frac{8}{3}\right)-\left(\frac{1}{2}-2+\frac{1}{3}\right)=6-\frac{8}{3}-\frac{1}{2}+2-\frac{1}{3}=\frac{9}{2}$$

[EX 4] Sketch the region enclosed by the curves and find its area.

(1) $x=2y^2,\ x=0$, and $y=3$

(2) $y^2=x$ and $x+y=2$

(3) $x-y^2=0$ and $x+2y^2=3$

(4) $x=y^2-4y$ and $x=-y^2+2y$

(5) $y=\sqrt{x-1}, y=x,\ y=0$ and $y=2$

(6) $y=\ln x, y=\sqrt{-x},\ y=0$ and $y=1$

(7) $y=e^x$, $y=\sqrt{x}$, $y=1$ and $y=e$ (8) $y=e^{-x}$, $y=\dfrac{1}{x}$, $y=1$ and $y=e$

6.3 Volume of Solids with Known Cross Sections

Let, the slices are taken perpendicular to the $x-$axis so that $\triangle V = A(x)\triangle x$, where $A(x)$ is the area of a cross section and $\triangle x$ is the thickness of the slice.

$$V=\int_a^b A(x)dx \qquad \text{where } a<b$$

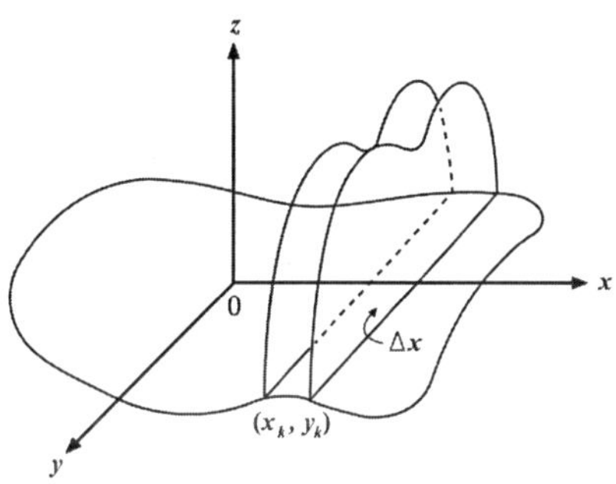

Figure 6.7

Example 1

A solid has as its base the circle $x^2+y^2=9$, and all cross sections parallel to the $y-$axis and the line $y-$axis are squares. Find the volume of the solid.

Solution

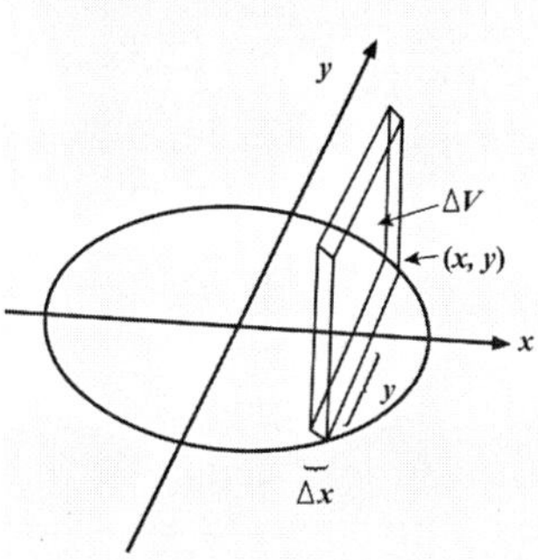

Figure 6.8

the element of volume is a square prism with sides of length $2y$ and thickness Δx, so

$$\triangle V = (2y)^2\triangle x = 4y^2\triangle x = 4(9-x^2)\triangle x.$$

Now, using symmetry across the y-axis, we find, the volume of the solid :

$$V=2\int_0^3 4(9-x^2)\,dx = 8\int_0^3 (9-x^2)dx = 8\left(9x-\frac{x^3}{3}\right)\Bigg|_0^3 = 144.$$

[EX 1] The base of a solid is the region enclosed by the parabola $x=y^2$ and the line $x=3$. For this solid, each cross section perpendicular to the $x-$axis is a square. Find the volume of the solid.

[EX 2] The solid lies between planes perpendicular to the $x-$axis at $x=-1$ and $x=1$. The cross-sections perpendicular to the $x-$axis are circular disks whose diameters run from the parabola $y=x^2$ to the parabola $y=2-x^2$.

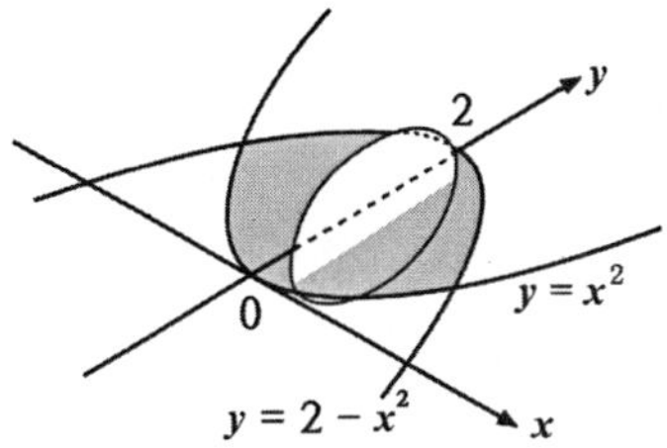

Figure 6.9

[EX 3] The base of a solid is the region between the curve $y=2\sqrt{\sin x}$ and the interval $[0,\pi]$ on the $x-$axis. The cross-sections perpendicular to the $x-$axis are equilateral triangles with bases running from the $x-$axis to the curve as shown in the accompanying figure.

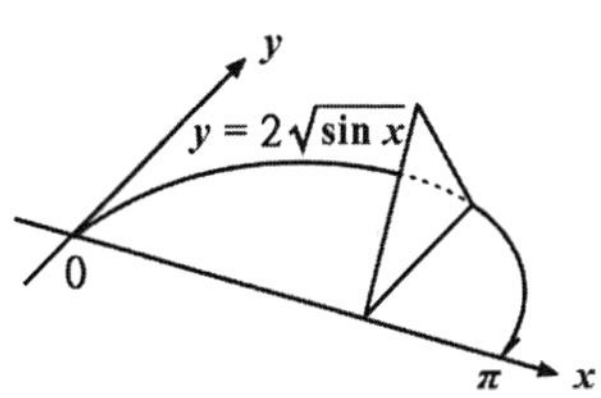

Figure 6.10

[EX 4] The solid lies between planes perpendicular to the $x-$axis at $x=-1$ and $x=1$. The cross-sections perpendicular to the $x-$axis between these planes are squares whose diagonals run from the semicircle $y=-\sqrt{1-x^2}$ to the semicircle $y=\sqrt{1-x^2}$.

[EX 5] The functions f and g are given by $f(x)=\sqrt{x}$ and $g(x)=6-x$. Let R be the region bounded by yhe $x-$axis and the graphs of f and g, as shown in the figure below.

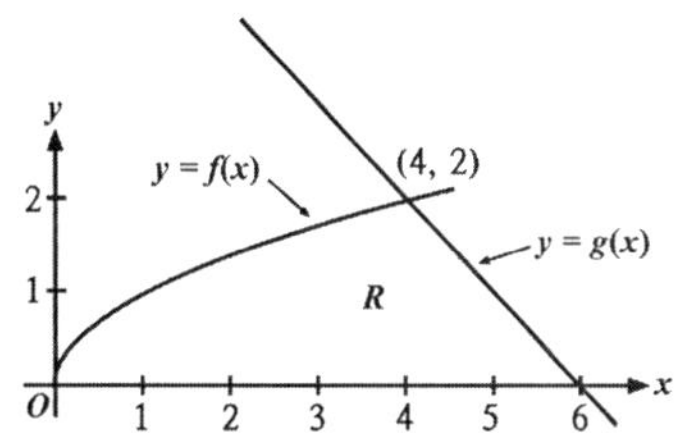

Figure 6.11

(1) Find the area of R

(2) The region R is the base of a solid. For each y. where $0 \le y \le 2$, <u>the cross section of the solid taken perpendicular to the $y-$axis is a rectangle whose base lies in R and whose height is $2y$</u>. Write, but do not evaluate, an integral expression that gives the volume of the solid.

(3) There is a point P on the graph of f at which th line tangent to the graph of f is perpendicular to the graph of g. Find the coordinates of point P

Let S be a solid bounded by $y=c$ and $y=d$. If the cross-sectional area of S (perpendicular to the $y-$axis) is $A(y)$, then the volume of S is

$$V=\int_c^d A(y)dy \quad \text{where } c<d$$

[EX 6] The solid lies between planes perpendicular to the $y-$axis at $y=0$ and $y=2$. The cross-sections perpendicular to the $y-$axis are cir-cular disks with diameters running from the $y-$axis to the parabola $x=\sqrt{5}y^2$.

[EX 7] The base of a solid is the region enclosed by the graphs of $y=x^2$ and $y=1$. For this solid. each cross section perpendicular to the <u>$y-$axis</u> is a <u>rectangle</u> whose height is 4 times the length of its base. Find the volume of the solid.

[EX 8] The base of the solid is the disk $x^2+y^2 \leq 1$. The cross-sections by planes perpendicular to the $y-$axis between $y=-1$ and $y=1$ are isosceles right triangles with one leg in the disk.

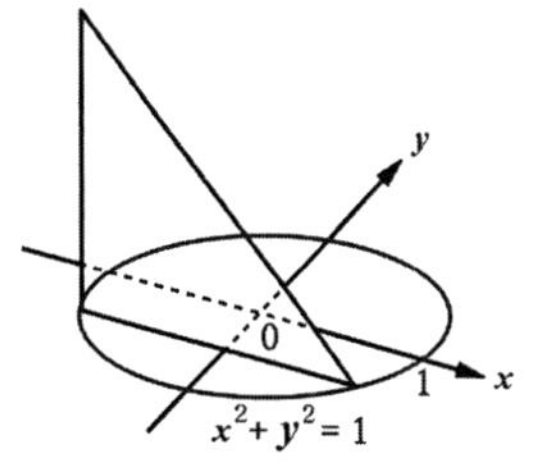

Figure 6.12

[EX 9] The base of a solid is the region enclosed by the graphs of $y=\sqrt{x}$ and $y=\frac{1}{2}x$. For this solid each cross section perpendicular to the <u>$y-$axis</u> is a <u>semicircle</u>. Find the volume of the solid.

6.4 Volume of Solids of Revolution : Disk, Washers and Cylindrical Shells

Disk

① Revolution about the $x-$axis : $V=\pi\int_a^b r^2 dx$

② Revolution about the $y-$axis : $V=\pi\int_c^d r^2 dy$

The region bounded by a curve and the $x-$axis is revolved around the $x-$axis, forming the solid of revolution seen in Figure 6.13, We think of the "rectangular strip" of the region at the left as generating the solid disk. $\triangle V$ (an element of the volume), shown at the right

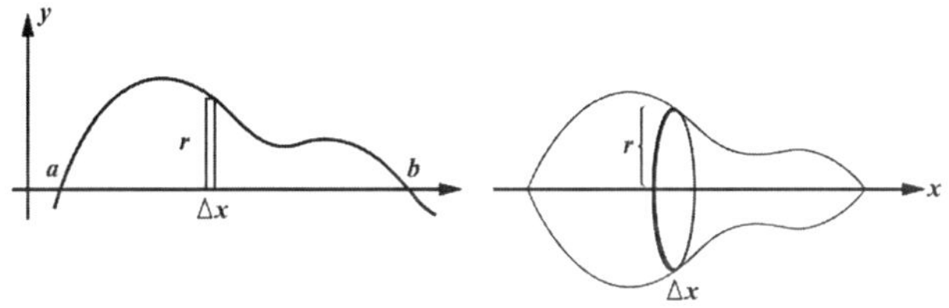

Figure 6.13

This disk is a cylinder whose radius, r, is the height of the rectangular strip, and whose height is the thickness of the strip, $\triangle x$, Thus

$$\triangle V=\pi r^2\triangle x \text{ and } V=\pi\int_a^b r^2 dx.$$

Example 1

Find the volume of the solid generated by revolving the region enclosed by $y=\sqrt{r^2-x^2}$ and $y=0$ about the $x-$axis.

Solution

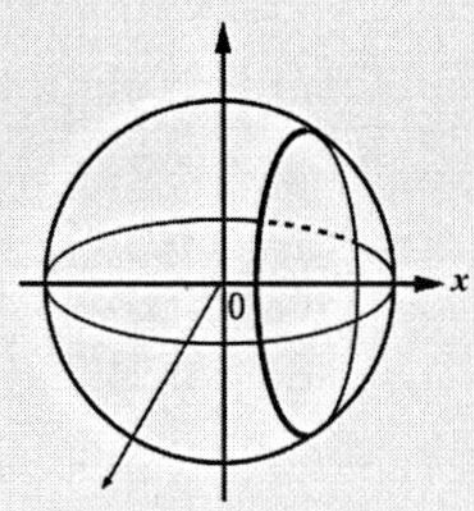

Figure 6.13

$$V=\int_{-r}^{r}\pi y^2dx=\pi\int_{-r}^{r}(r^2-x^2)dx=\pi\left[r^2x-\frac{x^3}{3}\right]\Big|_{-r}^{r}=\frac{4}{3}\pi r^3$$

Example 2

Find the volume of the solid generated when the region bounded by $y=x^2$, $x=2$, and $y=0$ is rotated about the line $x=2$ as shown in figure.

Solution

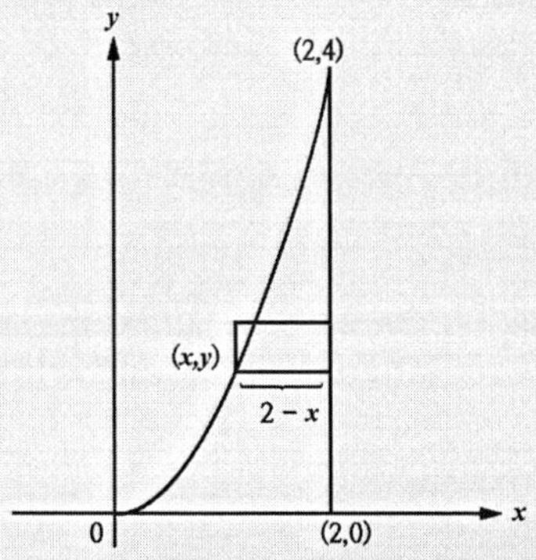

Figure 6.14

Disk.

$r=2-x.$

$\Delta V=\pi(2-x)^2\Delta y.$

$$V=\pi\int_0^4(2-x)^2dy$$

$$=\pi\int_0^4(2-\sqrt{y})^2dy$$

$$=\frac{8\pi}{3}.$$

[EX 1] Find the volume of the solid generated by revolving the region enclosed by the following graphs about the <u>$x-$axis.</u>

(1) $y=x^2$, $x=2$, and $y=0$

(2) $y=\sin x$, $x=0$, $x=\pi$, and $y=0$

(3) $y=e^{-x}$, $y=0$, $x=0$, $x=1$

(4) $y=x-x^2$, $y=0$

[EX 2] Find the volume of solid generated by revolving the region enclosed by the following graphs about the $y-$axis.

(1) $y=x^2$ and $y=2$

(2) $x=-y^2+2y$ and $x=0$

(3) $x=\sqrt{5}y^2$, $x=0$, $y=-1$, $y=1$

(4) $x=2/\sqrt{y+1}$, $x=0$, $y=0$, $y=3$

[EX 3] Find the volume of the solid generated by revolving the region enclosed by $y=\sqrt{\frac{\ln x}{x}}$, $x=1$, $x=e$, and $y=0$ about the $x-$axis.

[EX 4] Find the volume of the solid generated by revolving the region enclosed by the following graphs about the $y-$axis.

(1) $y=e^{-x^2}$ and $y=\frac{1}{e}$

(2) $\frac{x^2}{9}+\frac{y^2}{4}=1$ (ellipse)

[EX 5] Show that the volume of a right circular cone with radius r, height h is $\frac{1}{3}\pi r^2 h$ using the disk method. (Hint: The cone is generated by revolving the triangle about the $y-$axis.)

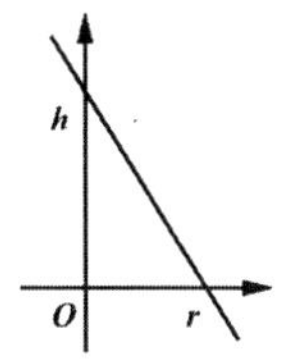

Figure 6.15

Washer

① Revolution about the x − axis : $V=\pi\int_a^b (R^2-r^2)dx = \pi\int_a^b [f(x)^2 - g(x)^2]\,dx$

② Revolution about the y − axis : $V=\pi\int_c^d (R^2-r^2)dy = \pi\int_c^d [f(y)^2 - g(y)^2]\,dy$

The region enclosed by $y=f(x)$, $y=g(x)$ $x=a$ and $x=b$ is revolved about the x − axis.

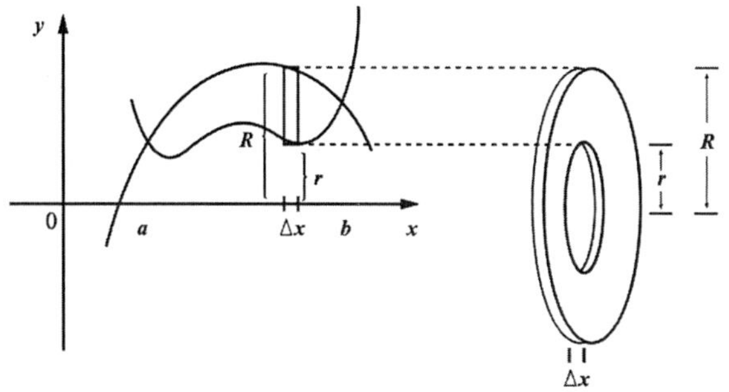

Washers

$A(x) = A = \pi R^2 - \pi r^2$

$V=\int_a^b \pi R^2 - \pi r^2 dx$

Figure 6.16

This washer's height is the thickness of the rectangular strip, Δx. The washer is a disk whose outer radius, R, is the distance to the top of the rectangular strip, with the disk of inner radius r(the distance to the bottom of the strip) removed. Thus:

$\Delta V=\pi R^2\Delta x - \pi r^2\Delta x = \pi(R^2-r^2)\Delta x$ and $V=\pi\int_a^b (R^2-r^2)\,dx$.

Example 3

Find the volume obtained when the region bounded by $y = x^2$ and $y = 2x$ is revolved about the $x-$axis.

Solution

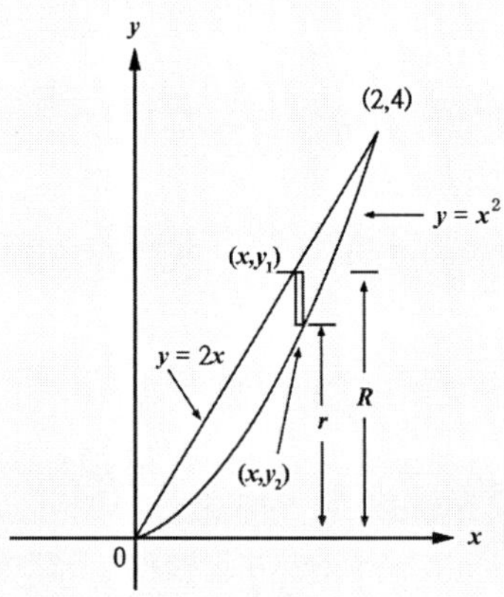

Figure 6.17

Washer.

$R = y_1 = 2x.$

$r = y_2 = x^2.$

$\Delta V = \pi({y_1}^2 - {y_2}^2)\Delta x.$

$$V = \pi\int_0^2 ({y_1}^2 - {y_2}^2)dx$$

$$= \pi\int_0^2 ((2x)^2 - (x^2)^2)dx$$

$$= \frac{64\pi}{15}.$$

[EX 6] Find the volume of the solid generated by revolving the region enclosed by the following graphs about the $x-$axis.

(1) $y = x^2 + 1,\ y = x + 3$

(2) $y = x^2,\ y^2 = 8x$

(3) $y = x^2,\ y = 4$

(4) $y = \dfrac{6}{\sqrt{x^2+1}},\ y = 3$

(5) $y = \sec x,\ y = \sqrt{2},\ -\pi/4 \le x \le \pi/4$

(6) $y = \sin x,\ \ y = \dfrac{1}{2},\ \ x = \dfrac{\pi}{6}$ and $x = \dfrac{\pi}{2}$

[EX 7] Find the volume of the solid generated by revolving the region enclosed by the following graphs about the $y-$axis.

(1) $y=x^2, x=2$, and $y=0$

(2) $y=\frac{1}{2}x^2,\ y=-x$

(3) $y=x^3,\ y^2=32x$

(4) $y=\ln x,\ y=\sqrt{x},\ y=0$ and $y=1$

[EX 8] Find the volume of the solid generated by revolving the region bounded by the parabola $y=x^2$ and the line $y=1$ about

(1) the line $y=1$

(2) the line $y=2$

(3) the line $y=-1$

[EX 9] Find the volume of the solid generated by revolving the triangular region bounded by the lines $y=2x,\ y=0$ and $x=1$ about

(1) the line $x=1$

(2) the line $x=2$

[EX 10] Let R be the region bounded by the graph of $y=e^{2x-x^2}$ and the horizontal line $y=2$, and Let S be region bounded by the graph of $y=e^{2x-x^2}$ and the horizontal line $y=1$ and $y=2$, as shown below.

(1) Find the area of R

(2) Find the area of S

(3) Write, but do not evaluate, an integral expression that gives the volume of the solid generated when R is rotated about the horizontal line $y=1$

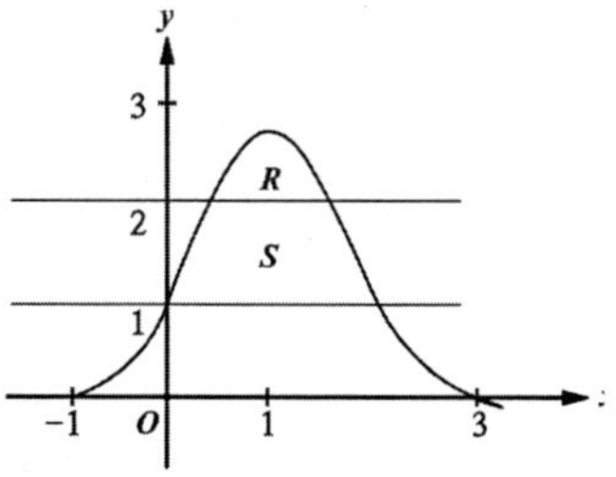

Figure 6.18

[EX 11] Let R be the region in the first quadrant bounded by the graph of $y=2\sqrt{x}$, the horizontal line $y=6$, and the $y-$axis, as shown in the figure below.

(1) Find the area of R

(2) Write, but do not evaluate, an intergral expression that gives the volume of the solid generated when R is rotated about the horizontal line $y=7$

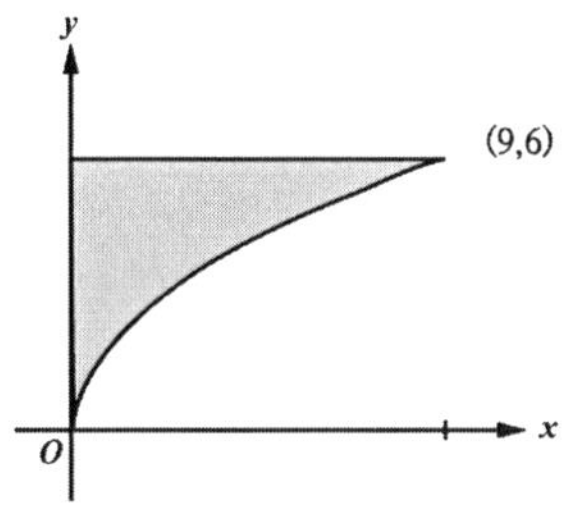

Figure 6.19

(3) Region R is the bse of a solid. For each y, where $0 \le y \le 6$, the cross section of the solid taken perpendicular to the $y-$axis is a rectangle whose height is 3 times the length of its base in region R. Write, but do evaluate, an integral expression that gives the volume of the solid.

Cylindrical Shell

① Revolution about the $y-$axis : $V=2\pi\int_a^b(Shell\,Radius)(Shell\,Height)dx$

② Revolution about the $x-$axis : $V=2\pi\int_a^b(Shell\,Radius)(Shell\,Height)\,dy$

Consider the volume of solid of revolution formed when the region bounded by the two curves seen in Figure 6.20 is revolved around the $y-$axis. We think of the rectangular strip of the region at the left as generating the shell, $\triangle V$(an element of the volume), shown at the right.

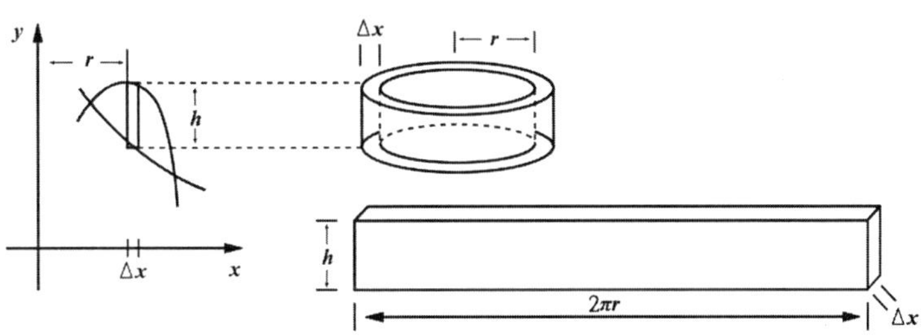

Figure 6.20

This shell's radius, r, is the distance from the axis to the rectangular strip, and its height is the height of the rectangular strip, h. When the shell is unwound and flattened to from a rectangular solid, the length of the solid is the circumference of the cylinder, $2\pi r$, its height of the cylinder, h, and its thickness is the thickness of the rectangular strip, Δx. Thus:

$$\Delta V = 2\pi r h \Delta x \text{ and } V = 2\pi \int_a^b rh\, dx.$$

Example 1

The region enclosed by the x − axis and the parabola $y = f(x) = 3x - x^2$ is revolved about the vertical line $x = -1$ to generate a solid (Figure). Find the volume of the solid.

Solution

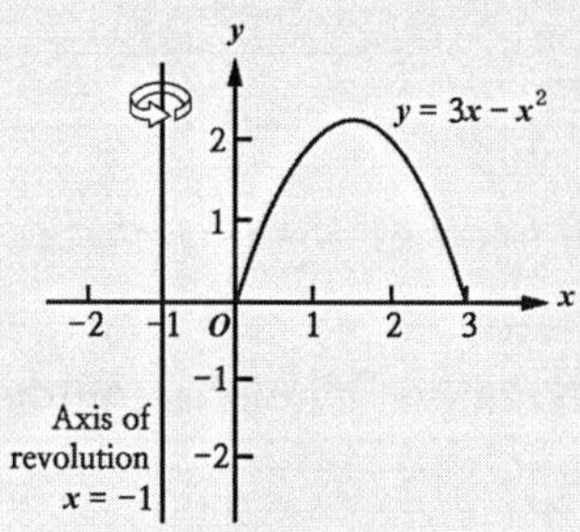

Figure 6.21

$$V = \int_0^3 2\pi(x+1)(3x - x^2)dx$$

$$= \int_0^3 2\pi(3x^2 + 3x - x^3 - x^2)dx$$

$$= 2\pi \int_0^3 (2x^2 + 3x - x^3)dx$$

$$= 2\pi \left[\frac{2}{3}x^3 + \frac{3}{2}x^2 - \frac{1}{4}x^4\right]_0^3 = \frac{45\pi}{2}.$$

Example 2

The region bounded by the curve $y = \sqrt{x}$, the $x-$axis, and the line $x = 4$ is revolved about the $x-$axis to generate a solid. Find the volume of the solid by the shell method.

Solution

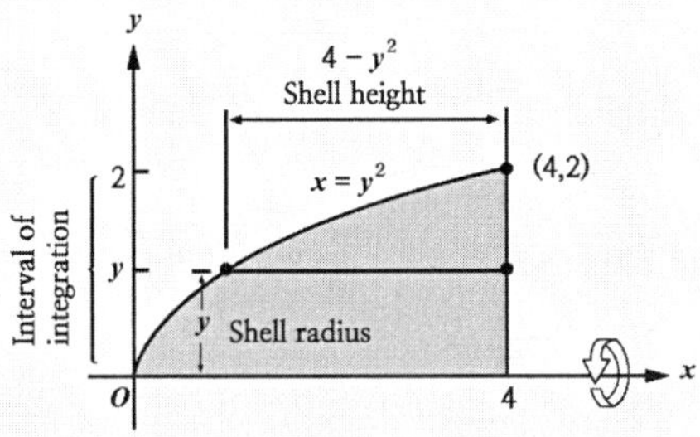

Figure 6.22

$$V = \int_a^b 2\pi \begin{pmatrix}\text{shell}\\ \text{radius}\end{pmatrix}\begin{pmatrix}\text{shell}\\ \text{height}\end{pmatrix} dy$$

$$= \int_0^2 2\pi(y)(4-y^2)dy$$

$$= 2\pi \int_0^2 (4y-y^3)dy$$

$$= 2\pi \left[2y^2 - \frac{y^4}{4}\right]_0^2 = 8\pi.$$

[EX 12] Use the cylindrical shell method to find volume of the solid generated by revolving the region enclosed by the following graphs about the $y-$axis.

(1) $y = x^2,\ y = 2-x,\ x = 0, x = 1$

(2) $y = x^2,\ x = 1,\ x = 2$ and $y = 0$

(3) $y = 2x-1,\ y = \sqrt{x},\ x = 0$

(4) $y = \sin x,\ x = 0,\ x = \pi,$ and $y = 0$

(5) $y = 3x - x^2$ and $y = x$

(6) $y = \sqrt{x}$ and $y = x$

[EX 13] Use the cylindrical shell method to find the volume of the solid generated by revolving the region enclosed by the following graphs about the $x-$axis.

(1) $y = \sqrt{x},\ y = 0,\ y = 2-x$

(2) $x = 2y - y^2,\ x = y$

(3) $y = x,\ y = 2x,\ y = 2$

(4) $x = 2y - y^2,\ x = 0$

[EX 14] Use the cylindrical shell method to find the volume of the solid generated by revolving the region enclosed by $y=x^3$, $y=8$, $x=0$ about the following lines.

(1) $x=3$

(2) $x=-2$

(3) $y=8$

(4) $y=-1$

[EX 15] Use the cylindrical shell method to find the volume of solid generated by revolving the region enclosed by $x=\frac{y^4}{4}-\frac{y^2}{2}$ and $x=\frac{y^2}{2}$ about the line $y=5$

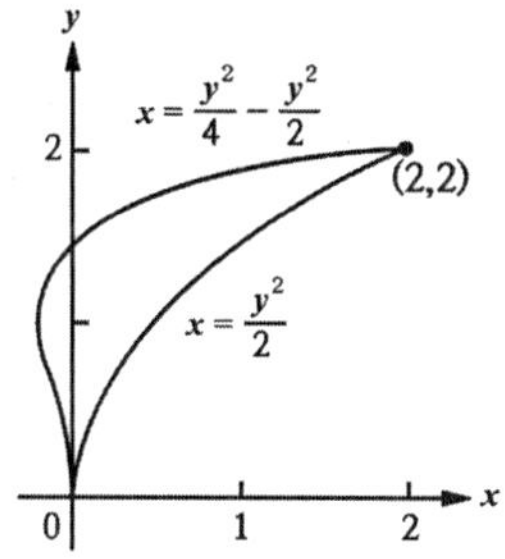

Figure 6.23

6.5 DEFINITE INTEGRAL OF A RATE IS NET CHANGE

If f is continuous and $f(t)=\frac{dF}{dt}$, then

$$\int_a^b f(t)dt = F(b)-F(a).$$

The definite integral of the rate of changs of a quantity over an interval is the net change or net accumulation of the quantity over that interval. Thus, $F(b)-F(a)$ is the net change in $F(t)$ as t varies from a to b

Example 1

Let $G(t)$ be the rate of growth of a population at time t. Then the increase in population between times $t=a$ and $t=b$ is given by $\int_a^b G(t)\,dt$. The population may consist of people, deer, fruit files, bacteria, and so on.

Solution

Example 2

If after t minutes a chemical is decomposing at the rate of $10e^{-t}$ grams per minute, then during the first 3 minutes the amount hat has decomposed is

Solution

$$\int_0^3 10e^{-t}\,dt \simeq 9.5\,g$$

[EX 1]

(1) A beach opens at 8 A.M. and people arrive at a rate of $R(t)=10+40t$ people per hour, where t represents the number of hours the beach has been open. Assuming no one leaves before noon, at what time will there be 100 people there?

(2) Oil is leaking from a tanker at the rate of $1000e^{-0.3t}$ gal/hr, where t is given in hours. the total number of gallons of oil that will leak out during the next 8 hr is approximately

(3) If a factory continuously dumps pollutants into a river at the rate of $\frac{\sqrt{t}}{180}$ tons per day, then the amount dumped after 7 weeks is approximately

(4) How long woll it take to release 9 tons of pollutant if the rate at which pollutant is being released is $te^{-0.3t}$ tons per week?

(5) If you deposit \$1000 today at 8% interest compounded continuously, it will grow at the rate of $80e^{0.08t}$ dollars per year. In 6 years it will be worth (in dollars)

(6) Water is leaking from a tank at the rate of $(-0.1t^2-0.3t+2)$ gal/hr. The total amount, in gallons, that will leak out in the next 3 hr is approximately

New Position = initial Position + Displacement

$$s(t)=s(a)+\int_a^t v(t)dt$$

New Velocity = initial Velocity + Change of Velocity

$$v(t)=v(a)+\int_a^t a(t)dt$$

Example 3

The velocity function of a particle is given by $v(t)=2t+3$ and its postion is 2 at time $t=1$. Find the position function of the particle.

Solution

(Method 1)

$s(t)=\int v(t)dt=\int (2t+3)dt$

$=t^2+3t+C$

$s(1)=2 : (1)^2+3(1)+C=2,$

$C=-2$

$\therefore s(t)=t^2+3t-2$

(Method 2)

$s(t)=s(1)+\int_1^t v(t)dt$

$=2+\int_1^t (2t+3)dt$

$=2+\left[t^2+3t\right]_1^t$

$=2+(t^2+3t)-(1+3)$

$=t^2+3t-2$

[EX 2] A particle moves along a straight line. Find the position of the particle at time t.

(1) $v(t)=3t^2+2t,\ s(0)=2$

(2) $v(t)=4\cos 2t,\ s(\pi)=-3$

(3) $v(t)=\dfrac{1}{\sqrt{1+t}},\ s(3)=1$

(4) $v(t)=\dfrac{\ln t}{t},\ s(1)=2$

(5) $a(t)=6t-3,\ v(1)=2,\ s(0)=2$

(6) $a(t)=2\sin 2t,\ v(0)=3,\ s(0)=-2$

[EX 3] Use a graphing calculator to evaluate the following values.

(1) Given $f'(x)=3\sin(x^2)$ and $f(1)=3$, find $f(3)$.

(2) Given $f'(x)=e^{-x^2}$ and $f(-1)=2$, find $f(2)$.

(3) A particle moves along the $x-$axis with velocity given by $v(t)=e^{\sin t}-1$ for $t\geq 0$.

The position of the particle is 2 at time $t=0$. Find the position of the particle at time $t=4$.

(4) A particle moves along the $x-$axis with acceleration given by $a(t)=2\cos(t+\ln t)$ for $t\geq 0$. The velocity of the particle is 3 at time $t=1$. Find the velocity of the paticle at time $t=5$

(5) The number of bacteria in a culture changes at a rate of $500e^{-0.02t}$ bacteria/hour at time t. Determine the increase in number of bacteria in the culture during the time interval $10\leq t\leq 30$.

[EX 4]

t	0	2	4	6	8	10	12
P(t)	0	46	53	57	60	62	63

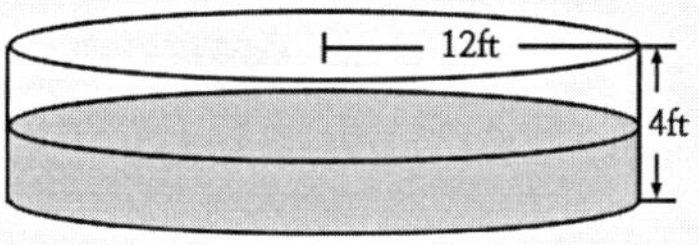

Figure 6.24

The figure above shows an aboveground swimming pool in the shape of a cylinder with a radius of 12 feet and a height of 4 feet. The pool contains 1000 cubic feet of water at time $t=0$. During the time interval $0 \le t \le 12$ hours, water is pumped into the pool at the rate $P(t)$ cubic feet per hour. The table above gives values of $P(t)$ for selected values of t. During the same time interval, water is leaking from the pool at the rate $R(t)$ cubic feet per hour, where $R(t)=25e^{-0.05t}$. (Note: The volume V of a cylinder with radius r and height h is given by $V=\pi r^2 h$.)

(1) Use a midpoint Riemann sum with three subintervals of equal length to approximate the total amount of water that was pumped into the pool during the time interval $0 \le t \le 12$ hours. Show the computations that lead to your answer.

(2) Calculate the total amount of water that leaked out of the pool during the time interval $0 \le t \le 12$ hours.

(3) Use the results from parts (a) and (b) to approximate the volume of water in the pool at time $t=12$ hours. Round your answer to the nearest cubic foot.

(4) Find the rate at which the volume of water in ther pool is increasing at time $t=8$ hours. How fast is the water level in the pool rising at $t=8$ hours? Indicate units of measure in both answers.

[EX 5]

t (minutes)	0	2	5	9	10
$H(t)$ (degrees Celsius)	66	60	52	44	43

As a pot of tea cools, the temperature of the tea is modeled by a differentiable function H for $0 \le t \le 10$, where time t is measured in minutes and temperature $H(t)$ is measured in degrees Celsius. Values of $H(t)$ at selected values of time t are shown in the table above.

(1) Use the date in the table to approximate the rate at which the temperature of the tea is changing at time $t = 3.5$. Show the computations that lead to your answer.

(2) Using correct units, explain the meaning of $\frac{1}{10}\int_0^{10} H(t)\,dt$ in the context of this problem. Use a trapezoidal sum with the four subintervals indicated by the tables to estimate $\frac{1}{10}\int_0^{10} H(t)\,dt$.

(3) Evaluate $\int_0^{10} H'(t)dt$, Using correct units, explain the meaning of the expression in the context of this problem.

(4) At time $t = 0$, biscuits with temperature $100\,^\circ\mathrm{C}$ were removed from an oven. The temperature of the biscuits at time t is modeled by a differentiable function B for which it is known that $B'(t) = -13.84e^{-0.173t}$. Using the given models, at time $t = 10$, how much cooler are the biscuits than the tea?

6.6 Motion Along a Straight Line (1-Dimension)

Displacement (Net Change in Position) from $t=a$ to $t=b$

$$\int_a^b v(t)dt = [s(t)]_a^b = s(b) - s(a)$$

Net Change in Velocity from $t=a$ to $t=b$

$$\int_a^b a(t)dt = [v(t)]_a^b = v(b) - v(a)$$

Distance from $t=a$ to $t=b$

$$\int_a^b |v(t)|\,dt$$

Example 1

The velocity function of a particle is given by $6t^2 - 18t + 12$. Find the displacement and the distance traveled by the particle during the interval $0 \le t \le 4$.

Solution

Displacement

$$= \int_0^4 v(t)dt = \int_0^4 (6t^2 - 18t + 12)dt = (2t^3 - 9t^2 + 12t)\Big|_0^4 = 128 - 144 + 48 = 32.$$

Since $v(t) = 6t^2 - 18t + 12 = 6(t-1)(t-2)$, we see that:

if $t < 1$, than $v > 0$;

if $1 < t < 2$, than $v < 0$;

if $2 < t$, than $v > 0$.

$$\text{Distance} = \int_0^4 |v(t)|\,dt = \int_0^1 v(t)dt - \int_1^2 v(t)dt + \int_2^4 v(t)dt.$$

$$= 34$$

[EX 1] A particle moves along a straight line. Find the displacement and the distance traveled by the particle during the given interval.

(1) $v(t) = 1 - \cos t,\ 0 \le t \le \frac{5\pi}{6}$

(2) $v(t) = 4t^3 - 4t,\ 0 \le t \le 2$

(3) $a(t) = 6t - 6,\ v(0) = 0,\ 0 \le t \le 3$

(4) $a(t) = e^{t-1},\ v(1) = 0,\ 0 \le t \le 2$

[EX 2]

A squirrel starts at building A at time $t = 0$ and travels along a straight wire connected to building B. For $0 \le t \le 18$, the squirrel's velocity is modeled by the piecewise-linear function defined by the graph above.

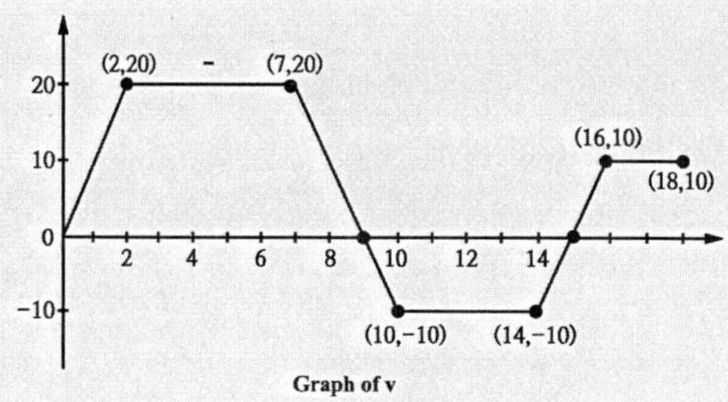

Figure 6.25

(1) At what times in the intervals $0 < t < 18$, if any, does the squirrel change direction? Give a reason for your answer.

(2) At What time in the intervals $0 \le t \le 18$ is the squirrel farthest from building A? How far from building A is the squirrel at this time?

(3) Find the total distance the squirrel travels during the time interval $0 \le t \le 18$.

(4) Write expressions for the squirrel's acceleration $a(t)$. velocity $v(t)$, and distance $x(t)$ from building A that are valid for the time interval $7 < t < 10$.

[EX 3]

Caren rides her bicycle along a straight road from home to school, starting at home at time $t = 0$ minutes and arriving at school at time $t = 12$ minutes. During the time interval $0 \leq t \leq 12$ minutes, her velocity $v(t)$, in miles per minutes, is modeled by the piecewise-linear function whose graph is shown above.

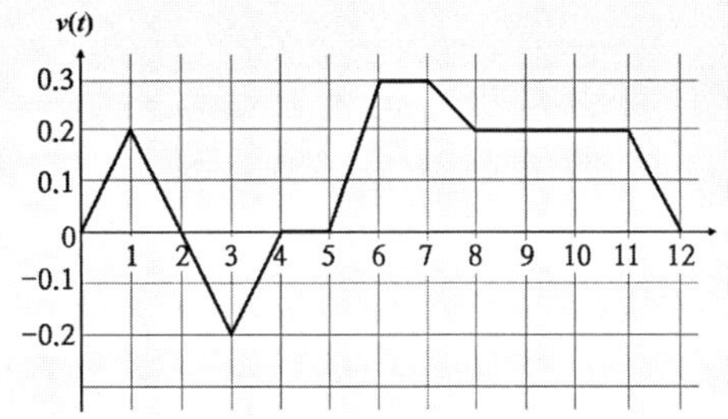

Figure 6.26

(1) Find the acceleration of Caren's bicycle at time $t = 7.5$ minutes. Indicate units of measure.

(2) Using correct units, explain the meaning of $\int_0^{12} |v(t)|\,dt$ in terms of Caren's trip.

Find the value of $\int_0^{12} |v(t)|\,dt$.

(3) Shortly after leaving home, Caren realizes she left her calculus homework at home, and she returns to get it. At what time does she turn around to go back home? Give a reason for your answer.

(4) Larry also rides his bicycle along a straight road from home to school in 12 minutes. His velocity is modeled by the function w given by $w(t) = \frac{\pi}{15}\sin\left(\frac{\pi}{12}t\right)$, where $w(t)$ is in miles per minute for $0 \leq t \leq 12$ minutes. Who lives closer to school : Caren or Larry? Show the work that leads to your answer.

[EX 4]

t (seconds)	0	10	40	60
$B(t)$ (meters)	100	136	9	49
$v(t)$ (meters per second)	2.0	2.3	2.5	4.6

Ben rides a unicycle back and forth along a straight east-west track.
The twice-differentiable function B models Ben's position on the track, measured in meters from the western end of the track, at time t, measured in seconds from the start of the ride. The table above gives values for $B(t)$ and Ben's velocity, $v(t)$, measured in meters per seconds, at selected times t.

(1) Use the data in the table to approximate Ben's acceleration at time $t = 5$ seconds. Indicate units of measure.

(2) Using correct units, interpret the meaning of $\int_0^{60} |v(t)|\,dt$ in the context of this problem. Approximate $\int_0^{60} |v(t)|\,dt$ using a left Riemann sum with the subintervals indicated by the data in the table.

(3) For $40 \le t \le 60$, must there be a time t when Ben's velocity is 2 meters per second? Justify your answer.

(4) A light is directly above the western end of the track. Ben rides so that at time t, the distance $L(t)$ between Ben and the light satisfies $(L(t))^2 = 12^2 + (B(t))^2$. At what rate is the distance between Ben and the light changing at time $t = 40$?

[EX 5] There is no snow on Janet's driveway when snow begins to fall at midnight. From midnight to 9 A.M., snow accumulates on the driveway at a rate modeled by $f(t)=7te^{\cos t}$ cubic feet per hour, where t is measured in hours since midnight, Janet starts removing snow at 6 A.M. $(t=6)$. The rate $g(t)$, in cubic feet per hour, at which Janet removes snow from the driveway at time t hours after midnight is modeled by

$$g(t)=\begin{cases} 0 & \text{for } 0\le t<6 \\ 125 & \text{for } 6\le t<7 \\ 108 & \text{for } 7\le t\le 9 \end{cases}$$

(1) How many cubic feet of snow have accumulated on the driveway by 6 A.M. ?

(2) Find the rate of change of the volume of snow on the driveway at 8 A.M.

(3) Let $h(t)$ represent the total amount of snow, in cubic feet, that Janet has removed from the driveway at time t hours after midnight. Express h as a piecewise-defined function with domain $0\le t\le 9$.

(4) How many cubic feet of snow are on the driveway at 9 A.M. ?

6.7 Motion Along a Plane Curve

If the motion of a particle P along a curve is given parametrically by the equations $x = x(t)$ and $y = y(t)$, then at time t the position vector R. The velocity vector v, and the acceleration vector a are:

$$R = xi + yj;$$

$$V = \frac{dR}{dt} = \frac{dx}{dt}i + \frac{dy}{dt}j = v_x i + v_y j;$$

$$a = \frac{d^2R}{dt^2} = \frac{dv}{dt} = \frac{d^2x}{dt^2}i + \frac{d^2y}{dt^2}j = a_x i + a_y j.$$

The components in the horizontal and vertical directions of R, v, and a are given respectively by the coefficients of i and j in the corresponding vector. The slope of v is $\frac{dy}{dx}$; its magnitude.

$$|v| = \sqrt{\left(\frac{dx}{dt}\right)^2 + \left(\frac{dy}{dt}\right)^2},$$

is the speed of the particle, and the velocity vector is tangent to the path. The slope of a is $\frac{d^2y}{dt^2} / \frac{d^2x}{dt^2}$, The distance the particle travels from t_1, to t_2, it given by

$$\int_{t_1}^{t_2} |v|\, dt = \int_{t_1}^{t_2} \sqrt{\left(\frac{dx}{dt}\right)^2 + \left(\frac{dy}{dt}\right)^2}\, dt.$$

Example 1

The particle $P(t)=\left((t+1)^2, \frac{t}{t+1}\right)$ in motion for 1 second, $0 \le t \le 1$. Find its position velocity, speed, and acceleration at $t=1$ and the distance it traveled.

Solution

From $P(t)$, we write the velocity vector: $v=\frac{dx}{dt}i+\frac{dy}{dt}j=2(t+1)i+\frac{1}{(t+1)^2}j$.

Hence, at $t=1$ the particle's velocity is $v=4i+\frac{1}{4}j$.

Speed is the magnitude of the velocity vector, so after 1 second the particle's speed is

$$|v|=\sqrt{4^2+\left(\frac{1}{4}\right)^2}\approx 4.008\text{ units/sec.}$$

The particle's acceleration vector at $t=1$ is $a=\frac{d^2x}{dt^2}i+\frac{d^2y}{dt^2}j=2i+\frac{-2}{(t+1)^3}j=2i-\frac{1}{4}j$.

On the interval $0 \le t \le 1$ the distance traveled by the particle is

$$\int_0^1 \sqrt{\left(\frac{dx}{dt}\right)^2+\left(\frac{dy}{dt}\right)^2}\,dt=\int_0^1 \sqrt{[2(t+1)]^2+\left[\frac{1}{(t+1)^2}\right]^2}\,dt\approx 3.057\text{ units.}$$

[EX 1] A particle $P(x,y)$ moves along a curve so that its acceleration is given by

$$a=-4\cos 2ti-2\sin tj\left(-\frac{\pi}{2}\leqq t\leqq\frac{\pi}{2}\right);$$

when $t=0$, the particle is at $(1,0)$ with $\frac{dx}{dt}=0$ and $\frac{dy}{dt}=2$.

(1) Find the position vector R at any time t.

(2) Find a Cartesian equation for the path of the particle, and identify the conic on which P moves.

[EX 2]

A diver leaps from the edge of diving platform into a pool below. The figure above shows the initial position of the diver and her position at a later time. At time t seconds after she leaps, the horizontal distance from the front edge of the platform to the diver's shoulders is given by $x(t)$, and the vertical distance from the water surface to her shoulders is given by $y(t)$, where $x(t)$ and $y(t)$ are measured in meters. <u>Suppose that the diver's shoulders are 11.4 meters above the water when she makes her leap and that</u>

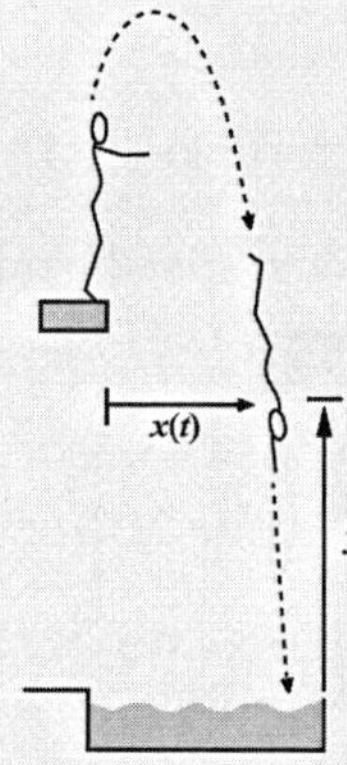

Note: Figure not drawn to scale.

Figure 6.27

$$\frac{dx}{dt} = 0.8 \text{ and } \frac{dy}{dt} = 3.6 - 9.8t,$$

for $0 \le t \le A$, where A is the time that the diver's shoulders enter the water.

(1) Find the maximum vertical distance from the water surface to the diver's shoulders.

(2) Find A, the time that the diver's shoulders enter the water.

(3) Find the total distance traveled by the diver's shoulders from the time she leaps from the platform until the time her shoulders enter the water.

(4) Find the angle θ, $0 \le \theta \le \frac{\pi}{2}$, between the path of the diver and the water at the instant the diver's shoulders enter the water.

[EX 3] A particle is moving along a curve so that its position at time t is $(x(t), y(t))$, where $x(t) = t^2 - 4t + 8$ and $y(t)$ is not explicitly given. Both x and y are measure in meters, and t is measured in seconds. It is known that $\frac{dy}{dt} = te^{t-3} - 1$.

(1) Find the speed of the particle at time $t = 3$ seconds.

(2) Find the total distance traveled by the particle for $0 \leq t \leq 4$ seconds.

(3) Find the time t, $0 \leq t \leq 4$, when the line tangent to the path of the particle is horizontal. Is the direction of motion of the paricle toward the left or toward the right at that time? Given a reason for your answer.

(4) There is a point with $x-$coordinate 5 through which the particle passes twice. Find each of the following.

(i) The two values of t when that occurs.

(ii) The slopes of the lines tangent to the particle's path at that point

(iii) The $y-$coordinate of that point, given $y(2) = 3 + \frac{1}{e}$

[EX 4] At time t, a particle moving in the $xy-$plane is at position $(x(t), y(t))$, where $x(t)$ and $y(t)$ are not explicitly given. For $t \geq 0$, $\frac{dx}{dt} = 4t + 1$ and $\frac{dy}{dt} = \sin(t^2)$. At time $t = 0$, $x(0) = 0$ and $y(0) = -4$.

(1) Find the speed of the particle at time $t = 3$, and find the acceleration vector of the particle at time $t = 3$.

(2) Find the slope of the line tangent to the path of the particle at time $t = 3$.

(3) Find the position of the particle at time $t = 3$.

(4) Find the total distance traveled by the particle over the time interval $0 \leq t \leq 3$.

6.8 Arc Length

If the derivative of function $y = f(x)$ is continuous on the interval $a \le x \le b$, then the length s of the arc of the curve of $y = f(x)$ from the point where $x = a$ to the point where $x = b$ is given by

$$s = \int_a^b \sqrt{1 + \left(\frac{dy}{dx}\right)^2}\, dx. \qquad (1)$$

Here a small piece of the curve is equal approximately to $\sqrt{1 + (f'(x))^2}\, \Delta x$.
As $\Delta x \to 0$, the sum of these pieces approaches the definite integral above.
If the derivative of the function $x = g(y)$ is continuous on the interval $c \le y \le d$, then the length s of the arc from $y = c$ to $y = d$ is given by

$$s = \int_c^d \sqrt{1 + \left(\frac{dx}{dy}\right)^2}\, dy. \qquad (2)$$

If a curve if defined parametrically by the equations $x = x(t)$ and $y = y(t)$, if the derivatives of the functions $x(t)$ and $y(t)$ are continuous on $[t_a, t_b]$, (and if the curve does not intersect ifself, then the length of the arc from $t = t_a$ to $t = t_b$ is given by

$$s = \int_{t_a}^{t_b} \sqrt{\left(\frac{dx}{dt}\right)^2 + \left(\frac{dy}{dt}\right)^2}\, dt. \qquad (3)$$

Example 1

Find the length, to three decimal places, of the arc of $y = x^{3/2}$ from $x = 1$ to $x = 8$.

Here $\dfrac{dy}{dx} = \dfrac{3}{2}x^{1/2}$, so, by (1),

Solution

$$s = \int_1^8 \sqrt{1 + \frac{9}{4}x}\, dx = 22.803.$$

[EX 1] Find the length, to three decimal placves, of the curve $(x-2)^2 = 4y^3$ from $y=0$ to $y=1$.

[EX 2] Find the length of the arc of the curve $y = x^{3/2}$ from $x=0$ to $x=4$

[EX 3] Find the length of the arc of $y = \ln \sec x$ from $x=0$ to $x = \frac{\pi}{3}$.

[EX 4] The position (x, y) of a particle at time t is given parametrically by $x = t^2$ and $y = \frac{t^3}{3} - t$. Find the distance the particle travels between $t=1$ and $t=2$.

[EX 5] Find the length one arc of the cycloid $x = t - \sin t$, $y = 1 - \cos t$.

[EX 6] Find the length of $x = e^t \cos t$, $y = e^t \sin t$ from $t=2$ to $t=3$.

[EX 7]

Let $f(x)=e^{2x}$. Let R be the region in the first quadrant bounded by the graph of f, the coordinate axes, and the vertical line $x=k$, where $k>0$. The region R is shown in the figure.

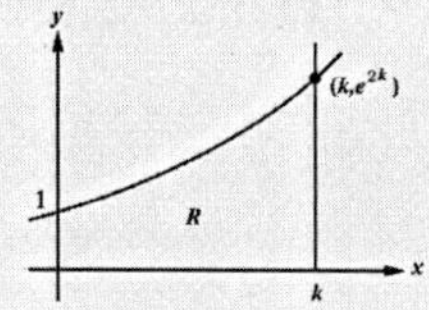

Figure 6.28

(1) Write, but do not evaluate, and expression involving an integral that gives the perimeter of R in terms of k.

(2) The region R is rotated about the x-axis to form a solid, Find the volume, V, of the solid in terms of k.

(3) The volume V, found in part (2), changes as k changes. If $\dfrac{dk}{dt}=\dfrac{1}{3}$, determine $\dfrac{dV}{dt}$ when $k=\dfrac{1}{2}$.

[EX 8]

A baker is creating a birthday cake. The base of the cake is the region R in the first quadrant under the graph of $y=f(x)$ for $0 \le x \le 30$, where $f(x)=20\sin\left(\dfrac{\pi x}{30}\right)$. Both x and y are measured in centimeters. The region R is shown in the figure. The derivative of f is $f'(x)=\dfrac{2\pi}{3}\cos\left(\dfrac{\pi x}{30}\right)$.

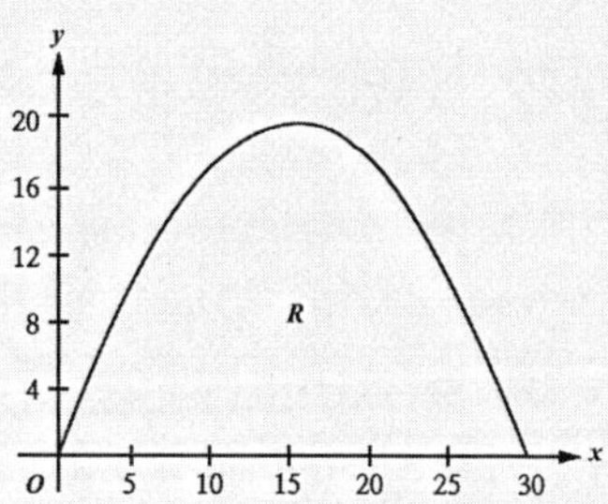

Figure 6.29

(1) The region R is cut out of a 30-centimeter-by-20-centimeter rectangular sheet of cardboard, and the remaining cardboard is discarded. Find the area of the discarded cardboard.

(2) The cake is a solid with base R. Cross sections of the cake perpendicular to the x-axis are semicircles. If the baker uses 0.05 gram of unsweetened chocolate for each cubic centimeter of cake, how many grams of unsweetened chocolate will be in the cake?

(3) Find the perimeter of the base of the cake.

6.9 Improper Integrals

There are two classes of improper integrals:

(1) those in wich at least one of the limits of integration is infinite (the interval is not bounded); and Illustrations of improper integrals of class (1) are:

$$\int_0^{\infty}\frac{dx}{\sqrt[3]{x+1}}\ ;\ \int_0^{\infty}\frac{dx}{x}\ ;\ \int_{-\infty}^{\infty}\frac{dx}{a^2+x^2}\ ;\ \int_{-\infty}^{0}e^{-x}dx\ ;\ \int_{-\infty}^{-1}\frac{dx}{x^n}\ (n\text{ a real number})\ ;$$

$$\int_{-\infty}^{0}\frac{dx}{e^x+e^{-x}}\ ;\ \int_0^{\infty}\frac{dx}{(4+x)^2}\ ;\ \int_{-\infty}^{0}e^{-x^2}dx\ ;\ \int_1^{\infty}\frac{e^{-x^2}}{x^2}dx$$

Improper integrals of class (1), where the interval is not bounded, are handled as limits:

$$\int_a^{\infty}f(x)dx=\lim_{b\to\infty}\int_a^b f(x)dx\ ,$$

where f is continuous on $[a, b]$. If the limit on the right exists, the improper integral on the left is said to converge to this limit; if the limit on the right fails to exist, we say that the improper integral diverges (or is meaningless).

Example 1

$$\int_1^{\infty}\frac{dx}{x^2}=$$

Solution

$\int_1^{\infty}\frac{dx}{x^2}=\lim_{b\to\infty}\int_1^b x^{-2}dx=\lim_{b\to\infty}-\frac{1}{x}\Big|_1^b=\lim_{b\to\infty}-\left(\frac{1}{b}-1\right)=1$. The given integral thus converges to 1.

Example 2

$$\int_{-\infty}^{0} e^{-x}dx=$$

Solution

$\int_{-\infty}^{0} e^{-x}dx=\lim_{b\to-\infty} -e^{-x}\Big|_b^0=\lim_{b\to-\infty} -(1-e^{-b})=+\infty$. Thus, this improper integral diverges.

(2) those of the type $\int_a^b f(x)dx$, where $f(x)$ has a point of discontinuity (becoming infinite) at $x=c$, $a\le c\le b$ (the function is not bounded).

The following improper integrals are of class (2):

$$\int_0^1 \frac{dx}{x}\ ;\ \int_1^2 \frac{dx}{(x-1)^n}\ \text{(n a real number)}\ ;\ \int_{-1}^1 \frac{dx}{1-x^2}\ ;\ \int_0^2 \frac{x}{\sqrt{4-x^2}}dx\ ;\ \int_1^2 \frac{dx}{1+\sin x}$$
$$;\ \int_{-1}^2 \frac{dx}{x(x-1)^2}\ ;\ \int_0^{2x} \frac{\sin x\,dx}{\cos x+1}\ ;\ \int_a^b \frac{dx}{(x-c)^n}\ \text{(n real; } a\le c\le b\text{)}\ ;\ \int_0^1 \frac{dx}{\sqrt{x+x^4}}\ ;$$
$$\int_{-2}^2 \sqrt{\frac{2+x}{2-x}}\,dx\ .$$

Improper integrals of class (2), where the function has an infinite discontinuity, are handled as follows.

To investigate $\int_a^b f(x)dx$, where f becomes infinite at $x=a$, we define $\int_a^b f(x)dx$ to be $\lim_{k\to a^+}\int_k^b f(x)dx$. The given integral then converges or diverges according to whether the limit does or does not exist. If f has its discontinuity at b, we define $\int_a^b f(x)dx$ to be $\lim_{k\to b^-}\int_a^k f(x)dx$; again, the given integral converges or diverges as the limit does or does not exist. When, finally, the integrand has a discontinuity at an interior point c on the interval of integration ($a<c<b$), we let

$$\int_a^b f(x)dx = \lim_{k \to c^-} \int_a^k f(x)dx + \lim_{m \to c^+} \int_m^b f(x)dx$$

Now the improper integral converges only if both of the limits exist. If either limit does not exist, the improper integral diverges.

Example 3

$$\int_0^1 \frac{dx}{\sqrt[3]{x}} =$$

Solution

$$\int_0^1 \frac{dx}{\sqrt[3]{x}} = \lim_{k \to 0^+} \int_k^1 x^{-1/3} dx = \lim_{k \to 0^+} \frac{3}{2} x^{2/3} \Big|_k^1 = \lim_{k \to 0^+} \frac{3}{2}(1 - k^{2/3}) = \frac{3}{2}$$

Example 4

$$\int_2^3 \frac{dt}{(3-t)^2} =$$

Solution

$$\int_2^3 \frac{dt}{(3-t)^2} = \lim_{k \to 3^-} - \int_2^k (3-t)^{-2}(-dt) = \lim_{k \to 3^-} \frac{1}{3-t} \Big|_2^k = +\infty$$

This integral diverges.

[EX 1] Evaluate the integrals

(1) $\int_1^{\infty} \frac{1}{\sqrt{x}} dx$

(2) $\int_0^{\infty} \frac{1}{x^2+9} dx$

(3) $\int_0^{\infty} \frac{dy}{e^y}$

(4) $\int_{-\infty}^{0} \frac{dz}{(z-1)^2}$

(5) $\int_0^{\infty} \cos x \, dx$

(6) $\int_0^{1} \frac{1}{x^3} dx$

(7) $\int_0^{2} \frac{dy}{\sqrt{4-y^2}}$

(8) $\int_0^{2} \frac{dx}{(x-1)^{2/3}}$

(9) $\int_{-2}^{2} \frac{dx}{x^2}$

(10) $\int_0^{\pi/2} \frac{\sin x}{\sqrt{1-\cos x}} dx$

[EX 2] Choose the alternative that gives the area, if it exists, of the region described.

(1) In the first quadrant under the curve of $y = e^{-x}$

(A) 1 (B) e (C) $\frac{1}{e}$ (D) 2 (E) none of these

(2) In the first quadrant under the curve of $y = xe^{-x^2}$

(A) 2 (B) $\frac{2}{e}$ (C) $\frac{1}{2}$ (D) $\frac{1}{2e}$ (E) none of these

(3) In the first quadrant above $y=1$, between by the y-axis and the curve $xy=1$.

(A) 1 (B) 2 (C) $\frac{1}{2}$ (D) 4 (E) none of these

(4) Between the curve $y=\frac{4}{1+x^2}$ and the x-axis.

(A) 2π (B) 4π (C) 8π (D) π (E) none of these

(5) Above the x-axis, between the curve $y=\frac{4}{\sqrt{1-x^2}}$ and its asymptotes.

(A) $\frac{\pi}{2}$ (B) π (C) 2π (D) 4π (E) none of these

[EX 3] Choose the alternative that gives the volume, if it exists, of the solid generated.

(1) $y=\frac{1}{x}$, at the right by $x=1$, and above by $y=0$; about the x-axis.

(A) $\frac{\pi}{2}$ (B) π (C) 2π (D) 4π (E) none of these

(2) The first quadrant region under $y=e^{-x}$; about the x-axis.

(A) $\frac{\pi}{2}$ (B) π (C) 2π (D) 4π (E) none of these

[EX 4] Let f and g be the functions defined by $f(x)=\frac{1}{x}$ and $g(x)=\frac{4x}{1+4x^2}$, for all $x>0$.

(1) Find the absolute maximum value of g on the open interval $(0, \infty)$ if the maximum exists. Find the absolute minimum value of g on the open interval $(0, \infty)$ if the minimum exists. Justify your answers.

(2) Find the area of the unbounded region in the first quadrant to the right of the vertical line $x=1$. below the graph of f, and above the graph of g.

CALCULUS

CHAPTER 07

Differential Equations

7.1 Slope Fields (Direction Fields)

A slope field (or direction field) is a graphical solution to a first order differential equation $\frac{dy}{dx} = f(x,y)$.

Step 1. Select a rectangular grid of points.

Step 2. Calculate the tangent slope at each point.

Step 3. Draw a short line segment through each point with a graphical representation of its slope.

Example

Consider the differential equation $y' = x + 1$

(1) On a grid, like that shown in the figure, sketch a slope field for the given differential equation at the 20 points indicated.

(2) Sketch the solution curve that passes through the point (-1,0).

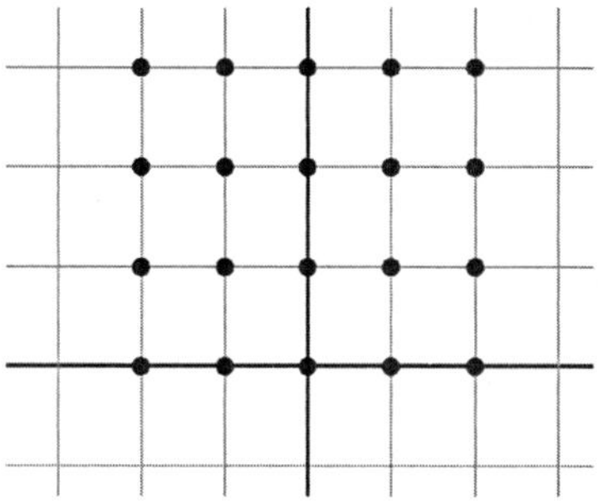

Figure 7.1

Solution

(1)

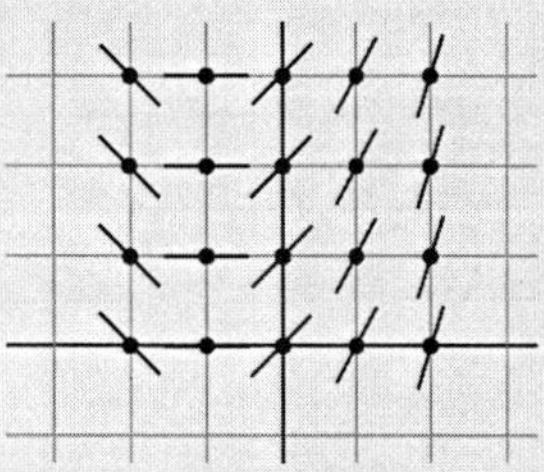

Figure 7.2

(2)

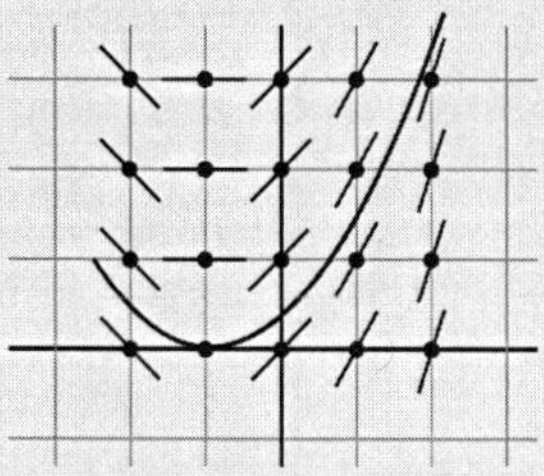

Figure 7.3

[EX 1] The slope field for the differential equation $\frac{dy}{dx} = \frac{1}{x}$ is shown below.

(1) Carefully draw the solution curve that passes through the point (1,0.5).

(2) Find the general solution for the equation.

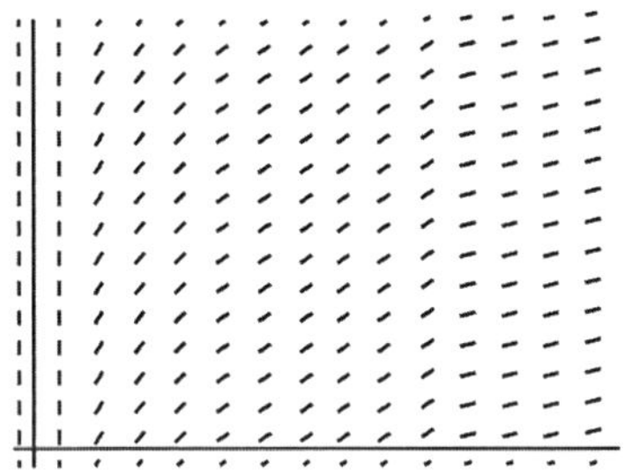

Figure 7.4

[EX 2]

(a) Sketch a slope field for each of the following differential equations.

(b) Sketch the solution curve that passes through the indicated point.

(1) $\frac{dy}{dx} = y + 1$, (0,1)

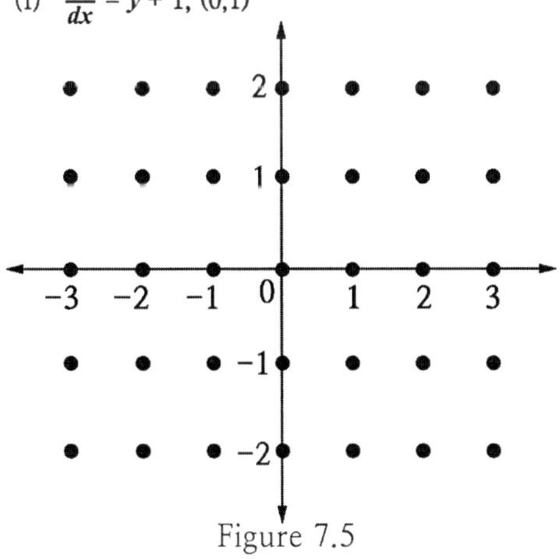

Figure 7.5

(2) $\frac{dy}{dx} = -1xy$, (1,1)

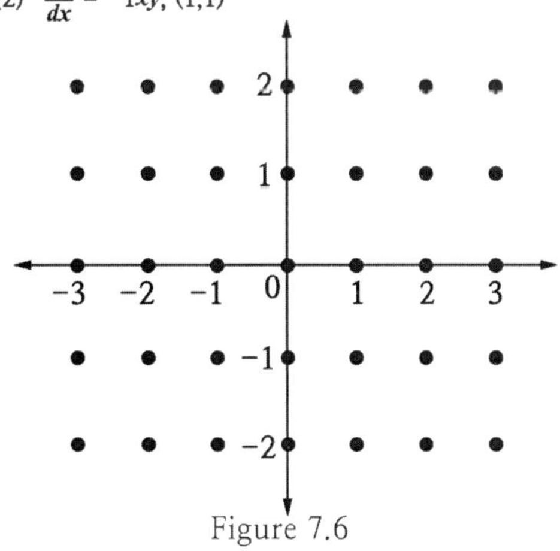

Figure 7.6

(3) $\frac{dy}{dx} = -\frac{y}{x}$, (−2,1)

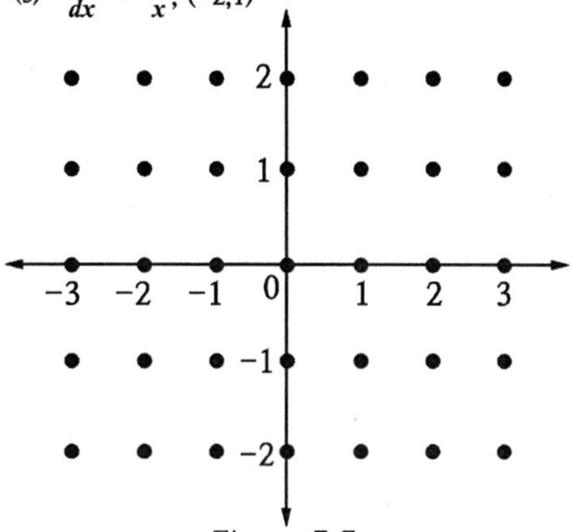

Figure 7.7

(4) $\frac{dy}{dx} = -\frac{x}{y}$, (−2,1)

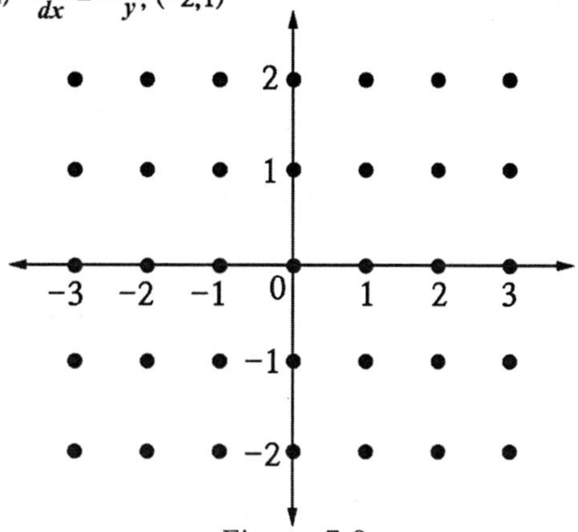

Figure 7.8

[EX 3] Match the slope fields with their differential equations.

(1) $y' = x + y$ (2) $y' = .5y$ (3) $y' = .5y - 1$ (4) $y' = -\frac{x}{y}$

a)

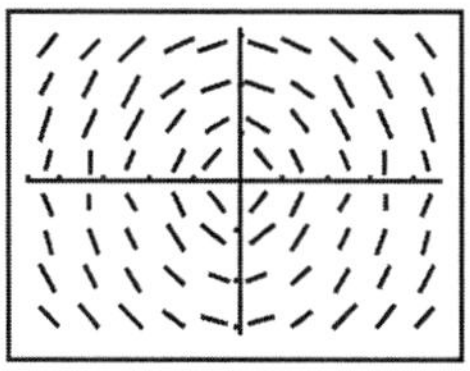

Figure 7.9

b)

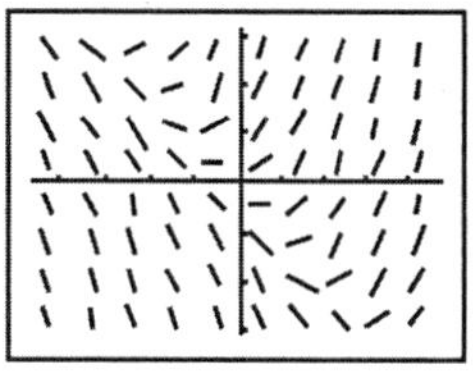

Figure 7.10

c)

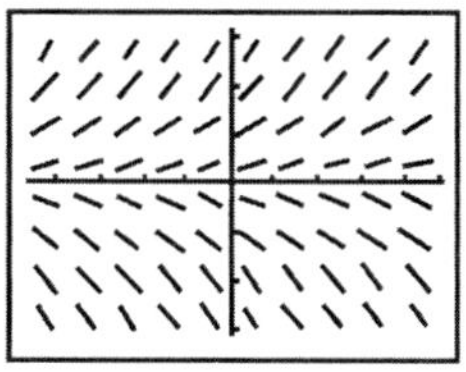

Figure 7.11

d)

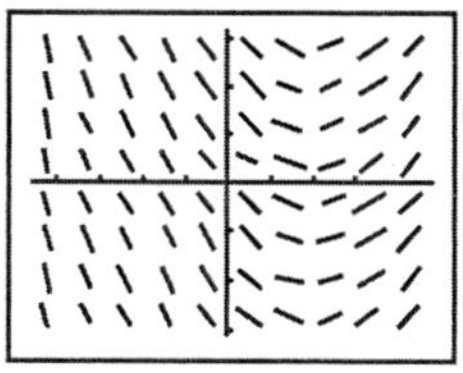

Figure 7.12

[EX 4] Match each differential equation with its slope field.

(1) $\frac{dy}{dx} = y(y+1)$

(2) $\frac{dy}{dx} = \sin 2x$

(3) $\frac{dy}{dx} = \cos 2x$

(4) $\frac{dy}{dx} = \frac{1}{1+x^2}$

(5) $\frac{dy}{dx} = 1 + y^2$

(6) $\frac{dy}{dx} = -xy$

(a) Figure 7.13

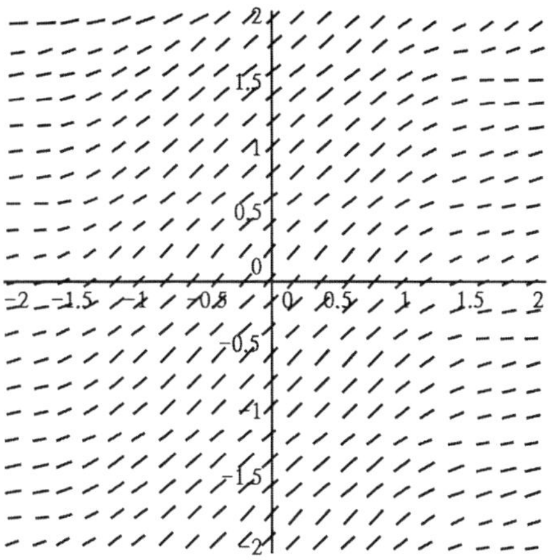

(b) Figure 7.14

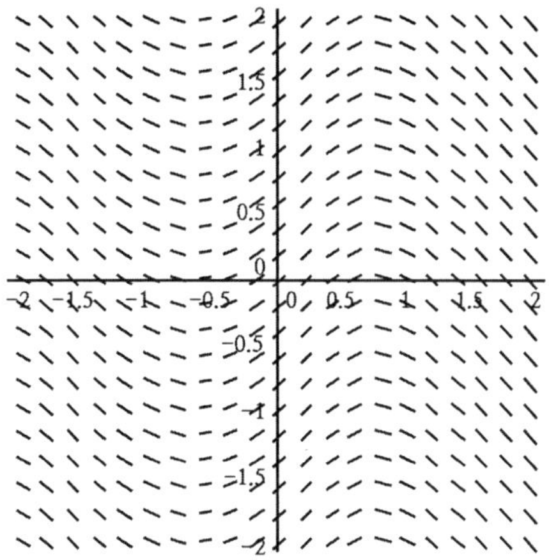

(c) Figure 7.15

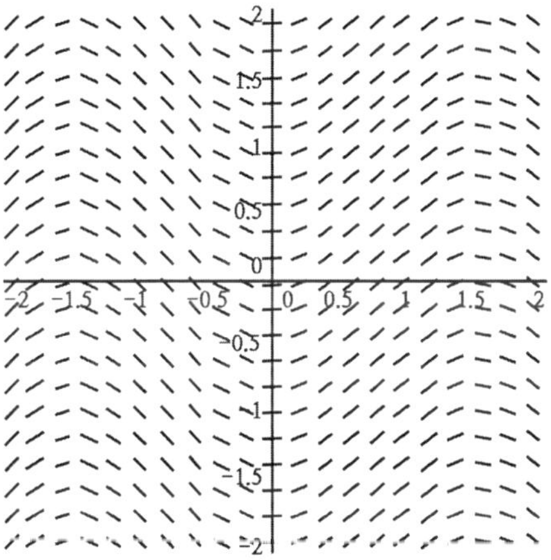

(d) Figure 7.16

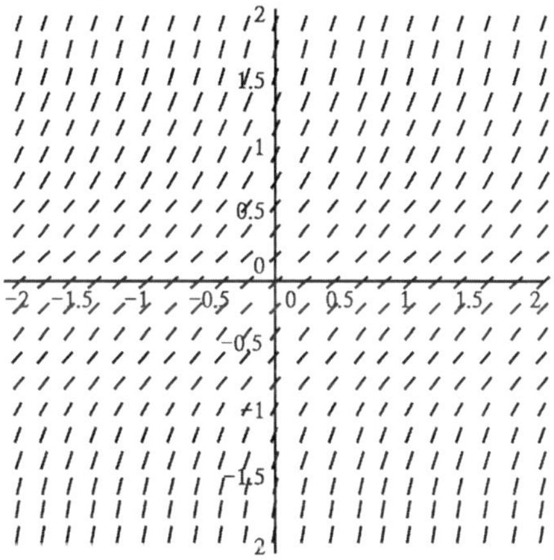

(e) Figure 7.17

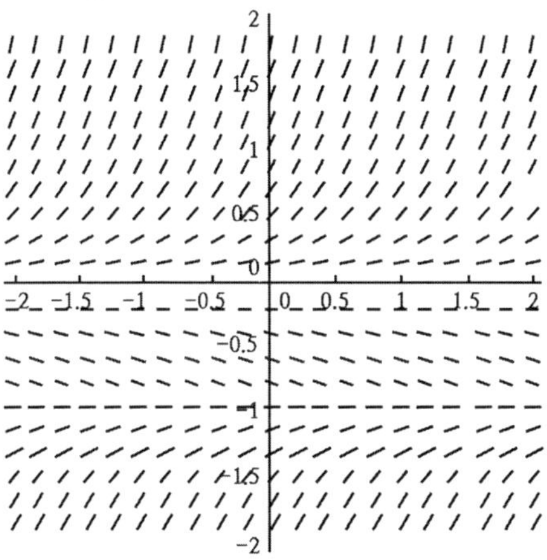

(d) Figure 7.18

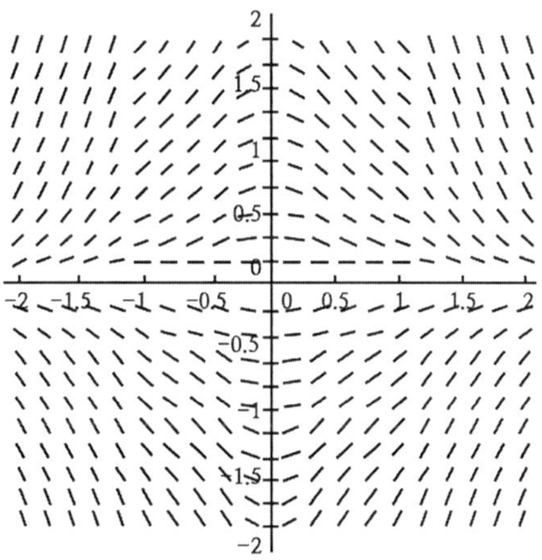

[EX 5] Match each differential equation with its slope field

(1) $\dfrac{dy}{dx} = \cos x$

(2) $\dfrac{dy}{dx} = 2x$

(3) $\dfrac{dy}{dx} = 3x^2 - 3$

(4) $\dfrac{dy}{dx} = -\dfrac{\pi}{2}$

(a)

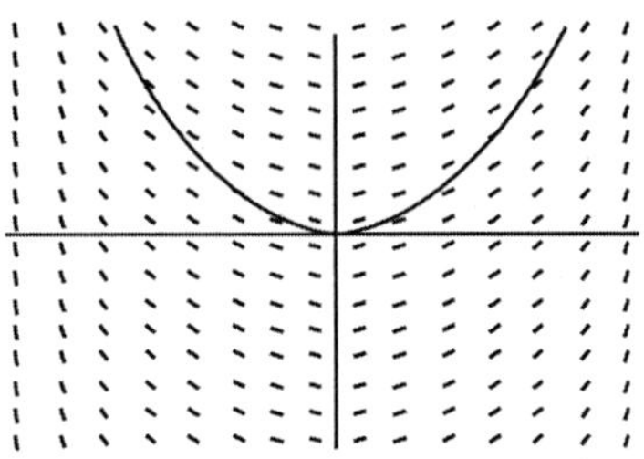

Figure 7.19

(b)

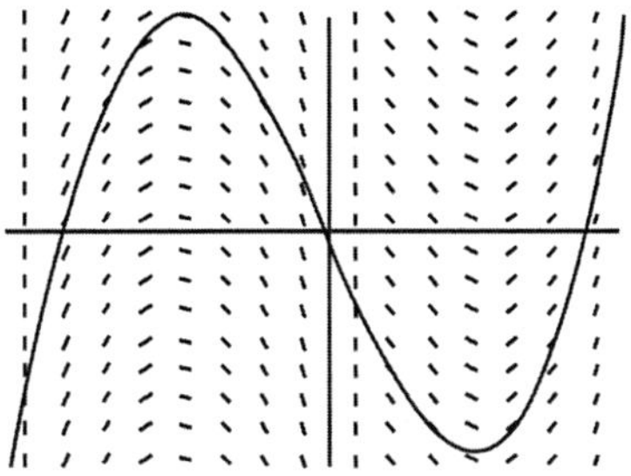

Figure 7.20

(c)

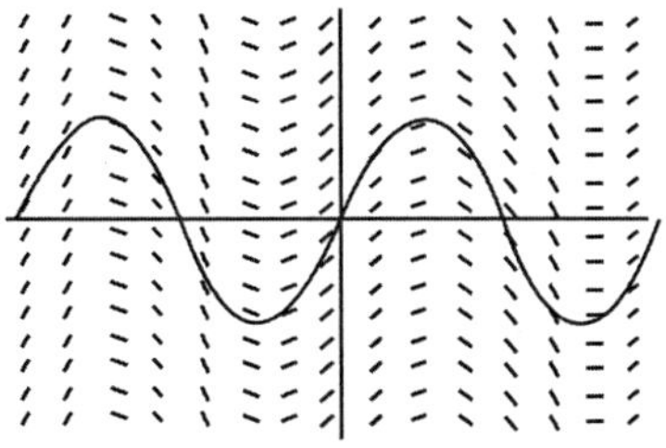

Figure 7.21

(d)

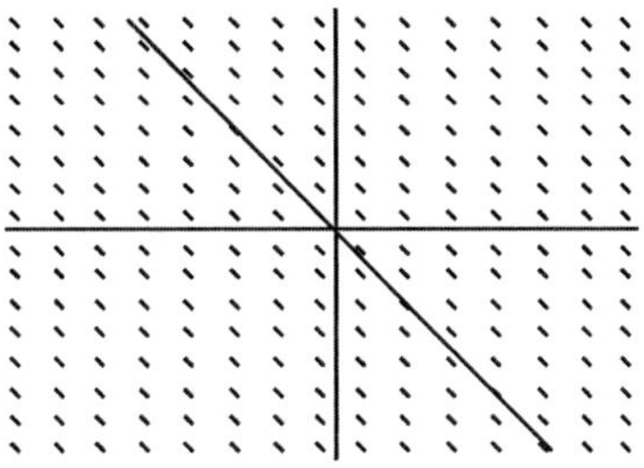

Figure 7.22

7.2 First Order Separable Differential Equations

A differential equation has variables separable if it is of the form

$$\frac{dy}{dx}=\frac{f(x)}{g(y)} \qquad \text{or} \qquad g(y)\,dy-f(x)\,dx=0.$$

The general solution is

$$\int g(y)\,dy-\int f(x)\,dx=C \quad \text{(C arbitrary).}$$

Example 1

Solve the d.e. $\frac{dy}{dx}=-\frac{x}{y}$, given the initial condition $y(0)=2$.

Solution

We rewrite the equation as $y\,dy=-x\,dx$. We then integrate, getting

$\int y\,dy=-\int x\,dx,$

$\frac{1}{2}y^2=-\frac{1}{2}x^2+k,$

$y^2+x^2=C \qquad (where\ C=2k).$

Since $y(0)=2$, we get $4+0=C$; the particular solution is therefore $x^2+y^2=4$.

[EX 1] Solve the differential equation.

(1) $\frac{dy}{dx}=\sqrt{xy},\ f(0)=1$

(2) $\frac{dy}{dx}=-2xy^3,\ f(0)=1$

(3) $\frac{dy}{dx}=e^{x-y},\ f(1)=1$

(4) $2x\frac{dy}{dx}-3y=0,\ y(1)=100$

[EX 2] Solve the differential equation.

(1) $\frac{dy}{dx}=\frac{y}{2\sqrt{x}}$, $f(4)=1$

(2) $y'=y^2\cos 2x$, $f(\pi/4)=-1$

(3) $(\ln y)\frac{dy}{dx}=\frac{y}{x}$, $f(1)=e$

(4) $y'=\frac{2+y}{x}$, $y(-1)=3$

(5) $\frac{dy}{dx}=e^y$, $f(1)=0$

(6) $\frac{dy}{dx}=\frac{2\sqrt{y}}{x^2+1}$, $y(0)=4$

[EX 3]

(1) Given $f(x)f'(x)=x$ for all real x and $f(0)=1$, find the value of $f(x)$

(2) Given $f'(x)=(f(x))^2+1$ for all real x and $f(0)=\sqrt{3}$, find the value of $f\left(\frac{\pi}{2}\right)$.

7.3 Exponential Growth and Decay

Exponential Growth and Decay

"A positive quantity y increases (or decreases) at a rate that at any time t is proportional to the amount present. " It follows that the quantity y satisfies the d.e.

$$\frac{dy}{dt}=ky$$

where $k>0$ if y is increasing and $k<0$ if y is decreasing.

proof

$\frac{dy}{y}=k\,dt,\ \int\frac{1}{y}dy=\int k\,dt,\ \ln y = kt+C$ (C a constant).

$y=e^{kt+C}=e^{kt}\cdot e^{C}$

$=ce^{kt}$ (where $c=e^{C}$).

Example 1

A radioactive substance decays at a rate proportional to the amount of substance present. Radium$-$226 is a radioactive substance with a half-life of 1612 years. How long will it take for 200 g of this substance to decay to 150g?

Solution

$\frac{dy}{dt}=ky$

$\int\frac{dy}{y}=\int k\,dt$

$\ln|y|=kt+C$

$\therefore y=Ae^{kt}$

When $t=0,\ y=200$

$y=200e^{kt}$

when $t=1612,\ y=100$

$100=200e^{k(1612)}$

$\frac{1}{2}=e^{1612k}$

$k=\frac{\ln\frac{1}{2}}{1612}=-0.00043$

$y=200\,e^{-0.00043t}$

Find when $y=150$

$150=200e^{-0.00043t}$

$\frac{3}{4}=e^{-0.00043t}$

$t=\frac{\ln\frac{3}{4}}{-0.00043}\approx 669\,yr$

[EX 1] The bacteria in a certain increase continuously at a rate proportional to the number present. (1) If the number triples in 6 hours, how many will there be in 12 hours? (2) In how many hours will the original number quadruple?

[EX 2] It is estimated that 20% of a certain radioactive substance decays in 45 hours. What is the half-life of the substance?

[EX 3] The population of a country is growing at a rate proportional to its population. If the growth rate per year is 4% of the current population, how long will it take for the population to double?

[EX 4] The rate of change of y is proportional to y. If $y = 4$ when $t = 0$ and $y = 10$ when $t = 2$, What is the value of y when $t = 5$?

Restricted Growth and Decay

Suppose the temperature of an object changes at a rate proportional to the difference between its temperature(T) and that of the surrounding medium (T_S).

If $T > T_S$: Newton's Law of Cooling

$\frac{dT}{dt} = -k(T - T_S) \Rightarrow T = T_S + Ae^{-kt}$ where k is the positive proportional constant.

If $T < T_S$

$\frac{dT}{dt} = k(T_S - T) \Rightarrow T = T_S - Ae^{-kt}$ where k is the positive proportional constant.

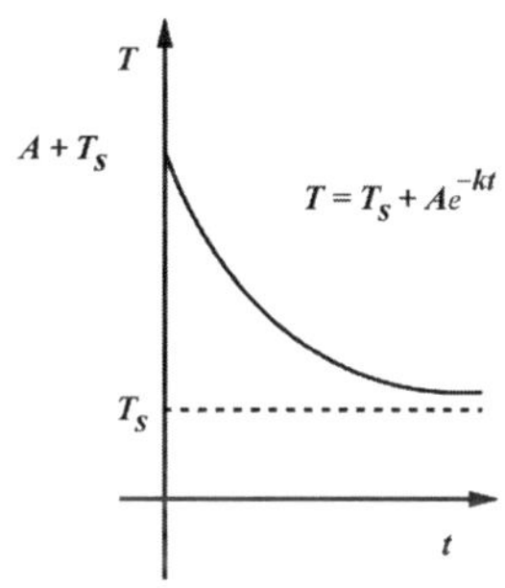

Figure 7.23

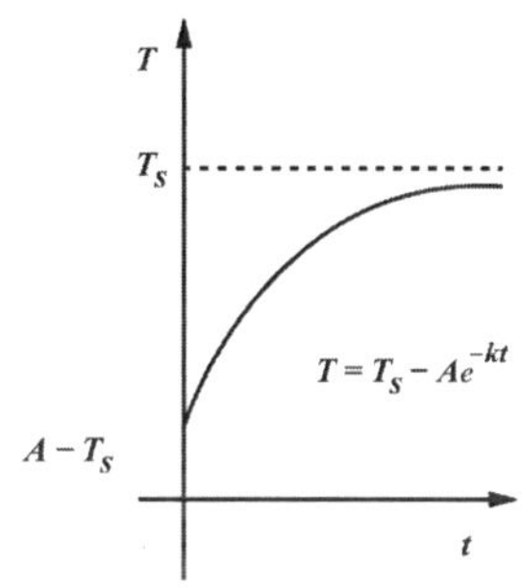

Figure 7.24

proof

$$\frac{dT}{dt} = -k(T - T_S)$$

$$\frac{dT}{T - T_S} = -kdt$$

$$\int \frac{dT}{T - T_S} = \int -kdt$$

$$\ln|T - T_S| = -kt + C \qquad where \quad T > T_S$$

$$|T - T_S| = e^{-kt + C}$$

$$T - T_S = \pm e^{-kt + C}$$

$$T - T_S = Ae^{-kt} \quad (where \ \ A = \pm e^C)$$

$$\therefore T = T_S + Ae^{-kt}$$

Example 2

A cup of coffee is initially 80℃ and is placed in a room with a temperature of 20℃. After 5 minutes, the temperature of the coffee drops to 60℃. How long will it take for the coffee to drop to 30℃ according to Newton's law of cooling?

Solution

$\frac{dT}{dt} = -k(T-20)$

$\int \frac{dT}{t-20} = \int -kdt$

$\ln|T-20| = -kt + C$

$\therefore\ T-20 = Ae^{-kt}$

When $t=0,\ T=80.$

$80-20 = Ae^0$

$A = 60$

$\therefore T-20 = 60e^{-kt}$

When $t=5,\ T=60.$

$40 = 60e^{-5k}$

$\frac{2}{3} = e^{-5k}$

$-5k = \ln\left(\frac{2}{3}\right)$

$k = -\frac{1}{5}\ln\left(\frac{2}{3}\right)$

$\therefore\ T-20 = 60e^{\frac{t}{5}\ln\left(\frac{2}{3}\right)}$

Find t when $T=30.$

$10 = 60e^{\frac{t}{5}\ln\left(\frac{2}{3}\right)}$

$\frac{1}{6} = e^{\frac{t}{5}\ln\left(\frac{2}{3}\right)}$

$\ln\left(\frac{1}{6}\right) = \frac{t}{5}\ln\left(\frac{2}{3}\right)$

$t = \frac{5\ln\left(\frac{1}{6}\right)}{\ln\left(\frac{2}{3}\right)}$

$= 22.095$ min

[EX 5] According to Newton's law of cooling, a hot object cools at a rate proportional to the difference between its own temperature and that of its environment. If a roast at room temperature 68°F is put into a 20°F freezer, and if, after 2hours, the temperature of the roast is 40°F: (a) What it its temperature after 5 hours? (b) How long will it take for the temperature of the roast to fall to 21°F?

[EX 6] A bottle of water initially has a temperature of 77°F. It is left in a refrigerator set to 41°F. How long will it take for the water to reach 45°F, if it drops to 68°F after 10 minutes?

[EX 7] Advertisers generally assume that the rate at which people hear about a product is proportional to the number of people who have not yet heard about it. Suppose that the size of a community is 15,000, that to begin with no one has heard about a product, but that after 6 days 1500 people know about it. How long will it take for 2700 people to have heard of it?

[EX 8] A frozen pizza with a temperature of -5℃ is put into an oven set to 250℃. How long would it take for the pizza to reach 200℃, if it is 100℃ after 7 minutes?

[EX 9] A murder victim was found at midnight. A detective established the corpse's body temperature to be 25℃. Two hours later it dropped to 23℃. Suppose the temperature of a living body is 37℃ and the temperature of the room was constant at 20℃. Estimate the time at which the murder occurred.

7.4 Euler's Method

Euler's Method Approximate values for the solution of the initial-value problem

$dy/dx = f(x, y),\ y(x_0) = y_0$ with step size dx, at $x_n = x_{n-1} + dx$ are

$$y_n = y_{n-1} + f(x_{n-1}, y_{n-1})dx,\ n = 1,2,3,\cdots$$

proof Given a differential equation $dy/dx = f(x,y)$ and an initial condition $y(x_0) = y_0$, we can approximate the solution $y = y(x)$ by its linearization

$$L(x) = y(x_0) + y'(x_0)(x - x_0) \quad \text{or} \quad L(x) = y_0 + f(x_0, y_0)(x - x_0).$$

The function $L(x)$ gives a good approximation to the solution $y(x)$ in a shor interval about x_0(Figure 7.25).

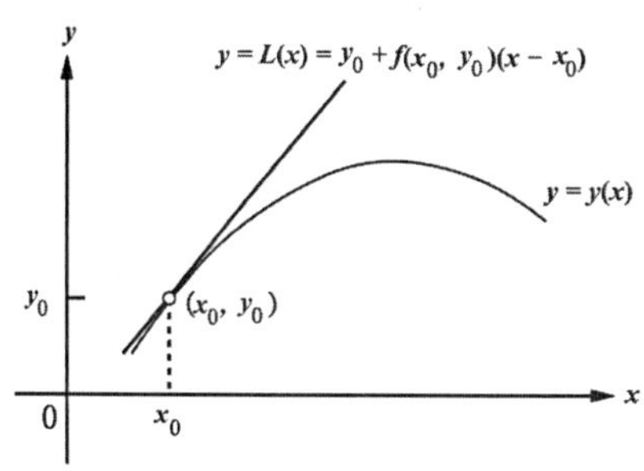

Figure 7.25 $y = y(x)$ at $x = x_0$.

We know the point (x_0, y_0) lies on the solution curve. Suppose that we specify a new value for the independent variable to be $x_1 = x_0 + dx$. (Recall that $dx = \Delta x$ in the definition of differentials.) If the increment dx is small, then

$$y_1 = L(x_1) = y_0 + f(x_0, y_0)\, dx$$

is a good approximation to the exact solution value $y = y(x_1)$. so from the point (x_0, y_0), which lies exactly on the solution curve.

We have obtained the point (x_1, y_1), which lies very close to the point $(x_1, y(x_1))$ on the solution curve (Figure 7.26).

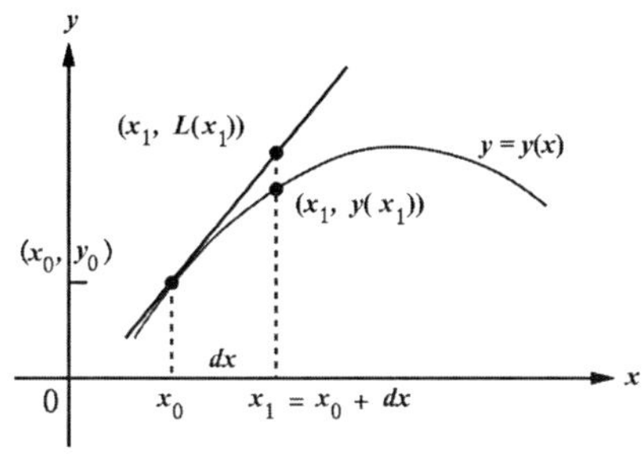

Figure 7.26 The first Euler step approximates $y(x_1)$ with $y_1 = L(x_1)$.

Using the point (x_1, y_1) and the slope $f(x_1, y_1)$ of the solution curve through (x_1, y_1), we take a second step.

Setting $x_2 = x_1 + dx$, we use the linearization of the solution curve through (x_1, y_1) to calculate

$$y_2 = y_1 + f(x_1, y_1)\,dx.$$

This gives the next approximation (x_2, y_2) to values along the solution curve $y = y(x)$ (Figure 7.27). Continuing in this fashion, we take a third step from the point (x_2, y_2) with slope $f(x_2, y_2)$ to obtain the third approximation

$$y_3 = y_2 + f(x_2, y_2)\,dx,$$

and so on. We are literally building an approximation to one of the solutions by following the direction of the slope field of the differential equation.

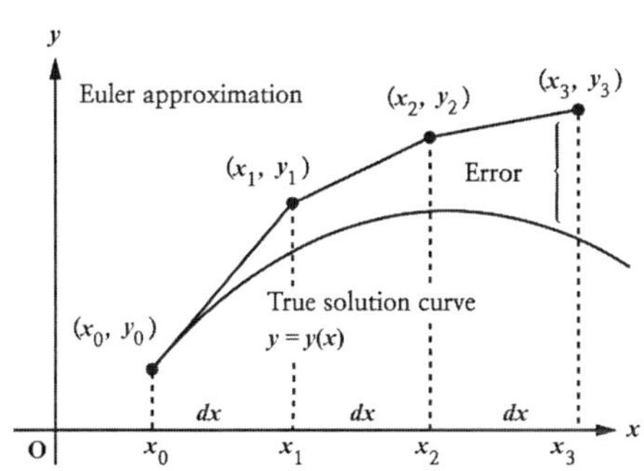

Figure 7.27 The steps in the Euler approximation to the solution of the initial value problem $y' = f(x,y)$, $y(x_0) = y_0$

Example 1

Find the first three approximations y_1, y_2, y_3 using Euler's method for the initial value problem

$$y' = 1 + y, \quad y(0) = 1,$$

stating at x_0 with $dx = 0.1$.

Solution

We have $x_0 = 0$, $y_0 = 1$, $x_1 = x_0 + dx = 0.1$, $x_2 = x_0 + 2dx = 0.2$, and $x_3 = x_0 + 3dx = 0.3$.

$$y_1 = y_0 + f(x_0, y_0)\,dx = y_0 + (1 + y_0)\,dx = 1 + (1+1)(0.1) = 1.2$$

$$y_2 = y_1 + f(x_1, y_1)\,dx = y_1 + (1 + y_1)\,dx = 1.2 + (1 + 1.2)(0.1) = 1.42$$

$$y_3 = y_2 + f(x_2, y_2)\,dx = y_2 + (1 + y_2)\,dx = 1.42 + (1 + 1.42)(0.1) = 1.662$$

[EX 1] Use Euler's method to calculate the first three approximations to the given intial value problem for the specified increment size.

(1) $y' = 1 - \frac{y}{x}$, $y(2) = -1$, $dx = 0.5$

(2) $y' = x(1 - y)$, $y(1) = 0$, $dx = 0.2$

(3) $y' = 2xy + 2y$, $y(0) = 3$, $dx = 0.2$

(4) $y' = y^2(1 + 2x)$, $y(-1) = 1$, $dx = 0.5$

[EX 2] Consider the differential equation $\frac{dy}{dx}=y^2(2x+2)$. $\leq t$ $y=f(x)$ be the particular solution to the differential equation with initial condition $f(0)=-1$.

(1) Find $\lim_{x\to 0}\frac{f(x)+1}{\sin x}$. Show the work that leads to your answer.

(2) Use Euler's method, stating at $x=0$ with two steps of equal size. to approximate $f\left(\frac{1}{2}\right)$.

(3) Find $y=f(x)$, the particular solution to the differential equation with initial condition $f(0)=-1$.

[EX 3] Consider the differential equation $\frac{dy}{dx}=1-y$. $\leq t$ $y=f(x)$ be the particular solution to this differential equation with the initial condition $f(1)=0$. For this particular solution, $f(x)<1$ for all values of x.

(1) Use Euler's method, starting at $x=1$ with two steps of equal size. to approximate $f(0)$. Show the work that leads to your answer.

(2) Find $\lim_{x\to 1}\frac{f(x)}{x^3-1}$. Show the work that leads to your answer.

(3) Find the particular solution $y=f(x)$ to the differential equation $\frac{dy}{dx}=1-y$ with the initial condition $f(1)=0$.

[EX 4] Consider the differential equation $\frac{dy}{dx} = 3x + 2y + 1$.

(1) Find $\frac{d^2y}{dx^2}$ in terms of x and y

(2) Find the values of the constants m, b and r for which $y = mx + b + e^{rx}$ is a solution to the differential equation.

(3) Let $y = f(x)$ be a particular solution to the differential equation with the initial condition $f(0) = -2$. Use Euler's method, stating at $x = 0$ with a step size of $\frac{1}{2}$, to approximate $f(1)$. Show the work that leads to your answer.

(4) $\leq t$ $y = g(x)$ be another solution to the differential equation with the initial condition $g(0) = k$, where k is a constants. Euler's method, starting at $x = 0$ with a step size of 1, gives the approximation $g(1) \approx 0$. Find the value of k.

7.5 Logistic Growth

The rate of change of a quantity (for example, a population) may be proportional both to the amount (size) of the quantity and to the difference between a fixed constants A and its amount (size). If $y = f(t)$ is the amount, then

$$y' = ky(A - y), \qquad (1)$$

where k and A are both positive. Equation (1) is called the logistic differential equation; it is used to model logistic growth.

The solution of the d.e (1) is

$$y = \frac{A}{1 + ce^{-Akt}} \qquad (2)$$

for some positive constant c.

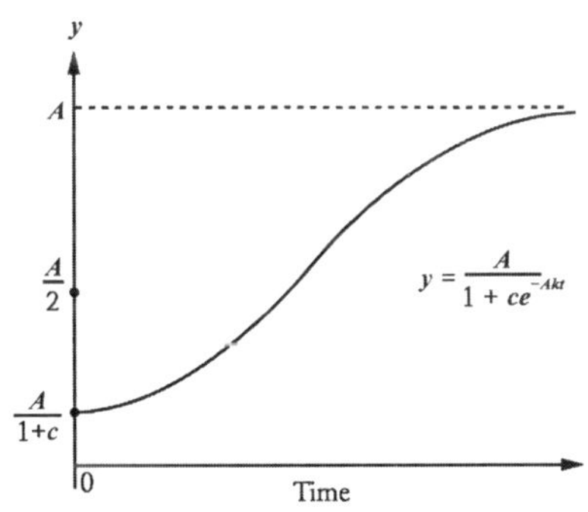

Figure 7.28

The S-shaped curve in (2) is often called a logistic curve. It shows that the rate of growth y' :

(1) increases slowly for a while; i.e., $y'' > 0$;

(2) attains a maximum when $y = A/2$, at the point of inflection;

(3) then decreases ($y'' < 0$), approaching 0 as y approaches its upper limit.

Example 1

A certain wild animal preserve can support no more than 250 lowland gorillas. twenty-eight gorillas were known to be in the preserve in 1970. Assume that the rate of growth of the population is

$$\frac{dy}{dt} = 0.0004(250 - y)y$$

where time t is in years.

Find a formula for the gorilla population in terms of t.

Solution

$k = 0.0004$ and $A = 250$, and $y = \dfrac{250}{1 + ce^{-0.1t}}$

initial condition : $y(0) = 28$, where $t = 0$ represents the year 1970.

$28 = \dfrac{250}{1+c} \Rightarrow 28(1+c) = 250 \Rightarrow c = 7.93$

Formula : $P = \dfrac{250}{1 + 7.93e^{-0.1t}}$

[EX 1] A 2000-gal tank can support no more than 150 guppies. Six guppies are introduced into the tank. Assume that the rate of growth of the population is

$$\frac{dy}{dt} = 0.0015(150 - y)y$$

where time t is in weeks.

(1) Find a formula for the guppy population in terms of t.

(2) How long will it take for the guppy population to be 100? 125?

[EX 2] Consider the logistic differential equation $\frac{dy}{dt}=\frac{y}{8}(6-y)$. $\leq t$ $y=f(t)$ be the particular solution to the differential equation with $f(0)=8$

(1) A slope field for this differential equation is given below. Sketch possible solution curves through the points (3,2) and (0,8).
(Note: Use the axes provided in the exam booklet.)

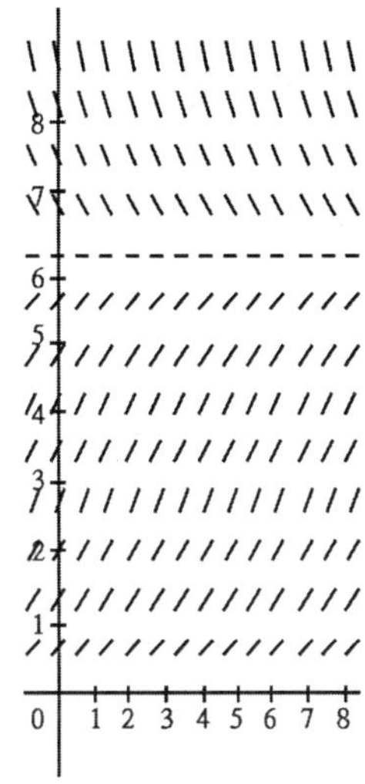

Figure 7.29

(2) Use Euler's method, starting at $t=0$ with two steps of equal size, to approximate $f(1)$

(3) Write the second-degree Taylor polynomial for f about $t=0$, and use it to approximate $f(1)$.

(4) What is the range of f for $t \geq 0$?

[EX 3] Suppose a flu-like virus is spreading through a population of 50,000 at a rate proportional both to the number of people already infected and to the number still uninfected. If 100 people were infected yesterday and 130 are infected today:

(1) write an expression for the number of people $N(t)$ infected after t days;

(2) determine how many will be infected a week from today.

CALCULUS

CHAPTER

08

Sequences and Series

8.1 Sequence

The sequence $\{a_n\}$ converges to the number L, If $\{a_n\}$ converges to L, we write $\lim_{n\to\infty} a_n = L$, or simply $a_n \to L$, and call L the limit of the sequence. If no such number L exists, say that $\{a_n\}$ diverges.

Let $\{a_n\}$ and $\{b_n\}$ be sequence of real numbers, and let A and B be real numbers. The following rules hold if $\lim_{n\to\infty} a_n = A$ and $\lim_{n\to\infty} b_n = B$.

1. *Sum Rule*: $\lim_{n\to\infty}(a_n + b_n) = A + B$
2. *Difference Rule* : $\lim_{n\to\infty}(a_n - b_n) = A - B$
3. Constant Multiple Rule : $\lim_{n\to\infty}(k \cdot b_n) = k \cdot B$ (any number k)
4. Product Rule: $\lim_{n\to\infty}(a_n \cdot b_n) = A \cdot B$
5. Quotient Rule : $\lim_{n\to\infty}\left(\frac{a_n}{b_n}\right) = \frac{A}{B}$ if $B \neq 0$

Example 1

Find the limit of each convergent sequence.

(1) $a_n = \frac{5}{n^2}$ (2) $a_n = \frac{4-7n^6}{n^6+3}$

Solution

(1) $a_n = \frac{5}{n^2}$

$$\lim_{n\to\infty}\frac{5}{n^2} = 5 \cdot \lim_{n\to\infty}\frac{1}{n} = 5 \cdot 0 \cdot 0 = 0$$

(2) $a_n = \frac{4-7n^6}{n^6+3}$

$$\lim_{n\to\infty}\frac{4-7n^6}{n^6+3} = \lim_{n\to\infty}\frac{(4/n^6)-7}{1+(3/n^6)} = \frac{0-7}{1+0} = -7$$

[EX 1] Find the limit of each convergent sequence.

(1) $a_n = 2 + (0.1)^n$

(2) $a_n = \dfrac{n^2 - 2n + 1}{n - 1}$

(3) $a_n = \left(\dfrac{n+1}{2n}\right)\left(1 - \dfrac{1}{n}\right)$

(4) $a_n = \sqrt{\dfrac{2n}{n+1}}$

(5) $a_n = \dfrac{\sin n}{n}$

(6) $a_n = \dfrac{\ln(n+1)}{\sqrt{n}}$

(7) $a_n = \left(\dfrac{3}{n}\right)^{\frac{1}{n}}$

(8) $a_n = \dfrac{\ln n}{n^{\frac{1}{n}}}$

(9) $a_n = \ln\left(1 + \dfrac{1}{n}\right)^n$

(10) $a_n = (3^n + 5^n)^{1/n}$

(11) $a_n = n \sin \dfrac{1}{n}$

(12) $a_n = \dfrac{\ln n}{n}$

(13) $a_n = \dfrac{4^{n+1} + 3^n}{4^n}$

(14) $a_n = \dfrac{2^n - 1}{3^n}$

[EX 2] Assume that each sequence convergent and find its limit.

(1) $a_1 = 1, \quad a_{n+1} = \dfrac{72}{1 + a_n}$

(2) $a_1 = -1, \quad a_{n+1} = \dfrac{a_n + 6}{a_n + 2}$

(3) $a_1 = -4, \quad a_{n+1} = \sqrt{8 + 2a_n}$

(4) $a_1 = 3, \quad a_{n+1} = 12 - \sqrt{a_n}$

8.2 Infinite series

Definition

If a_n is a sequence of real numbers, then and infinite series is an expression of the form

$$\sum_{k=1}^{\infty} a_k = a_1 + a_2 + a_3 + \cdots + a_n + \cdots$$

The elements in the sum are called *terms*; a_n is the nth or *general term* of the series. If there is a finite number S such that

$$\lim_{n\to\infty} \sum_{k=1}^{n} a_k = S$$

then we say that infinite series is *convergent*, or *converges to* S, or has the sum S and we write, in this case,

$$\sum_{k=1}^{\infty} a_k = S.$$

When there is no source of confusion, the infinite series (1) may be indicated simply by

$$\sum a_k \quad \text{or} \quad \sum a_n$$

Example 1

If a series converges, find its sum.

(1) $\sum_{n=1}^{\infty}\left(\frac{1}{n}-\frac{1}{n+1}\right)$ (2) $\sum_{n=1}^{\infty}(\ln\sqrt{n+1}-\ln\sqrt{n})$

Solution

(1) $\sum_{n=1}^{\infty}\left(\frac{1}{n}-\frac{1}{n+1}\right)$

$S_k=\left(1-\frac{1}{2}\right)+\left(\frac{1}{2}-\frac{1}{3}\right)+\left(\frac{1}{3}-\frac{1}{4}\right)+\cdots+\left(\frac{1}{k-1}-\frac{1}{k}\right)+\left(\frac{1}{k}-\frac{1}{k+1}\right)=1-\frac{1}{k+1}\Rightarrow\lim_{k\to\infty}S_k$

$=\lim_{k\to\infty}\left(1-\frac{1}{k+1}\right)=1$, series converges to 1.

(2) $\sum_{n=1}^{\infty}(\ln\sqrt{n+1}-\ln\sqrt{n})$

$S_k=(\ln\sqrt{2}-\ln\sqrt{1})+(\ln\sqrt{3}-\ln\sqrt{2})+(\ln\sqrt{4}-\ln\sqrt{3})+\cdots+(\ln\sqrt{k}-\ln\sqrt{k-1})+(\ln\sqrt{k+1}-\ln\sqrt{k})$

$=\ln\sqrt{k+1}-\ln\sqrt{1}=\ln\sqrt{k+1}\Rightarrow\lim_{k\to\infty}s_k=\lim_{k\to\infty}\ln\sqrt{k+1}=\infty$; series diverges

[EX 1] Find the sum of each series.

(1) $\sum_{n=1}^{\infty}\frac{4}{(4n-3)(4n+1)}$ (2) $\sum_{n=1}^{\infty}\frac{6}{(2n-1)(2n+1)}$

(3) $\sum_{n=1}^{\infty}\left(\frac{1}{\sqrt{n}}-\frac{1}{\sqrt{n+1}}\right)$ (4) $\sum_{n=1}^{\infty}\frac{2n+1}{n^2(n+1)^2}$

(5) $\sum_{n=1}^{\infty}\left(\frac{1}{\ln(n+2)}-\frac{1}{\ln(n+1)}\right)$ (6) $\sum_{n=1}^{\infty}(\tan^{-1}(n)-\tan^{-1}(n+1))$

Geometric Series

If $|r|<1$, the geometric series $a+ar+ar^2+\cdots+ar^{n-1}+\cdots$ converges to $a/(1-r)$:

$$\sum_{n=1}^{\infty} ar^{n-1}=\frac{a}{1-r},\quad |r|<1.$$

If $|r|\geq 1$, the series diverges.

Example 2

The geometric series with $a=1/9$ and $r=1/3$ is

Solution

$$\frac{1}{9}+\frac{1}{27}+\frac{1}{81}+\cdots=\sum_{n=1}^{\infty}\frac{1}{9}\left(\frac{1}{3}\right)^{n-1}=\frac{1/9}{1-(1/3)}=\frac{1}{6}.$$

[EX 2] Determine if the geometric series converges or diverges. If a series converges, find its sum.

(1) $\displaystyle\sum_{n=0}^{\infty}\left(\frac{1}{\sqrt{2}}\right)^n$

(2) $\displaystyle\sum_{n=1}^{\infty}(-1)^{n+1}\frac{3}{2^n}$

(3) $\displaystyle\sum_{n=0}^{\infty}\cos n\pi$

(4) $\displaystyle\sum_{n=0}^{\infty}e^{-2n}$

(5) $\displaystyle\sum_{n=1}^{\infty}\frac{2}{10^n}$

(6) $\displaystyle\sum_{n=0}^{\infty}\frac{2^n-1}{3^n}$

(7) $\displaystyle\sum_{n=1}^{\infty}\frac{2^n+3^n}{4^n}$

(8) $\displaystyle\sum_{n=0}^{\infty}\left(\frac{e}{\pi}\right)^n$

[EX 3] Find the values of x for which the given geometric series converges. Also, find the sum of the series for those values of x.

(1) $\sum_{n=0}^{\infty} 2^n x^n$

(2) $\sum_{n=0}^{\infty} (-1)^n (x+1)^n$

(3) $\sum_{n=0}^{\infty} \sin^n x$

(4) $\sum_{n=0}^{\infty} (\ln x)^n$

The nth–Term Test for Divergence

$\sum_{n=1}^{\infty} a_n$ diverges if $\lim_{n\to\infty} a_n$ fails to exist or is different from zero.

Example 3

The following are all examples of divergent series.

Solution

(1) $\sum_{n=1}^{\infty} n^2$ diverges because $n^2 \to \infty$.

(2) $\sum_{n=1}^{\infty} \frac{n+1}{n}$ diverges because $\frac{n+1}{n} \to 1$.

(3) $\sum_{n=1}^{\infty} (-1)^{n+1}$ diverges because $\lim_{n\to\infty}(-1)^{n+1}$ does not exist .

(4) $\sum_{n=1}^{\infty} \frac{-n}{2n+5}$ diverges because $\lim_{n\to\infty} \frac{-n}{2n+5} = -\frac{1}{2} \neq 0$.

[EX 4] Use the nth- Term Test for divergence to show that the series is divergent, or state that the test is inconclusive.

(1) $\sum_{n=1}^{\infty} \frac{n}{n+10}$

(2) $\sum_{n=1}^{\infty} \frac{n(n+1)}{(n+2)(n+3)}$

(3) $\sum_{n=0}^{\infty} \frac{1}{n+4}$

(4) $\sum_{n=1}^{\infty} \frac{n}{n^2+3}$

(5) $\sum_{n=1}^{\infty} \cos\frac{1}{n}$

(6) $\sum_{n=0}^{\infty} \frac{e^n}{e^n+n}$

(7) $\sum_{n=1}^{\infty} \ln\frac{1}{n}$

(8) $\sum_{n=0}^{\infty} \cos n\pi$

The Integral Test

If $f(x)$ is a continuous, positive, decreasing function and $f(n)=a_n$,

then $\sum a_n$ converges if and only if the improper integral $\int_1^{\infty} f(x)\,dx$ converges.

Example 4

Does $\sum \frac{n}{n^2+1}$ converge?

Solution

The associated improper integral is $\int_1^{\infty} \frac{x\,dx}{x^2+1}$ which equal $\lim_{b\to\infty} \frac{1}{2}\ln(x^2+1)\Big|_1^b = \infty$

The improper integral and the infinite series both diverge.

Example 5

Test the series $\sum \frac{n}{e^n}$ for convergence.

Solution

$$\int_1^{\infty} \frac{x}{e^x}dx = \lim_{b\to\infty}\int_1^b xe^{-x}dx = \lim_{b\to\infty} -e^{-x}(1+x)\Big|_1^b = -\lim_{b\to\infty}\left(\frac{1+b}{e^b} - \frac{2}{e}\right) = \frac{2}{e}$$

by an application of *L, Hopital's Rule*. Thus $\sum \frac{n}{e^n}$ converges.

The P–series Test

A p-series $\sum_{n=1}^{\infty} \frac{1}{n^p}$ converges if $p > 1$, but diverges if $p \le 1$.

This follows immediately from the Integral Test and the behavior of improper Integrals of the form $\int_1^{\infty} \frac{1}{x^n}dx$.

[EX 5] Use the Integral Test to determine if the series converges or diverges.

(1) $\sum_{n=1}^{\infty} \frac{1}{n^2}$

(2) $\sum_{n=1}^{\infty} \frac{1}{n^{0.2}}$

(3) $\sum_{n=2}^{\infty} \frac{1}{n^2+4}$

(4) $\sum_{n=1}^{\infty} \frac{1}{n+4}$

(5) $\sum_{n=1}^{\infty} e^{-2n}$

(6) $\sum_{n=2}^{\infty} \frac{1}{n(\ln n)^2}$

(7) $\sum_{n=1}^{\infty} \frac{n}{n^2+4}$

(8) $\sum_{n=2}^{\infty} \frac{\ln(n^2)}{n}$

The Comparision Test

Let $\sum a_n$ and $\sum d_n$ be series with nonnegative terms. Suppose that for some integer N

$$d_n \le a_n \le c_n \quad \text{for all} \quad n > N.$$

(a) If $\sum c_n$ converges, then $\sum a_n$ also converges.

(b) If $\sum d_n$ diverges, then $\sum a_n$ also diverges.

Example 6

We apply the comparison Test to several series.

Solution

(1) The series $\sum_{n=1}^{\infty} \frac{5}{5n-1}$ diverges because its nth term

$\frac{5}{5n-1} = \frac{1}{n-\frac{1}{5}} > \frac{1}{n}$ is greater than the nth term of the divergent harmonic series.

(2) The series $\sum_{n=0}^{\infty} \frac{1}{n!} = 1 + \frac{1}{1!} + \frac{1}{2!} + \frac{1}{3!} + \cdots$ converges because its terms are all positive and less than or equal to the corresponding terms of

$$1 + \sum_{n=0}^{\infty} \frac{1}{2^n} = 1 + 1 + \frac{1}{2} + \frac{1}{2^2} + \cdots$$

The geometric series on the left converges and we have

$$1 + \sum_{n=0}^{\infty} \frac{1}{2^n} = 1 + \frac{1}{1-(1/2)} = 3.$$

[EX 6] Use the Comparison Test to determine if each series converges or diverges.

(1) $\sum_{n=1}^{\infty} \frac{1}{n^2+30}$

(2) $\sum_{n=2}^{\infty} \frac{1}{\sqrt{n}-1}$

(3) $\sum_{n=1}^{\infty} \frac{\cos^2 n}{n^{3/2}}$

(4) $\sum_{n=1}^{\infty} \frac{n-1}{n^4+2}$

(5) $\sum_{n=2}^{\infty} \frac{n+2}{n^2-n}$

(6) $\sum_{n=1}^{\infty} \frac{1}{n3^n}$

The Limit Comparision Test

If $\lim_{n\to\infty} \frac{a_n}{b_n}$ is finite and nonzero, then $\sum a_n$, and $\sum b_n$ both converge or both diverge.

Example 7

Which of the following series converge, and which diverge?

(1) $\frac{3}{4}+\frac{5}{9}+\frac{7}{16}+\frac{9}{25}+\cdots=\sum_{n=1}^{\infty}\frac{2n+1}{(n+1)^2}=\sum_{n=1}^{\infty}\frac{2n+1}{n^2+2n+1}$

(2) $\frac{1}{1}+\frac{1}{3}+\frac{1}{7}+\frac{1}{15}+\cdots=\sum_{n=1}^{\infty}\frac{1}{2^n-1}$

(3) $\frac{1+2\ln 2}{9}+\frac{1+3\ln 3}{14}+\frac{1+4\ln 4}{21}+\cdots=\sum_{n=2}^{\infty}\frac{1+n\ln n}{n^2+5}$

Solution

(1) Let $a_n=(2n+1)/(n^2+2n+1)$. For large n, we expect a_n to behave like $2n/n^2=\frac{2}{n}$ since

the leading terms dominate for large n, so we let $b_n = 1/n$. Since

$$\sum_{n=1}^{\infty} b_n = \sum_{n=1}^{\infty} \frac{1}{n} \text{ diverges}$$

and

$$\lim_{n\to\infty} \frac{a_n}{b_n} = \lim_{n\to\infty} \frac{2n^2+n}{n^2+2n+1} = 2,$$

$\sum a_n$ diverges of the Limit Comparison Test.

(2) et $a_n = 1/(2^n - 1)$. For large n, we expect a_n to behave like $1/2^n$, so we let $b_n = 1/2^n$. Since

$$\sum_{n=1}^{\infty} b_n = \sum_{n=1}^{\infty} \frac{1}{2^n} \text{ converges}$$

and

$$\lim_{n\to\infty} \frac{a_n}{b_n} = \lim_{n\to\infty} \frac{2^n}{2^n - 1} = \lim_{n\to\infty} \frac{1}{1-(1/2^n)} = 1,$$

$\sum a_n$ converges of the Limit Comparison Test.

(3) Let $a_n = (1+n\ln n)/(n^2+5)$. For large n, we expect a_n to behave like $(n\ln n)/n^2 = (\ln n)/n$, which is greater that $1/n$ for $n \geq 3$, so we let $b_n = 1/n$. Since

$$\sum_{n=2}^{\infty} b_n = \sum_{n=2}^{\infty} \frac{1}{n} \text{ diverges}$$

and

$$\lim_{n\to\infty} \frac{a_n}{b_n} = \lim_{n\to\infty} \frac{n+n^2\ln n}{n^2+5} = \infty$$

$\sum a_n$ diverges of the Limit Comparison Test.

[EX 7] Use the Limit Comparison Test to determine if each series converges or diverges.

(1) $\displaystyle\sum_{n=1}^{\infty} \frac{n-2}{n^3 - n^2 + 3}$ (Hint : Limit comparison with $\displaystyle\sum_{n=1}^{\infty} (1/n^2)$)

(2) $\displaystyle\sum_{n=1}^{\infty} \sqrt{\frac{n+1}{n^2+2}}$ (Hint : Limit comparison with $\displaystyle\sum_{n=1}^{\infty} (1/\sqrt{n})$)

(3) $\sum_{n=2}^{\infty} \frac{n(n+1)}{(n^2+1)(n+1)}$

(4) $\sum_{n=1}^{\infty} \frac{2^n}{3+4^n}$

(5) $\sum_{n=2}^{\infty} \frac{1}{\ln n}$ (Hint : Limit comparison with $\sum_{n=2}^{\infty}(1/n)$)

(6) $\sum_{n=1}^{\infty} \ln\left(1+\frac{1}{n^2}\right)$ (Hint : Limit comparison with $\sum_{n=1}^{\infty}(1/n^2)$)

The Ratio Test

Let $\sum a_n$ be series with positive terms and suppose that

$$\lim_{n\to\infty} \frac{a_{n+1}}{a_n} = \rho.$$

Then (a) the series converges if $\rho < 1$, (b) the series diverges if $\rho > 1$ or ρ is infinite, (c) the test is inconclusive if $\rho = 1$.

Example 6

For $\sum \frac{n^n}{n!}$, we again use the Ratio Test :

Solution

$$\frac{a_{n+1}}{a_n} = \frac{(n+1)^{n+1}}{(n+1)!} \cdot \frac{n!}{n^n} = \frac{(n+1)^n}{n^n}$$

and

$\lim_{n\to\infty}\left(\frac{n+1}{n^n}\right) = \lim_{n\to\infty}\left(1+\frac{1}{n}\right)^n = e$. Since $e > 1$, $\sum \frac{n^n}{n!}$ diverges by the Ratio Test.

[EX 8] Use the Ratio Test to determine if each series converges or diverges

(1) $\sum_{n=1}^{\infty} \frac{2^n}{n!}$

(2) $\sum_{n=1}^{\infty} \frac{n+2}{3^n}$

(3) $\sum_{n=1}^{\infty} \frac{(n-1)!}{(n+1)^2}$

(4) $\sum_{n=1}^{\infty} \frac{2^{n+1}}{n3^{n-1}}$

(5) $\sum_{n=1}^{\infty} \frac{n^4}{4^n}$

(6) $\sum_{n=2}^{\infty} \frac{3^{n+2}}{\ln n}$

(7) $\sum_{n=1}^{\infty} \frac{n^2(n+2)!}{n!3^{2n}}$

The nth Root Test

Let $\lim_{n\to\infty} \sqrt[n]{a_n} = L$, if it exist, Then $\sum a_n$ converges if $L<1$ and diverges if $L>1$ If $L=1$ this test is inconclusive; try on of the other tests.

Note that the decision rule for this test is the same as that for Ratio Test.

Example 7

The series $\sum\left(\frac{n}{2n+1}\right)^n$ converges by the nth Root Test, since

$$\lim_{n\to\infty} \sqrt[n]{\left(\frac{n}{2n+1}\right)^n} = \lim_{n\to\infty} \frac{n}{2n+1} = \frac{1}{2}.$$

Solution

[EX 9] Use the Root test to determine if each series converges or diverges.

(1) $\sum_{n=1}^{\infty} \frac{7}{(2n+5)^n}$

(2) $\sum_{n=1}^{\infty} \frac{4^n}{(3n)^n}$

(3) $\sum_{n=1}^{\infty} \left(\frac{4n+3}{3n-5}\right)^n$

(4) $\sum_{n=1}^{\infty} \frac{8}{(3+(1/n))^{2n}}$

(5) $\sum_{n=1}^{\infty} \left(1-\frac{1}{n}\right)^{n^2}$

($Hint$: $\lim_{n\to\infty}(1+x/n)^n = e^x$)

(6) $\sum_{n=2}^{\infty} \frac{1}{n^{1+n}}$

The Alternating series test

An alternating series

$$\sum_{n-1}^{\infty} (-1)^{n+1} a_n = a_1 - a_2 + a_3 - \cdots \quad \text{converges if:}$$

(1) The a_n's are all positive.
(2) $a_{n+1} < a_n$ for all n, and
(3) $\lim_{n\to\infty} a_n = 0$

Example 8

The alternating harmonic series $\sum \frac{(-1)^{n+1}}{n}$ converges, since

(1) $\frac{1}{n+1} < \frac{1}{n}$ for all n and

(2) $\lim_{n\to\infty} \frac{1}{n} = 0$.

Solution

A series $\sum a_n$ converges absolutely (is absolutely convergent) if the corresponding series of absolute values, $\sum |a_n|$, *converges*.
A series that converges but does not converge absolutely converges conditionally.

Example 9

For $\sum_{n=1}^{\infty}(-1)^{n+1}\frac{1}{n^2}=1-\frac{1}{4}+\frac{1}{9}-\frac{1}{16}+\cdots$, the corresponding series of absolute values is the convergent series

Solution

$\sum_{n=1}^{\infty}\frac{1}{n^2}=1+\frac{1}{4}+\frac{1}{9}+\frac{1}{16}+\cdots$.
Then original series converges because it converges absolutely.

Example 10

Determine whether $\sum\frac{(-1)^{n+1}}{\sqrt[3]{n+1}}$ converges absolutely, converges conditionally, or diverges.

Solution

$\sum\frac{1}{\sqrt[3]{n+1}}$ is a p-series with $p=\frac{1}{3}$. so the nonnegative series diverges.
We see that

$$\frac{1}{\sqrt[3]{(n+1)+1}}<\frac{1}{\sqrt[3]{n+1}} \text{ and } \lim_{n\to\infty}\frac{1}{\sqrt[3]{n+1}}=0.$$

so the alternating series converges; hence $\sum\frac{(-1)^{n+1}}{\sqrt[3]{n+1}}$ is conditionally convergent.

[EX 10] Determine if the alternating series converges or diverges. Some of the series do not satisfy the conditions of the Alternating Series Test.

(1) $\sum_{n=1}^{\infty}(-1)^{n+1}\frac{1}{\sqrt{n}}$

(2) $\sum_{n=1}^{\infty}(-1)^{n+1}\frac{1}{n3^{n}}$

(3) $\sum_{n=1}^{\infty}(-1)^{n}\frac{n}{n^{2}+1}$

(4) $\sum_{n=1}^{\infty}(-1)^{n+1}\frac{2^{n}}{n^{2}}$

(5) $\sum_{n=1}^{\infty}(-1)^{n+1}\left(\frac{n}{10}\right)^{n}$

(6) $\sum_{n=1}^{\infty}(-1)^{n+1}\frac{\ln n}{n}$

[EX 11] Which of the series in Exercises converge absolutely, which converge, and which diverge? Give reasons for you answers.

(1) $\sum_{n-1}^{\infty}(-1)^{n+1}(0.1)^{n}$

(2) $\sum_{n-1}^{\infty}(-1)^{n}\frac{1}{\sqrt{n}}$

(3) $\sum_{n=1}^{\infty}(-1)^{n+1}\frac{n}{n^{3}+1}$

(4) $\sum_{n=1}^{\infty}(-1)^{n}\frac{1}{n+3}$

(5) $\sum_{n=1}^{\infty}(-1)^{n+1}\frac{3+n}{5+n}$

(6) $\sum_{n=1}^{\infty}(-1)^{n+1}(\sqrt[n]{10})$

Approximating the limit of an alternating series : error bound

If the alternating series $\sum_{n=1}^{\infty}(-1)^{n+1}a_n$ satisfies the three conditions then for $n \geq N$

$$s_n = a_1 - a_2 + \cdots + (-1)^{n+1}a_n$$

approximates the sum L of the series with an error whose absolute value is less than a_{n+1}, the absolute value of the first unused term. Furthermore, the sum L lies between any two successive partial sums s and s_{n+1},and the remainder, $L - s_n$, has the same sign as the first unused term.

Example 11

The series $\sum_{k=1}^{\infty}\frac{(-1)^{k+1}}{k}$ passes the Alternating Series Test; hence its sum differs from the sum

$$\left(1-\frac{1}{2}+\frac{1}{3}-\frac{1}{4}+\frac{1}{5}-\frac{1}{6}\right)$$

Solution

by less than $\frac{1}{7}$, which is the error bound.

Example 12

How many terms must be summed to approximate to three decimal places the value of

$$1-\frac{1}{4}+\frac{1}{9}-\frac{1}{16}+\cdots+\frac{(-1)^{n+1}}{n^2}+\cdots?$$

Solution

Since $\frac{1}{(n+1)^2} < \frac{1}{n^2}$ and $\lim_{n\to\infty}\frac{1}{n^2} = 0$ the series converges by the Alternating Series Test; therefore after summing a number of terms the remainder (error) will be less than the first omitted term.

We seek n such that $R_n = \dfrac{1}{(n+1)^2} < 0.001$. Thus n must satisfy $(n+1)^2 > 1000$. or $n > 31.623$.

Therefore 32 terms are needed for the desired accuracy.

[EX 12] Estimate the magnitude of the error involved in using the sum of the first four terms to approximate the sum of the entire series.

(1) $\sum_{n=1}^{\infty}(-1)^{n+1}\dfrac{1}{n}$

(2) $\sum_{n=1}^{\infty}(-1)^{n+1}\dfrac{1}{10^n}$

(3) $\sum_{n=1}^{\infty}(-1)^{n+1}\dfrac{(0.01)^n}{n}$

(4) $\dfrac{1}{1+t} = \sum_{n=0}^{\infty}(-1)^n t^n, \quad 0 < t < 1$

[EX 13] Determine how many terms should be sued to estimate the sum of the entire series with an error of less than 0.001.

(1) $\sum_{n=1}^{\infty}(-1)^{n}\dfrac{1}{n^2+3}$

(2) $\sum_{n=1}^{\infty}(-1)^{n+1}\dfrac{n}{n^2+1}$

(3) $\sum_{n=1}^{\infty}(-1)^{n+1}\dfrac{1}{(n+3\sqrt{n})^3}$

8.3 Power Series

A power series about $x=0$ is a series of the form,

$$\sum_{n=0}^{\infty} c_n x^n = c_0 + c_1 x + c_2 x^2 + \cdots + c_n x^n + \cdots. \quad (1)$$

A power series about $x=a$ is a series of the form,

$$\sum_{n=0}^{\infty} c_n (x-a)^n = c_0 + c_1 (x-a) + c_2 (x-a)^2 + \cdots + c_n (x-a)^n + \cdots. \quad (2)$$

in which the center a and the coefficients $c_0, c_1, c_2, \cdots, c_n \cdots$ are constants

Example 1

Taking all the coefficients to be 1 in Equation (1) gives the geometric power series.

$$\sum_{n=0}^{\infty} x^n = 1 + x + x^2 + \cdots + x^n + \cdots$$

Solution

This is the geometric series with first term 1 and ratio x. It converges to $1/(1-x)$ for $|x|<1$.

We express this fact by writing

$$\frac{1}{1-x} = 1 + x + x^2 + \cdots + x^n + \cdots, \qquad -1 < x < 1. \quad (3)$$

Example 2

The power series

$$1 - \frac{1}{2}(x-2) + \frac{1}{4}(x-2)^2 + \cdots + \left(-\frac{1}{2}\right)^n (x-2)^n + \cdots \quad (4)$$

Solution

matches Equation (2) with $a=2,\ c_0=1, c_1=-1/2, c_2=1/4, \cdots, c_n=(-1/2)^n$. This is a geometric series with first term 1 and ratio $r=-\dfrac{x-2}{2}$. The series converges for $\left|\dfrac{x-2}{2}\right|<1$ or $0<x<4$.

The sum is

$$\frac{1}{1-r}=\frac{1}{1+\dfrac{x-2}{2}}=\frac{2}{x},$$

so

$$\frac{2}{x}=1-\frac{(x-2)}{2}+\frac{(x-2)^2}{4}-\cdots+\left(-\frac{1}{2}\right)^n(x-2)^n+\cdots,\quad 0<x<4.$$

Radius and interval of convergence

If power series (1) converges when $|x|<r$ and diverges when $|x|>r$, then r is called the radius of convergence, Similarly, r is the radius of convergence of power series (2) if (2) converges when $|x-a|<r$ and diverges when $|x-a|>r$.

Example 3

For what x does $\displaystyle\sum_{n=1}^{\infty}\frac{(-1)^{n-1}x^{n-1}}{n+1}$ converge?

Solution

$$\lim_{n\to\infty}\left|\frac{a_{n+1}}{a_n}\right|=\lim_{n\to\infty}\left|\frac{x^n}{n+2}\cdot\frac{n+1}{x^{n-1}}\right|=\lim_{n\to\infty}|x|=|x|<1.$$

The radius of convergence is 1. when $x=1$, we have $\dfrac{1}{2}-\dfrac{1}{3}+\dfrac{1}{4}-\dfrac{1}{5}+\cdots$,

and alternating convergent series; when $x=-1$, the series is $\dfrac{1}{2}+\dfrac{1}{3}+\dfrac{1}{4}+\cdots$,

which diverges. Thus, the series converges if $-1<x\le 1$.

Example 4

Find all x for which the following series converges:

$$1+\frac{x-2}{2^1}+\frac{(x-2)^2}{2^2}+\cdots+\frac{(x-2)^{n-1}}{2^{n-1}}+\cdots \qquad (5)$$

Solution

$$\lim_{n\to\infty}\left|\frac{a_{n+1}}{a_n}\right|=\lim_{n\to\infty}\left|\frac{(x-2)^n}{2^n}\cdot\frac{2^{n-1}}{(x-2)^{n-1}}\right|=\lim_{n\to\infty}\frac{|x-2|}{2}=\frac{|x-2|}{2},$$

which is less than 1 if $|x-2|<2$, that is, if $0<x<4$ series converges on this interval and diverges
if $|x-2|>2$, that is, if $x<0$ or $x>4$. when $x=0$,(5) is $1-1+1-1+\cdots$ and diverges. when $x=4$,(5) is $1+1+1\cdots$ and diverges. Thus, the interval of convergence is $0<x<4$

[EX 1] For what value of x do the following power series converge?

(1) $\sum_{n=1}^{\infty}(-1)^{n-1}\frac{x^n}{n}=x-\frac{x^2}{2}+\frac{x^3}{3}-\cdots$

(2) $\sum_{n=0}^{\infty}n!(x-3)^n$

(3) $\sum_{n=1}^{\infty}(-1)^{n-1}\frac{x^{2n-1}}{2n-1}=x-\frac{x^3}{3}+\frac{x^5}{5}-\cdots$

(4) $x+\frac{x^2}{2}+\frac{x^3}{3}+\cdots+\frac{x^n}{n}+\cdots$

(5) $\sum_{n=0}^{\infty}\frac{x^n}{n!}=1+x+\frac{x^2}{2!}+\frac{x^3}{3!}+\cdots$

(6) $(x+1)-\frac{(x+1)^2}{2!}+\frac{(x+1)^3}{3!}-\frac{(x+1)^4}{4!}+\cdots$

(7) $(x-2)+\frac{(x-2)^2}{4}+\frac{(x-2)^3}{9}+\frac{(x-2)^4}{16}+\cdots$

(8) $\sum_{n=0}^{\infty}n!x^n=1+x+2!x^2+3!x^3+\cdots$

Taylor and Maclaurin series

Let f be function with derivatives of all orders throughout some interval containing a as an interior point. Then the Taylor series generated by f at $x=a$ is

$$\sum_{k=0}^{\infty}\frac{f^{(k)}(a)}{k!}(x-a)^k = f(a)+f'(a)(x-a)+\frac{f''(a)}{2!}(x-a)^2 + \cdots + \frac{f^{(n)}(a)}{n!}(x-a)^n+\cdots,$$

The Maclaurin series generated by f is

$$\sum_{k=0}^{\infty}\frac{f^{(k)}(0)}{k!}x^k = f(0)+f'(0)x+\frac{f''(0)}{2!}x^2+\cdots+\frac{f^{(n)}(0)}{n!}x^n+\cdots$$

the Taylor series generated by f at $x=0$.

Example 5

Find the Maclaurin expansion for $f(x)=\sin x$

Solution

$$\begin{aligned}
f(x) &= \sin x; & f(0) &= 0;\\
f'(x) &= \cos x; & f'(0) &= 1;\\
f''(x) &= -\sin x; & f''(0) &= 0;\\
f^{(3)}(x) &= -\cos x; & f^{(3)}(0) &= -1;\\
f^{(4)}(x) &= \sin x & f^{(4)}(0) &= 0;\\
&\vdots & &\vdots
\end{aligned}$$

Thus

$$\sin x = x-\frac{x^3}{3!}+\frac{x^5}{5!}-\cdots+(-1)^{n-1}\frac{x^{2n-1}}{(2n-1)!}+\cdots.$$

Example 6

Find the Taylor series for the function $f(x) = \ln x$ about $x = 1$.

Solution

$$f(x) = \ln x; \qquad f(1) = \ln 1 = 0;$$

$$f'(x) = \frac{1}{x}; \qquad f'(1) = 1;$$

$$f''(x) = -\frac{1}{x^2}; \qquad f''(1) = -1$$

$$f^{(3)}(x) = \frac{2}{x^3}; \qquad f^{(3)}(1) = 2;$$

$$f^{(4)}(x) = \frac{-3!}{x^4}; \qquad f^{(4)}(1) = -3!;$$

$$\vdots \qquad \vdots$$

$$f^{(n)}(x) = \frac{(-1)^{n-1}(n-1)!}{x^n}; \qquad f^{(n)}(1) = (-1)^{n-1}(n-1)!$$

$$\ln x = (x-1) - \frac{(x-1)^2}{2} + \frac{(x-1)^3}{3} - \frac{(x-1)^4}{4} + \cdots + \frac{(-1)^{n-1}(x-1)^n}{n} + \cdots$$

Taylor Polynomial

Let f be a function with derivatives of order k for $k = 1, 2, \cdots, N$ in some interval containing a as an interior point. Then for any integer n from 0 through N, the Taylor polynomial of order n generated by f at $x = a$ is the polynomial

$$P_n(x) = f(a) + f'(a)(x-a) + \frac{f''(a)}{2!}(x-a)^2 + \cdots + \frac{f^{(k)}(a)}{k!}(x-a)^k + \cdots + \frac{f^{(n)}(a)}{n!}(x-a)^n.$$

Example 7

Find the Taylor series and the Taylor polynomials generated by $f(x)=e^x$ at $x=0$

Solution

Since $f^{(n)}(x)=e^x$ and $f^{(n)}(0)=1$ for every $n=0,1,2,\cdots$, the Taylor series generated by f at $x=0$ is

$$f(0)+f'(0)x+\frac{f''(0)}{2!}x^2+\cdots+\frac{f^{(n)}(0)}{n!}x^n+\cdots=1+x+\frac{x^2}{2!}+\cdots+\frac{x}{n!}+\cdots=\sum_{k=0}^{\infty}\frac{x^k}{k!}$$

The Taylor polynomial of order n at $x=0$ is $P_n(x)=1+x+\frac{x^2}{2!}+\cdots+\frac{x^n}{n!}$.

[EX 2] Find the Taylor polynomials of orders 3 generated by f at a.

(1) $f(x)=e^{2x},\ a=0$

(2) $f(x)=\ln x,\ a=1$

(3) $f(x)=1/x,\ a=2$

(4) $f(x)=\sin x,\ a=\pi/4$

(5) $f(x)=\sqrt{x},\ a=4$

(6) $f(x)=\tan x,\ a=\pi/4$

[EX 3] Find the Maclaurin series for the functions.

(1) e^{-x}

(2) xe^x

(3) $\frac{1}{1+x}$

(4) x^4-2x^3-5x+4

[EX 4] Find the Taylor series generated by f at $x=a$.

(1) $f(x)=x^3-2x+4,\ a=2$

(2) $f(x)=1/x^2,\ a=1$

(3) $f(x)=e^x,\ a=2$

(4) $f(x)=2^x,\ a=1$

[EX 5] A function f has derivatives of all orders at $x=0$, Let $P_n(x)$ denote the n th-degree Taylor polynomial for f about $x=0$.

(1) It is known that $f(0)=-4$ and that $P_1\left(\frac{1}{2}\right)=-3$. Show that $f'(0)=2$.

(2) It is known that $f''(0)=-\frac{2}{3}$ and $f'''(0)=\frac{1}{3}$. Find $P_3(x)$.

(3) The function h has first derivative given by $h'(x)=f(2x)$. It is known that $h(0)=7$. Find the third-degree Taylor polynomial for h about $x=0$.

[EX 6] The function g has derivatives of all orders, and the Maclaurin series for g is

$$\sum_{n=0}^{\infty}(-1)^n\frac{x^{2n+1}}{2n+3}=\frac{x}{3}-\frac{x^3}{5}+\frac{x^5}{7}-\cdots$$

(1) Using the ratio test, determine the interval of convergence of the Maclaurin series for g.

(2) The Maclaurin series for g evaluated at $x=\frac{1}{2}$ is an alternating series whose terms decrease in absolute value to 0. The approximation for $g\left(\frac{1}{2}\right)$ using the first two nonzero terms of this series is $\frac{17}{120}$. Show that this approximation differs from $g\left(\frac{1}{2}\right)$ by less than $\frac{1}{200}$.

(3) Write the first three nonzero terms and the general term of the Maclaurin series for $g'(x)$.

[EX 7] The function f is defined by the power series

$$f(x)=1+(x+1)+(x+1)^2+\cdots+(x+1)^n+\cdots=\sum_{n=0}^{\infty}(x+1)^n$$

for all real numbers x for which the series converges.

(1) Find the interval of convergence of the power series for f. Justify your answer.

(2) The power series above is the Taylor series for f about $x=-1$. find the sum of the series for f.

(3) Let g be the function defined by $g(x)=\int_{-1}^{x} f(t)dt$. Find the value of $g\left(-\frac{1}{2}\right)$, if it exist, or explain why $g\left(-\frac{1}{2}\right)$ cannot be determined.

(4) Let h be the function defined by $h(x)=f(x^2-1)$. Find the first three nonzero terms and the general term of the Taylor series for h about $x=0$, and find the value of $h\left(\frac{1}{2}\right)$.

Taylor's Formular with Remainder ; Lagrange Error Bound.

If a function f and its first $(n+1)$ derivatives are continuous on the interval $|x-a|<r$, then for each x in this interval

$$f(x)=f(a)+f'(a)(x-a)+\frac{f''(a)}{2!}(x-a)^2+\cdots+\frac{f^{(n)}(a)}{n!}(x-a)^n+R_n(x),$$

where

$$R_n(x)=\frac{f^{(n+1)}(c)(x-a)^{n+1}}{(n+1)!}$$

and c is some number between a and x. $R_n(x)$ is called the Lagrange remainder.

$$R_n(x)<\max\left|\frac{f^{(n+1)}(c)(x-a)^{n+1}}{(n+1)!}\right|$$

the Lagrange error bound.

Example 8

Estimate the error in using the Maclaurin series generated by e^x to approximate the value of e.

Solution

we known that $f(x)=e^x$ generates the Maclaurin series

$$e^x=1+x+\frac{x^2}{2!}+\cdots+\frac{x^n}{n!}+\cdots$$

The Lagrange error bound $R_n(x)<\max\left|\frac{e^c(x)^{n+1}}{(n+1)!}\right|(0<c<x)$.

To estimate e, We use $x=1$. For $0<c<1$, the maximum value of e^c is e. Thus:

$$e\approx 1+1+\frac{1}{2!}+\frac{1}{3!}+\cdots+\frac{1}{n!}$$ with error less than $\frac{e}{(n+1)!}$.

[EX 8] Find the third-degree Maclaurin polynomial for $f(x)=\cos\left(x+\frac{\pi}{4}\right)$, and determine the upper bound on the error in estimating $f(0.1)$.

[EX 9] Let $f(x)=\sin(x^2)+\cos x$. The graph of $y=|f^{(5)}(x)|$ is shown above.

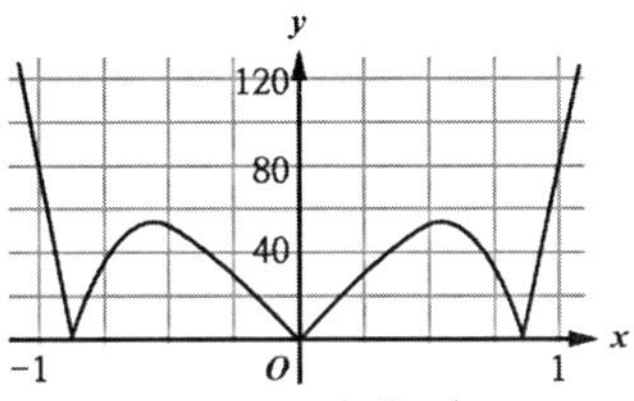

Figure 8.1

(1) Write the first four nonzero terms of the Taylor series for $\sin x$ about $x=0$, and write the first four nonzero terms of the Taylor series for $\sin(x^2)$ about $x=0$.

(2) Write the first four nonzero terms of the Taylor series for $\cos x$ about $x=0$. Use this series and the series for $\sin(x^2)$, found in part(a), to write the first four nonzero terms of the Taylor series for f about $x=0$.

(3) Find the value of $f^{(6)}(0)$.

(4) Let $P_4(x)$ be the fourth-degree Taylor polynomial for f about $x=0$. Using information from the graph of $y=|f^{(5)}(x)|$ shown above, show that $\left|P_4\left(\frac{1}{4}\right)-f\left(\frac{1}{4}\right)\right|<\frac{1}{3000}$.

[EX 10] Find the Maclaurin series for $\ln(1+x)$ and the associated Lagrange error bound.

Computations with Power Series.

The power series expansions of functions may be as any other functions for values of x that lie within their intervals of convergence. They may be added, subtracted multiplied, divided (with division by zero to be avoided), differentiated or integrated. These properties provide a valuable approach for many otherwise difficult computations, Indeed, power series are often very useful for approximating value of functions evaluating indeterminate forms of limits, and estimating definite integrals.

Example 9

Using known series, find the first few terms of the Taylor series for the given function using power series operations.

(1) $\frac{1}{3}(2x + x\cos x)$ (2) $e^x \cos x$

Solution

(1) $$\frac{1}{3}(2x+x\cos x)=\frac{2}{3}x+\frac{1}{3}x\left(1-\frac{x^2}{2!}+\frac{x^4}{4!}-\cdots+(-1)^k\frac{x^{2k}}{(2k)!}+\cdots\right)$$
$$=\frac{2}{3}x+\frac{1}{3}x-\frac{x^3}{3!}+\frac{x^5}{3\cdot 4!}-\cdots=x-\frac{x^3}{6}+\frac{x^5}{72}-\cdots$$

(2) $$e^x\cos x=\left(1+x+\frac{x^2}{2!}+\frac{x^3}{3!}+\frac{x^4}{4!}+\cdots\right)\cdot\left(1-\frac{x^2}{2!}+\frac{x^4}{4!}-\cdots\right)$$
$$=\left(1+x+\frac{x^2}{2!}+\frac{x^3}{3!}+\frac{x^4}{4!}+\cdots\right)-\left(\frac{x^2}{2!}+\frac{x^3}{2!}+\frac{x^4}{2!2!}+\frac{x^5}{2!3!}+\cdots\right)+\left(\frac{x^4}{4!}+\frac{x^5}{4!}+\frac{x^6}{2!4!}+\cdots\right)+\cdots$$
$$=1+x-\frac{x^3}{3}-\frac{x^4}{6}+\cdots$$

[EX 11] Find substitution to find Taylor series at $x=0$ of the functions.

(1) e^{-5x} (2) $\sin\left(\frac{\pi x}{2}\right)$

(3) $\ln(1+x^2)$ (4) $\frac{1}{2-x}$

[EX 12] Use power series operations to find the Taylor series at $x=0$ for the functions in Exercises (1)-(4).

(1) xe^x

(2) $x\cos\pi x$

(3) $x\ln(1+2x)$

(4) $x\tan^{-1}x^2$

Example 10

Evaluate $\lim_{x\to 1}\frac{\ln x}{x-1}$.

Solution

We represent $\ln x$ as a Taylor series in powers of $x-1$. This can be accomplished by calculating the Taylor series generated by $\ln x$ at $x=1$ directly of by replacing x by $x-1$ in the series for $\ln(1+x)$ in Example 6. Either way, we obtain

$$\ln x=(x-1)-\frac{1}{2}(x-1)^2+\frac{(x-1)^3}{3}-\cdots+\frac{(-1)^{n-1}(x-1)^n}{n}+\cdots$$

from which we find that

$$\lim_{x\to 1}\frac{\ln x}{x-1}=\lim_{x\to 1}\left(1-\frac{1}{2}(x-1)+\cdots\right)=1.$$

Example 11

Express $\int \sin x^2\,dx$ as a power series.

Solution

From the series from $\sin x$ we substitute x^2 for x to obtain

$$\sin x^2=x^2-\frac{x^6}{3!}+\frac{x^{10}}{5!}-\frac{x^{14}}{7!}+\frac{x^{18}}{9!}-\cdots.$$

Therefore,

$$\int \sin x^2 dx= C+\frac{x^3}{3}-\frac{x^7}{7\cdot 3!}+\frac{x^{11}}{11\cdot 5!}-\frac{x^{15}}{15\cdot 7!}+\frac{x^{19}}{19\cdot 9!}-\cdots.$$

[EX 13] Use series to estimate the integrals values with an error of magnitude less than 10^{-3}. (The answer section gives the integrals values rounded to five decimal places.)

(1) $\int_0^{0.2} \sin x^2 dx$

(2) $\int_0^{0.2} \frac{e^{-x}-1}{x} dx$

[EX 14] Use series to approximate the values of the integrals in Exercises (1-3) with an error of magnitude less than 10^{-8}.

(1) $\int_0^{0.1} \frac{\sin x}{x} dx$

(2) $\int_0^{0.1} e^{-x^2} dx$

(3) $\int_0^{1} \frac{1-\cos x}{x^2} dx$

[EX 15] Use series to evaluate the limits

(1) $\lim_{x \to 0} \frac{e^x-(1+x)}{x^2}$

(2) $\lim_{t \to 0} \frac{1-\cos t-(t^2/2)}{t^4}$

(3) $\lim_{y \to 0} \frac{y-\tan^{-1} y}{y^3}$

(4) $\lim_{x \to 0} \frac{\ln(1+x^2)}{1-\cos x}$

[EX 16] Let $f(x)=\ln(1+x^3)$

(1) The Maclaurin series for $\ln(1+x)$ is $x-\frac{x^2}{2}+\frac{x^3}{3}-\frac{x^4}{4}+\cdots+(-1)^{n+1}\cdot\frac{x^n}{n}+\cdots$. Use the series to write the first four nonzero terms and the general term of the Maclaurin series for f.

(2) The radius of convergence of the Maclaurin series for f is 1. Determine the interval of convergence. Show the work that leads to your answer.

(3) Write the first four nonzero terms of the Maclaurin series for $f'(t^2)$.
If $g(x)=\int_0^x f'(t^2)dt$, use the first two nonzero terms of the Maclaurin series for g to approximate $g(1)$.

(4) The Maclaurin series for g , evaluated at $x=1$, is a convergent alternating series with individual terms that decrease in absolute value to 0. Show that your approximation in *part*(3) must differ from $g(1)$ by less than $\frac{1}{5}$.

[EX 17] Let f be the function given by $f(x)=e^{-x^2}$

(1) Write the first four nonzero terms and the general term of the Taylor series for f about $x=0$.

(2) Use your answer to part (1) to find $\lim_{x\to 0}\frac{1-x^2-f(x)}{x^4}$.

(3) Write the first four nonzero terms of the Taylor series for $\int_0^x e^{-t^2} dt$ about $x=0$. Use the first two terms of your answer to estimate $\int_0^{1/2} e^{-t^2} dt$.

(4) Explain why the estimate found in part (3) differs from the actual value of $\int_0^{1/2} e^{-t^2} dt$ by less than $\frac{1}{200}$.

[EX 18] Let f be the function given by $f(x)=\frac{2x}{1+x^2}$.

(1) Write the first four nonzero terms and the general term of the Taylor series for f about $x=0$.

(2) Does the series found in part (1), when evaluated at $x=1$, converge to $f(1)$? Explain why or why not.

(3) The derivative of $\ln(1+x^2)$ is $\frac{2x}{1+x^2}$. Write the first four nonzero terms of the Taylor series for $\ln(1+x^2)$ about $x=0$.

(4) Use the series found in part (3) to find a rational number A such that $\left|A-\ln\left(\frac{5}{4}\right)\right|<\frac{1}{100}$. Justify your answer.

Frequently used Taylor series

$$\frac{1}{1-x}=1+x+x^2+\cdots+x^n+\cdots=\sum_{n=0}^{\infty}x^n,\quad |x|<1$$

$$\frac{1}{1+x}=1-x+x^2-\cdots+(-x)^n+\cdots=\sum_{n=0}^{\infty}(-1)^n x^n,\quad |x|<1$$

$$e^x=1+x+\frac{x^2}{2!}+\cdots+\frac{x^n}{n!}+\cdots=\sum_{n=0}^{\infty}\frac{x^n}{n!},\quad |x|<\infty$$

$$\sin x=x-\frac{x^3}{3!}+\frac{x^5}{5!}-\cdots+(-1)^n\frac{x^{2n+1}}{(2n+1)!}+\cdots=\sum_{n=0}^{\infty}\frac{(-1)^n x^{2n+1}}{(2n+1)!},\quad |x|\le\infty$$

$$\cos x=1-\frac{x^2}{2!}+\frac{x^4}{4!}-\cdots+(-1)^n\frac{x^{2n}}{(2n)!}+\cdots=\sum_{n=0}^{\infty}\frac{(-1)^n x^{2n}}{(2n)!},\quad |x|\le\infty$$

$$\ln(1+x)=x-x^2/2+x^3/3-\cdots+(-1)^{n-1}x^n/n+\cdots=\sum_{n=0}^{\infty}(-1)^{n+1}x^n/n,\quad -1<x\le 1$$

$$\tan^{-1}x=x-\frac{x^3}{3}+\frac{x^5}{5}-\cdots+(-1)^n\frac{x^{2n+1}}{(2n+1)}+\cdots=\sum_{n=0}^{\infty}\frac{(-1)^n x^{2n+1}}{(2n+1)},\quad |x|\le 1$$

CALCULUS

CHAPTER 09

Parametric Equations and Polar Coordinates

9.1 Parametric Equations

If x and y are given as functions

$$x = f(t), \quad y = g(t)$$

over an interval I of t-values, then the set of points $(x,y) = (f(t), g(t))$ defined by these equations is a parametric curve. The equations are parametric equations for the curve.

Example 1

Identify geometrically the curve defined by the parametric equations $x = t^2, \quad y = t+1$ $-\infty < x < \infty$ by eliminating the parameter t and obtaining an algebraic equation in x and y.

Solution

We solve the equation $y=t+1$ for the parameter t and substitute the result into the parametric equation for$x$.
This procedure gives $t=y-1$ and $x=t^2=(y-1)^2=y^2-2y+1$.

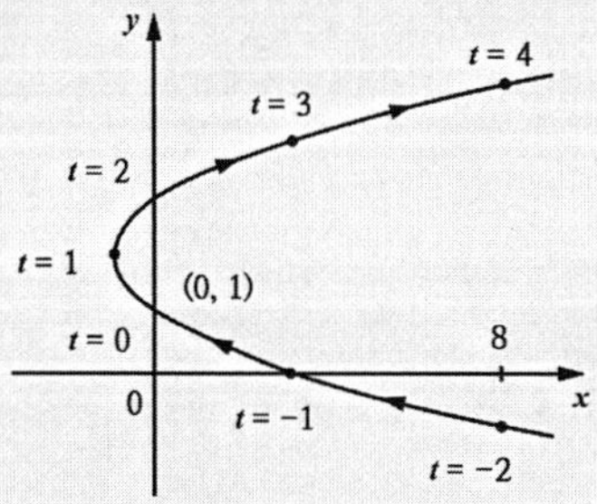

Figure 9.1

t	x	y
−2	8	−1
−1	3	0
0	0	1
1	−1	2
2	0	3
3	3	4
4	8	5

[EX 1] Identify the particle's path by finding a Cartesian equation for it. Graph the Cartesian equation.

(1) $x = -\sqrt{t},\ y = t,\ t \geq 0$

(2) $x = 3 - 3t,\ y = 2t,\ 0 \leq t \leq \pi$

(3) $x = \cos 2t,\ y = \sin 2t,\ 0 \leq t \leq \pi$

(4) $x = 4\cos t,\ y = 2\sin t,\ 0 \leq t \leq 2\pi$

(5) $x = \sin t,\ y = \cos 2t,\ -\frac{\pi}{2} \leq t \leq \frac{\pi}{2}$

(6) $x = 1 + \sin t,\ y = \cos t - 2,\ 0 \leq t \leq \pi$

(7) $x = t,\ y = \sqrt{1 - t^2},\ -1 \leq t \leq 0$

(8) $x = \sec^2 t - 1,\ y = \tan t,\ -\pi/2 < t < \pi/2$

parametric Formula for $\frac{dy}{dx}$, $\frac{d^2y}{dx^2}$

$$\frac{dy}{dx} = \frac{dy/dt}{dx/dt},\quad \frac{d^2y}{dx^2} = \frac{dy'/dt}{dx/dt}$$

Example 2

Find the slope of the tangent to the cycloid $\begin{cases} x = \theta - \sin\theta \\ y = 1 - \cos\theta \end{cases}$ at the point $\theta = \frac{2}{3}\pi$.

Solution

We have $\frac{dx}{d\theta} = 1 - \cos\theta$ and $\frac{dy}{d\theta} = \sin\theta$.

Therefore, the derivative $\frac{dy}{dx}$ is given by $\frac{dy}{dx} = \frac{\sin\theta}{1 - \cos\theta}$.

When $\theta = \frac{2}{3}\pi$, $\frac{dy}{dx} = \frac{\sqrt{3}/2}{3/2} = \frac{\sqrt{3}}{3} \approx 0.57735027$.

Example 3

For the cycloid in (example 2) determine the concavity on the interval $(0, 2\pi)$.

Solution

$\dfrac{dx}{d\theta} = 1 - \cos\theta$ and $\dfrac{dy}{d\theta} = \sin\theta$, we have $\dfrac{dy}{dx} = \dfrac{\sin\theta}{1 - \cos\theta}$.

Thus, $\dfrac{d\left(\dfrac{dy}{dx}\right)}{d\theta} = \dfrac{(1-\cos\theta)(\cos\theta) - (\sin\theta)(\sin\theta)}{(1-\cos\theta)^2} = \dfrac{-1}{(1-\cos\theta)}$

and $\dfrac{d^2y}{dx^2} = \dfrac{\dfrac{d\left(\dfrac{dy}{dx}\right)}{d\theta}}{\dfrac{dx}{d\theta}} = \dfrac{-\dfrac{1}{1-\cos\theta}}{1-\cos\theta} = -\dfrac{1}{(1-\cos\theta)^2}$

∴ concave down

[EX 2] Find an equation for the line tangent to the curve at the point defined by the given value of t. Also, find the value of d^2y/dx^2 at this point.

(1) $x = 2\cos t,\ y = 2\sin t,\ t = \pi/4$

(2) $x = 4\sin t,\ y = 2\cos t,\ t = \pi/4$

(3) $x = t,\ y = \sqrt{t},\ t = 1/4$

(4) $x = \sec t,\ y = \tan t,\ t = \pi/6$

(5) $x = 2t^2 + 3,\ y = t^4,\ t = -1$

(6) $x = 1/t,\ y = -2 + \ln t,\ t = 1$

(7) $x = \cos t,\ y = 1 + \sin t,\ t = \pi/2$

(8) $x = t + e^t,\ y = 1 - e^t,\ t = 0$

Region Bounded by a parametric Curve

If x and y are given parametrically, say by $x = f(\theta)$, $y = g(\theta)$, then to evaluate $\int_a^b y\,dx$, we express y, dx, and the limits a and b in terms of θ and $d\theta$, then integrate. Remember that we define dx to be $x'(\theta)d\theta$, or $f'(\theta)d\theta$.

Example 4

Find the area enclosed by the ellipse $x = 2\cos\theta$, $y = 3\sin\theta$

Solution

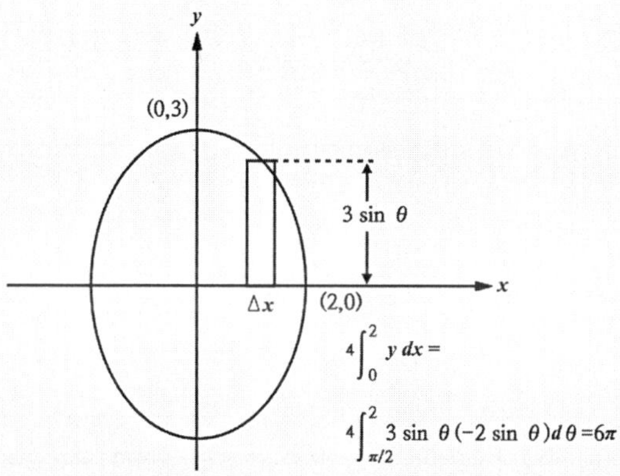

Figure 9.2

[EX 3] Find the area enclosed by one arch of the cycloid with parametric equations
$x = \theta - \sin\theta$, $y = 1 - \cos\theta$

[EX 4] Find the area enclosed by the astroid (Figure)

$x = \cos^3 t.\ y = \sin^3 t,\ \ 0 \le t \le 2\pi.$

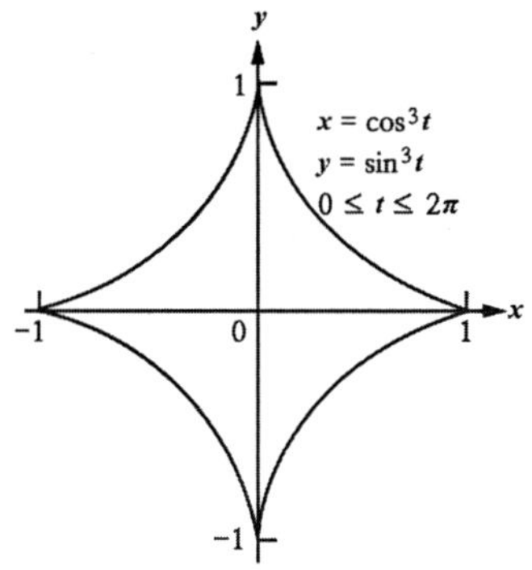

Figure 9.3

The length L of an arc defined by parametric equations $x = f(t), y = g(t)$ on the interval $[a,b]$

$$L = \int_a^b \sqrt{[f'(t)]^2 + [g'(t)]^2}\, dt$$

Example 5

Find the length of the curve C defined for $0 \le t \le \pi$ by $\begin{cases} x = e^t \cos t \\ y = e^t \sin t \end{cases}$

Solution

We have $\dfrac{dx}{dt} = e^t \cos t - e^t \sin t = e^t(\cos t - \sin t)$

$\dfrac{dy}{dt} = e^t(\sin t) + e^t \cos t = e^t(\cos t + \sin t)$

Then $\left(\dfrac{dx}{dt}\right)^2 + \left(\dfrac{dy}{dt}\right)^2 = e^{2t}(\cos t - \sin t)^2 + e^{2t}(\cos t + \sin t)^2$

$= e^{2t}(\cos^2 t - 2\cos t \sin t + \sin^2 t + \cos^2 t + 2\cos t \sin t + \sin^2 t)$

$= 2e^{2t}.$

Thus, we obtain

$$\int_0^\pi \sqrt{[f'(t)]^2 + [g'(t)]^2}\, dt = \int_0^\pi \sqrt{2e^{2t}}\, dt = \int_0^\pi \sqrt{2}\, e^t dt = \sqrt{2}\left[e^t\right]_0^\pi = \sqrt{2}(e^\pi - 1).$$

[EX 5] Find the length of the curves

(1) $x = t^2$
$y = t^3 \quad 0 \le t \le 1$

(2) $x = r\cos t,\ y = r\sin t,\ 0 \le t \le 2\pi$

(3) $x = e^{-t}\cos t,\ y = e^{-t}\sin t,\ 0 \le t \le \pi$

(4) $x = \cos^3 t,\ y = \sin^3 t,\ 0 \le t \le 2\pi$

9.2 Polar Coordinates

Let $\overrightarrow{OA}$ be any ray of the plane (called the polar axis, with point O called the pole), and let P be any point of the plane. If $\angle AOP = \theta$(measured in radians) and using the convention that $\theta > 0$ if $\angle AOP$ is measured counterclockwise while $\theta < 0$ if ⟨AOP is measured clockwise and if $OP = r$, then one way of giving polar coordinates to P is to call them (r, θ).

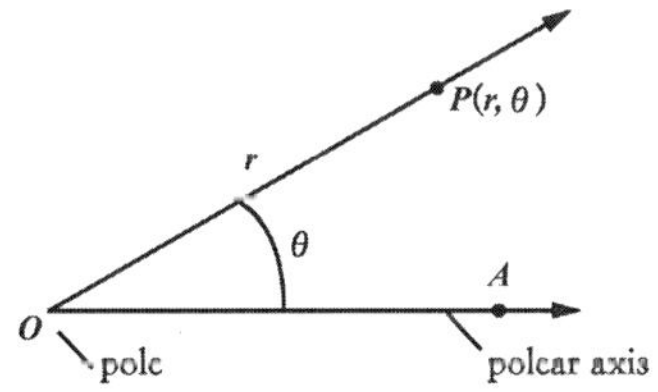

Figure 9.4

Example 1

Find the points in the plane with polar coordinates $P\left(2, \frac{\pi}{6}\right)$, $Q\left(1, \frac{2\pi}{3}\right)$, $R\left(-1, \frac{\pi}{2}\right)$, and $S\left(-\frac{3}{2}, \frac{3\pi}{4}\right)$.

Solution

The points are plotted in the graph.

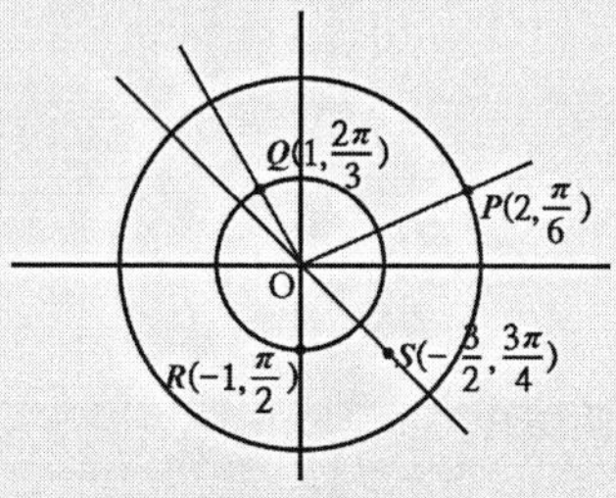

Figure 9.5

[EX 1] Plot the points with the following polar coordinates

(1) $(3, \pi/4)$

(2) $\left(-3, \frac{\pi}{4}\right)$

(3) $(3, -\pi/4)$

(4) $(-3, -\pi/4)$

(5) $(2, \pi/2)$

(6) $(2, 0)$

(7) $(-2, \pi/2)$

(8) $(-2, 0)$

Equations Relating polar and coordinates

P has polar coordinates (r, θ), the Cartesian coordinates of P will be given by

$(x, y) = (r\cos\theta, r\sin\theta)$ or $\begin{cases} x = r\cos\theta \\ y = r\sin\theta \end{cases}$

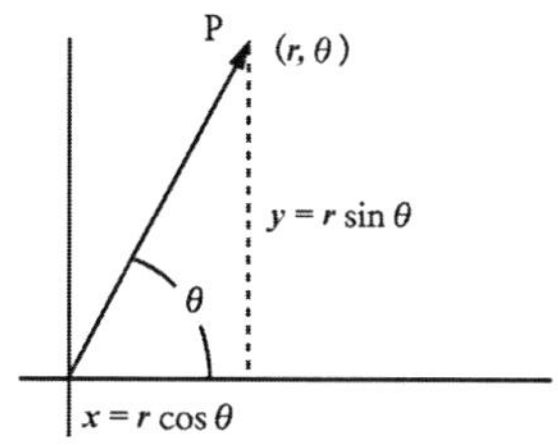

Figure 9.6

A conversion from Cartesian to polar coordinates is somewhat more difficult. From the fact that $x = r\cos\theta$, and $y = r\sin\theta$, we find that $r^2 = x^2 + y^2$ and $\tan\theta = \dfrac{y}{x}$.

Example 2

Find polar coordinates of the point with Cartesian coordinates $(-2, 2\sqrt{3})$.

Solution

From the equations above, $r^2 = 4 + 12 = 16$ and $\tan\theta = -\sqrt{3}$. Since the point $(-2, 2\sqrt{3})$ is in the second quadrant, $\tan\theta = -\sqrt{3} \Rightarrow \theta = \dfrac{2\pi}{3}$. With $r = 4$, one set of polar coordinates is $(r, \theta) = \left(4, \dfrac{2\pi}{3}\right)$. We can also use the coordinates $(r, \theta) = \left(-4, -\dfrac{\pi}{3}\right)$, making use of a ray in the fourth quadrant along with a negative r.

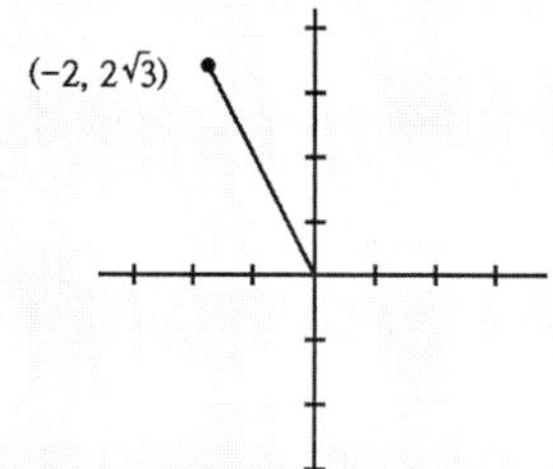

Figure 9.7

[EX 2] For each of the points given by the following Cartesian coordinates, find the corresponding polar coordinates.

(1) $(2, 2\sqrt{3})$

(2) $(\sqrt{3}, -1)$

(3) $(-4, -4)$

(4) $(\sqrt{3}, 1)$

(5) $(-3, 3)$

(6) $(-2, 2\sqrt{3})$

Example 3

Convert the point $(2, \pi/3)$ from polar to Cartesian coordinates.

Solution

Since $r=2$ and $\theta = \pi/3$, Equations 3 give

$$x = r\cos\theta = 2\cos\frac{\pi}{3} = 2 \cdot \frac{1}{2} = 1$$

$$y = r\sin\theta = 2\sin\frac{\pi}{3} = 2 \cdot \frac{\sqrt{3}}{2} = \sqrt{3}$$

Therefore the point is $(1, \sqrt{3})$ in Cartesian coordinates.

[EX 3] Find the Cartesian coordinates of the following points (given in polar coordinates).

(1) $(\sqrt{2}, \pi/4)$

(2) $(1, 0)$

(3) $(0, \pi/2)$

(4) $(-\sqrt{2}, \pi/4)$

(5) $(-3, 5\pi/6)$

(6) $(5, \tan^{-1}(4/3))$

(7) $(-1, 7\pi)$

(8) $(2\sqrt{3}, 2\pi/3)$

Example 4

Find a polar equation for the circle $x^2+(y-3)^2=9$

Solution

We apply the equations relating polar and Cartesian coordinates:

$$x^2+(y-3)^2=9$$

$$x^2+y^2-6y+9=9$$

$$x^2+y^2-6y=0$$

$$r^2-6r\sin\theta=0$$

$$r=0 \text{ or } r-6\sin\theta=0$$

$$r=6\sin\theta$$

[EX 4] Transform the following equations of graphs from Cartesian coordinates to polar coordinates

(1) $x=7$

(2) $x=y$

(3) $x^2-y^2=1$

(4) $\dfrac{x^2}{9}+\dfrac{y^2}{4}=1$

(5) $y^2=4x$

(6) $x^2+(y-2)^2=4$

(7) $(x-3)^2+(y+1)^2=4$

Example 5

Replace the following polar equations by equivalent Cartesian equations and identify their graphs.

(1) $r\cos\theta = -4$　　(2) $r^2 = 4r\cos\theta$　　(3) $r = \dfrac{4}{2\cos\theta - \sin\theta}$

Solution

We use the substitutions $r\cos\theta = x$, $r\sin\theta = y$, $r^2 = x^2 + y^2$

(1) $r\cos\theta = -4$　The Cartesian equation: $r\cos\theta = -4$

$$x = -4$$

The graph: Vertical line through $x = -4$ on the x-axis

(2) $r^2 = 4r\cos\theta$　The Cartesian equation : $r^2 = 4r\cos\theta$

$$x^2 + y^2 = 4x$$

$$x^2 - 4x + y^2 = 0$$

$$x^2 - 4x + 4 + y^2 = 4$$

$$(x-2)^2 + y^2 = 4$$

The graph: Circle, radius 2, center $(h,k) = (2,0)$

(3) $r = \dfrac{1}{2\cos\theta - \sin\theta}$　The Cartesian Equation: $r(2\cos\theta - \sin\theta) = 4$

$$2r\cos\theta - r\sin\theta = 4$$

$$2x - y = 4$$

$$y = 2x - 4$$

The graph: Line, slope $m = 2$, $y-$ intercept $b = -4$

[EX 5] Transform the following equations of graphs from polar coordinates to Cartesian coordinates

(1) $r\cos\theta = 2$　　(2) $r\sin\theta = 0$

(3) $r = 4\csc\theta$　　(4) $r\cos\theta + r\sin\theta = 1$

(5) $r = \dfrac{5}{\sin\theta - 2\cos\theta}$

(6) $r = \cot\theta\csc\theta$

(7) $r = \csc\theta\, e^{r\cos\theta}$

(8) $r^2 = 4r\cos\theta$

(9) $r = 8\sin\theta$

(10) $r = 2\cos\theta + 2\sin\theta$

Graphing polar equations

If we have an equation involving r and θ, its graph consists of all points in the plane having at least on set of polar coordinates satisfying the equation. If the equation happens to specify r as a function of θ, as many do, then various inputs θ are taken and the corresponding output r computed in order to find the points.
We began with the basic point-plotting method, which was then enhanced by sketching aids such as symmetry, intercepts, asymptotes, periods, and shifts.

Example 6

Sketch a graph of the set of points with polar coordinates equation $r = 2\cos\theta$.

Solution

θ	r
0	2
$\frac{\pi}{6}$	$\sqrt{3}$
$\frac{\pi}{3}$	1
$\frac{\pi}{2}$	0
$\frac{2\pi}{3}$	-1
$\frac{5\pi}{6}$	$-\sqrt{3}$
π	-2

θ	r
$\frac{7\pi}{6}$	$-\sqrt{3}$
$\frac{4\pi}{3}$	-1
$\frac{3\pi}{2}$	0
$\frac{5\pi}{3}$	1
$\frac{11\pi}{6}$	$\sqrt{3}$
2π	2

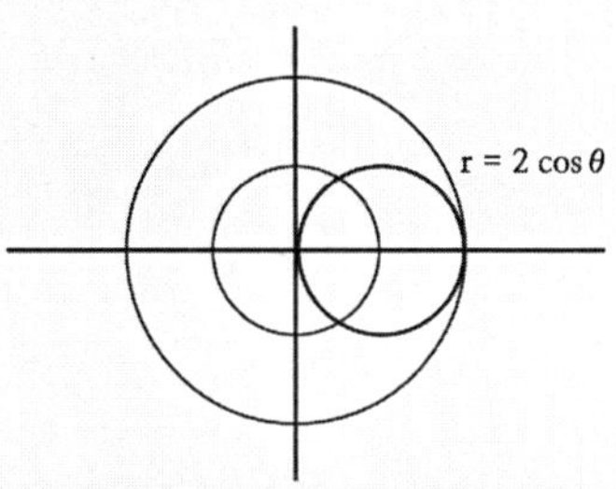

Figure 9.8

We make a table inputs θ and outputs $r = 2\cos\theta$ and use the pairs to plot points

Symmetry

Three tests for symmetry in polar graphs are useful. For example, a polar curve is

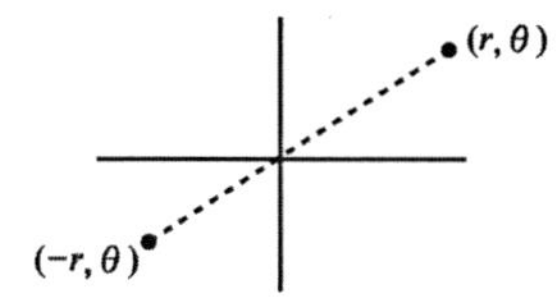

Figure 9.9

a) symmetric about the origin if substitution of $-r$ for r produces an equivalent equation;

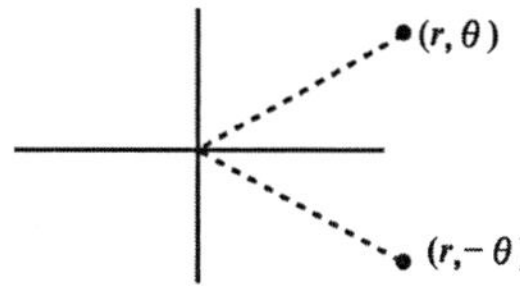

Figure 9.10

b) symmetric about x-axis if substitution for $-\theta$ for θ produces an equivalent equation;

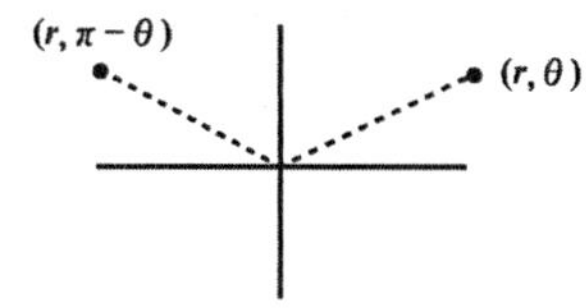

Figure 9.11

c) symmetric about the $y-$axis if substitution of $\pi-\theta$ for θ produces an equivalent equation

---

Example 7

Sketch a graph of the polar curve $r=1-\cos\theta$

Solution

Since $\cos(-\theta)=\cos\theta$, the equation is unchanged when $-\theta$ is substituted for θ and the graph is symmetric about the $x-$axis. Therefore we focus on the portion of the graph in the first and second quadrants to determine the shape of the curve for $0 \le x \le \pi$. Plotting a few points and using the symmetry about the $x-$axis we have

θ	r
0	0
$\frac{\pi}{6}$	$1-\frac{\sqrt{3}}{2}=13$
$\frac{\pi}{3}$	$\frac{1}{2}$
$\frac{\pi}{2}$	1
$\frac{2\pi}{3}$	$\frac{3}{2}$
$\frac{5\pi}{6}$	$1+\frac{\sqrt{3}}{2}=1.87$
π	2

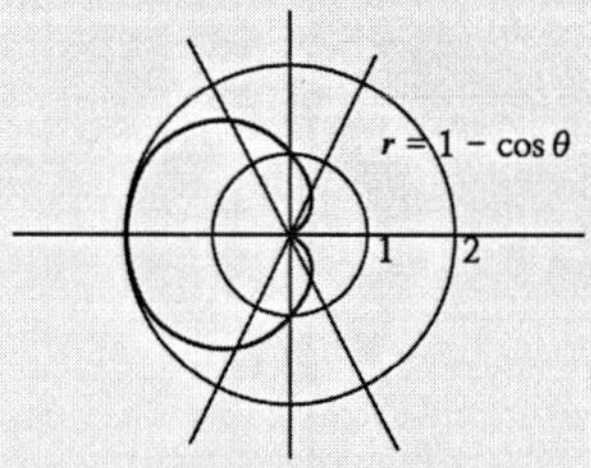

Figure 9.12

This heart- shaped curve is called a cardioid, from the Greek (kordia), meaning heart.

[EX 6] Sketch the curves

(1) $\theta = \frac{\pi}{3}$

(2) $r = 1 - \sin\theta$

(3) $r = 4\sin\theta$

(4) $r = 1 + 2\sin\theta$

(5) $r = \cos 2\theta$

(6) $r = 2\sin 3\theta$

Slope of the curve $f(\theta)$

$$\left.\frac{dy}{dx}\right|_{(r,\theta)} = \frac{f'(\theta)\sin\theta + f(\theta)\cos\theta}{f'(\theta)\cos\theta - f(\theta)\sin\theta},$$

provided $dx/d\theta \neq 0 \; at \; (r, \theta)$.

Example 8

Find the horizontal and vertical tangent lines of $r = \sin\theta,\ 0 \leq \theta \leq \pi$.

Solution

Begin by writing the equation in parametric form.

$$x = r\cos\theta = \sin\theta\cos\theta$$

and

$$y = r\sin\theta = \sin\theta\sin\theta = \sin^2\theta$$

Next, differentiate x and y with respect to θ and set each derivative equal to 0.

$$\frac{dx}{d\theta} = \cos^2\theta - \sin^2\theta = \cos 2\theta = 0 \Rightarrow \theta = \frac{\pi}{4}, \frac{3\pi}{4}$$

$$\frac{dy}{d\theta} = 2\sin\theta\cos\theta = \sin 2\theta = 0 \Rightarrow \theta = 0, \frac{\pi}{2}$$

So, the graph has vertical tangent lines at $(\sqrt{2}/2, \pi/4)$ and $(\sqrt{2}/, 3\pi/4)$, and it has horizontal tangent lines at $(0,0)$ and $(1,\pi/2)$, as shown in **Figure.**

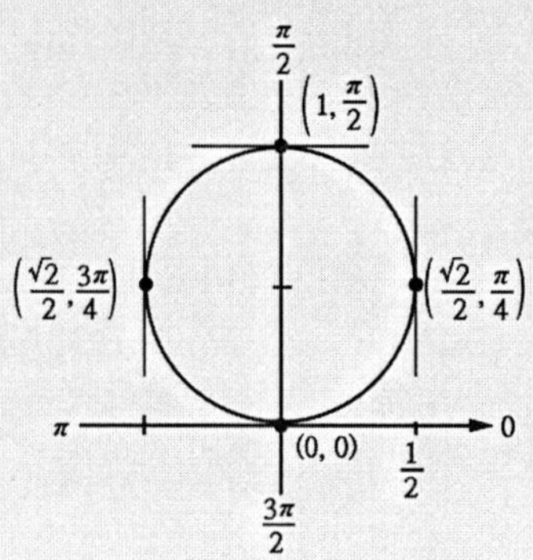

Figure 9.13

[EX 7] The polar curve r is given by $r(\theta) = 2(1+\cos\theta)$, where $0 \le \theta \le 2\pi$. See Figure

(1) Find the slope of the cardioid $r = 2(1+\cos\theta)$ at $\theta = \frac{\pi}{6}$. See Figure 9.14.

(2) where is the tangent to the curve horizontal?

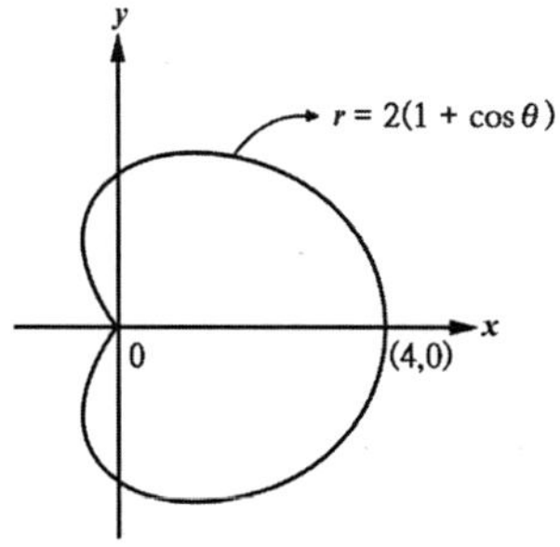

Figure 9.14

[EX 8] The polar curve r is given by $r(\theta)=3\theta+\sin\theta$, where $0 \le \theta \le 2\pi$.

(1) For $\frac{\pi}{2} \le \theta \le \pi$, there is one point P on the polar curve r with $x-$ coordinate -3. Find the angle θ that corresponds to point P. Find the $y-$ coordinate of point P. Show the work that leads to your answers.

(2) A particle is traveling along the polar curve r so that its position at time t is $(x(t), y(t))$ and such that $\frac{d\theta}{dt}=2$, Find $\frac{dy}{dt}$ at the instant that $\theta=\frac{2\pi}{3}$, and interpret the meaning of your answer in the context of the problem.

[EX 9] The graph of the polar curve $r=1-2\cos\theta$ for $0 \le \theta \le \pi$ is shown above. Let S be the shaded region in the third quadrant bounded by the curve and the $x-$axis.

(1) Write expressions for $\frac{dx}{d\theta}$ and $\frac{dy}{d\theta}$ in terms of θ.

(2) Write an equation in terms of x and y for the line tangent to the graph of the polar curve at the point where $\theta=\frac{\pi}{2}$. Show the computations that lead to your answer.

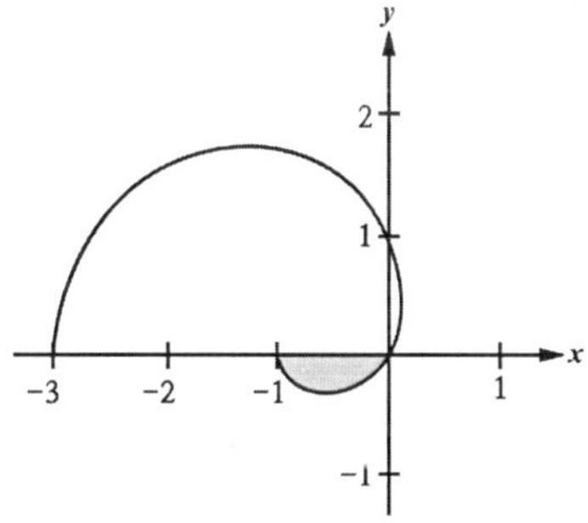

Figure 9.15

Area of the Region $r = f(\theta)$, $\alpha \leq \theta \leq \beta$

$$A = \int_{\alpha}^{\beta} \frac{1}{2} r^2 d\theta$$

Example 9

Find the area enclosed by the cardioid $r = 2(1 + \cos\theta)$. We graphed the cardioid on out calculator, using polar mode, in the window $[-2.5] \times [-3.3]$ with θ in $[0, 2\pi]$.

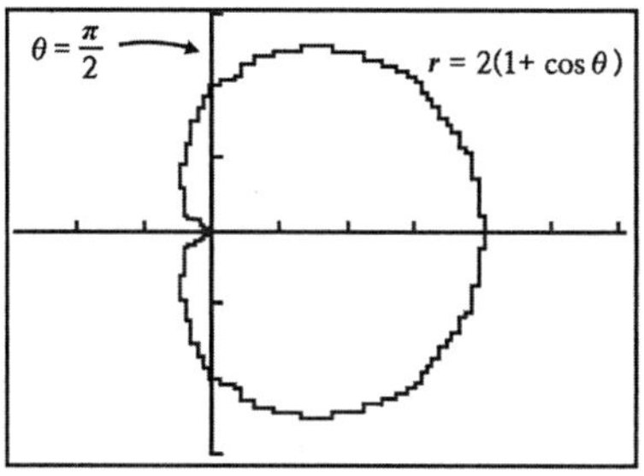

Figure 9.16

Solution

See Figure 9.16. We use the symmetry of the curve with respect to the polar axis and write

$$A = 2 \cdot \frac{1}{2} \int_0^{\pi} r^2 d\theta = 4 \int_0^{\pi} (1 + \cos\theta)^2 d\theta$$

$$= 4 \int_0^{\pi} (1 + 2\cos\theta + \cos^2\theta) d\theta$$

$$= 4 \int_0^{\pi} \left(1 + 2\cos\theta + \frac{1}{2} + \frac{\cos 2\theta}{2}\right) d\theta$$

$$= 4 \left[\theta + 2\sin\theta + \frac{\theta}{2} + \frac{\sin 2\theta}{4}\right]_0^{\pi} = 6\pi.$$

Area of the Region $0 \le r_1(\theta) \le r \le r_2(\theta), \alpha \le \theta \le \beta$

$$A = \int_{\alpha}^{\beta} \frac{1}{2} r_2^2 d\theta - \int_{\alpha}^{\beta} \frac{1}{2} r_1^2 d\theta = \int_{\alpha}^{\beta} \frac{1}{2}\left(r_2^2 - r_1^2\right) d\theta \qquad (1)$$

Example 10

Find the area of the region that lies inside the circle $r=1$ and outside the cardioid $r = 1 - \cos\theta$.

Solution

We sketch the region to determine its boundaries and find the limits of integration (Figure). The outer curve is $r_2 = 1$, the inner curve is $r_1 = 1 - \cos\theta$, and θ runs from $-\pi/2$ to $\pi/2$. The area is

$$A = \int_{-\pi/12}^{\pi/2} \frac{1}{2}(r_2^2 - r_1^2) d\theta$$

$$= 2\int_{0}^{\pi/2} \frac{1}{2}\left(r_2^2 - r_1^2\right) d\theta$$

$$= \int_{0}^{\pi/2} (2\cos\theta - \cos^2\theta) = \int_{0}^{\pi/2} \left(2\cos\theta - \frac{1+\cos 2\theta}{2}\right) d\theta$$

$$= \left[2\sin\theta - \frac{\theta}{2} - \frac{\sin 2\theta}{4}\right]_0^{\pi/2} = 2 - \frac{\pi}{4}.$$

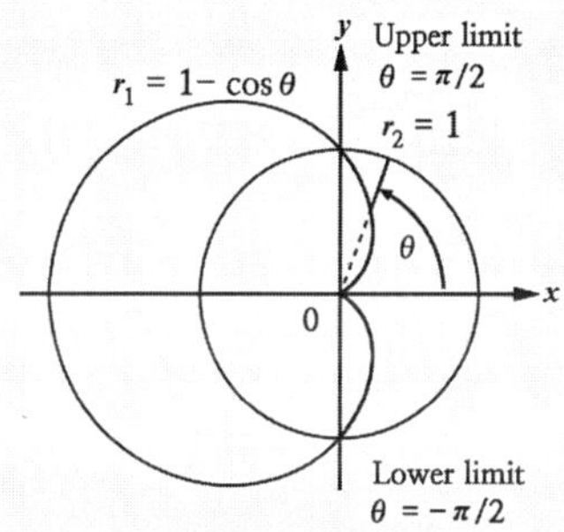

Figure 9.17

[EX 10] Find the area of the regions

(1) Bounded by the spiral $r = \theta$ for $0 \le \theta \le \pi$

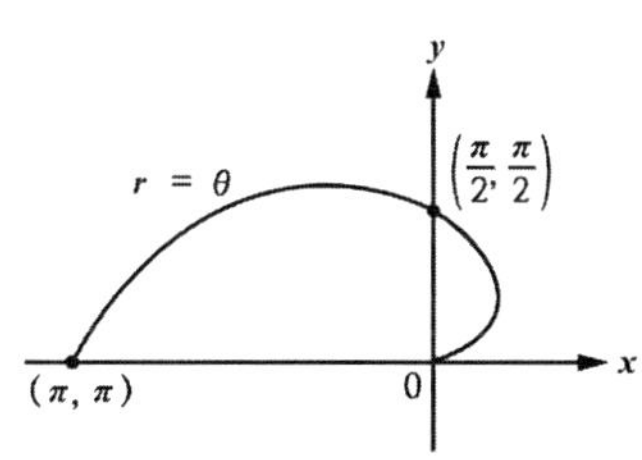

Figure 9.18

(2) Bounded by the circle $r=2\sin\theta$ for $\pi/4 \le \theta \le \pi/2$

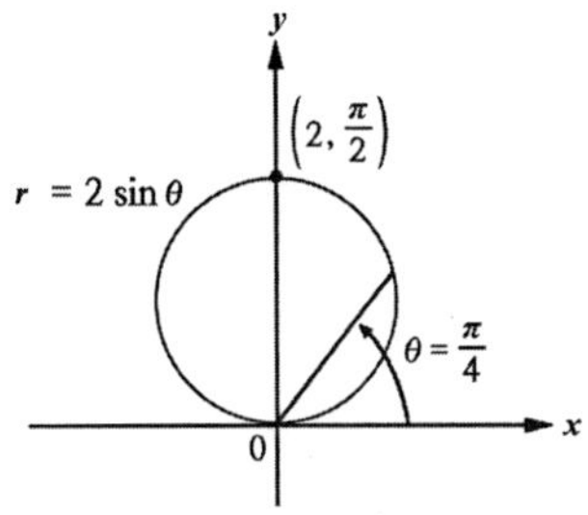

Figure 9.19

(3) Inside one leaf of the four-leaved rose $r=\cos 2\theta$

(4) Inside one leaf of the tree leaved rose $r=\cos 3\theta$

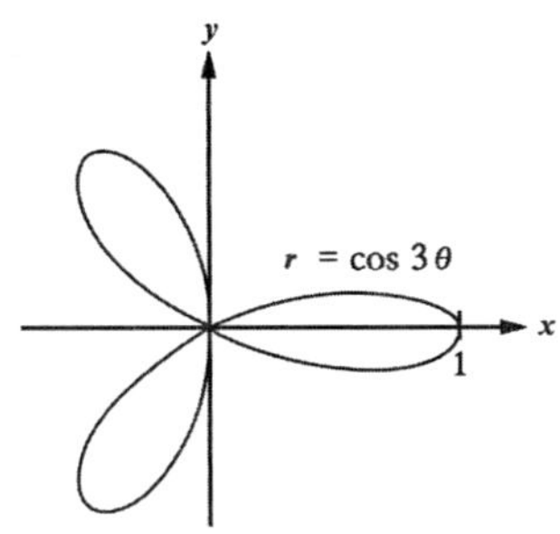

Figure 9.20

[EX 11] Find the area of the regions in Exercises (1)-(6).

(1) Shared by the circles $r=2\cos\theta$ and $r=2\sin\theta$

(2) Shared by the circles $r=1$ and $r=2\sin\theta$

(3) Shared by the circle $r=2$ and the cardioid $r=2(1-\cos\theta)$

(4) Inside the circle $r=-2\cos\theta$ and outside the circle $r=1$

(5) Inside the circle $r=6$ above the line $r=3\csc\theta$

(6) Inside the circle $r=4\cos\theta$ and to the right of the vertical line $r=\sec\theta$

[EX 12] The graphs of the polar curves $r=3$ and $r=4-2\sin\theta$ are shown in the figure above. The curves intersect when $\theta=\dfrac{\pi}{6}$ and $\theta=\dfrac{5\pi}{6}$.

(1) Let S be the shaded region that is inside the graph of $r=3$ and also inside the graph of $r=4-2\sin\theta$. Find the area of S.

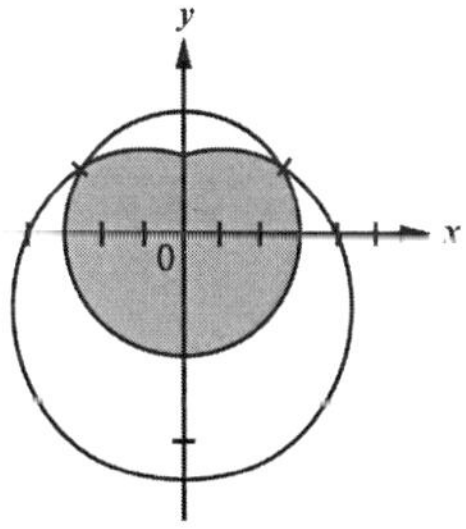

Figure 9.21

(2) A particle moves along the polar curve $r=4-2\sin\theta$ so that at time t seconds, $\theta=t^2$. Find the time t in the interval $1\le t\le 2$ for which the $x-$coordinate of the particle's position is -1.

(3) For the particle described in part (2), find the position vector in terms of t. Find the velocity vector at time $t=1.5$.

[EX 13] The graphs of the polar curves $r=2$ and $r=3+2\cos\theta$ are shown n the figure. The curves intersect when $\theta=\dfrac{2\pi}{3}$ and $\theta=\dfrac{4\pi}{3}$.

(1) Let R be the region that is inside the graph of $r=2$ and also inside the graph of $r=3+2\cos\theta$, as shaded in the figure above. Find the area of R.

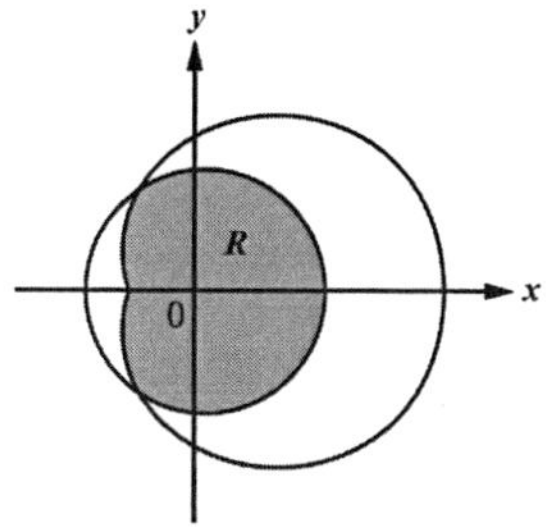

Figure 9.22

(2) A particle moving with nonzero velocity along the polar curve given by $r=3+2\cos\theta$ has position $(x(t), y(t))$ at time t, with $\theta=0$ when $t=0$.
This particle moves along the curve so that $\dfrac{dr}{dt}=\dfrac{dr}{d\theta}$. Find the value of $\dfrac{dr}{dt}$ at $\theta=\dfrac{\pi}{3}$ and interpret your answer in terms of the motion of the particle.

(3) For the particle described in part(2), $\dfrac{dy}{dt}=\dfrac{dy}{d\theta}$. Find the value of $\dfrac{dy}{dt}$ at $\theta=\dfrac{\pi}{3}$ and interpret your answer in terms of the motion of the particle.

memo

CALCULUS

CHAPTER 10

Vectors and the Geometry of Space

10.1 Vectors

The term vector is used by scientists to indicate a quantity (such as displacement or velocity or force) that has both magnitude and direction. A vector is often represented by an arrow or a directed line segment. The length of the arrow represents the magnitude of the vector and the arrow points in the direction of the vector. We denote a vector by printing a letter in boldface (v) or by putting an arrow above the letter $(\vec{v})$

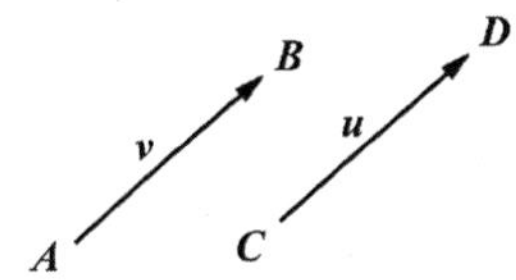

Figure 10.1

For instance, suppose a particle moves along a line segment from point A to B. The corresponding vector v, shown in Figure 10.1. has initial point A and terminal point B and we indicate this by writing $\mathrm{v}=\overrightarrow{AB}$. Notice that the vector $\mathrm{u}=\overrightarrow{CD}$ has the same length and the same direction as v even though it is in a different position. We say that u and v are equivalent (or equal) and we write $\mathrm{u}=\mathrm{v}$. The zero vector, denoted by 0, has length 0. It is the only vector with no specific direction.

Definition of Vector Addition If u and v are vectors positioned so the initial point of v is at the terminal point of u, then the sum $\mathrm{u}+\mathrm{v}$ is the vector from the initial point of u to the terminal point of v.

The definition of vector addition is illustrated in Figure 10.2. You can see why the definition is sometimes called the Triangle Law.

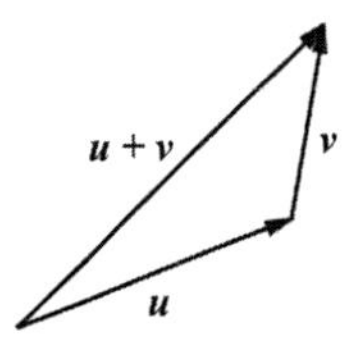

Figure 10.2

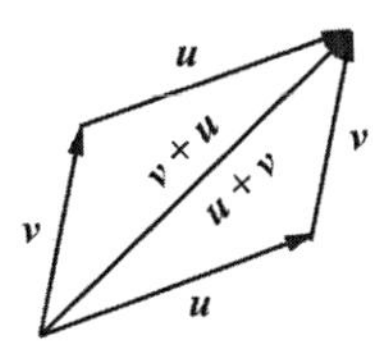

Figure 10.3

In Figure 10.3 we start with the same vectors u and v as in Figure 10.2 and draw another copy of v with the same initial point as u. Completing the parallelogram, we see the

$u+v=v+u$. This also gives another way to construct the sum: If we place u and v. they start at the same point, then $u+v$ lies along the diagonal of the parallelogram with u and v as sides. (This is called the Parallelogram Law)

Example 1

Draw the sum of the vectors u and v shown in Figure 10.4

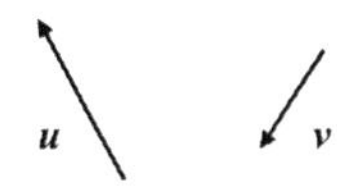

Figure 10.4

Solution

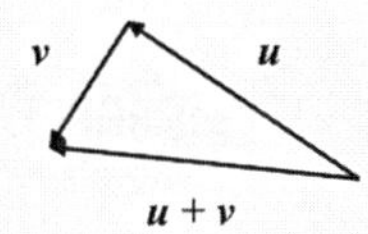

Figure 10.5

Definition of Scalar Multiplication If c is a scalar and v is a vector, then the scalar multiple cv is the vector whose length is $|c|$ times the length of v and whose direction is the same as v if $c>0$ and is opposite to v if $c<0$. if $c=0$ or $v=\mathbf{0}$, then $cv=0$.

Example 2

If a and b are the vectors shown in Figure 10.6, draw $2b-a$.

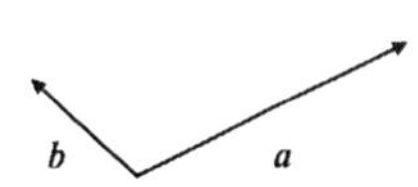

Figure 10.6

Solution

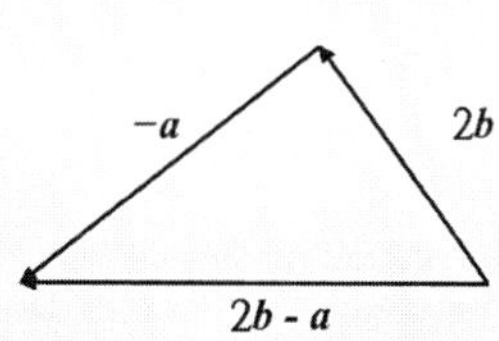

Figure 10.7

[EX 1] Copy the vectors in the figure and use them to draw the following vectors.

(1) $u+v$

(2) $u-v$

(3) $v+u+w$

(4) $u-w-v$

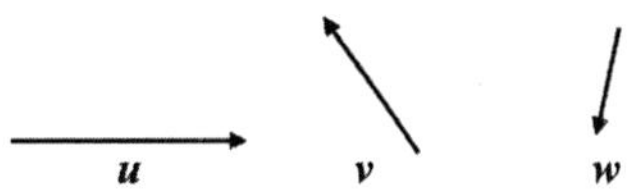

Figure 10.8

[EX 2] Copy the vectors in the figure and use them to draw the following vectors.

(1) $a-b$

(2) $\frac{1}{2}a$

(3) $-3b$

(4) $a+2b$

Figure 10.9

Definition of Component Form of a Vector in the plane

If v is a vector in the plane whose initial point is the origin and whose terminal point is (v_1, v_2), then the component form of v is given by

$$\mathrm{v} = \langle v_1, v_2 \rangle.$$

The coordinates v_1 and v_2 are called the components of v. If both the initial point and the terminal point lie at the origin, then v is called the zero vector and is denoted by

$$0 = \langle 0, 0 \rangle.$$

The following procedures can be used to convert directed line segments to component form or vice versa.

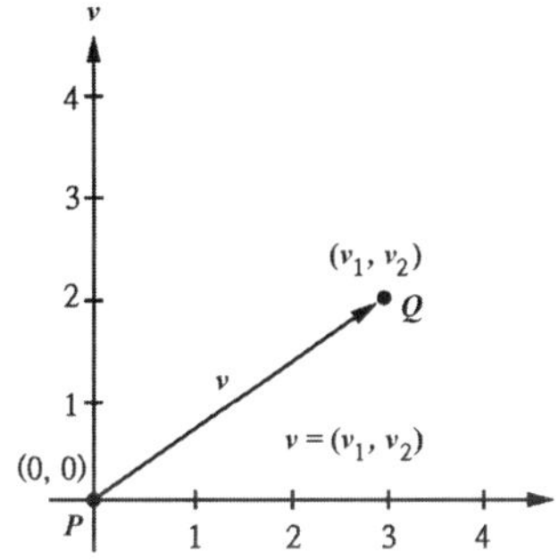

Figure 10.10

1. If $P(p_1, p_2)$ and $Q(q_1, q_2)$ are the initial and terminal points of a directed segment, the component from of the vector v represented by $\overrightarrow{PQ}$ is $\langle v_1, v_2 \rangle = \langle q_1 - p_1, q_2 - p_2 \rangle$. Moreover, the length (or magnitude) of v is

$$\begin{aligned} \|\mathrm{v}\| &= \sqrt{(q_1 - p_1)^2 + (q_2 - p_2)^2} \\ &= \sqrt{{v_1}^2 + {v_2}^2} \end{aligned}$$

2. If $\mathrm{v} = \langle v_1, v_2 \rangle$, v can be represented by the directed line segment, in standard position, from $P(0, 0)$ to $Q(v_1, v_2)$

The length of v is also called the norm of v.

Example 3

Find the component form and length of the vector v that has intial point (2, 5) and terminal point (-3, -1)

Solution

Let $P(2, 5) = (p_1, p_2)$ and $Q(-3, -1) = (q_1, q_2)$.

Then the component of $v = \langle v_1, v_2 \rangle$ are $v_1 = q_1 - p_1 = -3 - 2 = -5$

$$v_2 = q_2 - p_2 = -1 - 5 = -6$$

So, as shown in Figure 10.11, $v = \langle -5, -6 \rangle$, and the length of v is $\|v\| = \sqrt{(-5)^2 + (-6)^2} = \sqrt{61}$

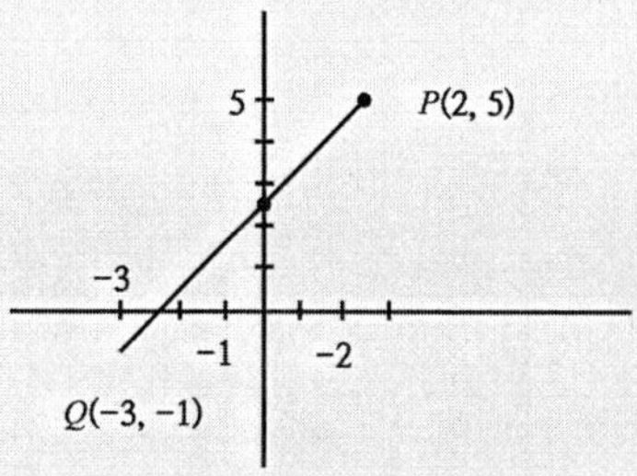

Figure 10.11

[EX 3] The initial and terminal points of a vector are given. (a) Sketch the given directed line segment, (b) write the vector in component form, and (c) find the magnitude of v.

	Initial Point	Terminal Point
(1)	(1, 2)	(5, 5)
(2)	(10, 2)	(6, −1)
(3)	(6, 2)	(6, 6)
(4)	$\left(\frac{3}{2}, \frac{4}{3}\right)$	$\left(\frac{1}{2}, 3\right)$

Definitions of Vector Addition and Scalar Multiplication

Let u= $\langle u_1, u_2 \rangle$ and v= $\langle v_1, v_2 \rangle$ be vector and let c be a scalar.

1. The vector sum of u and v is the vector $\mathrm{u}+\mathrm{v}=\langle u_1+v_1, u_2+v_2 \rangle$
2. The scalar multiple of c and u is the vector $cu=\langle cu_1, cu_2 \rangle$
3. The negative of v is vector
$$-\mathrm{v}=(-1)\mathrm{v}=\langle -v_1, -v_2 \rangle$$
4. The difference of u and v is
$$\mathrm{u}-\mathrm{v}=\mathrm{u}+(-\mathrm{v})=\langle u_1-v_1, u_2-v_2 \rangle$$

Example 4

Given $\mathrm{u}=\langle 4, 9 \rangle$ and $\mathrm{v}=\langle 2, -5 \rangle$, find (1) $\frac{2}{3}\mathrm{u}$, (2) $\mathrm{v}-\mathrm{u}$, and (3) $2\mathrm{u}+5\mathrm{v}$

Solution

(1) $\frac{2}{3}\mathrm{u}=\frac{2}{3}\langle 4, 9 \rangle=\left\langle \frac{8}{3}, 6 \right\rangle$

(2) $\mathrm{v}-\mathrm{u}=\langle 2, -5 \rangle-\langle 4, 9 \rangle=\langle -2, -14 \rangle$

(3) $2\mathrm{u}+5\mathrm{v}=2\langle 4, 9 \rangle+5\langle 2, -5 \rangle=\langle 18, -7 \rangle$

[EX 4] Given $\mathrm{u}=\langle -3, -8 \rangle$ and $\mathrm{v}=\langle 8, 25 \rangle$, find each of the vectors.

(1) $\frac{2}{3}\mathrm{u}$ (2) $\mathrm{v}-\mathrm{u}$ (3) $2\mathrm{u}+5\mathrm{v}$

THEOREM 10.1 Length of a Scalar Multiple

Let $\mathbf{v}$ be a vector and let c be a scalar. Then
$\|c\mathbf{v}\| = |c|\,\|\mathbf{v}\|$

THEOREM 10.2 Unit Vector in the Direction of $\mathbf{v}$

If $\mathbf{v}$ is a nonzero vector in the plane, then the vector

$$\mathbf{u} = \frac{\mathbf{v}}{\|\mathbf{v}\|} = \frac{1}{\|\mathbf{v}\|}\mathbf{v}$$

has length 1 and the same direction as $\mathbf{v}$.

Example 5

Find a unit vector in the direction of $\mathbf{v} = \langle -3, 4 \rangle$

Solution

From Theorem,

the unit vector in the direction of $\mathbf{v}$ is $\dfrac{\mathbf{v}}{\|\mathbf{v}\|} = \dfrac{\langle -3, 4\rangle}{\sqrt{(-3)^2+4^2}} = \dfrac{1}{5}\langle -3, 4\rangle = \left\langle -\dfrac{3}{5}, \dfrac{4}{5}\right\rangle$

[EX 5] Find the unit vector in the direction of $\mathbf{u}$ and verify that it has length 1.

(1) $\mathbf{u} = \langle 3, 12\rangle$

(2) $\mathbf{u} = \langle 5, 15\rangle$

(3) $\mathbf{u} = \left\langle \dfrac{3}{2}, \dfrac{5}{2}\right\rangle$

(4) $\mathbf{u} = \langle -6.2, 3.4\rangle$

Standard Unit Vectors

The unit vectors $\langle 1, 0 \rangle$ and $\langle 0, 1 \rangle$ are called the standard unit vectors in the plane and are denoted by

$$\mathrm{i} = \langle 1, 0 \rangle \text{ and } \mathrm{j} = \langle 0, 1 \rangle$$

as shown in Figure 10.12. These vectors can be used to represent any vector unique as follows.

$$\mathrm{v} = \langle v_1, v_2 \rangle = \langle v_1, 0 \rangle + \langle 0, v_2 \rangle = v_1 \langle 1, 0 \rangle + v_2 \langle 0, 1 \rangle = v_1 \mathrm{i} + v_2 \mathrm{j}$$

The vector $\mathrm{v} = v_1\mathrm{i} + v_2\mathrm{j}$ is called a linear combination of i and j. The scalar v_1 and v_2 are called the horizontal and vertical components of v.

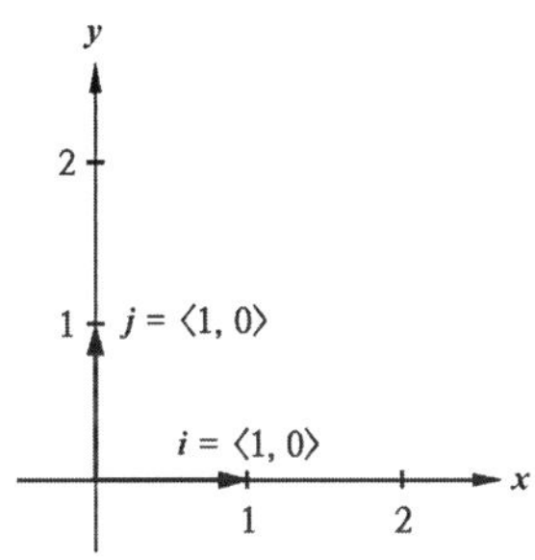

Figure 10.12

Example 6

Let u be the vector with initial point (2, -5) and terminal point (-1, 3), and let $\mathrm{v} = 2\mathrm{i} - \mathrm{j}$. Write each vector as a linear combination of i and j.

(1) u (2) $\mathrm{w} = 2\mathrm{u} - 3\mathrm{v}$

Solution

(1) $\mathrm{u} = \langle q_1 - p_1, q_2 - p_2 \rangle = \langle -1-2, 3-(-5) \rangle = \langle -3, 8 \rangle = -3\mathrm{i} + 8\mathrm{j}$

(2) $\mathrm{w} = 2\mathrm{u} - 3\mathrm{v} = 2(-3\mathrm{i} + 8\mathrm{j}) - 3(2\mathrm{i} - \mathrm{j}) = -6\mathrm{i} + 16\mathrm{j} - 6\mathrm{i} + 3\mathrm{j} = -12\mathrm{i} + 19\mathrm{j}$

Direction Angles

If u is a unit vector and θ is the angle (measured counterclockwise) form the positive x-axis to u, then the terminal point of u lies on the unit circle, and you have

$$u = \langle \cos\theta, \sin\theta \rangle = \cos\theta i + \sin\theta j$$

as shown in Figure 10.13. Moreover, it follows that any other nonzero vector v making an angle θ with the positive x-axis has the same direction as u, and you can write

$$v = \| v \| \langle \cos\theta, \sin\theta \rangle = \| v \| \cos\theta i + \| v \| \sin\theta j$$

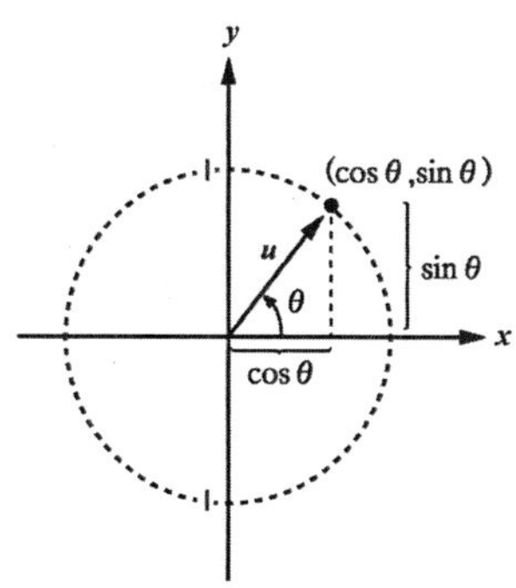

Figure 10.13

Example 7

The vector v has a magnitude of 5 and makes an angle of $30^\circ = \pi/6$ with the positive x-axis. Write v as a linear combination of the unit vectors i and j.

Solution

Because the angle between v and the positive x-axis is $\theta = \pi/6$, you can write the following.

$$v = \|v\|\cos\theta i + \|v\|\sin\theta j = 5\cos\frac{\pi}{6} i + 5\sin\frac{\pi}{6} j = \frac{5\sqrt{3}}{2} i + \frac{5}{2} j$$

[EX 6] Find the component form of v given its magnitude and the angle it makes with the positive $x-$axis.

(1) $\|v\| = 3,\ \theta = 0°$ (2) $\|v\| = 5,\ \theta = 120°$

(3) $\|v\| = 2,\ \theta = 150°$ (4) $\|v\| = 1,\ \theta = 3.5°$

Applications of vectors

Vectors are useful in many aspects of physics and engineering. Here we look at forces. A force is represented by a vector because it has both a magnitude (measured in pounds or newtons) and a direction. If several forces are acting on an object, the resultant force experienced by the object is the vector sum of these forces.

Example 8

The magnitude of a velocity vector is called speed. Suppose that a wind is blowing from the direction N45° W at a speed of 50 km/h. (This means that the direction from which the wind blows is 45° west of the northerly direction.) A pilot is steering a plane in the direction N60° E at an airspeed (speed in still air) of 250 km/h. The true course, or track, of the plane is the direction of the resultant of the velocity vectors of the plane and the wind. The ground speed of the plane is the magnitude of the resultant. Find the true course and the ground speed of the plane.

Solution

Set up the coordinate axes so that north is the positive $y-$direction, and east is the positive $x-$direction. The wind is blowing at 50 km/h from the direction N45° W, so that its velocity vector is 50 km/h S45° E, which can be written as $v_{wind} = 50(\cos 45° i - \sin 45° j)$. With respect to the still air, the velocity vector of the plane is 250 km/h N60° E, or equivalently $v_{plane} = 250(\cos 30° i + \sin 30° j)$. The velocity of the plane relative to ground is

$$v = v_{wind} + v_{plane} = (50\cos 45° + 250\cos 30°)i + (-50\sin 45° + 250\sin 30°)j$$
$$= (25\sqrt{2} + 125\sqrt{3})i + (125 - 25\sqrt{2})j \approx 251.9i + 89.6j$$

The ground speed is $|v| \approx \sqrt{(251.9)^2 + (89.6)^2} \approx 267$km/h. The angle the velocity vector makes with the $x-$axis is $\theta \approx \tan^{-1}\left(\frac{89.6}{251.9}\right) \approx 20°$. Therefore, the true course of the plane is about N(90-20)° E = N70° E.

[EX 7] A plane files at a constant groundspeed of 400 miles per hour due east and encounters a 50-mile-per hour wind from the northwest. Find the airspeed and compass direction that will allow the plane to maintain its groundspeed and eastward direction.

[EX 8] Consider a 100-N weight suspended by two wires as shown in the accompanying figure. Find the magnitudes and components of the force vectors F_1 and F_2.

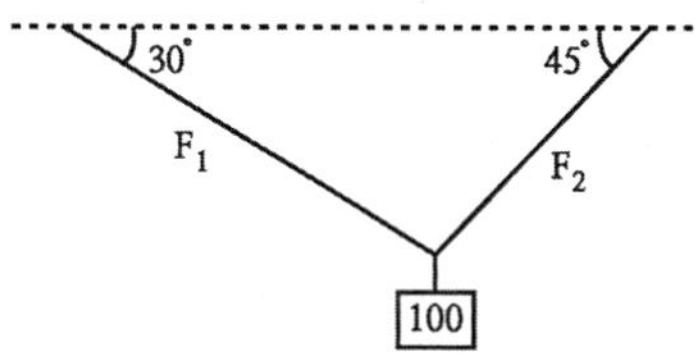

Figure 10.14

[EX 9] A straight river flows east at a speed of 10 mi/h. A boater starts at the north shore of the river and heads in a direction 60 ° from the shore (see the figure). The motorboat has a speed of 20 mi/h relative to the water.

(1) Express the velocity of the river as a vector in component form.

(2) Express the velocity of the motorboat relative to the water as a vector in component form.

(3) Find the true velocity of the motorboat.

(4) Find the true speed and direction of the motorboat.

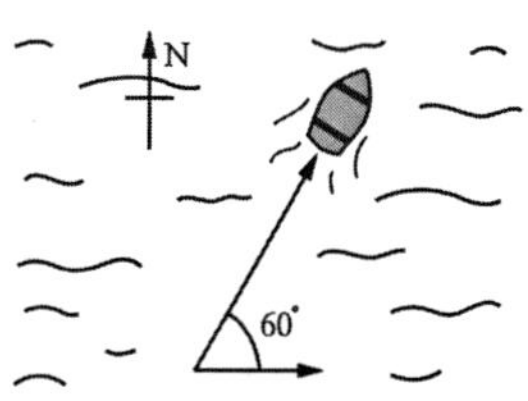

Figure 10.15

10.2 Space Coordinate and Vectors in Space

Coordinates in Space

The point $P(a, b, c)$ determines a rectangular box as in Figure 10.16. If we drop a perpendicular form P to the xy-plane, we get a point Q with coordinates $(a, b, 0)$ called the projection of P onto the xy-plane. Similarly, $R(0, b, c)$ and $S(a, 0, c)$ are the projections of P onto the yz-plane and xy-plane, respectively.

As numerical illustrations, the points (-4, 3, -5) and (3, -2, -6) are plotted in Figure 10.16

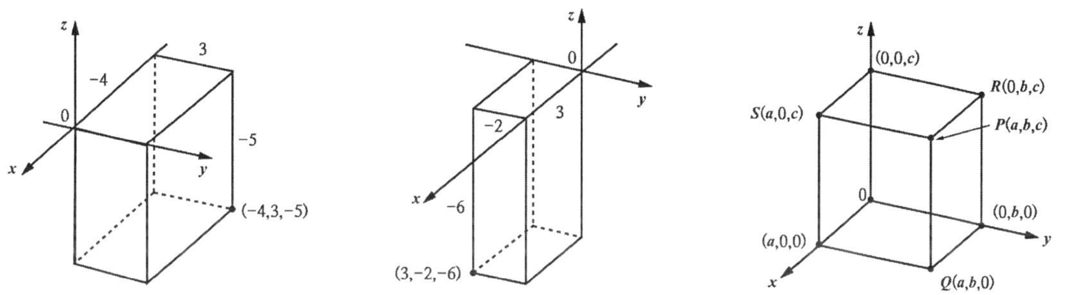

Figure 10.16

The Cartesian product $R \times R \times R = \{(x, y, z) | x, y, z \in R\}$ is the set of all ordered triples of real numbers and is denoted by R^3. We have given a one-to-one correspondence between points P in space and ordered triples (a, b, c) in R^3. It is called a three-dimensional rectangular coordinate system. Notice that, in terms of coordinates, the first octant can be described as the set of points whose coordinates are all positive.

Example 9

Sketch the points (0, 5, 2), (4, 0, -1), (2, 4, 6), and (1, -1, 2) on a single set of coordinate axes.

Solution

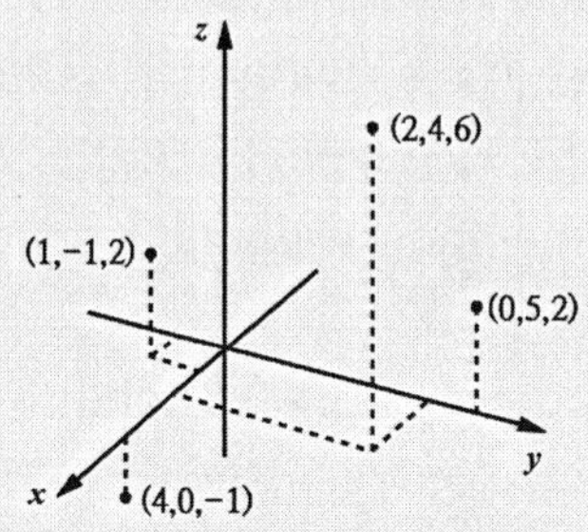

Figure 10,17

[EX 10] plot the points on the same three-dimen-sional coordinate system.

(1) (a) (5, -2, 2)   (b) (5, -2, -2)

(2) (a) (0, 4, -5)   (b) (4, 0, 5)

Distance Formula in Three Dimensions

The distance $|P_1P_2|$ between the points $P_1(x_1, y_1, z_1)$ and $P_2(x_2, y_2, z_2)$ is

$$|P_1P_2| = \sqrt{(x_2 - x_1)^2 + (y_2 - y_1)^2 + (z_2 - z_1)^2}$$

Example 10

The distance between $P_1(2, 1, 5)$ and $P_2(-2, 3, 0)$ is

Solution

$|P_1P_2| = \sqrt{(-2-2)^2 + (3-1)^2 + (0-5)^2} = \sqrt{16+4+25} = \sqrt{45} \approx 6.708$

[EX 11] Find the distance between points P_1 and P_2.

(1) $P_1(1, 1, 1)$, $P_2(3, 3, 0)$

(2) $P_1(-1, 1, 5)$, $P_2(2, 5, 0)$

(3) $P_1(1, 4, 5)$, $P_2(4, -2, 7)$

(4) $P_1(3, 4, 5)$, $P_2(2, 3, 4)$

The Standard Equation for the Sphere of Radius a and Center (x_0, y_0, z_0)

$$(x - x_0)^2 + (y - y_0)^2 + (z - z_0)^2 = a^2$$

We can use the distance formula to write equations for spheres in space (Figure 10.18) A point $P(x, y, z)$ lies on the sphere of radius a centered at $P_0(x_0, y_0, z_0)$ precisely when $|P_0P| = a$ or

$$(x - x_0)^2 + (y - y_0)^2 + (z - z_0)^2 = a^2$$

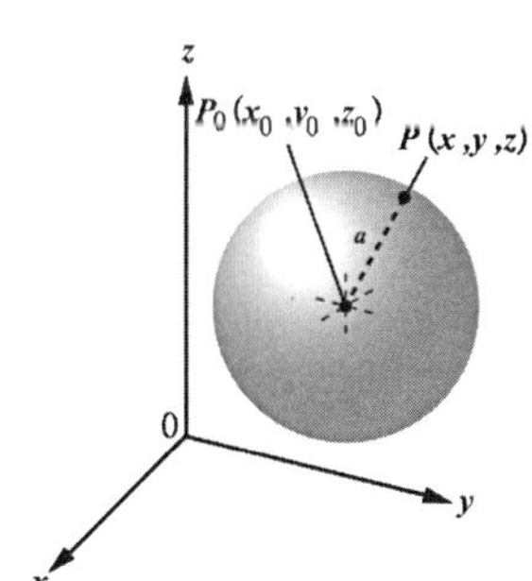

Figure 10,18

Example 11

Show that $x^2+y^2+z^2+4x-6y+2z+6=0$ is the equation of a sphere, and find its center and radius.

Solution

We can rewrite the given equation in the form of an equation of a sphere if we complete squares :

$(x^2+4x+4)+(y^2-6y+9)+(z^2+2z+1)=-6+4+9+1$

$(x+2)^2+(y-3)^2+(z+1)^2=8$

[EX 12] Show that the equation represents a sphere, and find its center and radius.

(1) $x^2+y^2+z^2-2x-4y+8z=15$

(2) $x^2+y^2+z^2+8x-6y+2z+17=0$

(3) $2x^2+2y^2+2z^2=8x-24z+1$

(4) $3x^2+3y^2+3z^2=10+6y+12z$

[EX 13] Find to each of the following

(1) Find an equation of the sphere with center (1, -4, 3) and radius 5. What is the intersection of this sphere with the xz-plane?

(2) Find an equation of the sphere with center (2, -6, 4) and radius 5. Describe its intersection with each of the coordinate planes.

(3) Find an equation of the sphere that passes through the point (4, 3, -1) and has center (3, 8, 1).

(4) Find an equation of the sphere that passes through the origin and whose center is (1, 2, 3).

Vectors in Space

In space, vectors are denoted by ordered triples v= $\langle v_1, v_2, v_3\rangle$. The zero vector is denoted by 0= $\langle 0,0,0\rangle$. Using the unit vectors i= $\langle 1,0,0\rangle$, j= $\langle 0,1,0\rangle$ and k = $\langle 0,0,1\rangle$ in the direction of the positive z-axis, the standard unit vector notation for v is

$$\mathrm{v}= v_1\mathrm{i}+ v_2\mathrm{j}+ v_3\mathrm{k}$$

as shown in Figure 10.19. If v is represented by the directed line segment form $P(p_1, p_2, p_3)$ to $Q(q_1, q_2, q_3)$, as shown in Figure 10.20. The component form of v is given by subtracting the coordinates of the initial point from coordinates of the terminal point. as follows.

$$\mathrm{v}= \langle v_1, v_2, v_3\rangle = \langle q_1 - p_1, q_2 - p_2, q_3 - p_3\rangle$$

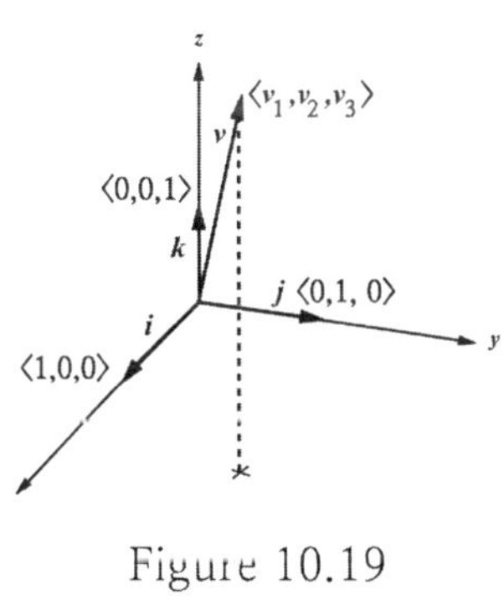

Figure 10.19

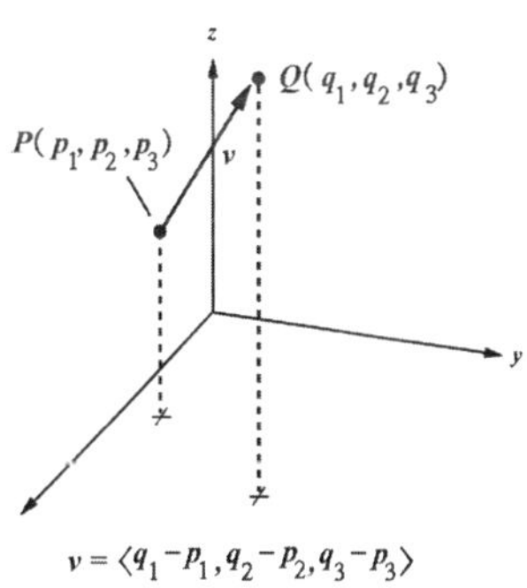

Figure 10.20

Vectors in Space

Let u= $\langle u_1, u_2, u_3\rangle$ and v= $\langle v_1, v_2, v_3\rangle$ be vectors in space and let c be a scalar.

1. Equality of Vectors : u = v if and only if $u_1 = v_1$, $u_2 = v_2$ and $u_3 = v_3$.
2. Component Form : If v is represented by the directed line segment from $P(p_1, p_2, p_3)$ to $Q(q_1, q_2, q_3)$, then v= $\langle v_1, v_2, v_3\rangle = \langle q_1 - p_1, q_2 - p_2, q_3 - p_3\rangle$

3. Length : $\| \mathrm{v} \| = \sqrt{\mathrm{v}_1{}^2 + \mathrm{v}_2{}^2 + \mathrm{v}_3{}^2}$

4. Unit Vector in the Direction of v : $\dfrac{\mathrm{v}}{\| \mathrm{v} \|} = \left(\dfrac{1}{\| \mathrm{v} \|} \right) \langle v_1, v_2, v_3 \rangle$, $\mathrm{v} \neq 0$

5. Vector Addition : $\mathrm{v} + \mathrm{u} = \langle v_1 + u_1, v_2 + u_2, v_3 + u_3 \rangle$

6. Scalar Multiplication : $c\mathrm{v} = \langle cv_1, cv_2, cv_3 \rangle$

Example 12

Find the (1) component form and (2) length of the vector with initial point $P(-3, 4, 1)$ and terminal point $Q(-5, 2, 2)$.

Solution

(1) The standard position vector v representing $\overrightarrow{PQ}$ has components
$v_1 = x_2 - x_1 = -5 - (-3) = -2$, $v_2 = y_2 - y_1 = 2 - 4 = -2$,
and $v_3 = z_2 - z_1 = 2 - 1 = 1$
The component form of $\overrightarrow{PQ}$ is $\mathrm{v} = \langle -2, -2, 1 \rangle$

(2) The length or magnitude of $\mathrm{v} = \overrightarrow{PQ}$ is $|\mathrm{v}| = \sqrt{(-2)^2 + (-2)^2 + (1)^2} = \sqrt{9} = 3$

Example 13

Find the component form and magnitude of the vector v having initial point (-2, 3, 1) and terminal point (0, -4, 4). Then find a unit vector in the direction of v.

Solution

The component form of v is

$$\mathrm{v} = \langle q_1 - p_1, q_2 - p_2, q_3 - p_3 \rangle = \langle 0 - (-2), -4 - 3, 4 - 1 \rangle = \langle 2, -7, 3 \rangle$$

which implies that its magnitude is $\| \mathrm{v} \| = \sqrt{2^2 + (-7)^2 + 3^2} = \sqrt{62}$

The unit vector in the direction of v is $\mathrm{u} = \dfrac{\mathrm{v}}{\| \mathrm{v} \|} = \dfrac{1}{\sqrt{62}} \langle 2, -7, 3 \rangle$

[EX 14] Find

a. unit vector in the direction of $\overrightarrow{P_1P_2}$ and b. the midpoint of line segment P_1P_2.

(1) $P_1(-1, 1, 5)$ $\qquad$ $P_2(2, 5, 0)$

(2) $P_1(1, 4, 5)$ $\qquad$ $P_2(4, -2, 7)$

(3) $P_1(3, 4, 5)$ $\qquad$ $P_2(2, 3, 4)$

(4) $P_1(0, 0, 0)$ $\qquad$ $P_2(2, -2, -2)$

10.3 The Dot Product of Two Vectors

Definition of Dot Product

The dot product of $\mathrm{u} = \langle u_1, u_2 \rangle$ and $\mathrm{v} = \langle v_1, v_2 \rangle$ is

$$\mathrm{u} \cdot \mathrm{v} = u_1v_1 + u_2v_2$$

The dot product of $\mathrm{u} = \langle u_1, u_2, u_3 \rangle$ and $\mathrm{v} = \langle v_1, v_2, v_3 \rangle$ is

$$\mathrm{u} \cdot \mathrm{v} = u_1v_1 + u_2v_2 + u_3v_3$$

THEOREM 10.3 Properties of the Dot Product

Let u, v, and w be vectors in the plane or in space and let c be a scalar.

$\mathrm{u} \cdot \mathrm{u} = |\mathrm{u}|^2$ $\qquad$ 2. $\mathrm{u} \cdot \mathrm{v} = \mathrm{v} \cdot \mathrm{u}$

3. $\mathrm{u} \cdot (\mathrm{v} + \mathrm{w}) = \mathrm{u} \cdot \mathrm{v} + \mathrm{u} \cdot \mathrm{w}$ $\qquad$ 4. $(c\mathrm{u}) \cdot \mathrm{v} = c(\mathrm{u} \cdot \mathrm{v}) = \mathrm{u} \cdot (c\mathrm{v})$

5. $0 \cdot \mathrm{u} = 0$

proof To prove the second property, let $\mathrm{u} = \langle u_1, u_2, u_3 \rangle$ and $\mathrm{v} = \langle v_1, v_2, v_3 \rangle$. Then

$$\mathrm{u} \cdot \mathrm{v} = u_1v_1 + u_2v_2 + u_3v_3 = v_1u_1 + v_2u_2 + v_3u_3 = \mathrm{v} \cdot \mathrm{u}$$

For the first property, let $\mathrm{v} = \langle v_1, v_2, v_3 \rangle$. Then

$$\mathrm{u} \cdot \mathrm{v} = v_1^2 + v_2^2 + v_3^2 = (\sqrt{v_1^2 + v_2^2 + v_3^2})^2 = \|\mathrm{v}\|^2$$

Proofs of the other properties are left to you.

Example 14

Given $u=\langle 2, -2\rangle$, $v=\langle 5, 8\rangle$ and $w=\langle -4, 3\rangle$ find each of the following.

(1) $u \cdot v$ (2) $(u \cdot v)w$

(3) $u \cdot (2v)$ (3) $\|w\|^2$

Solution

(1) $u \cdot v = \langle 2, -2\rangle \cdot \langle 5, 8\rangle = 2(5)+(-2)(8) = -6$

(2) $(u \cdot v)w = -6\langle -4, 3\rangle = \langle 24, -18\rangle$

(3) $u \cdot (2v) = 2(u \cdot v) = 2(-6) = -12$

(4) $\|w\|^2 = w \cdot w = \langle -4, 3\rangle \cdot \langle -4, 3\rangle$

$= (-4)(-4)+(3)(3)$

$= 25$

[EX 15] Find (a) $u \cdot v$, (b) $\|u\|^2$, (c) $(u \cdot v)v$ and (d) $u \cdot (2v)$.

(1) $u=\langle 5, -1\rangle$, $v=\langle -3, 2\rangle$

(2) $u=\langle 2, -3, 4\rangle$, $v=\langle 0, 6, 5\rangle$

(3) $u=2i-j+k$, $v=i-k$

THEOREM 10.4 Angle Between Two Vectors

If θ is the angle between two nonzero vectors u and v, then

$$\cos\theta = \frac{u \cdot v}{\|u\|\|v\|}$$

proof Consider the triangle determined by vectors u, v and v − u as shown in Figure 10.21. By the Law of Cosines, you can write

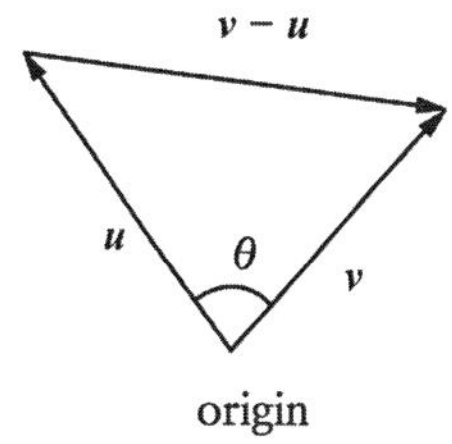

Figure 10.21

$$\|v-u\|^2 = (v-u)\cdot(v-u)$$

$$= (v-u)\cdot v - (v-u)\cdot u$$

$$= v\cdot v - u\cdot v - v\cdot u + u\cdot u$$

$$= \|v\|^2 - 2u\cdot v + \|u\|^2$$

and substitution back into the Law of Cosines yields

$$\|v\|^2 - 2u\cdot v + \|u\|^2 = \|u\|^2 + \|v\|^2 - 2\|u\|\|v\|\cos\theta$$

$$-2u\cdot v = -2\|u\|\|v\|\cos\theta$$

$$\cos\theta = \frac{u\cdot v}{\|u\|\|v\|}$$

Definition of Orthogonal Vectors

The vectors u and v are orthogonal if $u\cdot v = 0$

Example 15

For $u=\langle 3, -1, 2\rangle$, $v=\langle -4, 0, 2\rangle$, $w=\langle 1, -1, 2\rangle$, and $z=\langle 2, 0, -1\rangle$ find the angle between each pair of vectors.

(1) u and v (2) u and w (3) v and z

Solution

(1) $\cos\theta=\dfrac{u \cdot v}{\|u\|\|v\|}=\dfrac{-12+0+4}{\sqrt{14}\sqrt{20}}=\dfrac{-8}{2\sqrt{14}\sqrt{5}}=\dfrac{-4}{\sqrt{70}}$

Because $u \cdot v<0$, $\theta=\arccos\dfrac{-4}{\sqrt{70}}\approx 2.069$ radians.

(2) $\cos\theta=\dfrac{u \cdot v}{\|u\|\|w\|}=\dfrac{3+1-4}{\sqrt{14}\sqrt{6}}=\dfrac{0}{\sqrt{84}}=0$

Because $u \cdot w=0$, u and w are orthogonal. So, $\theta=\pi/2$.

(3) $\cos\theta=\dfrac{v \cdot z}{\|v\|\|z\|}=\dfrac{-8+0-2}{\sqrt{20}\sqrt{5}}=\dfrac{-10}{\sqrt{100}}=-1$

Consequently, $\theta=\pi$. Note that v and z are parallel, with $v=-2z$.

[EX 16] Find the angle θ between the vectors.

(1) $u=3i+j$, $v=-2i+4j$

(2) $u=\cos\left(\dfrac{\pi}{6}\right)i+\sin\left(\dfrac{\pi}{6}\right)j$, $v=\cos\left(\dfrac{3\pi}{4}\right)i+\sin\left(\dfrac{3\pi}{4}\right)j$

(3) $u=\langle 1,1,1\rangle$
$v=\langle 2,1,-1\rangle$

(4) $u=3i+2j+k$
$v=2i-3j$

Direction Angle and Direction Cosines

The direction angles of a nonzero vector are the angles α, β and γ (in the interval $[0, \pi]$) that u makes with the positive $x-$, $y-$, and $z-$axes. (See Figure 10.22)

The cosines of these direction angles, $\cos\alpha$, $\cos\beta$ and $\cos\gamma$, are called the direction cosines of the vector u.

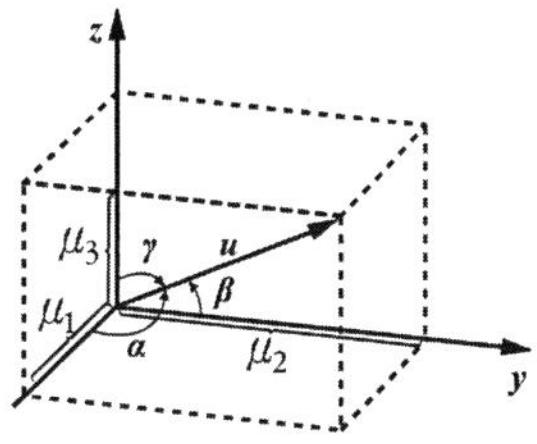

Figure 10.22

we obtain

$$\cos\alpha = \frac{u \cdot i}{|u||i|} = \frac{\mu_1}{|u|}$$

(This can also be seen directly from Figure 10.22)

Similarly, we also have $\cos\beta = \dfrac{\mu_2}{|u|}$ $\cos\gamma = \dfrac{\mu_3}{|u|}$

By squaring the expressions $\cos^2\alpha + \cos^2\beta + \cos^2\gamma = 1$

We can also write $\langle \mu_1, \mu_2, \mu_3 \rangle = \langle |u|\cos\alpha,\ |u|\cos\beta,\ |u|\cos\gamma \rangle$

$$= |u| \langle \cos\alpha,\ \cos\beta,\ \cos\gamma \rangle$$

Example 16

Find the direction cosines and angles for the vector $v = 2i + 3j + 4k$, and show that $\cos^2\alpha + \cos^2\beta + \cos^2\gamma = 1$.

Solution

Because $\|v\| = \sqrt{2^2 + 3^2 + 4^2} = \sqrt{29}$, you can write the following.

$\cos\alpha = \dfrac{v_1}{\|v\|} = \dfrac{2}{\sqrt{29}} \Rightarrow \alpha \approx 68.2^\circ$ Angle between v and i

$\cos\beta = \dfrac{v_2}{\|v\|} = \dfrac{3}{\sqrt{29}} \Rightarrow \beta \approx 56.1^\circ$ Angle between v and j

$\cos\gamma = \dfrac{v_3}{\|v\|} = \dfrac{4}{\sqrt{29}} \Rightarrow \gamma \approx 42.0^\circ$ Angle between v and k

Furthermore, the sum of the squares of the direction cosines is

$$\cos^2\alpha + \cos^2\beta + \cos^2\gamma = \frac{4}{29} + \frac{9}{29} + \frac{16}{29} = \frac{29}{29} = 1$$

[EX 17] Find the direction angles of the vector.

(1) $u = 3i + 2j - 2k$ (2) $u = \langle -2, 6, 1 \rangle$

Definition of Projection and Vector Components

Let u and v be nonzero vectors. Moreover, let $u = w_1 + w_2$, where w_1 is parallel to v and w_2 is orthogonal to v, as shown in Figure 10.23

1. Scalar projection of u onto v : $\text{comp}_v u = \dfrac{u \cdot v}{|u|}$
2. Vector projection of u onto v : $w_1 = \text{proj}_v u = \left(\dfrac{u \cdot v}{|v|}\right)\dfrac{v}{|v|} = \dfrac{u \cdot v}{|v|^2}v$
3. Scalar component of u orthogonal to v : $w_2 = u - w_1 +$

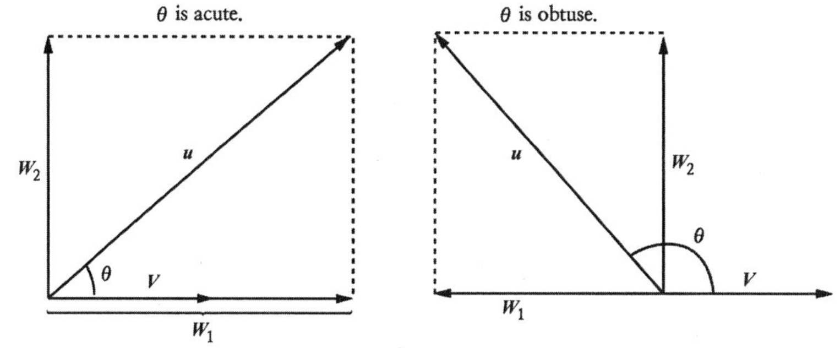

Figure 10.23

$w_1 = \text{proj}_v u =$ projection of u onto v = vector component of u along v

w_2 = vector component of u orthogonal to v

Example 17

Find the projection of u onto v and the vector component of u orthogonal to v for the vectors $u = 3i - 5j + 2k$ and $v = 7i + j - 2k$ shown in Figure 10.24.

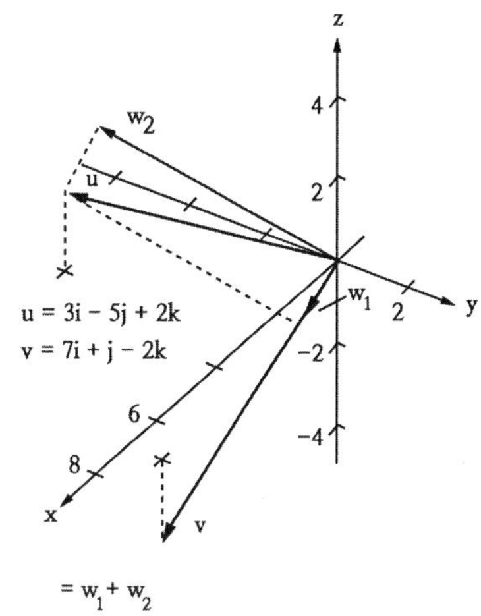

Figure 10.24

Solution

The projection of u onto v is

$$w_1 = \left(\frac{u \cdot v}{\|v\|^2}\right)v = \left(\frac{12}{54}\right)(7i + j - 2k) = \frac{14}{9}i + \frac{2}{9}j - \frac{4}{9}k$$

The vector component of u orthogonal to v is the vector

$$w_2 = u - w_1 = (3i - 5j + 2k) - \left(\frac{14}{9}i + \frac{2}{9}j - \frac{4}{9}k\right) = \frac{13}{9}i - \frac{47}{9}j + \frac{22}{9}k$$

[EX 18] (a) find the projection of u onto v, and (b) find the vector component of u orthogonal to v,

(1) $u = \langle 2,3 \rangle$, $v = \langle 5,1 \rangle$

(2) $u = \langle 2, -3 \rangle$, $v = \langle 3,2 \rangle$

(3) $u = \langle 2,1,2 \rangle$, $v = \langle 0,3,4 \rangle$

(4) $u = \langle 1,0,4 \rangle$, $v = \langle 3,0,2 \rangle$

Work

The work W done by the constant force F acting along the line of motion of an object is given by

$$W = (\text{magnitude of force})(\text{distance}) = \|\mathbf{F}\|\,\|\overrightarrow{PQ}\|$$

as shown in Figure 10.25 (a). If the constant force F is not directed along the line motion, you can see from Figure 10.25 (b) that the work W done by the force is

$$W = \|\text{proj}_{PQ}\mathbf{F}\|\,\|\overrightarrow{PQ}\| = (\cos\theta)\|\mathbf{F}\|\,\|\overrightarrow{PQ}\| = \mathbf{F}\cdot\overrightarrow{PQ}$$

This notion of work is summarized in the following definition.

Definition of Work

The work W done by a constant force F as its point of application moves along the vector $\overrightarrow{PQ}$ is given by either of the following.

1. $W = \|\text{proj}_{PQ}\mathbf{F}\|\,\|\overrightarrow{PQ}\|$ Projection form
2. $W = \mathbf{F}\cdot\overrightarrow{PQ}$ Dot product form

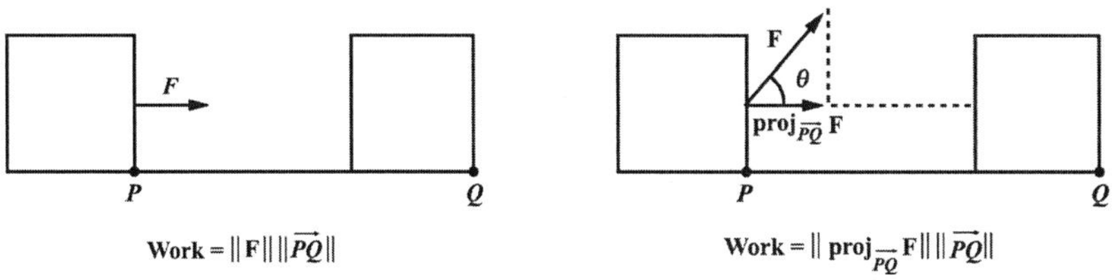

(a) Force acts along the line of motion.

(b) Force acts at angle θ with the line of motion.

Figure 10.25

Example 18

A wagon is pulled a distance of 100 m along a horizontal path by a constant force of 70 N. The handle of the wagon is held at an angle of 35° above the horizontal. Find the work done by the force.

Solution

If F and D are the force and displacement vectors, as pictured in Figure 10.26 then the work done is

$$W = \mathrm{F} \cdot \mathrm{D} = |\mathrm{F}||\mathrm{D}|\cos 35°$$

$$= (70)(100)\cos 35° \approx 5734\,\mathrm{N} \cdot \mathrm{m} = 5734\,\mathrm{J}$$

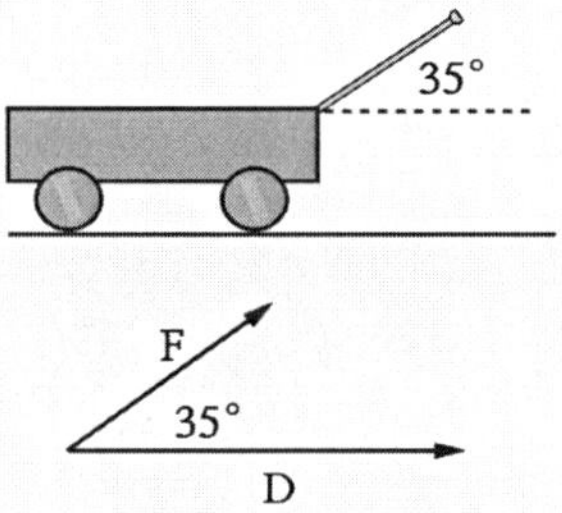

Figure 10.26

Example 19

A force is given by a vector $\mathrm{F} = 3\mathrm{i} + 4\mathrm{j} + 5\mathrm{k}$ and moves a particle from the point $P(2, 1, 0)$ to the point $Q(4, 6, 2)$. Find the work done.

Solution

The displacement vector is $\mathrm{D} = \overrightarrow{PQ} = \langle 2,5,2 \rangle$, so the work done is

$$W = \mathrm{F} \cdot \mathrm{D} = \langle 3, 4, 5 \rangle \cdot \langle 2, 5, 2 \rangle$$

$$= 6 + 20 + 10 = 36$$

If the unit of length is meters and the magnitude of the force is measured in newtons, then the work done is 36 J.

[EX 19]

(1) Find the work done by a force $F = 8i - 6j + 9k$ that moves an object from the point (0, 10, 8) to the point (6, 12, 20) along a straight line. The distance is measured in meters and the force in newtons.

(2) A tow truck drags a stalled car along a road. The chain makes an angle of $30°$ with the road and the tension in the chain is 1500 N. How much work is done by the truck in pulling the car 1 km?

(3) A woman exerts a horizontal force of 140 N on a crate as the pushes it up a ramp that is 4 m long and inclined at an angle of $20°$ above the horizontal. Find the work done on the box.

(4) Find the work done by a force of 100 N acting in the direction $N50°W$ in moving an object 5 m due west.

10.4 The Cross Product

Definition

If $\mathrm{u} = \langle u_1, u_2, u_3 \rangle$ and $\mathrm{v} = \langle v_1, v_2, v_3 \rangle$, then the cross product of u and v in the vector $\mathrm{u} \times \mathrm{v} = \langle u_2v_3 - u_3v_2,\ u_3v_1 - u_1v_3,\ u_1v_2 - u_2v_1 \rangle$

A convenient way to calculate $\mathrm{u} \times \mathrm{v}$ is to use the following determinant form with cofactor expansion.

$$\begin{aligned}\mathrm{u} \times \mathrm{v} &= \begin{vmatrix} \mathrm{i} & \mathrm{j} & \mathrm{k} \\ u_1 & u_2 & u_3 \\ v_1 & v_2 & v_3 \end{vmatrix} \\ &= \begin{vmatrix} u_2 & u_3 \\ v_2 & v_3 \end{vmatrix}\mathrm{i} - \begin{vmatrix} u_1 & u_3 \\ v_1 & v_3 \end{vmatrix}\mathrm{j} + \begin{vmatrix} u_1 & u_2 \\ v_1 & v_2 \end{vmatrix}\mathrm{k} \\ &= (u_2v_3 - u_3v_2)\mathrm{i} - (u_1v_3 - u_3v_1)\mathrm{j} + (u_1v_2 - u_2v_1)\mathrm{k}\end{aligned}$$

Example 20

Given $\mathrm{u} = \mathrm{i} - 2\mathrm{j} + \mathrm{k}$ and $\mathrm{v} = 3\mathrm{i} + \mathrm{j} - 2\mathrm{k}$, find each of the following.

(1) $\mathrm{u} \times \mathrm{v}$ (2) $\mathrm{v} \times \mathrm{u}$ (3) $\mathrm{v} \times \mathrm{v}$

Solution

(1) $$\mathrm{u} \times \mathrm{v} = \begin{vmatrix} \mathrm{i} & \mathrm{j} & \mathrm{k} \\ 1 & -2 & 1 \\ 3 & 1 & -2 \end{vmatrix} = \begin{vmatrix} -2 & 1 \\ 1 & -2 \end{vmatrix}\mathrm{i} - \begin{vmatrix} 1 & 1 \\ 3 & -2 \end{vmatrix}\mathrm{j} + \begin{vmatrix} 1 & -2 \\ 3 & 1 \end{vmatrix}\mathrm{k}$$

$$= (4-1)\mathrm{i} - (-2-3)\mathrm{j} + (1+6)\mathrm{k}$$

$$= 3\mathrm{i} + 5\mathrm{j} + 7\mathrm{k}$$

(2) $$\mathrm{v} \times \mathrm{u} = \begin{vmatrix} \mathrm{i} & \mathrm{j} & \mathrm{k} \\ 3 & 1 & -2 \\ 1 & -2 & 1 \end{vmatrix} = \begin{vmatrix} 1 & -2 \\ -2 & 1 \end{vmatrix}\mathrm{i} - \begin{vmatrix} 3 & -2 \\ 1 & 1 \end{vmatrix}\mathrm{j} + \begin{vmatrix} 3 & 1 \\ 1 & -2 \end{vmatrix}\mathrm{k}$$

$$= (1-4)\mathrm{i} - (3+2)\mathrm{j} + (-6-1)\mathrm{k}$$

$$= -3\mathrm{i} - 5\mathrm{j} - 7\mathrm{k}$$

Note that this result is the negative of that in part(1).

(3) $$\mathrm{v} \times \mathrm{v} = \begin{vmatrix} \mathrm{i} & \mathrm{j} & \mathrm{k} \\ 3 & 1 & -2 \\ 3 & 1 & -2 \end{vmatrix} = 0$$

[EX 20] Find the cross product $u \times v$

(1) $a = \langle 1, 1, -1 \rangle$, $b = \langle 2, 4, 6 \rangle$

(2) $a = i + 3j - 2k$, $b = -i + 5k$

(3) $a = j + 7k$, $b = 2i - j + 4k$

(4) $a = ti + \cos tj + \sin tk$, $b = i - \sin tj + \cos tk$

THEOREM 10.5 The vector $u \times v$ is orthogonal to both u and v

proof In order to show that $u \times v$ is orthogonal to u, we compute their dot product at follows:

$$(u \times v) \cdot u = \begin{vmatrix} \mu_2 & \mu_3 \\ v_2 & v_3 \end{vmatrix} \mu_1 - \begin{vmatrix} \mu_1 & \mu_3 \\ v_1 & v_3 \end{vmatrix} \mu_2 + \begin{vmatrix} \mu_1 & \mu_2 \\ v_1 & v_2 \end{vmatrix} \mu_3$$

$$= \mu_1(\mu_2 v_3 - \mu_3 v_2) - \mu_2(\mu_1 v_3 - \mu_3 v_1) + \mu_3(\mu_1 v_2 - \mu_2 v_1)$$

$$= \mu_1\mu_2 v_3 - \mu_1 v_2 \mu_3 - \mu_1\mu_2 v_3 + v_1\mu_2\mu_3 + \mu_1 v_2 \mu_3 - v_1 \mu_2 \mu_3$$

$$= 0$$

Example 21

find $u \times v$ and show that it in orthogonal to both u and v.

$u = \langle 2, -3, 1 \rangle$, $v = \langle 1, -2, 1 \rangle$

Solution

$u = \langle 2, -3, 1 \rangle$, $v = \langle 1, -2, 1 \rangle$

$$u \times v = \begin{vmatrix} i & j & k \\ 2 & -3 & 1 \\ 1 & -2 & 1 \end{vmatrix} = -i - j - k = \langle -1, -1, -1 \rangle$$

$u \cdot (u \times v) = 2(-1) + (-3)(-1) + (1)(-1) = 0 \Rightarrow u \perp u \times v$

$v \cdot (u \times v) = 1(-1) + (-2)(-1) + (1)(-1) = 0 \Rightarrow v \perp u \times v$

[EX 21] Find $\mathrm{u} \times \mathrm{v}$ and show that it is orthogonal to both u and v.

(1) $\mathrm{u} = \langle 12, -3, 0 \rangle$
$\mathrm{v} = \langle -2, 5, 0 \rangle$

(2) $\mathrm{u} = \mathrm{i} + \mathrm{j} + \mathrm{k}$
$\mathrm{v} = 2\mathrm{i} + \mathrm{j} - \mathrm{k}$

THEOREM 10.6

If θ is the angle between u and v (so $0 \le \theta \le \pi$), then $|\mathrm{u} \times \mathrm{v}| = |\mathrm{u}||\mathrm{v}|\sin\theta$

proof From the definitions of the cross product and length of vector, we have

$$
\begin{aligned}
|\mathrm{u} \times \mathrm{v}|^2 &= (u_2v_3 - u_3v_2)^2 + (u_3v_1 - u_1v_3)^2 + (u_1v_2 - u_2v_1)^2 \\
&= u_2^2v_3^2 - 2u_2u_3v_2v_3 + u_3^2v_2^2 + u_3^2v_1^2 - 2u_1u_3v_1v_3 + u_1^2v_3^2 \\
&\quad + u_1^2v_2^2 - 2u_1u_2v_1v_2 + u_2^2v_1^2 \\
&= (u_1^2 + u_2^2 + u_3^2)(v_1^2 + v_2^2 + v_3^2) - (u_1v_1 + u_2v_2 + u_3v_3)^2 \\
&= |\mathrm{u}|^2|\mathrm{v}|^2 - (\mathrm{u} \cdot \mathrm{v})^2 \\
&= |\mathrm{u}|^2|\mathrm{v}|^2 - |\mathrm{u}|^2|\mathrm{v}|^2\cos^2\theta \\
&\equiv |\mathrm{u}|^2|\mathrm{v}|^2(1 - \cos^2\theta) \\
&\equiv |\mathrm{u}|^2|\mathrm{v}|^2\sin^2\theta
\end{aligned}
$$

Taking square roots and observing that $\sqrt{\sin^2\theta} = \sin\theta$ because $\sin\theta \ge 0$ when $0 \le \theta \le \pi$, we have

$$|\mathrm{u} \times \mathrm{v}| = |\mathrm{u}||\mathrm{v}|\sin\theta$$

THEOREM 10.7

The length of the cross product $u \times v$ is equal to the area of the parallelogram determined by u and v.

The geometric interpretation of Theorem 10.7 can be seen by looking at Figure 10.27. If u and v are represented by directed line segments with the same initial point, then they determine a parallelogram with base $|u|$, altitude $|v| \sin\theta$, and area

$$A = |u|(|v|\sin\theta) = |u \times v|$$

Thus we have the following way of interpreting the magnitude of a cross product.

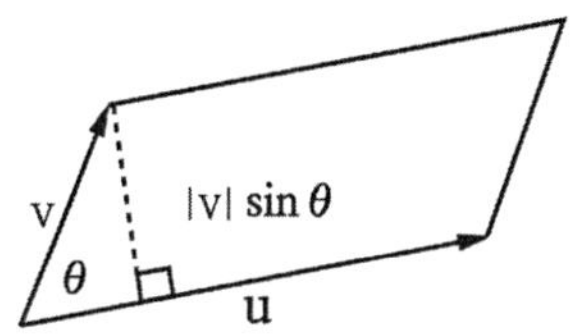

Figure 10.27

Example 22

Find a vector perpendicular to the plane that passes through the points $P(1,4,6)$, $Q(-2,5,-1)$, and $R(1,-1,1)$.

Solution

The vector $\overrightarrow{PQ} \times \overrightarrow{PR}$ is perpendicular to both $\overrightarrow{PQ}$ and $\overrightarrow{PR}$ and is therefore perpendicular to the plane through P, Q, and R. We Know that

$$\overrightarrow{PQ} = (-2-1)i + (5-4)j + (-1-6)k = -3i + j - 7k$$

$$\overrightarrow{PR} = (1-1)i + (-1-4)j + (1-6)k = -5j - 5k$$

We compute the cross product of these vectors:

$$\overrightarrow{PQ} \times \overrightarrow{PR} = \begin{vmatrix} i & j & k \\ -3 & 1 & -7 \\ 0 & -5 & -5 \end{vmatrix}$$

$$= (-5-35)i - (15-0)j + (15-0)k = -40i - 15j + 15k$$

So the vector $\langle -40, -15, 15 \rangle$ is perpendicular to the given plane. Any nonzero scalar multiple of this vector, such as $\langle -8, -3, 3 \rangle$, is also perpendicular to the plane.

Example 23

Find the area of the triangle with vertices $P(1,4,6)$, $Q(-2,5,-1)$, and $R(1,-1,1)$.

Solution

In Example 22 we computed that $\overrightarrow{PQ}\times\overrightarrow{PR}=\langle -40, -15, 15\rangle$. The area of the parallelogram with adjacent sides PQ and PR is the length of this cross product:

$$|\overrightarrow{PQ}\times\overrightarrow{PR}|=\sqrt{(-40)^2+(-15)^2+15^2}=5\sqrt{82}$$

The area A of the triangle PQR is half the area of this parallelogram, that is, $\frac{5}{2}\sqrt{82}$.

[EX 22] find the area of the parallelogram that has the given vectors as adjacent sides, Use a computer algebra system or a graphing utility to verify your result.

(1) $u=j$
$v=j+k$

(2) $u=i+j+k$
$v=j+k$

(3) $u=\langle 3,2,-1\rangle$
$v=\langle 1,2,3\rangle$

(4) $u=\langle 2,-1,0\rangle$
$v=\langle -1,2,0\rangle$

[EX 23] find the area of the triangle with the given vertices. (Hint: $\frac{1}{2}\|u\times v\|$ is the area of the triangle having u and v as adjacent sides.)

(1) $(0,0,0)$, $(1,2,3)$, $(-1,0,0)$

(2) $(2,-3,4)$, $(0,1,2)$, $(-1,2,0)$

(3) $(2,-7,3)$, $(-1,5,8)$, $(4,6,-1)$

(4) $(1,2,0)$, $(-2,1,0)$, $(0,0,0)$

THEOREM 10.8

Let u, v, and w be vectors in space, and let c be a scalar.

1. $\mathrm{u}\times\mathrm{v}=-(\mathrm{v}\times\mathrm{u})$
2. $\mathrm{u}\times(\mathrm{v}+\mathrm{w})=(\mathrm{u}\times\mathrm{v})+(\mathrm{u}\times\mathrm{w})$
3. $c(\mathrm{u}\times\mathrm{v})=(c\mathrm{u})\times\mathrm{v}=\mathrm{u}\times(c\mathrm{v})$
4. $\mathrm{u}\times 0=0\times\mathrm{u}=0$
5. $\mathrm{u}\times\mathrm{u}=0$
6. $\mathrm{u}\cdot(\mathrm{v}\times\mathrm{w})=(\mathrm{u}\times\mathrm{v})\cdot\mathrm{w}$

proof To prove Property 1, let $\mathrm{u}=u_1\mathrm{i}+u_2\mathrm{j}+u_3\mathrm{k}$ and $\mathrm{v}=v_1\mathrm{i}+v_2\mathrm{j}+v_3\mathrm{k}$

Then,

$$\mathrm{u}\times\mathrm{v}=(u_2v_3-u_3v_2)\mathrm{i}-(u_1v_3-u_3v_1)\mathrm{j}+(u_1v_2-u_2v_1)\mathrm{k}$$

and

$$\mathrm{v}\times\mathrm{u}=(v_2u_3-v_3u_2)\mathrm{i}-(v_1u_3-v_3u_1)\mathrm{j}+(v_1u_2-v_2u_1)\mathrm{k}$$

which implies that $\mathrm{u}\times\mathrm{v}=-(\mathrm{v}\times\mathrm{u})$. Proofs of Properties 2, 3, 5, and 6 are left as exercises.

THEOREM 10.9 The Triple Scalar Product

For $\mathrm{u}=u_1\mathrm{i}+u_2\mathrm{j}+u_3\mathrm{k}$, $\mathrm{v}=v_1\mathrm{i}+v_2\mathrm{j}+v_3\mathrm{k}$, and $\mathrm{w}=w_1\mathrm{i}+w_2\mathrm{j}+w_3\mathrm{k}$, the triple scalar product is given by

$$\mathrm{u}\cdot(\mathrm{v}\times\mathrm{w})=\begin{vmatrix} u_1 & u_2 & u_3 \\ v_1 & v_2 & v_3 \\ w_1 & w_2 & w_3 \end{vmatrix}$$

THEOREM 10.10 Geometric Property of Triple Scalar Product

The volume V of a parallelepiped with vectors u, v, and w as adjacent edges is given by $V=|\mathrm{u}\cdot(\mathrm{v}\times\mathrm{w})|$.

proof In Figure 10.28, note that

$$\|\mathrm{v}\times\mathrm{w}\| = \text{area of base and } \|\mathrm{proj}_{\mathrm{v}\times\mathrm{w}}\mathrm{u}\| = \text{height of parallelepiped.}$$

Therefore, the volume is

$$V = (\text{height})(\text{area of base}) = \| \text{proj}_{v \times w}\, u \| \| v \times w \|$$

$$= \left| \frac{u \cdot (v \times w)}{\| v \times w \|} \right| \| v \times w \|$$

$$= | u \cdot (v \times w) |$$

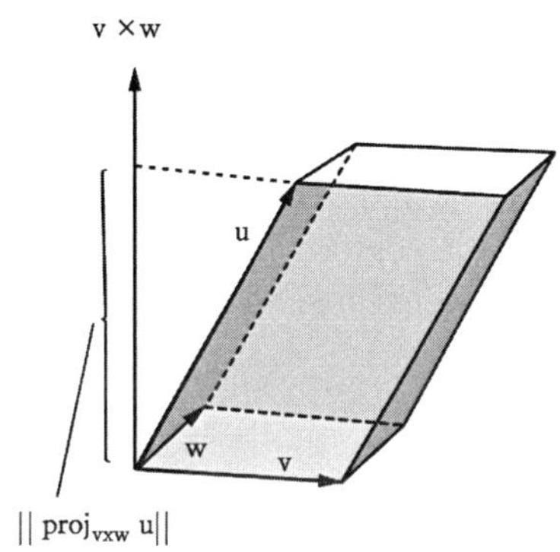

Area of base = ||v × w|| volume of

parallelepiped = | u • (v × w)|

Figure 10.28

Example 24

Find the volume of the parallelepiped shown in Figure 10.28 having $u = 3i - 5j + k$, $v = 2j - 2k$, and $w = 3i + j + k$ as adjacent edges.

Solution

By Theorem 10.10, you have

$$V = | u \cdot (v \times w) |$$

$$= \begin{vmatrix} 3 & -5 & 1 \\ 0 & 2 & -2 \\ 3 & 1 & 1 \end{vmatrix}$$

$$= 3\begin{vmatrix} 2 & -2 \\ 1 & 1 \end{vmatrix} - (-5)\begin{vmatrix} 0 & -2 \\ 3 & 1 \end{vmatrix} + (1)\begin{vmatrix} 0 & 2 \\ 3 & 1 \end{vmatrix}$$

$$= 3(4) + 5(6) + 1(-6)$$

$$= 36$$

[EX 24] Use the triple scalar product. find the volume of the parallelepiped having adjacent edges u, v, and w.

(1) $\mathrm{u} = \mathrm{i} + \mathrm{j}$
$\mathrm{v} = \mathrm{j} + \mathrm{k}$
$\mathrm{w} = \mathrm{i} + \mathrm{k}$

(2) $\mathrm{u} = \langle 1, 3, 1 \rangle$
$\mathrm{v} = \langle 0, 6, 6 \rangle$
$\mathrm{w} = \langle -1, 0, -4 \rangle$

[EX 25] find the volume of the parallelepiped with the given vertices.

(1) (0, 0, 0), (3, 0, 0), (0, 5, 1), (3, 5, 1)
(2, 0, 5), (5, 0, 5), (2, 5, 6), (5, 5, 6)

(2) (0, 0, 0), (1, 1, 0), (1, 0, 2), (0, 1, 1)
(2, 1, 2), (1, 1, 3), (1, 2, 1), (2, 2, 3)

10.5 Equations of Lines and Planes

THEOREM 10.11 Parametric Equations of a Line in Space

A line L parallel to the vector $\mathrm{v} = \langle a, b, c \rangle$ and passing through the point $P(x_1, y_1, z_1)$ is represented by the parametric equations

$$x = x_1 + at, \quad y = y_1 + bt, \quad \text{and} \quad z = z_1 + ct.$$

If the direction numbers a, b, and c are all nonzero, you can eliminate the parameter t to obtain symmetric equations of the line.

$$\frac{x - x_1}{a} = \frac{y - y_1}{b} = \frac{z - z_1}{c} \qquad \text{Symmetric equations}$$

In Figure 10.29, consider the line L through the point $P(x_1, y_1, z_1)$ and parallel to the vector $\mathrm{v} = \langle a, b, c \rangle$. The vector v is a direction vector for the line L, and a, b, and c are direction numbers. One way of describing the line L is to say that it consists of all points $Q(x, y, z)$ for which the vector $\overrightarrow{PQ}$ is parallel to v. This means that $\overrightarrow{PQ}$ is a scalar multiple of v, and you can write $\overrightarrow{PQ} = t\mathrm{v}$, where t is a scalar (a real number).

$$\overrightarrow{PQ} = \langle x - x_1, y - y_1, z - z_1 \rangle = \langle at, bt, ct \rangle = t\mathrm{v}$$

By equating corresponding components, you can obtain parametric equations of a line in space.

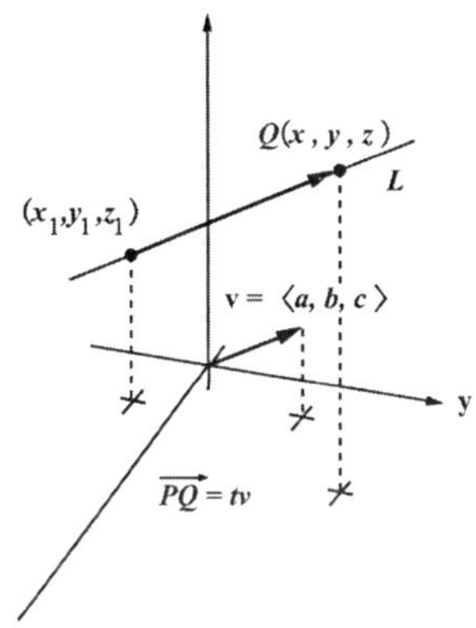

Figure 10.29

Example 25

Find parametric and symmetric equations of the line L that passes through the point (1, -2, 4) and is parallel to $\mathrm{v} = \langle 2, 4, -4 \rangle$.

Solution

To find a set of parametric equations of the line, use the coordinates $x_1 = 1$, $y_1 = -2$, and $z_1 = 4$ and direction numbers $a = 2$, $b = 4$, and $c = -4$

$$x = 1 + 2t, \quad y = -2 + 4t, \quad z = 4 = 4t$$

Because a, b, and c are all nonzero, a set of symmetric equations is

$$\frac{x-1}{2} = \frac{y+2}{4} = \frac{z-4}{-4}$$

[EX 26] Find sets of (a) parametric equations and (b) symmetric of the line through the point parallel to the given vector or line. (For each line, write the direction numbers as integers.)

	Point	Parallel to
(1)	(0, 0, 0)	$\mathrm{v} = \langle 1,2,3 \rangle$
(2)	(0, 0, 0)	$\mathrm{v} = \langle -2, \frac{5}{2}, 1 \rangle$
(3)	(–2, 0, 3)	$\mathrm{v} = 2\mathrm{i} + 4\mathrm{j} - 2\mathrm{k}$
(4)	(–3, 0, 2)	$\mathrm{v} = 6\mathrm{j} + 3\mathrm{k}$
(5)	(1, 0, 1)	$x = 3 + 3t,\ \ y = 5 - 2t,\ \ z = -7 + t$
(6)	(–3, 5, 4)	$\frac{x-1}{3} = \frac{y+1}{-2} = z - 3$

Example 26

Find a set of parametric equations of the line that passes through the points (-2,1,0) and (1,3,5).

Solution

Begin by using the points $P(-2,1,0)$ and $Q(1,3,5)$ to find a direction vector for the line passing through P and Q. given by

$$\mathrm{v} = \overrightarrow{PQ} = \langle 1-(-2), 3-1, 5-0 \rangle = \langle 3,2,5 \rangle = \langle a,b,c \rangle$$

Using the direction numbers $a=3$, $b=2$, and $c=5$ with the point $P(-2, 1, 0)$, you can obtain the parametric equations

$$x = -2+3t,\ \ y = 1+2t,\ \text{ and }\ z = 5t$$

[EX 27] Find sets of (a) parametric equations and (b) symmetric equations of the line through the two points. (For each line, write the direction numbers as integers.)

(1) $(5, -3, -2), \left(-\frac{2}{3}, \frac{2}{3}, 1\right)$

(2) $(2, 0, 2), (1, 4, -3)$

(3) $(2, 3, 0), (10, 8, 12)$

(4) $(0, 0, 25), (10, 10, 0)$

THEOREM 10.12 Standard Equation of a Plane in Space

The plane containing the point (x_1, y_1, z_1) and having a normal vector $\mathbf{n} = \langle a, b, c \rangle$ can be represented, in standard form, by the equation

$$a(x - x_1) + b(y - y_1) + c(z - z_1) = 0$$

Consider the plane containing the point $P(x_1, y_1, z_1)$ having a nonzero normal vector $\mathbf{n} = \langle a, b, c \rangle$, as shown in Figure 10.30. This plane consists of all point $Q(x, y, z)$ for which vector $\overrightarrow{PQ}$ is orthogonal to $\mathbf{n}$. Using the dot product, you can write the following.

$$\mathbf{n} \cdot \overrightarrow{PQ} = 0$$

$$\langle a, b, c \rangle \cdot \langle x - x_1, y - y_1, z - z_1 \rangle = 0$$

$$a(x - x_1) + b(y - y_1) + c(z - z_1) = 0$$

The third equation of the plane is said to be in standard form.

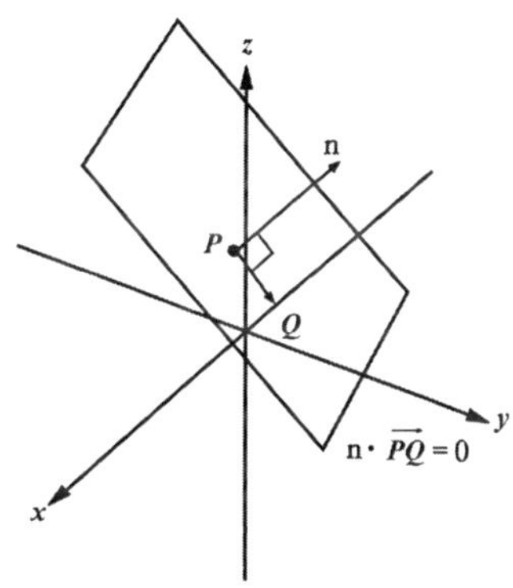

Figure 10.30

Example 27

Find an equation for the plane through $P_0(-3, 0, 7)$ perpendicular to $\mathbf{n} = 5\mathbf{i} + 2\mathbf{j} - \mathbf{k}$.

Solution

The component equation is $5(x-(-3)) + 2(y-0) + (-1)(z-7) = 0$

Simplifying, we obtain $5x + 15 + 2y - z + 7 = 0$

$$5x + 2y - z = -22$$

[EX 28] Find an equation of the plane passing through the point perpendicular to the given or line.

	Point	Parallel to
(1)	(2, 1, 2)	$\mathrm{n}=\mathrm{i}$
(2)	(1, 0, −3)	$\mathrm{n}=\mathrm{k}$
(3)	(3, 2, 2)	$\mathrm{n}=2\mathrm{i}+3\mathrm{j}-\mathrm{k}$
(4)	(0, 0, 0)	$\mathrm{n}=-3\mathrm{i}+2\mathrm{k}$
(5)	(0, 0, 6)	$x=1-t,\ \ y=2+t,\ \ z=4-2t$
(6)	(3, 2, 2)	$\dfrac{x-1}{4}=y+2=\dfrac{z+3}{-3}$

Example 28

Find an equation for the plane through $A(0,0,1)$, $B(2,0,0)$, and $C(0,3,0)$.

Solution

We find a vector normal to the plane and use it with one of the points (it does not matter which) to write an equation for the plane.
The cross product

$$\overrightarrow{AB}\times\overrightarrow{AC}=\begin{vmatrix} \mathrm{i} & \mathrm{j} & \mathrm{k} \\ 2 & 0 & -1 \\ 0 & 3 & -1 \end{vmatrix}=3\mathrm{i}+2\mathrm{j}+6\mathrm{k}$$

is normal to the plane. We substitute the components of this vector and the coordinates of $A(0,0,1)$ into the component form of the equation to obtain

$$3(x-0)+2(y-0)+6(z-1)=0$$

$$3x+2y+6z=6$$

[EX 29] Find an equation of the plane.

(1) The plane passes through (0, 0, 0), (1, 2, 3), and (−2, 3, 3).

(2) The plane passes through (2, 3, −2), (3, 4, 2), and (1, −1, 0).

(3) The plane passes through (1, 2, 3), (3, 2, 1), and (−1, −2, 2).

(4) The plane passes through the point (1, 2, 3) and is parallel to the yz−plane.

Lines of Intersection

Just as lines are parallel if and only if they have the same direction, two planes are parallel if and only if their normals are parallel, or $\mathbf{n}_1 = k\mathbf{n}_2$ for some scalar k. Two planes that are not parallel intersect in a line.

Example 29

Find a vector parallel to the line of intersection of the planes $3x-6y-2z=15$ and $2x+y-2z=5$

Solution

The line of intersection of two planes is perpendicular to both planes' normal vectors $\mathbf{n}_1$ and $\mathbf{n}_2$ (Figure 10.31) and therefore parallel to $\mathbf{n}_1 \times \mathbf{n}_2$. Turning this around, $\mathbf{n}_1 \times \mathbf{n}_2$ is a vector parallel to the planes' line of intersection. In our case,

$$\mathbf{n}_1 \times \mathbf{n}_2 = \begin{vmatrix} \mathbf{i} & \mathbf{j} & \mathbf{k} \\ 3 & -6 & -2 \\ 2 & 1 & -2 \end{vmatrix} = 14\mathbf{i} + 2\mathbf{j} + 15\mathbf{k}$$

Any nonzero scalar multiple of $\mathbf{n}_1 \times \mathbf{n}_2$ will do as well.

Example 30

Find parametric equations for the line in which the planes $3x-6y-2z=15$ and $2x+y-2z=5$ intersect.

Solution

We find a vector parallel to the line and a point on the line Example 29 identifies $\mathrm{v}=14\mathrm{i}+2\mathrm{j}+15\mathrm{k}$ as a vector parallel to the line. To find a point on the line, we can take any point common to the two planes. Substituting $z=0$ in the plane equations and solving for x and y simultaneously identifies one of these points as $(3,\,-1,\,0)$. The line is

$$x=3+14t,\quad y=-1+2t,\quad z=15t$$

The choice $z=0$ is arbitrary and we could have chosen $z=1$ or $z=-1$ just as well. Or we could have let $x=0$ and solved for y and z. The different choices would simply give different parametrizations of the same line.

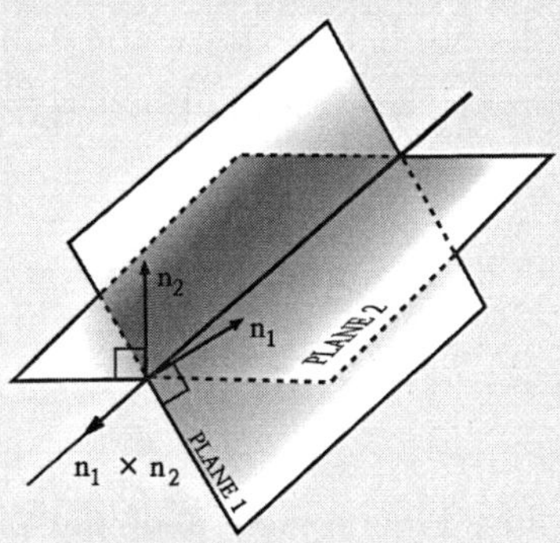

Figure 10.31

[EX 30] Find parametrizations for the lines in which the planes in (1)-(4) intersect.

(1) $x+y+z=1,\ \ x+y=2$

(2) $3x-6y-2z=3,\ \ 2x+y-2z=2$

(3) $x-2y+4z=2,\ \ x+y-2z=5$

(4) $5x-2y=11,\ \ 4y-5z=-17$

The Distance form a Point to a Plane

If P is a point on a plane with normal n, then the distance from any point S to the plane is the length of the vector projection of $\overrightarrow{PS}$ onto n. That is, the distance from S to the plane is

$$d=\left|\overrightarrow{PS}\cdot\frac{\mathrm{n}}{|\mathrm{n}|}\right|$$

where $\mathrm{n}=A\mathrm{i}+B\mathrm{j}+C\mathrm{k}$ is normal to the plane.

Example 31

Find the distance from $S(1,1,3)$ to the plane $3x+2y+6z=6$.

Solution

We find a point P in the plane and calculate the length of the vector projection of $\overrightarrow{PS}$ onto a vector n normal to the plane (Figure 10.32). The coefficients in the equation $3x+2y+6z=6$ give

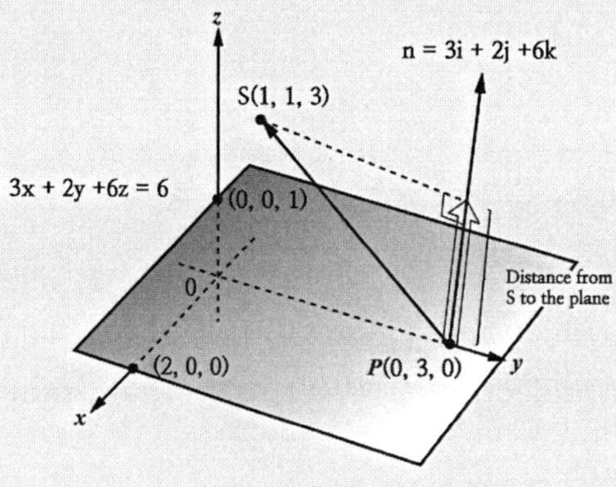

Figure 10.32

The points on the plane easiest to find form the plane's equation are the intercepts. If we take P to be the y−intercept (0, 3, 0), then

$$\overrightarrow{PS}=(1-0)\mathrm{i}+(1-3)\mathrm{j}+(3-0)\mathrm{k}$$

$$=\mathrm{i}-2\mathrm{j}+3\mathrm{k}$$

$$|\mathrm{n}|=\sqrt{(3)^2+(2)^2+(6)^2}=\sqrt{49}=7$$

The distance from S to the plane is

$$d=\left|\overrightarrow{PS}\cdot\frac{\mathrm{n}}{|\mathrm{n}|}\right|$$

$$=\left|(\mathrm{i}-2\mathrm{j}+3\mathrm{k})\cdot\left(\frac{3}{7}\mathrm{i}+\frac{2}{7}\mathrm{j}+\frac{6}{7}\mathrm{k}\right)\right|$$

$$=\left|\frac{3}{7}-\frac{4}{7}+\frac{18}{7}\right|=\frac{17}{7}$$

[EX 31] Find the distance from the point to the plane.

(1) $(2, -3, 4)$, $x+2y+2z=13$

(2) $(0, 1, 1)$, $4y+3z=-12$

(3) $(0, -1, 0)$, $2x+y+2z=4$

(4) Find the distance from the plane $x+2y+6z=1$ to the plane $x+2y+6z=10$

The Distance from a Point to a Line in Space

To find the distance form a point S to a line that passes through a point P parallel to a vector v, we find the absolute value of the scalar component of $\overrightarrow{PS}$ in the direction of a vector normal to the line (Figure 10.33). In the notation of the figure, the absolute value of the scalar component is $|\overrightarrow{PS}|\sin\theta$, which is $\dfrac{|\overrightarrow{PS}\times\mathrm{v}|}{|\mathrm{v}|}$.

Distance from a Point S to a Line Through P Parallel to v

$$d=\frac{|\overrightarrow{PS}\times\mathrm{v}|}{|\mathrm{v}|}$$

Example 32

Find the distance from the point $S(1, 1, 5)$ to the line

$L: \quad x=1+t, \quad y=3-t, \quad z=2t$

Solution

We see from the equations for L passes through $P(1, 3, 0)$ parallel to $\mathbf{v}=\mathbf{i}-\mathbf{j}+2\mathbf{k}$. With

$$\overrightarrow{PS}=(1-1)\mathbf{i}+(1-3)\mathbf{j}+(5-0)\mathbf{k}=-2\mathbf{j}+5\mathbf{k}$$

and

$$\overrightarrow{PS}\times\mathbf{v}=\begin{vmatrix} \mathbf{i} & \mathbf{j} & \mathbf{k} \\ 0 & -2 & 5 \\ 1 & -1 & 2 \end{vmatrix}=\mathbf{i}+5\mathbf{j}+2\mathbf{k}$$

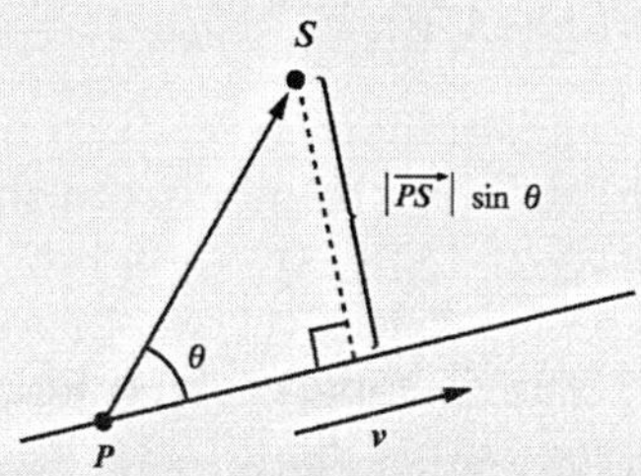

Figure 10.33

Equation gives

$$d=\frac{|\overrightarrow{PS}\times\mathbf{v}|}{|\mathbf{v}|}=\frac{\sqrt{1+25+4}}{\sqrt{1+1+4}}=\frac{\sqrt{30}}{\sqrt{6}}=\sqrt{5}$$

[EX 32] Find the distance from the point to the line.

(1) $(0, 0, 12)$; $\quad x=4t, \quad y=-2t, \quad z=2t$

(2) $(2, 1, 3)$; $\quad x=2+2t, \quad y=1+6t, \quad z=3$

(3) $(3, -1, 4)$; $\quad x=4-t, \quad y=3+2t, \quad z=-5+3t$

(4) $(-1, 4, 3)$; $\quad x=10+4t, \quad y=-3, \quad z=4t$

Angles Between Planes

The angle between two intersecting planes is defined to be the acute angle between their normal vectors (Figure 10.34).

Example 33

Find the angle between the planes $3x-6y-2z=15$ and $2x+y-2z=5$.

Solution

The vectors

$$\mathrm{n}_1 = 3\mathrm{i}-6\mathrm{j}-2\mathrm{k}, \quad \mathrm{n}_2 = 2\mathrm{i}+\mathrm{j}-2\mathrm{k}$$

are normals to the planes. The angle between them is

$$\begin{aligned}\theta &= \cos^{-1}\left(\frac{\mathrm{n}_1 \cdot \mathrm{n}_2}{|\mathrm{n}_1||\mathrm{n}_2|}\right)\\ &= \cos^{-1}\left(\frac{4}{21}\right)\\ &\approx 1.38 \text{ radians.}\end{aligned}$$

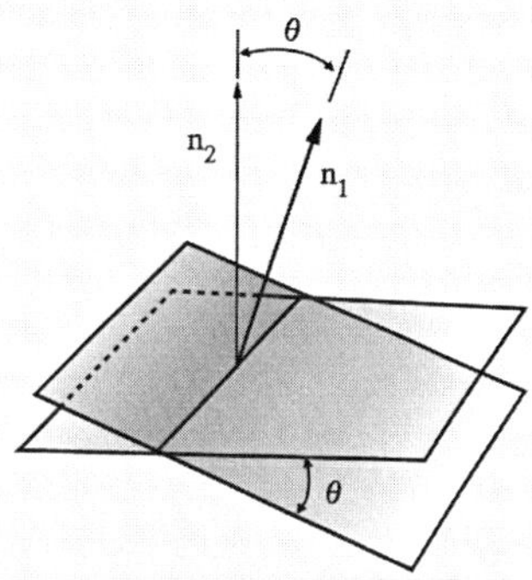

Figure 10.34

[EX 33] Find the angles between the planes

(1) $x+y=1, \quad 2x+y-2z=2$

(2) $5x+y-z=10, \quad x-2y+3z=-1$

[EX 34] Find an equation of the plane

(1) The plane contains the lines given by

$$\frac{x-1}{-2} = y-4 = z \quad \text{and} \quad \frac{x-2}{-3} = \frac{y-1}{4} = \frac{z-2}{-1}$$

(2) The plane passes through the point $(2,\ 2,\ 1)$ and contains the line given by

$$\frac{x}{2} = \frac{y-4}{-1} = z$$

(3) The plane passes through the points $(2, 2, 1)$ and $(-1, 1, -1)$ and is perpendicular to the plane $2x-3y+z=3$

(4) The plane passes through the points $(3, 2, 1)$ and $(3, 1, -5)$ and is perpendicular to the plane $6x+7y+2z=10$

(5) The plane passes through the points $(1, -2, -1)$ and $(2, 5, 6)$ and is parallel to the $x-$axis

(6) The plane passes through the points $(4, 2, 1)$ and $(-3, 5, 7)$ and is parallel to the $z-$axis

ANSWERS AND EXPLANATIONS

CALCULUS

CHAPTER 02. Limits and Continuity

2.1 Limit Existence:

EX 1

(1) 1 (2) 0 (3) 0 (4) 0
(5) DNE (6) 0 (7) 0 (8) DNE
(9) DNE (10) 1

EX 2

(1) (a) -1 (b) 1 (c) DNE (d) DNE
(2) (a) $+\infty$ (b) 2 (c) DNE (d) -2
(3) (a) $-\infty$ (b) $-\infty$ (c) $-\infty$ (d) 1
(4) (a) 0 (b) 0 (c) 0 (d) 0

EX 3

(1)

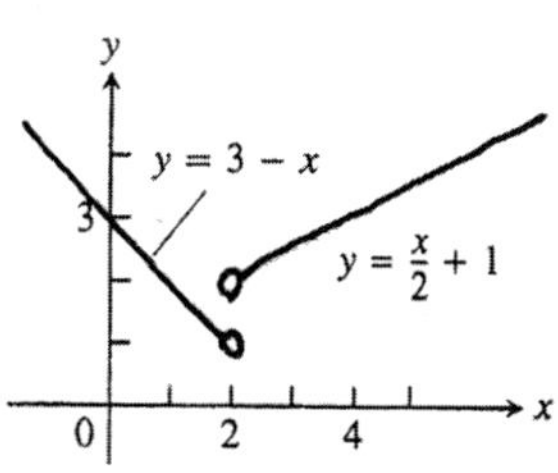

(a) 2
(b) 1
(c) DNE
(d) 3
(e) 3
(f) 3

(2)

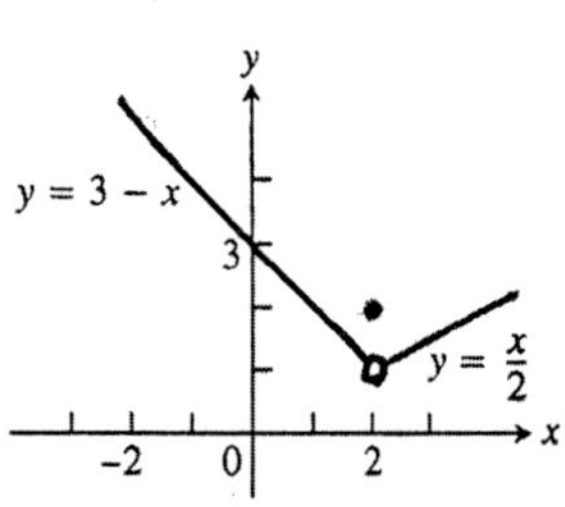

(a) 1
(b) 1
(c) 2
(d) 1
(e) 4
(f) 4
(g) 4

(3)

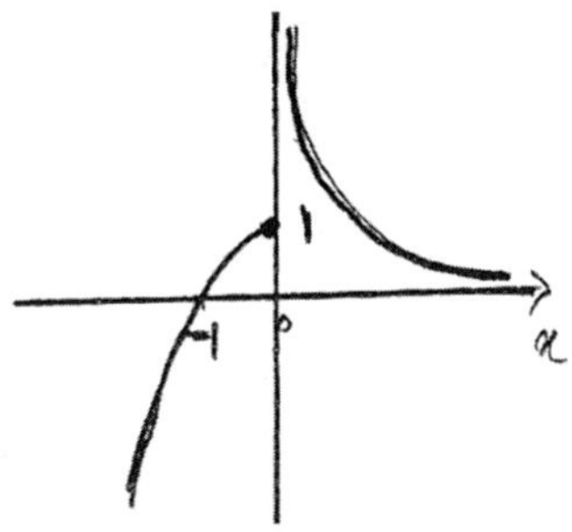

(a) 1
(b) $+\infty$
(c) DNE
(d) 1

(4)

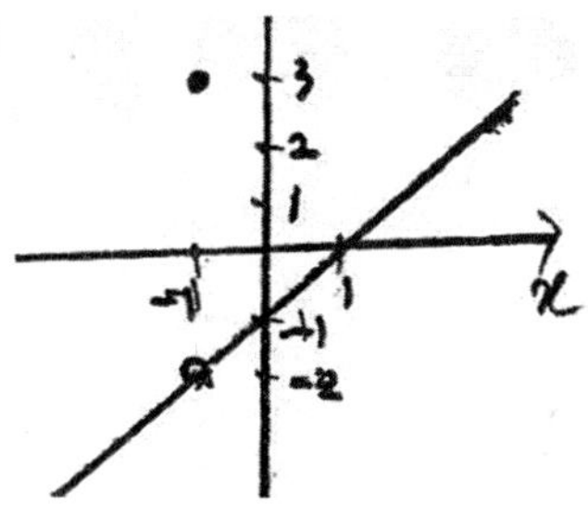

(a) -2
(b) -2
(c) -2
(d) 3

(5)

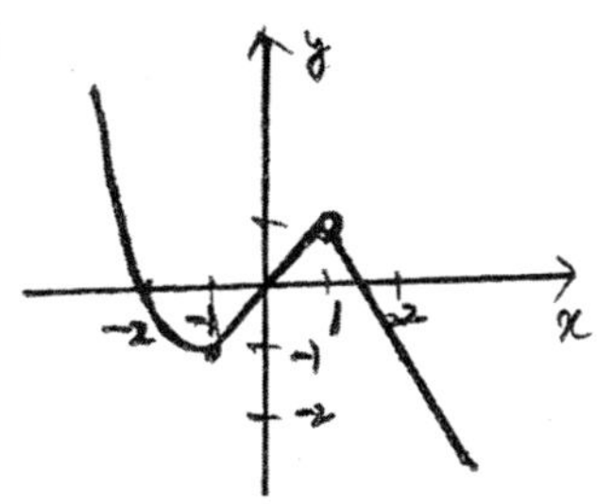

(a) -1
(b) -1
(c) -1
(d) 1
(e) 0
(f) DNE

(6)

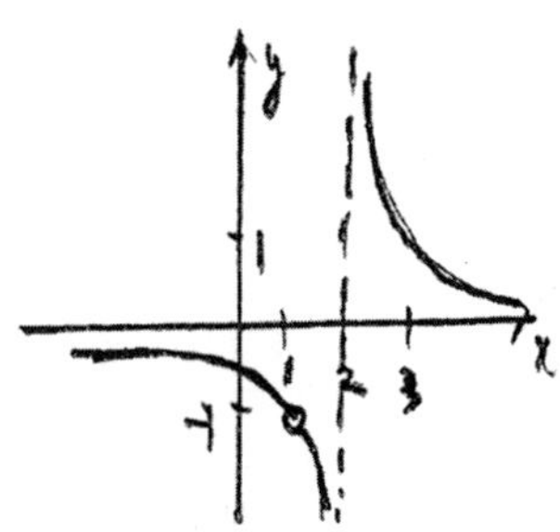

(a) -1
(b) -1
(c) -1
(d) 1

EX 4

(1) c^3+4c^2-3

(2) $\lim_{x\to c}\frac{x^4+x^2-1}{x^2+5}=\frac{\lim_{x\to c}(x^4+x^2-1)}{\lim_{x\to c}(x^2+5)}$

$=\frac{\lim_{x\to c}x^4+\lim_{x\to c}x^2-\lim_{x\to c}1}{\lim_{x\to c}x^2+\lim_{x\to c}5}$

$=\frac{c^4+c^2-1}{c^2+5}$

(3) $\frac{1}{2}$

(4) $\lim_{x\to -2}\sqrt{4x^2-3}=\sqrt{\lim_{x\to -2}(4x^2-3)}$

$=\sqrt{\lim_{x\to -2}4x^2-\lim_{x\to -2}3}=\sqrt{4(-2)^2-3}$

$=\sqrt{16-3}=\sqrt{13}$

EX 5

(1) -10

(2) $\lim_{x\to c}2f(x)g(x)=2\left[\lim_{x\to c}f(x)\right]\left[\lim_{x\to c}g(x)\right]$

$=2(5)(-2)=-20$

(3) -1

(4) $\lim_{x\to c}\frac{f(x)}{f(x)-g(x)}=\frac{\lim_{x\to c}f(x)}{\lim_{x\to c}f(x)-\lim_{x\to c}g(x)}$

$=\frac{5}{5-(-2)}=\frac{5}{7}$

EX 6

(1) $=\frac{3}{2}$

(2) $=\lim_{x\to 4}\frac{x(x-4)}{(x-1)(x-4)}=\lim_{x\to 4}\frac{x}{(x-1)}=\frac{4}{3}$

(3) $=\frac{3}{8}$

(4) $=\lim_{x\to 2}\frac{(x-2)(x^2+2x+4)}{(x-2)(x-1)}=\lim_{x\to 2}\frac{x^2+2x+4}{x-1}$

$=\frac{4+4+4}{1}=12$

(5) $=-\frac{1}{2}$

(6) $=\lim_{x\to 2}\frac{(x-2)(x^2-4)}{x^2(x-2)}=\lim_{x\to 2}\frac{(x^2-4)}{x^2}=\frac{0}{4}=0$

(7) $=4$

(8) $=\lim_{x\to 1}\frac{x^3+3x^2+3x+1-8}{x-1}$

$=\lim_{x\to 1}\frac{x^3+3x^2+3x-7}{x-1}$

$=\lim_{x\to 1}\frac{(x-1)(x^2+4x+7)}{x-1}=\lim_{x\to 1}(x^2+4x+7)$

$=1+4+7=12$

EX 7

(1) $=\frac{1}{2}$

(2) $=\lim_{x\to 4}\frac{(x-4)(2+\sqrt{x})}{(2-\sqrt{x})(2+\sqrt{x})}$

$=\lim_{x\to 4}\frac{(x-4)(2+\sqrt{x})}{4-x}=\lim_{x\to 4}\frac{(2+\sqrt{x})}{-1}$

$=-4$

(3) $=4$

(4) $-\lim_{x\to 3}\frac{(\sqrt{x+1}-2)(\sqrt{x+1}+2)(\sqrt{x-2}+1)}{(\sqrt{x-2}-1)(\sqrt{x+1}+2)(\sqrt{x-2}+1)}$

$=\lim_{x\to 3}\frac{(x+1-4)(\sqrt{x-2}+1)}{(x-2-1)(\sqrt{x+1}+2)}$

$=\lim_{x\to 3}\frac{(x-3)(\sqrt{x-2}+1)}{(x-3)(\sqrt{x+1}+2)}$

$=\lim_{x\to 3}\frac{(\sqrt{x-2}+1)}{(\sqrt{x+1}+2)}=\frac{1+1}{2+2}=\frac{1}{2}$

(5) $=-3$

EX 8

(1) = -2

(2) $= \lim_{h\to 0}\dfrac{\left[\dfrac{1}{(h+3)^2}-\dfrac{1}{9}\right]\cdot 9(h+3)^2}{h\cdot 9(h+3)^2}$

$= \lim_{h\to 0}\dfrac{9-(h+3)^2}{h\cdot 9(h+3)^2}$

$= \lim_{h\to 0}\dfrac{9-h^2-6h-9}{h\cdot 9(h+3)^2} = \lim_{h\to 0}\dfrac{-h^2-6h}{9h(h+3)^2}$

$= \lim_{h\to 0}\dfrac{-h-6}{9(h+3)^2} = -\dfrac{6}{81} = -\dfrac{2}{27}$

(3) $= -\dfrac{1}{54}$

(4) $= \lim_{x\to 2}\dfrac{\left(\dfrac{1}{\sqrt{x+2}}-\dfrac{1}{2}\right)\cdot 2\sqrt{x+2}}{(x^2-4)\cdot 2\sqrt{x+2}}$

$= \lim_{x\to 2}\dfrac{2-\sqrt{x+2}}{(x^2-4)\cdot 2\sqrt{x+2}}$

$= \lim_{x\to 2}\dfrac{(2-\sqrt{x+2})\cdot(2+\sqrt{x+2})}{(x^2-4)\cdot 2\sqrt{x+2}\cdot(2+\sqrt{x+2})}$

$= \lim_{x\to 2}\dfrac{4-(x+2)}{(x^2-4)\cdot 2\sqrt{x+2}\cdot(2+\sqrt{x+2})}$

$= \lim_{x\to 2}\dfrac{2-x}{(x+2)(x-2)\cdot 2\sqrt{x+2}\cdot(2+\sqrt{x+2})}$

$= \lim_{x\to 2}\dfrac{-1}{(x+2)2\sqrt{x+2}(2+\sqrt{x+2})}$

$= \dfrac{-1}{(4)(2)(2)(2+2)} = -\dfrac{1}{64}$

EX 9

(1) (a)

$\lim_{x\to -2^+}(x+3)\dfrac{|x+2|}{x+2} = \lim_{x\to -2^+}(x+3)\dfrac{(x+2)}{(x+2)}$

$(|x+2| = x+2 \text{ for } x > -2)$

$= \lim_{x\to -2^+}(x+3) = (-2)+3 = 1$

(b)

$\lim_{x\to -2^-}(x+3)\dfrac{|x+2|}{x+2} = \lim_{x\to -2^-}(x+3)\left[\dfrac{-(x+2)}{(x+2)}\right]$

$(|x+2| = -(x+2) \text{ for } x < -2)$

$= \lim_{x\to -2^-}(x+3)(-1) = -(-2+3) = -1$

(2) (a)

$\lim_{x\to 1^+}\dfrac{\sqrt{2}(x+1)}{|x+1|} = \lim_{x\to 1^+}\dfrac{\sqrt{2}(x+1)}{(x+1)}$

$(|x-1| = x-1 \text{ for } x > 1) = \lim_{x\to 1^+}\sqrt{2} = \sqrt{2}$

(b) $\lim_{x\to 1^-}\dfrac{\sqrt{2}(x-1)}{|x-1|} = \lim_{x\to 1^-}\dfrac{\sqrt{2}(x-1)}{-(x-1)}$

$(|x-1| = -(x-1) \text{ for } x < 1)$

$= \lim_{x\to 1^-} -\sqrt{2} = -\sqrt{2}$

(3) (a) $\lim_{\theta\to 3^+}\dfrac{\lfloor\theta\rfloor}{\theta} = \dfrac{3}{3} = 1$

(b) $\lim_{\theta\to 3^-}\dfrac{\lfloor\theta\rfloor}{\theta} = \dfrac{2}{3}$

(4) (a) $\lim_{t\to 4^+}(t-\lfloor t\rfloor) = 4-4 = 0$

(b) $\lim_{t\to 4^-}(t-\lfloor t\rfloor) = 4-3 = 1$

EX 10

(1) $= \dfrac{3}{0^-} = -\infty$

(2) $= \dfrac{-1}{0^-} = +\infty$

(3) $= \dfrac{-16}{0^+} = -\infty$

(4) DNE $\left(\lim_{x\to 2^-}\dfrac{1}{x-2} = -\infty, \lim_{x\to 2^+}\dfrac{1}{x-2} = +\infty\right)$

(5) $= \infty$

(6) $= \dfrac{-1}{0^+} = -\infty$

(7) $= +\infty$

(8) $= \dfrac{1}{0^+} = +\infty$

(9) $= +\infty$

(10) $= -\infty$

(11) $= +\infty$

(12) $= -\infty$

EX 11

(1) $\lim_{x \to -1^-} f(x) = \lim_{x \to -1^-} (x+2) = 1,\ \lim_{x \to -1^+} f(x)$

$= \lim_{x \to -1^+} (x^2+1) = 2$

$\lim_{x \to -1^-} f(x) \neq \lim_{x \to -1^+} f(x)$

$\therefore \lim_{x \to -1} f(x) = DNE$

(2) $\lim_{x \to 3^-} f(x) = \lim_{x \to 3^-} 4\cos\left(\frac{\pi}{9}x\right) = 4\cos\left(\frac{\pi}{3}\right) = 4\left(\frac{1}{2}\right) = 2$

, $\lim_{x \to 3^+} f(x) = \lim_{x \to 3^+} \sqrt{x^2-5} = \sqrt{9-5} = 2$

$\lim_{x \to 3^-} f(x) = \lim_{x \to 3^+} f(x) = 2$

$\therefore \lim_{x \to 3} f(x) = 2$

(3) $\lim_{x \to \ln 3} f(x) = 6$

EX 12

(1) $a = 4, b = 0$

(2) $\lim_{x \to 1} \frac{x^2-a}{x-1}$

must me the indeterminate form $\left(\frac{0}{0}\right)$

because the limit exists.

$\lim_{x \to 1} (x^2-a) = 0 \quad 1^2 - a = 0,\ \therefore a = 1$

$\lim_{x \to 1} \frac{x^2-1}{x-1} = b$

$\lim_{x \to 1} (x+1) = b$

$\therefore b = 2$

(3) $a = 4, b = 4\sqrt{2}$

(4) $\lim_{x \to 2} \frac{2-x}{ax^2+x-6}$

must be the indeterminate form $\left(\frac{0}{0}\right)$

because the limit exists and is not equal to zero.

$\lim_{x \to 2} (ax^2+x-6) = 0$

$a(2)^2 + (2) - 6 = 0$

$4a = 4$

$\lim_{x \to 2} \frac{2-x}{x^2+x-6} = b$

$\lim_{x \to 2} \frac{2-x}{(x-2)(x+3)} = b$

$\lim_{x \to 2} \frac{-1}{x+3} = b$

$\therefore a = 1 \qquad \therefore b = -\frac{1}{5}$

2.2 Squeezing Theorem (Sandwich Theorem)

EX 1

(1) $-1 \le \cos\left(\frac{1}{x}\right) \le 1$

$-x^2 \le x^2\cos\left(\frac{1}{x}\right) \le x^2$

$\lim_{x \to 0}(-x^2) = \lim_{x \to 0} x^2 = 0.$

By the Squeezing Theorem,

$\lim_{x \to 0} x^2\cos\left(\frac{1}{x}\right) = 0.$

(2) $-1 \le \sin\left(\frac{1}{x^2}\right) \le 1$

$-x \le x\sin\left(\frac{1}{x^2}\right) \le x$

$\lim_{x \to 0}(-x) = \lim_{x \to 0}(x) = 0$

By the Squeezing Theorem,

$\lim_{x \to 0} x\sin\left(\frac{1}{x^2}\right) = 0.$

(3) $-1 \le \sin x \le 1$

$$-\frac{1}{x} \le \frac{\sin x}{x} \le \frac{1}{x}$$

$$\lim_{x\to\infty}\left(-\frac{1}{x}\right)=\lim_{x\to\infty}\left(\frac{1}{x}\right)=0$$

By the Squeezing Theorem,

$$\lim_{x\to\infty}\frac{\sin x}{x}=0.$$

(4) $-1 \le \cos 2x \le 1$

$$-\frac{1}{x^2} \le \frac{\cos x}{x^2} \le \frac{1}{x^2}$$

$$\lim_{x\to\infty}\left(-\frac{1}{x^2}\right)=\lim_{x\to\infty}\left(\frac{1}{x^2}\right)=0$$

By the Squeezing Theorem,

$$\lim_{x\to\infty}\frac{\cos 2x}{x^2}=0.$$

EX 2

(1) 1

(2) $\lim_{x\to -1}(-6x-4)=\lim_{x\to -1}(x^4+x^2)=2$

By the Squeezing Theorem $\lim_{x\to -1} f(x)=2$

(3) $\sqrt{5}$

2.3 Limit of Trigonometric Functions

EX 1

(1) 1

(2) $=\lim_{x\to 0}\frac{\sin 2x}{2x}\cdot\frac{2}{3}$

$=1\cdot\frac{2}{3}=\frac{2}{3}$

(3) $\frac{3}{4}$

(4) $=\lim_{x\to 0}\frac{\frac{\sin 5x}{5x}\cdot 5x}{\frac{\sin 2x}{2x}\cdot 2x}$

$=\lim_{x\to 0}\frac{5x}{2x}=\frac{5}{2}$

(5) $=\left(\lim_{x\to 0}\frac{\sin 2x}{2x}\right)^3$

$=\left(\lim_{x\to 0}\frac{\sin 2x}{2x}\cdot 2\right)^3$

$=(1\cdot 2)^3=8$

(6) $=\lim_{x\to 0}\left(\frac{\sin 3x}{\sin 2x}\right)^2$

$=\lim_{x\to 0}\left(\frac{\frac{\sin 3x}{3x}\cdot 3x}{\frac{\sin 2x}{2x}\cdot 2x}\right)^2$

$=\left(\lim_{x\to 0}\frac{3x}{2x}\right)^2=\left(\frac{3}{2}\right)^2=\frac{9}{4}$

(7) 2

(8) $=\lim_{x\to 0}\frac{\tan 3x}{3x}\cdot\frac{3}{2}$

$1\cdot\frac{3}{2}=\frac{3}{2}$

(9) $\frac{3}{8}$

(10) $=\left(\lim_{x\to 0}\frac{\tan 3x}{2x}\right)^3$

$=\left(\lim_{x\to 0}\frac{\tan 3x}{3x}\cdot\frac{3}{2}\right)^3=\left(\frac{3}{2}\right)^3=\frac{27}{8}$

(11) $\frac{2}{3}$

(12) $=\lim_{x\to 0}\frac{\sin^2 x}{x^2}\cdot\sin x$

$=1\cdot 0=0$

(13) 0

(14) $=\lim_{x\to 0^+}\frac{\sin x\sqrt{x}\sqrt{x}}{\sin\sqrt{x}\cdot x}$

$=\lim_{x\to 0^+}\sqrt{x}$

$=0$

(15) 1

(16) 0

(17) $=\lim_{x\to 0}\frac{\sin(1-\cos x)}{(1-\cos x)}\cdot\frac{(1-\cos x)}{x^2}$

$=\lim_{x\to 0}\frac{(1-\cos x)(1+\cos x)}{x^2(1+\cos x)}$

$=\lim_{x\to 0}\frac{\sin^2 x}{x^2(1+\cos x)}$

$=\lim_{x\to 0}\left(\frac{\sin x}{x}\right)^2\cdot\frac{1}{1+\cos x}$

$=1\cdot\frac{1}{1+1}=\frac{1}{2}$

2.4 Limit of Exponential and Logarithmic Functions

EX 1

(1) e^3

(2) $=\left[\lim_{x\to\infty}\left(1+\frac{3}{x}\right)^{\frac{x}{3}}\right]^3=e^3$

(3) e^3

(4) $=\left[\lim_{x\to-\infty}\left(1+\frac{2}{3x}\right)^{\frac{3x}{2}}\right]^{\frac{2}{3}}=e^{\frac{2}{3}}$

(5) $e^{\frac{3}{2}}$

(6) $=\lim_{n\to\infty}P\left[\left(1+\frac{r}{n}\right)^{\frac{n}{r}}\right]^{rt}=Pe^{rt}$

(7) e^3

(8) $=\left[\lim_{x\to 0}\left(1+\frac{x}{2}\right)^{\frac{2}{x}}\right]^{\frac{1}{4}}=e^{\frac{1}{4}}$

(9) e^3

(10) $=\left[\lim_{x\to 0}\left(1+\frac{x}{-2}\right)^{\frac{-2}{x}}\right]^{-\frac{1}{2}}=e^{-\frac{1}{2}}$

(11) e^6

(12) $=\lim_{x\to\infty}\left(2+\frac{1}{x}\right)^{3x}=(2+0)^{\infty}=\infty$

EX 2

(1) $=\lim_{x\to 0}\frac{1}{x}\cdot\ln(x+1)$

$=\lim_{x\to 0}\ln(x+1)^{\frac{1}{x}}$

$=\ln e=1$

(2) Let $e^x-1=t$

$e^x=1+t$

$x=\ln(1+t)$

As $x\to 0, t\to 0$

$=\lim_{x\to 0}\frac{e^x-1}{x}=\lim_{t\to 0}\frac{t}{\ln(1+t)}$

$=\lim_{t\to 0}\frac{1}{\frac{1}{t}\ln(1+t)}$

$=\lim_{t\to 0}\frac{1}{\ln(1+t)^{\frac{1}{t}}}=\frac{1}{\ln e}=1$

(3) $\log_a e$

(4) $\ln a$

2.5 Limits Involving Infinity

EX 1

(1) ∞

(2) $= \lim_{x\to\infty} \dfrac{\frac{3x^3+4}{x^3}}{\frac{2x^3-3x+5}{x^3}} = \lim_{x\to\infty} \dfrac{3+\frac{4}{x^3}}{2-\frac{3}{x^2}+\frac{5}{x^3}}$

$= \dfrac{3+0}{2-0+0} = \dfrac{3}{2}$

(3) 0

(4) $= \lim_{x\to\infty} \dfrac{\frac{3x^2+1}{x^2}}{\frac{5x^2-2}{x^2}} = \lim_{x\to\infty} \dfrac{3+\frac{1}{x^2}}{5-\frac{2}{x^2}}$

$= \dfrac{3+0}{5-0} = \dfrac{3}{5}$

(5) 1

(6) $= \lim_{x\to-\infty} \dfrac{\frac{2x^4-4x}{x^5}}{\frac{x^5+9x-5}{x^5}} = \lim_{x\to-\infty} \dfrac{\frac{2}{x}-\frac{4}{x^4}}{1+\frac{9}{x^4}-\frac{5}{x^5}}$

$= \dfrac{0-0}{1+0-0} = 0$

(7) ∞

(8) 1

(9) -1

(10) $\dfrac{1}{2}$

(11) 3

EX 2

(1) $y=\dfrac{2}{5}$

(2) $y=-\dfrac{1}{2}$

(3) no horizontal asymptote

(4) $y=0$

(5) $y=0$

(6) $y=-2$

(7) $\lim_{x\to\pm\infty}\left(\dfrac{7}{x+4}+5\right)=(0+5)=5$

$\therefore y=5$

(8) $\lim_{x\to\pm\infty}\left(\dfrac{2x}{x^2+4}+2\right)=(0+2)=2$

$\therefore y=2$

(9) $\lim_{x\to\pm\infty}\left(-\dfrac{4x}{2x-3}+5\right)=(-2+5)=3$

$\therefore y=3$

EX 3

(1) (a) $x=3$

(b) $\lim_{x\to3^-} y = -\infty$

$\lim_{x\to3^+} y = +\infty$

(2) (a) $x=-2$

(b) $\lim_{x\to-2^-} y = \lim_{x\to-2^-}\left(\dfrac{x^2}{x+2}\right) = \dfrac{4}{0^-} = -\infty$

$\lim_{x\to-2^+} y = \lim_{x\to-2^+}\left(\dfrac{x^2}{x+2}\right) = \dfrac{4}{0^+} = +\infty$

(3) (a) $x=0$

(b) $\lim_{x\to0^-} y = +\infty$

$\lim_{x\to0^+} y = +\infty$

(4) (a) $x=\pm1$

(b) $\lim_{x\to-1^-}\left(\dfrac{x^2}{1-x^2}\right)=\dfrac{1}{0^-}=-\infty$

$\lim_{x\to-1^+}\left(\dfrac{x^2}{1-x^2}\right)=\dfrac{1}{0^+}=+\infty$

$\lim_{x\to1^-}\left(\dfrac{x^2}{1-x^2}\right)=\dfrac{1}{0^+}=+\infty$

$\lim_{x\to1^+}\left(\dfrac{x^2}{1-x^2}\right)=\dfrac{1}{0^-}=-\infty$

(5) (a) $x=\pm 2$

(b) $\lim_{x\to -2^-} y=-\infty$

$\lim_{x\to -2^+} y=+\infty$

$\lim_{x\to 2^-} y=-\infty$

$\lim_{x\to 2^+} y=+\infty$

(6) (a) $x=2$

(b) *Domain* : $x<2$

$\lim_{x\to 2^-}\left(\frac{x}{\sqrt{2-x}}+3\right)$

$=\frac{2}{0^+}+3=+\infty$

EX 4

(1) (a) $\therefore y=2$(to the right),

$y=-2$(to the left).

(b) $\therefore |x|-1=0 \Rightarrow \therefore x=\pm 1$

(2) $|x|=\begin{cases} x & ,\text{if } x\ge 0 \\ -x & ,\text{if } x\le 0 \end{cases}$

(a) $\lim_{x\to\infty} y=\lim_{x\to\infty}\left(\frac{x-1}{2|x|-1}\right)=\lim_{x\to\infty}\frac{x-1}{2x-1}=\frac{1}{2}$

$\lim_{x\to -\infty} y=\lim_{x\to -\infty}\left(\frac{x-1}{2|x|-1}\right)=\lim_{x\to -\infty}\frac{x-1}{-2x-1}$

$=-\frac{1}{2}$

$\therefore y=\frac{1}{2}$ (to the right),

$y=-\frac{1}{2}$ (to the left)

(b) no vertical asymptote

(3) (a) $\therefore y=5$ (to the right),

$y=-2$ (to the left)

(b) $4^x-1=0$

$4^x=1$

$\therefore x=0$

(4) (a) $\lim_{x\to\infty} y=\lim_{x\to\infty}\left(\frac{3(2)^x+1}{(2)^x-3}\right)=\lim_{x\to\infty}\left(\frac{3+\frac{1}{2^x}}{1-\frac{3}{2^x}}\right)$

$=\frac{3+0}{1-0}=3$

$\lim_{x\to -\infty} y=\lim_{x\to -\infty}\left(\frac{3(2)^x+1}{(2)^x-3}\right)=\frac{0+1}{0-3}$

$=-\frac{1}{3}$

$\therefore y=3$ (to the right),

$y=-\frac{1}{3}$ (to the left)

(b) $2^x-3=0$

$2^x=3$

$\therefore x=\log_2 3$

EX 5

(1) $\lim_{x\to\left(\frac{\pi}{2}\right)^-} \tan x=+\infty$

(2) $\lim_{x\to\left(\frac{\pi}{2}\right)^+} \sec x=-\infty$

(3) $\lim_{x\to 0^-} \csc x=-\infty$

(4) $\lim_{x\to 0^+} \cot x=+\infty$

(5) $\lim_{x\to 0^+} \ln x=-\infty$

(6) $\lim_{x\to 2^-} \log_3(2-x)=-\infty$

EX 6

(1) 0

(2) $+\infty$

(3) $=e^{-\infty}=0$

(4) $=(0.9)^{+\infty}=0$

(5) $=e^{+\infty}=+\infty$

(6) $=\frac{1}{\ln(0^+)}=\frac{1}{-\infty}=0$

(7) $=\left(\dfrac{3^{-\infty}+1}{3^{-\infty}-1}\right)=\dfrac{0+1}{0-1}=-1$

(8) $-\dfrac{\pi}{2}$

EX 7

(1) DNE

(2) $-1\le \sin x\le 1 \Rightarrow -\dfrac{1}{x}\le \dfrac{\sin x}{x}\le \dfrac{1}{x}$

$\Rightarrow \lim\limits_{x\to\infty}\left(-\dfrac{1}{x}\right)=\lim\limits_{x\to\infty}\left(\dfrac{1}{x}\right)=0 \Rightarrow$ By the Squeezing Theorem, $\lim\limits_{x\to\infty}\dfrac{\sin x}{x}=0$.

$-1\le \sin\left(\dfrac{1}{x}\right)\le 1 \Rightarrow -x\le x\sin\left(\dfrac{1}{x}\right)\le x$

$\Rightarrow \lim\limits_{x\to\infty}(-x)=\lim\limits_{x\to\infty}(x)=0 \Rightarrow$ By the Squeezing Theorem, $\lim\limits_{x\to\infty}x\sin\left(\dfrac{1}{x}\right)=0$.

(3) DNE

(4) $\lim\limits_{x\to\infty}x\sin\dfrac{1}{x}=\lim\limits_{x\to\infty}\dfrac{\sin\left(\dfrac{1}{x}\right)}{\dfrac{1}{x}}=1$

$\left(\because \text{As } x\to\infty, \dfrac{1}{x}\to 0\right)$

EX 8

(1) $\dfrac{3}{\sqrt{2}}$

(2) $=\lim\limits_{x\to\infty}\dfrac{\dfrac{x+3}{\sqrt{x^2}}}{\dfrac{\sqrt{x^2+1}+2}{\sqrt{x^2}}}$

$=\lim\limits_{x\to\infty}\dfrac{\dfrac{x}{|x|}+\dfrac{3}{|x|}}{\sqrt{1+\dfrac{1}{x}}+\dfrac{2}{|x|}}$

$=\lim\limits_{x\to\infty}\dfrac{\dfrac{x}{x}+0}{\sqrt{1+0}+0}=\dfrac{1}{1}=1$

(3) 1

(4) $=\lim\limits_{x\to\infty}\dfrac{\dfrac{\sqrt{x^2-1}+3}{\sqrt{x^2}}}{\dfrac{2x}{\sqrt{x^2}}}$

$=\lim\limits_{x\to\infty}\dfrac{\sqrt{1-\dfrac{1}{x^2}}+\dfrac{3}{|x|}}{\dfrac{2x}{|x|}}$

$=\dfrac{\sqrt{1-0}+0}{\lim\limits_{x\to\infty}\dfrac{2x}{x}}=\dfrac{1}{2}$

EX 9

(1) $\dfrac{5}{2}$

(2) $=\lim\limits_{x\to-\infty}\dfrac{(\sqrt{x^2+x}+x)(\sqrt{x^2+x}-x)}{(\sqrt{x^2+x}-x)}$

$=\lim\limits_{x\to-\infty}\dfrac{x^2+x-x^2}{\sqrt{x^2+x}-x}=\lim\limits_{x\to-\infty}\dfrac{x}{\sqrt{x^2+x}-x}$

$=\dfrac{1}{-\sqrt{1+0}-1}=-\dfrac{1}{2}$

(3) $-\dfrac{3}{4}$

(4) $=\lim\limits_{x\to\infty}\dfrac{(2x-\sqrt{4x^2+x})(2x+\sqrt{4x^2+x})}{(2x+\sqrt{4x^2+x})}$

$=\lim\limits_{x\to\infty}\dfrac{4x^2-(4x^2+x)}{2x+\sqrt{4x^2+x}}=\lim\limits_{x\to\infty}\dfrac{-x}{2x+\sqrt{4x^2+x}}$

$=\lim\limits_{x\to\infty}\dfrac{\dfrac{-x}{\sqrt{x^2}}}{\dfrac{2x}{\sqrt{x^2}}+\dfrac{\sqrt{4x^2+x}}{\sqrt{x^2}}}=\lim\limits_{x\to\infty}\dfrac{\dfrac{-x}{|x|}}{\dfrac{2x}{|x|}+\sqrt{4+\dfrac{1}{x}}}$

$=\dfrac{-1}{2+\sqrt{4+0}}=-\dfrac{1}{4}$

EX 10

(1) 2	(2) -3	(3) 1	(4) DNE
(5) $+\infty$	(6) $+\infty$	(7) $+\infty$	(8) $+\infty$
(9) $-\infty$	(10) DNE	(11) 0	(12) -1

EX 11

(1)

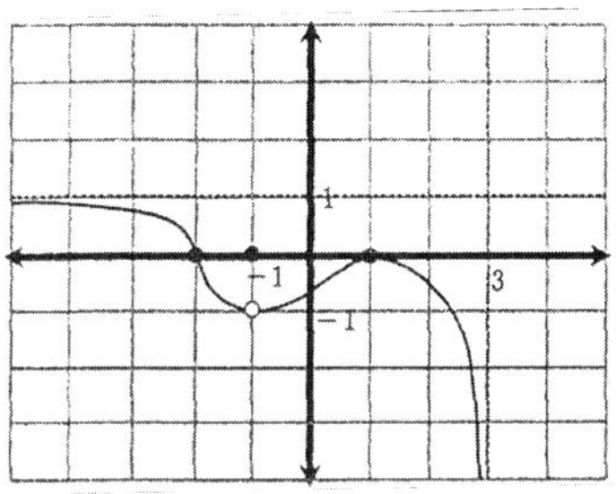

(2)

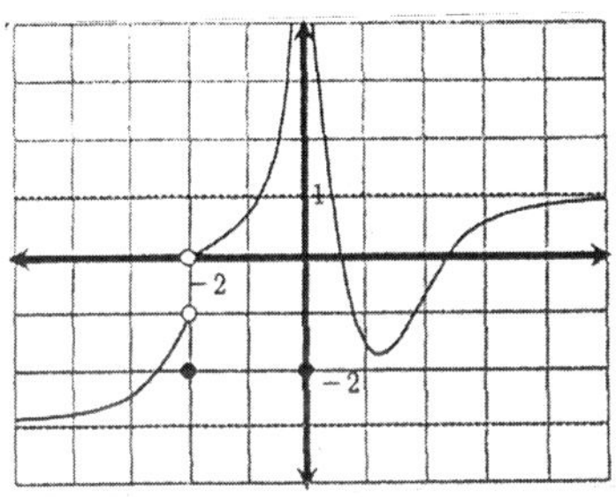

2.6 Continuity

EX 1

(1) (a) $\lim_{x\to -3} f(x) \neq f(-3)$

(b) removable discontinuity

(2) (a) $\lim_{x\to -1^-} f(x) \neq \lim_{x\to -1^+} f(x)$

(b) jump discontinuity

(3) (a) $f(1) = DNE$

(b) removable discontinuity

(4) (a) $\lim_{x\to 3^-} f(x) \neq \lim_{x\to 3^+} f(x)$

(b) jump discontinuity

(5) (a) $\lim_{x\to 5} f(x) = DNE,\ f(5) = DNE$

(b) infinite discontinuity

EX 2

(1)

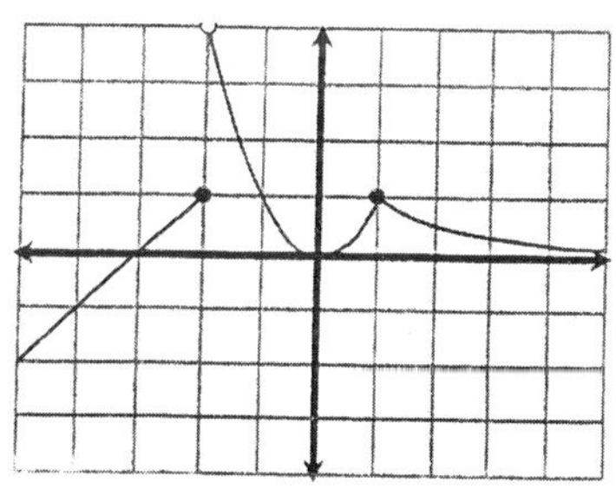

discontinuous at $x = -2$

(2)

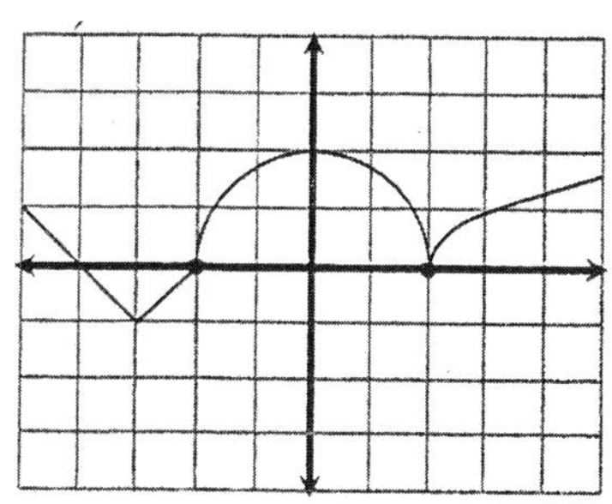

continuous everywhere

(3)

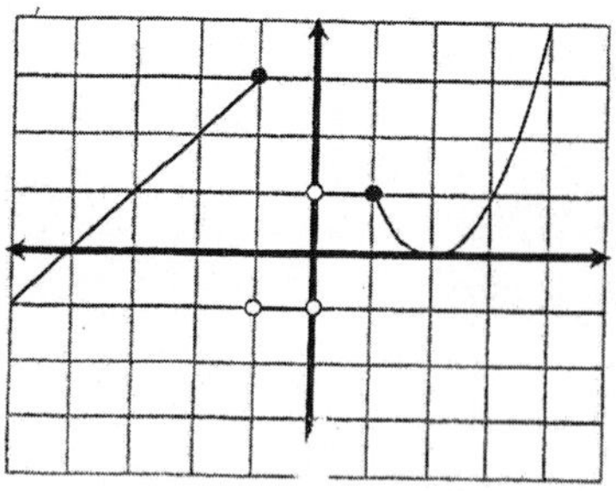

discontinuous at $x=-1, 0$

(4)

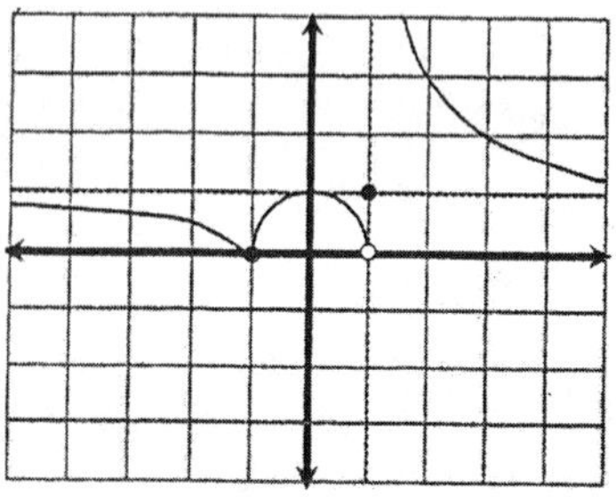

discontinuous at $x=1$

EX 3

(1) 6 (2) 7

(3) $\frac{3}{2}$ (4) $\frac{8}{5}$

EX 4

(1) 1. $\lim_{x\to-1^-} f(x)=\lim_{x\to-1^-}(x+3)=2,$

$\lim_{x\to-1^+} f(x)=\lim_{x\to-1^+}\sqrt{x+5}=2$

$\therefore \lim_{x\to-1} f(x)=2$

2. $f(-1)=-1+3=2$

3. $\lim_{x\to-1} f(x)=f(-1)$

$\therefore f$ is continuous at $x=-1$

(2) 1. $\lim_{x\to1^-} f(x)=\lim_{x\to1^-}(3-\ln x)=3-\ln 1=3,$

$\lim_{x\to1^+} f(x)=\lim_{x\to1^+}(2e^{\sin(\pi x)}+1)=2e^0+1=3$

$\therefore \lim_{x\to1} f(x)=3$

2. $f(1)=2e^{\sin(\pi)}+1=2e^0+1=3$

3. $\lim_{x\to1} f(x)=f(1)$

$\therefore f$ is continuous at $x=1$

EX 5

(1) $k=4$

(2) $\lim_{x\to0} f(x)=f(0)$

$\lim_{x\to0}\frac{x^2-2x}{x}=k$

$\lim_{x\to0}(x-2)=k$

$\therefore k=-2$

(3) $k=0$

(4) $\lim_{x\to0} f(x)=f(0)$

$\lim_{x\to0}\frac{\sin x}{x}=k$

$\therefore k=1$

EX 6

(1) 3 or -2

(2) -2 or 0

(3) $a=\frac{5}{2}, b=-\frac{1}{2}$

(4) $a=-\frac{3}{2}, b=-\frac{3}{2}$

2.7 Average Rate of Change and Instantaneous Rate of Change

EX 1

(1) a^2+ab+b^2

(2) $\dfrac{f(4)-f(2)}{4-2}=\dfrac{24-8}{2}=8$

(3) $-\dfrac{4}{\pi}$

(4) $\dfrac{f(5)-f(1)}{5-1}=\dfrac{3-1}{4}=\dfrac{1}{2}$

(5) $\dfrac{-1}{a(a+h)}$

(6) $\dfrac{f\left(\frac{\pi}{3}\right)-f\left(-\frac{\pi}{6}\right)}{\frac{\pi}{3}-\left(-\frac{\pi}{6}\right)}=\dfrac{\left(\frac{\sqrt{3}}{2}+\frac{1}{2}\right)-\left(-\frac{1}{2}+\frac{\sqrt{3}}{2}\right)}{\frac{\pi}{2}}$

$=\dfrac{1}{\frac{\pi}{2}}=\dfrac{2}{\pi}$

EX 2

(1) (a) $f'(1)=4$

(b) $y-3=4(x-1)$

(2) (Method 1)

(a) $f'(2)=\lim\limits_{x\to2}\dfrac{f(x)-f(2)}{x-2}$

$=\lim\limits_{x\to2}\dfrac{x^2-6x+8}{x-2}$

$=\lim\limits_{x\to2}\dfrac{(x-2)(x-4)}{(x-2)}$

$=-2$

(b) $y+8=-2(x-2)$

(Method 2)

$f'(2)=\lim\limits_{h\to0}\dfrac{f(2+h)-f(2)}{h}$

$=\lim\limits_{h\to0}\dfrac{(2+h)^2-6(2+h)-(-8)}{h}$

$=\lim\limits_{h\to0}\dfrac{4+4h+h^2-12-6h+8}{h}$

$=\lim\limits_{h\to0}\dfrac{h^2-2h}{h}=\lim\limits_{h\to0}(h-2)=-2$

(3) (a) $f'(-1)=3$

(b) $y+1=3(x+1)$

(4) (Method 1)

(a) $f'(1)=\lim\limits_{x\to1}\dfrac{f(x)-f(1)}{x-1}$

$=\lim\limits_{x\to1}\dfrac{\left(\frac{1}{x+1}-\frac{1}{2}\right)\cdot2(x+1)}{(x-1)\cdot2(x+1)}$

$=\lim\limits_{x\to1}\dfrac{2-(x+1)}{x-1\cdot2(x+1)}$

$=\lim\limits_{x\to1}\dfrac{1-x}{(x-1)\cdot2(x+1)}$

$=\lim\limits_{x\to1}\dfrac{-1}{2(x+1)}=-\dfrac{1}{4}$

(b) $y-\dfrac{1}{2}=-\dfrac{1}{4}(x-1)$

(Method 2)

$f'(1)=\lim\limits_{h\to0}\dfrac{f(1+h)-f(1)}{h}$

$=\lim\limits_{h\to0}\dfrac{\left(\frac{1}{h+2}-\frac{1}{2}\right)\cdot2(h+2)}{h\cdot2(h+2)}$

$=\lim\limits_{h\to0}\dfrac{2-(h+2)}{h\cdot2(h+2)}$

$=\lim\limits_{h\to0}\dfrac{-h}{h\cdot2(h+2)}=-\dfrac{1}{4}$

(5) (a) $f'(6)=\dfrac{1}{4}$

(b) $y-2=\dfrac{1}{4}(x-6)$

(6) (Method 1)

(a) $f(3)=\dfrac{1}{\sqrt{4}}=\dfrac{1}{2}$, $\therefore (3,\dfrac{1}{2})$

$$f'(3)=\lim_{x\to3}\frac{f(x)-f(3)}{x-3}$$

$$=\lim_{x\to3}\frac{\left(\frac{1}{\sqrt{x+1}}-\frac{1}{2}\right)\cdot2\sqrt{x+1}}{(x-3)\cdot2\sqrt{x+1}}$$

$$=\lim_{x\to3}\frac{(2-\sqrt{x+1})\cdot(2+\sqrt{x+1})}{(x-3)\cdot2\sqrt{x+1}\cdot(2+\sqrt{x+1})}$$

$$=\lim_{x\to3}\frac{4-(x+1)}{(x-3)\cdot2\sqrt{x+1}\cdot(2+\sqrt{x+1})}$$

$$=\lim_{x\to3}\frac{(3-x)}{(x-3)\cdot2\sqrt{x+1}\cdot(2+\sqrt{x+1})}$$

$$=\lim_{x\to3}\frac{-1}{2\sqrt{x+1}\cdot(2+\sqrt{x+1})}$$

$$=\frac{-1}{2(2)(2+2)}=-\frac{1}{16}$$

(b) $y-\frac{1}{2}=-\frac{1}{16}(x-3)$

(Method 2)

$$f'(3)=\lim_{h\to0}\frac{f(3+h)-f(3)}{h}$$

$$=\lim_{h\to0}\frac{\left(\frac{1}{\sqrt{4+h}}-\frac{1}{2}\right)\cdot2\sqrt{4+h}}{h\cdot2\sqrt{4+h}}$$

$$=\lim_{h\to0}\frac{(2-\sqrt{4+h})\cdot(2+\sqrt{4+h})}{h\cdot2\sqrt{4+h}\cdot(2+\sqrt{4+h})}$$

$$=\lim_{h\to0}\frac{4-(4+h)}{h\cdot2\sqrt{4+h}\cdot(2+\sqrt{4+h})}$$

$$=\lim_{h\to0}\frac{-h}{h\cdot2\sqrt{4+h}\cdot(2+\sqrt{4+h})}$$

$$=\lim_{h\to0}\frac{-1}{2\sqrt{4+h}\cdot(2+\sqrt{4+h})}$$

$$=\frac{-1}{2(2)(2+2)}=-\frac{1}{16}$$

2.8 Intermediate Value Theorem (I.V.T.)

EX 1

(1) $x^3-3x-8=0$

Let $f(x)=x^3-3x-8$

Since f is a polynomial function, it is ontinuous everywhere.

$f(2)=8-6-8=-6<0$

$f(3)=27-9-8=10>0$

$\therefore f(2)<0<f(3)$

By I.V.T. there must be at least one solution c, $2<c<3$, such that $f(c)=0$.

(2) $x^3+x^2-2x-6=0$

Let $f(x)=x^3+x^2-2x-6=0$

Since f is a polynomial function, it is continuous everywhere.

$f(1)=1+1-2-6=-6<0$

$f(2)=8+4-4-6=2>0$

$\therefore f(1)<0<f(2)$

By I.V.T. , there must be at least one solution c, $1<c<2$, such that $f(c)=0$.

EX 2

(1) $v(5)=-1<0$, $v(10)=5>0$

$v(5)<0<v(10)$

$v(t)$ is continuous on the interval $[5, 10]$.

By I.V.T. there must be a value t for $5<t<10$, such that $v(t)=0$.

(2) $a(10)=2<4$, $a(20)=5>4$

$a(10)<4<a(20)$

$a(t)$ is continuous on the interval $[10, 20]$.

By I.V.T. , there must be a value t for $10<t<20$, such that $a(t)=4$.

EX 3

(1) $M > 0$

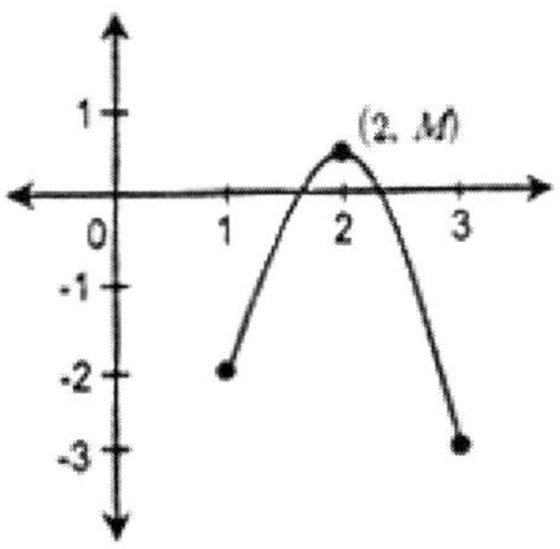

(2) $M < 0$

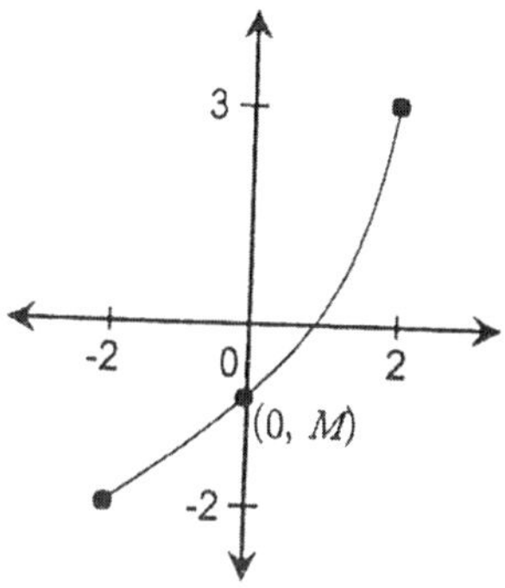

CHAPTER 03. Derivatives

3.1 Definition for the Derivative :

EX 1

(1) $f'(x) = \dfrac{-3}{2(3x-2)\sqrt{3x-2}}$

(2) $= 3x^2 - 2$

(3) $f'(x) = 1 + \dfrac{1}{x^2}$

(4) $f'(x) = \lim\limits_{h\to 0} \dfrac{f(x+h)-f(x)}{h}$

$= \lim\limits_{h\to 0} \dfrac{\left[\dfrac{1}{(x+h)^2} - \dfrac{1}{x^2}\right] \cdot x^2(x+h)^2}{h \cdot x^2(x+h)^2}$

$= \lim\limits_{h\to 0} \dfrac{x^2-(x+h)^2}{h \cdot x^2(x+h)^2}$

$= \lim\limits_{h\to 0} \dfrac{-2xh-h^2}{h \cdot x^2(x+h)^2}$

$= \lim\limits_{h\to} \dfrac{-2x-h}{x^2(x+h)^2}$

$= \dfrac{-2x}{x^2(x^2)} = \dfrac{-2}{x^3}$

(5) $f'(x) = \dfrac{3}{(x+1)^2}$

(6) $= \dfrac{2x}{2\sqrt{x^2+1}} = \dfrac{x}{\sqrt{x^2+1}}$

3.2 Differentiability

EX 1

(1) (a) The function is differentiable on its domain $-3 \le x \le 2$ (it is smooth)

(b) none

(c) none

(2) (a) The function is differentiable on its domain

$-2 \le x \le 3$ (it is smooth)

(b) none

(c) none

(3) (a) The function is differentiable on $-3 \le x \le 0$ and $0 < x \le 3$

(b) none

(c) The function is neither continuous nor differentiable at $x=0$ since $\lim\limits_{x\to 0^-} f(x) \ne \lim\limits_{x\to 0^+} f(x)$

(4) (a) f is differentiable on $-2 \le x < 1$, $-1 < x < 0, 0 < x < 2$, and $2 < x \le 3$.

(b) f is continuous but not differentiable at $x=-1$; $\lim\limits_{x\to -1} f(x)=0$ exists but there is a corner at $x=-1$ since

$\lim\limits_{h\to 0^-} \dfrac{f(-1+h)-f(-1)}{h} = -3$ and

$\lim\limits_{h\to 0+} \dfrac{f(-1+h)-f(-1)}{h} = 3 \Rightarrow f'(-1)$

does not exist.

(c) f is neither continuous nor differentiable at $x=0$ and $x=2$: at $x=0$, $\lim_{x\to0^-}f(x)=3$ but $\lim_{x\to0^+}f(x)=0 \Rightarrow \lim_{x\to0}f(x)$ does not exist : at $x=2$, $\lim_{x\to2}f(x)$ exists but $\lim_{x\to2}f(x) \neq f(2)$

(5) (a) f is differentiable on $-1 \leq x < 0$ and $0 < x \leq 2$

(b) f is continuous but not differentiable at $x=0$: $\lim_{x\to0}f(x)=0$ exists but there is a cusp at $x=0$,

so $f'(0)=\lim_{h\to0}\dfrac{f(0+h)-f(0)}{h}$

does not exist

(6) (a) f is differentiable on $-3 \leq x < 2$, $-2 < x < 2$, and $2 < x \leq 3$

(b) f is continuous but not differentiable at $x=-2$ and $x=2$: there are corners at those points.

(c) none

EX 2

(1) T : left-hand limit = right-hand limit = 2

(2) T : f is continuous at $x=-2$

(3) F : left-hand tangent slope=1
right-hand tangent slope=-2

(4) T : left-hand limit = right-hand limit=-2

(5) F : f is discontinuous at $x=0$
$\lim_{x\to0}f(x)=-2 \neq f(0)=-1$

(6) F : left-hand tangent slope=-2
right-hand tangent slope=2

(7) T

(8) F : f is discontinuous at x=1
$\therefore f'(1)$ does not exist

(9) T : right-hand tangent slope
$\lim_{x\to0^+}f'(x)=2$

(10) T : f is discontinuous at $x=1$
$\therefore f'(1)$ does not exist

(11) T : left-hand tangent slope=2
right-hand tangent slope=2

(12) T : left-hand limit=2
right-hand limit=1

(13) F : left-hand limit=2=$f(2)$

(14) T : The graph of f is smooth in the interval(2, 6)
$\Rightarrow f$ is continuous at $x=3$ and left-hand slope = right-hand slope
$\therefore f'(3)$ exists

(15) T : f is continuous at $x=4$

(16) F : $f'(4)=0$ (horizontal slope)

(17) T : f is continuous in the interval (2,6)
$\therefore$ The limit exists for every number in the interval (2,6)

(18) T : The graph of f is smooth in the interval (2,6)
$\therefore$ left-hand slope = right-hand slope for every number in the interval (2,6)

(19) T : tangent slope=1

(20) F : tangent slope= -2

(21) T : $f'(3) \approx -1$, $f'(4)=0$
$f'(5) \approx 1$, $f'(5.5) \approx 3$

EX 3

(1) $y=f'(x)$

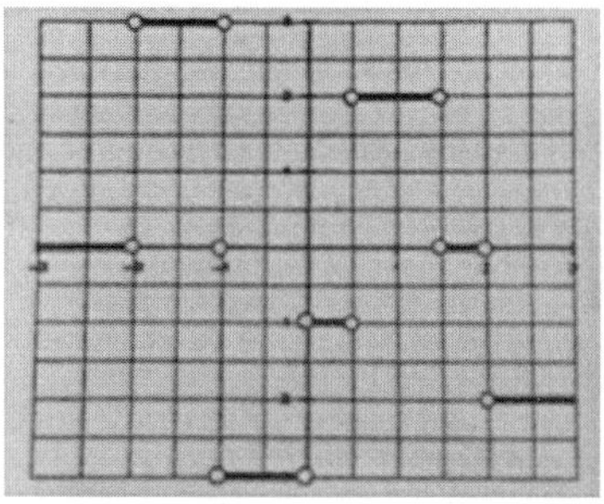

(2) $y=f'(x)$

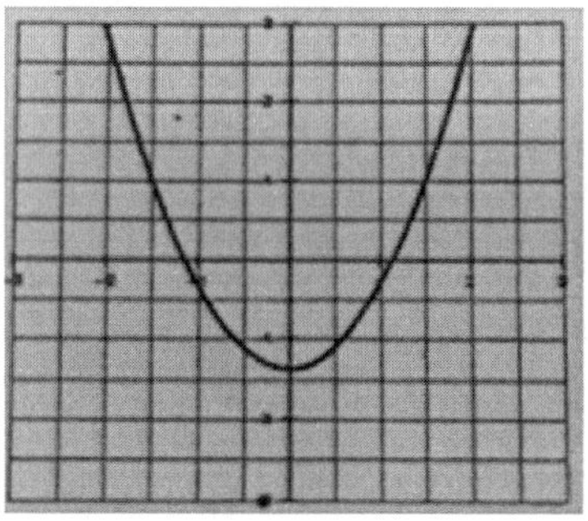

(3) $y=f'(x)$

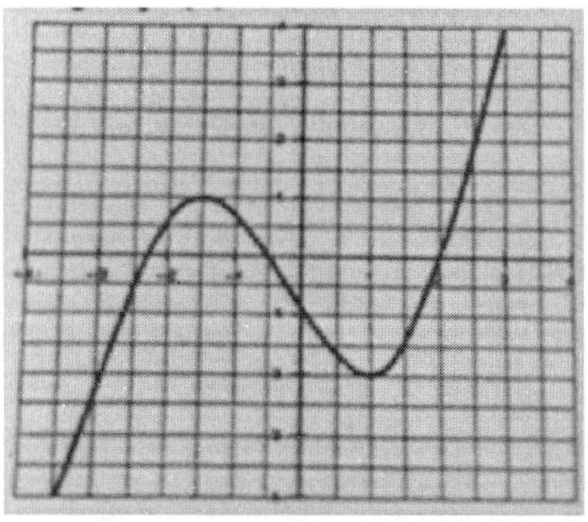

(4) $y=f'(x)$

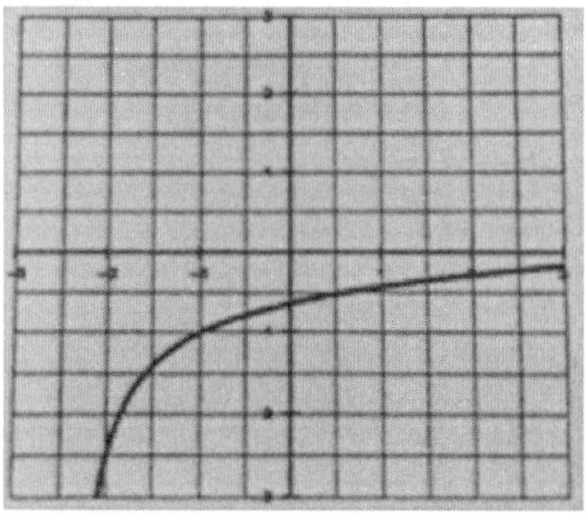

3.3 Power Rule

EX 1

(1) $y'=0$

(2) $y'=0$

(3) $y'=-20x^3$

(4) $y=x^{\frac{4}{5}}$

$y'=\frac{4}{5}x^{-\frac{1}{5}}$

(5) $y'=6x^{-4}$

(6) $y=2x^{-\frac{1}{2}}$

$y'=-x^{-\frac{3}{2}}=-\frac{1}{x^{3/2}}$

EX 2

(1) $y'=6x-5$

(2) $y'=15x^2-\frac{1}{\sqrt{x}}$

(3) $y=1+\frac{1}{x}+2x^{-2}$

$y'=0-\frac{1}{x^2}-4x^{-3}$

$=-\frac{1}{x^2}-\frac{4}{x^3}$

(4) $y=2\sqrt{x}+\frac{2}{x}$

$y'=\frac{1}{\sqrt{x}}-\frac{2}{x^2}$

(5) $y'=2x^{-2}-8x^{-3}$

(6) $y'=-12x^{-2}+12x^{-4}-4x^{-5}$

EX 3

(1) $y=1+5x^{-1}-x^{-2}$

$y'=-5x^{-2}+2x^{-3}$

(2) $y=2x^{5/2}-3x-x^{-1/2}$

$y'=5x^{3/2}-3+\frac{1}{2}x^{-3/2}$

$=5x^{3/2}-3+\frac{1}{2x^{3/2}}$

(3) $y = 3x^{-1} + 2x - x^2$

$y' = -3x^{-2} + 2 - 2x$

$= -\dfrac{3}{x^2} + 2 - 2x$

(4) $f(x) = 12x^{5/3} - 6x^{2/3}$

$f'(x) = 20x^{2/3} - 4x^{-1/3}$

$= 20x^{2/3} - \dfrac{4}{x^{1/3}}$

(5) $f(x) = x^{\frac{3}{2}}(x - 3x^2)$

$= x^{5/2} - 3x^{7/2}$

$f'(x) = \dfrac{5}{2}x^{3/2} - \dfrac{21}{2}x^{5/2}$

EX 4

(1) $y - 3 = 2(x + 2)$

(2) $y + 2 = 5(x - 1)$

EX 5

(1) $y = 2x + 5$

(2) $f(1) = 4 - 1 = 3 \Rightarrow (1,3)$

$f'(x) = 4 \cdot \dfrac{1}{2\sqrt{x}} + \dfrac{1}{x^2}$

$= \dfrac{2}{\sqrt{x}} + \dfrac{1}{x^2}$

$f'(1) = 3$

$\therefore y - 3 = 3(x - 1)$

(3) $g(4) = \dfrac{8}{16} + \dfrac{2}{2} = \dfrac{1}{2} + 1 \Rightarrow (4, \dfrac{3}{2})$

$g(x) = 8x^{-2} + 2x^{-1/2}$

$g'(x) = -16x^{-3} - x^{-3/2}$

$= \dfrac{-16}{x^3} - \dfrac{1}{x\sqrt{x}}$

$g'(4) = -\dfrac{16}{64} - \dfrac{1}{4\sqrt{4}} = -\dfrac{1}{4} - \dfrac{1}{8} = -\dfrac{3}{8}$

$\therefore y - \dfrac{3}{2} = -\dfrac{3}{8}(x - 4)$

EX 6

(1) $y = x^3 - 3x - 2 \Rightarrow y' = 3x^2 - 3$.
For the tangent to be horizontal, we need $m = y' = 0$
$\Rightarrow 0 = 3x^2 - 3 \Rightarrow 3x^2 = 3 \Rightarrow x = \pm 1$. When $x = -1, y = 0 \Rightarrow$ the tangent line has equation $y = 0$. The line perpendicular to this line at $(-1,0)$ is $x = -1$. When $x = 1, y = -4 \Rightarrow$ the tangent line has equation $y = -4$. The line perpendicular to this line at $(1,-4)$ is $x = 1$.

(2) $8x - 2y = 1 \Rightarrow y = 4x - \dfrac{1}{2} \Rightarrow m = 4;$

$g(x) = \dfrac{1}{3}x^3 - \dfrac{3}{2}x^2 + 1$

$\Rightarrow g'(x) = x^2 - 3x; x^2 - 3x = 4$
$\Rightarrow x = 4$ or $x = -1$

$\Rightarrow g(4) = \dfrac{1}{3}(4)^3 - \dfrac{3}{2}(4)^2 + 1 = -\dfrac{5}{3}$

$g(-1) = \dfrac{1}{3}(-1)^3 - \dfrac{3}{2}(-1)^2 + 1 = -\dfrac{5}{6}$

$\Rightarrow (4, -\dfrac{5}{3})$ or $(-1, -\dfrac{5}{6})$

EX 7

(1) $y=ax^2+bx+c$ passes through (0,0)

$\Rightarrow 0=a(0)+b(0)+c \Rightarrow c=0$; $y=ax^2+bx$ passes through (1,2)

$\Rightarrow 2=a+b$; $y'=2ax+b$

and since the curve is tangent to $y=x$ at the origin, its slope is 1 at $x=0 \Rightarrow y'=1$ when $x=0 \Rightarrow 1=2a(0)+b \Rightarrow b=1$.

Then $a+b=2$

$\Rightarrow a=1$. In summary $a=b=1$ and $c=0$ so the curve is $y=x^2+x$.

(2) $y=cx-x^2$ passes through

$(1,0) \Rightarrow 0=c(1)-1$

$\Rightarrow c=1$

$\Rightarrow$ the curve is $y=x-x^2$. For this curve, $y'=1-2x$ and $x=1 \Rightarrow y'=1-2=-1$. Since $y=x-x^2$ and $y=x^2+ax+b$ have common tangents at $x=0$, $y=x^2+ax+b$ must also slope -1 at $x=1$. Thus $y'=2x+a \Rightarrow$

$-1=2 \cdot 1+a \Rightarrow a=-3 \Rightarrow y=x^2-3x+b$.

Since this last curve passes through (1,0) we have $0=1-3+b \Rightarrow b=2$. In summary $a=-3$, $b=2$ and $c=1$ so the curves are $y=x^2-3x+2$ and $y=x-x^2$

EX 8

(1) $g'(x)=\begin{cases}2x-3 & x>0\\ a & x<0\end{cases}$, since g is differentiable at $x=0 \Rightarrow \lim_{x\to 0^-}(2x-3)=-3$ and

$\lim_{x\to 0^-} a=a \Rightarrow a=-3$

(2) $f'(x)=\begin{cases}a & x>-1\\ 2bx & x<-1\end{cases}$, since f is differentiable at $x=-1 \Rightarrow \lim_{x\to -1^+} a$ and $\lim_{x\to -1^-}(2bx)=-2b$

and $\Rightarrow a=-2b$, since f is continuous at $x=-1 \Rightarrow \lim_{x\to -1^+}(ax+b)=-a+b$ and

$\lim_{x\to -1^-}(bx^2-3)=b-3 \Rightarrow -a+b=b-3$

$\Rightarrow a=3 \Rightarrow 3=-2b \Rightarrow b=-\frac{3}{2}$

(3) $a=-\frac{1}{4}$, $b=-3$

(4) $a=-\frac{1}{2}$, $b=1$

3.4 Product Rule

EX 1

(1) $y'=-5x^4+12x^2-2x-3$

(2) $y'=\frac{3}{4}x^{-\frac{1}{4}}+3x^{-4}+\frac{11}{4}x^{\frac{7}{4}}+x^{-2}$

(3) $y'=-2\sqrt{x}+(3-2x)\frac{1}{2\sqrt{x}}$

$=\frac{-4x+3-2x}{2\sqrt{x}}=\frac{3-6x}{2\sqrt{x}}$

EX 2

(1) $f(-1)=(1-2)(1+3-1)=-3$

$\Rightarrow \therefore (-1,-3)$

$y'=(2x)(x^2-3x-1)+(x^2-2)(2x-3)$

$f'(-1)=-2(1+3-1)+(1-2)(-2-3)$

$=-2(3)+(-1)(-5)=-1$

$\therefore y+3=-(x+1)$

(2)

$f(4)=0 \Rightarrow \therefore (4,0)$

$y'=\left(1-\frac{1}{\sqrt{x}}\right)(x^2-6x+2)+(x-2\sqrt{x})(2x-6)$

$f'(4)=\left(1-\frac{1}{2}\right)(16-24+2)+0$

$=\frac{1}{2}(-6)=-3$

$\therefore y=-3(x-4)$

EX 3

(1) 13 (2) -7 (3) $\frac{7}{25}$ (4) 20

EX 4

$$\frac{d}{df}(f \cdot g \cdot h)=\frac{d}{dx}[(f \cdot g) \cdot h]$$
$$=\frac{d}{dx}(f \cdot g) \cdot h+(f \cdot g) \cdot \frac{d}{dx}h$$
(by Product rule)
$$=[f' \cdot g+f \cdot g'] \cdot h+f \cdot g \cdot h'$$
$$=f' \cdot g \cdot h+f \cdot g' \cdot h+f \cdot g \cdot h'$$

3.5 Quotient Rule

EX 1

$$y'=\frac{2 \cdot (x+1)-2x \cdot 1}{(x+1)^2}$$
$$=\frac{2x+2-2x}{(x+1)^2}$$
$$=\frac{2}{(x+1)^2}$$

EX 2

(1) $\frac{-19}{(3x-2)^2}$

(2) $y'=\frac{-3(3x^2+x)-(4-3x)(6x+1)}{(3x^2+x)^2}$
$$=\frac{9x^2-24x-4}{(3x^2+x)^2}$$

(3) $\frac{-6(x^2-2)}{(x-1)^2(x-2)^2}$

(4) $y'=\frac{\frac{1}{2\sqrt{x}}(\sqrt{x}+1)-(\sqrt{x}-1)\frac{1}{2\sqrt{x}}}{(\sqrt{x}+1)^2}$
$$=\frac{1}{\sqrt{x}(\sqrt{x}+1)^2}$$

EX 3

(1) $y'=\frac{1 \cdot (2x-1)-(1+x) \cdot 2}{(2x-1)^2}$
$$=\frac{-3}{(2x-1)^2}$$

$f'(1)=-3$

$\therefore y-2=-3(x-1)$

(2) $f(4)=\frac{4}{2-3}=-4 \Rightarrow (4,-4)$

$$y'=\frac{1 \cdot (\sqrt{x}-3)-x \cdot \frac{1}{2\sqrt{x}}}{(\sqrt{x}-3)^2}$$

$$f'(4)=\frac{(2-3)-4 \cdot \frac{1}{4}}{(2-3)^2}=\frac{-1-1}{1}=-2$$

The normal line is perpendicular to the tangent line.

$\therefore\ y+4=\frac{1}{2}(x-4)$

EX 4

(1) $y'=\frac{4(x^2+1)-4x \cdot 2x}{(x^2+1)^2}$
$$=\frac{4x^2+4-8x^2}{(x^2+1)^2}$$
$$=\frac{4\quad 4x^2}{(x^2+1)^2}$$

$y'=0 \Rightarrow 4-4x^2=0$
$x^2=1$
$x=\pm 1$

$f(1)=\frac{4}{1+1}=2,$
$f(-1)=\frac{-4}{1+1}=-2$

$\therefore$ (1,2),(−1,−2)

EX 5

(1) $\frac{2}{3}$ (2) $2\pi+5$ (3) $15-8\pi$ (4) $\frac{37}{6}$

EX 6

$$\left(\frac{f}{g}\right)(x)=\frac{f(x)}{g(x)}$$

$$\left(\frac{f}{g}\right)'(x)=\frac{f'(x)g(x)-f(x)g'(x)}{g(x)^2}$$

$$\left(\frac{f}{g}\right)'(3)=\frac{f'(3)g(3)-f(3)g'(3)}{[g(3)^2]}$$

$$=\frac{4(-2)-2(-5)}{(-2)^2}=\frac{2}{4}=\frac{1}{2}$$

EX 7

(1) $y'=\frac{4(x^2+1)-4x(2x)}{(x^2+1)^2}=\frac{-4x^2+4}{(x^2+1)^2}$

$y'|_{x=1}=\frac{0}{4}=0\Rightarrow y-2=0(x-1)$

$\therefore\ y=2$

$y'|_{x=0}=4\Rightarrow y-0=4(x-0)$

$y=4x$

(2) $y'=\frac{0-8(2x)}{(x^2+4)^2}=\frac{-16x}{(x^2+4)^2}$

$y'|_{x=2}=\frac{-32}{64}=-\frac{1}{2}$

$y-1=-\frac{1}{2}(x-2)$

$\therefore y=-\frac{1}{2}x+2$

3.6 Chain Rule

EX 1

(1) $y'=-27(4-3x)^8$

(2) $y'=4(3x^2-\sqrt{x})^3\left(6x-\frac{1}{2\sqrt{x}}\right)$

(3) $y'=4(\frac{x^2}{8}+x-\frac{1}{x})(\frac{x}{4}+1+\frac{1}{x^2})$

(4) $y'=\frac{2-2x}{3(2x-x^2)^{2/3}}$

EX 2

(1) $y'=\frac{12}{(2-3x)^5}$

(2) $y'=3(2x+1)^2(3x-4)^4(16x-3)$

(3) $y'=2(2-5x)^3(3x^2+1)^2(-75x^2+18x-10)$

(4) $y'=-\frac{2(x+1)(x+2)}{(2x+1)^4}$

EX 3

(1) $y'=\frac{x-2}{\sqrt{x^2-4x}}$

(2) $y'=6x\sqrt{x^2+1}+3x^2\cdot\frac{1}{2\sqrt{x^2+1}}\cdot 2x$

$$=6x\sqrt{x^2+1}+\frac{3x^3}{\sqrt{x^2+1}}$$

$$=\frac{6x(x^2+1)+3x^3}{\sqrt{x^2+1}}=\frac{9x^3+6x}{\sqrt{x^2+1}}$$

(3) $y'=\frac{-4x-4}{(4x+1)\sqrt{4x+1}}$

(4)

$$y'=\frac{\frac{1}{2\sqrt{x^2+1}}\cdot 2x\cdot(x-1)-\sqrt{x^2+1}\cdot 1}{(x-1)^2}$$

$$=\frac{\left[\frac{x(x-1)}{\sqrt{x^2+1}}-\sqrt{x^2+1}\right]\cdot\sqrt{x^2+1}}{(x-1)^2\sqrt{(x^2+1)}}$$

$$=\frac{x(x-1)-(x^2+1)}{(x-1)^2\sqrt{x^2+1}}$$

$$=\frac{x^2-x-x^2-1}{(x-1)^2\sqrt{x^2+1}}$$

$$=\frac{-x-1}{(x-1)^2\sqrt{x^2+1}}$$

(5) $y'=\frac{-25(x+2)^4}{(2x-1)^6}$

(6)

$$y' = 4\left(\frac{\sqrt{x}-1}{\sqrt{x}+1}\right)^3 \cdot \frac{\frac{1}{2\sqrt{x}} \cdot (\sqrt{x}+1) - (\sqrt{x}-1) \cdot \frac{1}{2\sqrt{x}}}{(\sqrt{x}+1)^2}$$

$$= 4\left(\frac{\sqrt{x}-1}{\sqrt{x}+1}\right)^3 \cdot \frac{\left[\frac{\sqrt{x}+1}{2\sqrt{x}} - \frac{\sqrt{x}-1}{2\sqrt{x}}\right] \cdot 2\sqrt{x}}{(\sqrt{x}+1)^2 \cdot 2\sqrt{x}}$$

$$= 4\left(\frac{\sqrt{x}-1}{\sqrt{x}+1}\right)^3 \cdot \frac{\sqrt{x}+1-(\sqrt{x}-1)}{2(\sqrt{x}+1)^2\sqrt{x}}$$

$$= 4\left(\frac{\sqrt{x}-1}{\sqrt{x}+1}\right)^3 \cdot \frac{2}{2(\sqrt{x}+1)^2\sqrt{x}} = \frac{4(\sqrt{x}-1)^3}{(\sqrt{x}+1)^5\sqrt{x}}$$

EX 4

(1) $y = \sqrt{x^2 - x + 7}$ and $x = 2$

$\Rightarrow y = \sqrt{(2)^2 - 2 + 7} = \sqrt{9} = 3.$

$$y' = \frac{1}{2}(x^2 - x + 7)^{-1/2}(2x - 1)$$
$$= \frac{2x-1}{2\sqrt{x^2 - x + 7}}$$

$$y'|_{x=2} = \frac{2(2)-1}{2\sqrt{(2)^2 - (2) + 7}} = \frac{3}{6} = \frac{1}{2}$$

$\Rightarrow y - 3 = \frac{1}{2}(x-2)$

$\Rightarrow y = \frac{1}{2}x + 2$

(2) $y = \left(\frac{x-1}{x+1}\right)^2$ and $x = 0 \Rightarrow y = \left(\frac{0-1}{0+1}\right)^2$

$= (-1)^2 = 1.$

$$y' = 2\left(\frac{x-1}{x+1}\right) \cdot \frac{(x+1) \cdot 1 - (x-1) \cdot 1}{(x+1)^2}$$
$$= 2\left(\frac{x-1}{x+1}\right)\frac{2}{(x+1)^2} = \frac{4(x-1)}{(x+1)^3}$$

$$y'|_{x=0} = \frac{4(0-1)}{(0+1)^3} = \frac{-4}{1^3} = -4$$

$\Rightarrow y - 1 = -4(x - 0)$

$\Rightarrow y = -4x + 1$

EX 5

(1) $y' = 3[f(x)]^2 \cdot f'(x)$

(2) $y' = f'(x^3) \cdot 3x^2$

$= 3x^2 f'(x^3)$

(3) $y' = 3[f(x^3)]^2 \cdot f'(x^3) \cdot 3x^2$

$= 9x^2[f(x^3)]^2 f'(x^3)$

(4) $y' = \dfrac{1}{2\sqrt{f(x) - 3x}}(f'(x) - 3)$

(5) $y = [f(x)]^{-\frac{1}{2}}$

$y' = -\frac{1}{2}[f(x)]^{-\frac{3}{2}} \cdot f'(x)$

(6)

$$y' = 6x \cdot f(g(\sqrt{x})) + 3x^2 \cdot f'(g(\sqrt{x})) \cdot g'(\sqrt{x}) \cdot \frac{1}{2\sqrt{x}}$$
$$= 6xf(g(\sqrt{x})) + \frac{3}{2}x^{3/2}f'(g(\sqrt{x}))g'(\sqrt{x})$$

(7) $y' = 12[f(x^2) + 2x]^3 \cdot [f'(x^2) \cdot 2x + 2]$

$= 24[f(x^2) + 2x]^3[xf'(x^2) + 1]$

EX 6

(1) 1 (2) 6 (3) 1 (4) $-\frac{1}{9}$

(5) $-\frac{40}{3}$ (6) $-\frac{1}{3}$ (7) $\frac{4}{9}$

3.7 Implicit Differentiation

EX 1

(1) $9y^2 \cdot y'$

(2) $-\dfrac{1}{2\sqrt{y}} \cdot y'$

(3) $-\dfrac{4}{y^2} \cdot y'$

(4) $= 1 \cdot y^2 + x \cdot 2y \cdot y'$

$= y^2 + 2xyy'$

(5) $= \dfrac{7(1 \cdot y - x \cdot y')}{y^2}$

$= 7(\dfrac{y - xy'}{y^2})$

EX 2

(1) $y' = -\dfrac{x^2}{y^2}$

(2) $\dfrac{d}{dx}(x^{\frac{2}{3}} + y^{\frac{2}{3}}) = \dfrac{d}{dx}(1)$

$\dfrac{2}{3}x^{-\frac{1}{3}} + \dfrac{2}{3}y^{-\frac{1}{3}} \cdot y' = 0$

$y' = \dfrac{-x^{-\frac{1}{3}}}{y^{-\frac{1}{3}}} = -\dfrac{y^{\frac{1}{3}}}{x^{\frac{1}{3}}}$

$\therefore \dfrac{dy}{dx} = -\dfrac{\sqrt[3]{y}}{\sqrt[3]{x}}$

(3) $\dfrac{d}{dx}(x^2 - xy + y^2) = \dfrac{d}{dx}(6)$

$2x - (1 \cdot y + x \cdot y') + 2y \cdot y' = 0$

$y'(2y - x) = y - 2x$

$\therefore \dfrac{dy}{dx} = \dfrac{y - 2x}{2y - x}$

(4) $y' = \dfrac{6y - x^2}{y^2 - 6x}$

(5) $y' = \dfrac{\sqrt{y}}{\sqrt{y} + 1}$

(6)

$\dfrac{d}{dx}(x^2y^3 - x^3y^2) = \dfrac{d}{dx}(-5)$

$2x \cdot y^3 + x^2 \cdot 3y^2y' - (3x^2 \cdot y^2 + x^3 \cdot 2yy') = 0$

$2xy^3 + 3x^2y^2y' - 3x^2y^2 - 2x^3yy' = 0$

$y'(3x^2y^2 - 2x^3y) = 3x^2y^2 - 2xy^3$

$\therefore \dfrac{dy}{dx} = \dfrac{xy^2(3x - 2y)}{x^2y(3y - 2x)} = \dfrac{y(3x - 2y)}{x(3y - 2x)}$

EX 3

(1) $x^2 + xy - y^2 = 1$

$\Rightarrow 2x + y + xy' - 2yy' = 0$

$\Rightarrow (x - 2y)y' = -2x - y$

$\Rightarrow y' = \dfrac{2x + y}{2y - x} \Rightarrow y'|_{(2,3)} = \dfrac{7}{4}$

(2) $(x^2 + y^2)^2 = (x - y)^2$ at $(1,0)$ at $(1,-1)$

$\Rightarrow 2(x^2 + y^2)(2x + 2y\dfrac{dy}{dx}) = 2(x - y)(1 - \dfrac{dy}{dx})$

$\Rightarrow \dfrac{dy}{dx}[2y(x^2 + y^2) + (x - y)]$

$= -2x(x^2 + y^2) + (x - y)$

$\Rightarrow \dfrac{dy}{dx} = \dfrac{-2x(x^2 + y^2) + (x - y)}{2y(x^2 + y^2) + (x - y)}$

$\Rightarrow \dfrac{dy}{dx}\Big|_{(1.0)} = -1$ and $\dfrac{dy}{dx}\Big|_{(1,-1)} = 1$

(3)

$$\frac{1}{2\sqrt{4x-y^2}}(4-2yy')-4\cdot\frac{1}{2\sqrt{xy}}(y+xy')=0$$

At (2,2),

$$\frac{1}{4}(4-4y')-4\cdot\frac{1}{4}(2+2y')=0$$

$$1-y'-2-2y'=0$$

$$-1-3y'=0$$

$$y'=-\frac{1}{3} \qquad \therefore \left.\frac{dy}{dx}\right|_{(2,2)}=-\frac{1}{3}$$

(4) $1=2yy'\sqrt{y-1}+y^2\cdot\frac{1}{2\sqrt{y-1}}\cdot y'$

At(4,2),

$$1=4y'+4\cdot\frac{1}{2}\cdot y'$$

$$1=6y'$$

$$y'=\frac{1}{6} \qquad \therefore \left.\frac{dy}{dx}\right|_{(4,2)}=\frac{1}{6}$$

EX 4

(1) slope: $y'=\frac{5}{4}$

$$y-2=\frac{5}{4}(x-4)$$

(2) slope ; $y'=1$,
$normal; y-1=-1(x-1)$

$$y=-x+2$$

$$\rightarrow x^2=2x(-x+2)-3(-x+2)^2=0$$

$$\rightarrow x^2-4x+3=0$$

$$\rightarrow x=1, x=3$$

$$\rightarrow x=3 \ \therefore (3,-1)$$

EX 5

(1) *Horizontal Tangent* $\Rightarrow \frac{dy}{dx}=0$

$$\frac{d}{dx}(xy^2-x^2y)=\frac{d}{dx}(16)$$

$$1\cdot y^2+x\cdot 2y\cdot y'-(2x\cdot y+x^2\cdot y')=0$$

$$y^2+2xyy'-2xy-x^2y'=0$$

$$(2xy-x^2)y'=2xy-y^2$$

$$\frac{dy}{dx}=\frac{y(2x-y)}{2xy-x^2}$$

$$\frac{dy}{dx}=0 \ \Rightarrow y(2x-y)=0$$

$$\therefore y=0 \text{ or } y=2x$$

if $y=0$

$$0-0=16$$

no solution

if $y=2x$

$$x(2x)^2-x^2(2x)=16$$

$$4x^3-2x^3=16$$

$$2x^3=16$$

$$x^3=8$$

$$x=2 \ \Rightarrow y=2(2)=4 \qquad \therefore (2,4)$$

(2) *Vertical Tangent* $\Rightarrow \frac{dy}{dx}=undefined$

$$\frac{d}{dx}(x^2y-xy^3)=\frac{d}{dx}(6)$$
$$2xy+x^2y'-y^3-3xy^2y'=0$$
$$(x^2-3xy^2)y'=y^3-2xy$$
$$\frac{dy}{dx}=\frac{y^3-2xy}{x^2-3xy^2}=\frac{y(y^2-2x)}{x(x-3y^2)}$$

$$\frac{dy}{dx}=undefined \ \Rightarrow x(x-3y^2)=0$$

$$\therefore x=0 \text{ or } x=3y^2$$

if $x=0$
$0-0=6$
no solution

if $x=3y^2$
$(3y^2)^2y-3y^2y^3=6$
$9y^5-3y^5=6$
$6y^5=6$
$y=1 \Rightarrow x=3(1)^2=3$
$\therefore (3,1)$

3.8 Higher Order Derivatives

EX 1

(1) $y=(2x-1)^{-1}$

$y' = -(2x-1)^{-2} \cdot 2$
$= -2(2x-1)^{-2}$
$y'' = 4(2x-1)^{-3} \cdot 2$
$= \dfrac{8}{(2x-1)^3}$

(2)

$y'=6(x^2-2x)^5(2x-2)$
$y''=30(x^2-2x)^4(2x-2) \cdot (2x-2)+6(x^2-2x)^5 \cdot 2$
$=120(x^2-2x)^4(x-1)^2+12(x^2-2x)^5$
$=12(x^2-2x)^4[10(x-1)^2+(x^2-2x)]$
$=12(x^2-2x)^4(11x^2-22x+10)$

(3) $y'=\dfrac{1}{2\sqrt{x^2+2}} \cdot 2x=\dfrac{x}{\sqrt{x^2+2}}$

$$y''=\frac{1 \cdot \sqrt{x^2+2}-x \cdot \dfrac{1}{2\sqrt{x^2+2}} \cdot 2x}{(x^2+2)}$$
$$=\frac{\left[\sqrt{x^2+2}-\dfrac{x^2}{\sqrt{x^2+2}}\right] \cdot \sqrt{x^2+2}}{(x^2+2) \cdot \sqrt{x^2+2}}$$
$$=\frac{x^2+2-x^2}{(x^2+2)\sqrt{x^2+2}}=\frac{2}{(x^2+2)\sqrt{x^2+2}}$$

(4) $2x-2yy'=0$

$y'=\dfrac{x}{y}$

$y''=\dfrac{1 \cdot y-x \cdot y'}{y^2}$

$$=\frac{y-x\left(\dfrac{x}{y}\right)}{y^2}=\frac{\left[y-x\left(\dfrac{x}{y}\right)\right] \cdot y}{y^2 \cdot y}$$
$$=\frac{y^2-x^2}{y^3}\left(\text{or } -\frac{4}{y^3}\right)$$

EX 2

(1) $y'=\dfrac{3(2x+3)-(3x-2) \cdot 2}{(2x+3)^2}$

$=\dfrac{13}{(2x+3)^2}=13(2x+3)^{-2}$

$y''=-26(2x+3)^{-3} \cdot 2$

$=\dfrac{-52}{(2x+3)^3}$

$f''(-2)=\dfrac{-52}{(-1)^3}=52$

(2) $y=(1-2x)^{-\frac{1}{2}}$

$y'=-\dfrac{1}{2}(1-2x)^{-\frac{3}{2}} \cdot (-2)=(1-2x)^{-\frac{3}{2}}$

$y''=-\dfrac{3}{2}(1-2x)^{-\frac{5}{2}} \cdot (-2)=\dfrac{3}{(1-2x)^{5/2}}$

$f''(0)=3$

(3)

$g'(x)=5(3-2x^2)^4(-4x)$

$=-20x(3-2x^2)^4$

$g''(x)=-20(3-2x^2)^4-20x \cdot 4(3-2x^2)^3(-4x)$

$g''(1)=-20(1)-20 \cdot 4(1)(-4)$

$=300$

(4) Let $y=x^n-1$

$y'=nx^{n-1}$

$y''=n(n-1)x^{n-2}$

$y'''=n(n-1)(n-2)x^{n-3}$

$f'''(1)=n(n-1)(n-2)$

EX 3

(1)

$G(x)=f(x^2)+xf'(x^2)\cdot 2x$

$=f(x^2)+2x^2f'(x^2)$

$G'(x)=f'(x^2)\cdot 2x+4xf'(x^2)+2x^2f''(x^2)\cdot 2x$

$=6xf'(x^2)+4x^3f''(x^2)$

(2) $F'(x)=2[f(x)]\cdot f'(x)$

$F''(x)=2f'(x)f'(x)+2f(x)f''(x)$

$F''(2)=2f'(2)f'(2)+2f(2)f''(2)$

$=2(-2)(-2)+2(-1)(3)=2$

(3) $\leq t\ y=[f(x^3)]$

$y'=f'(x^3)\cdot 3x^2$

$=3x^2f'(x^3)$

$y''=6xf'(x^3)+3x^2\cdot f''(x^3)\cdot 3x^2$

$=6xf'(x^3)+9x^4\cdot f''(x^3)$

$y''(1)=6f'(1)+9f''(1)$

$=6\cdot 3+9\cdot(-2)=18-18$

$=0$

EX 4

(1)

$2yy'+y+xy'=0$

$y'(2y+x)=-y$

$y'=\dfrac{-y}{x+2y},\quad \therefore \left.\dfrac{dy}{dx}\right|_{(1,1)}=\dfrac{1}{1+2}=-\dfrac{1}{3}$

$y''=\dfrac{-y'(x+2y)-(-y)(1+2y')}{(x+2y)^2}$

$\therefore \left.\dfrac{d^2y}{dx^2}\right|_{(1,1)}=\dfrac{-\left(-\frac{1}{3}\right)(3)-(-1)\left(1-\frac{2}{3}\right)}{(1+2)^2}=\dfrac{1+\frac{1}{3}}{9}$

$=\dfrac{4}{27}$

(2) $2x-(y+xy')+2yy'=0$

$2x-y-xy'+2yy'=0$

$y'(-x+2y)=-2x+y$

$y'=\dfrac{-2x+y}{-x+2y}=\dfrac{2x-y}{x-2y},$

$\therefore \left.\dfrac{dy}{dx}\right|_{(1,-1)}=\dfrac{2+1}{1+2}=1$

$y''=\dfrac{(2-y')(x-2y)-(2x-y)(1-2y')}{(x-2y)^2}$

$\therefore \left.\dfrac{d^2y}{dx^2}\right|_{(1,-1)}=\dfrac{(2-1)(1+2)-(2+1)(1-2)}{(1+2)^2}$

$=\dfrac{3+3}{9}=\dfrac{2}{3}$

EX 5

(1) $g(1)=e^{f(1)}=e^2$

$g'(x)=e^{f(x)}f'(x),\ g'(1)=e^{f(1)}f'(1)=-4e^2$

The tangent line is given by

$y=e^2-4e^2(x-1)$.

(2) $g'(x)=e^{f(x)}f'(x)$

$e^{f(x)}>0$ for all x

So, g' changes from positive to negative only when f' changes from positive to negative. This occurs at $x=-1$ only. Thus, g has a local maximum at $x=-1$.

(3) $g''(-1)=e^{f(-1)}[(f'(-1))^2+f''(-1)]$

$e^{f(-1)}>0$ and $f'(-1)=0$

Since, f' is decreasing on a neighborhood of -1, $f''(-1)<0$. Therefore, $g''(-1)<0$.

(4) $\dfrac{g'(3)-g'(3)}{3-1}=\dfrac{e^{f(3)}f'(3)-e^{f(1)}f'(1)}{2}=2e^2$

3.9 Derivatives of Trigonometric Functions

EX 1

(1) $y'=3\cos(3x+1)$

(2) $y'=-\sin(-3x+2)\cdot(-3)$
$=3\sin(-3x+2)$

(3) $y'=2\cos\left(x^3-\dfrac{1}{x}\right)\cdot\left(3x^2+\dfrac{1}{x^2}\right)$
$=2\left(3x^2+\dfrac{1}{x^2}\right)\cos\left(x^3-\dfrac{1}{x}\right)$

(4) $y'=-\sin(\sin x)\cdot\cos x$

(5) $y'=10\sec^2(10x-5)$

(6) $y'=-\csc^2(3x^2+2x)\cdot(6x+2)$
$=-(6x+2)\csc^2(3x^2+2x)$

(7) $y'=\sec\left(\dfrac{1}{2}x\right)\tan\left(\dfrac{1}{2}x\right)\cdot\dfrac{1}{2}$
$=\dfrac{1}{2}\sec\left(\dfrac{1}{2}x\right)\tan\left(\dfrac{1}{2}x\right)$

(8) $y'=\dfrac{1}{5}\csc(5x)\cot(5x)\cdot 5$
$=\csc(5x)\cot(5x)$

(9) $y'=\sec(\tan x)\tan(\tan x)\cdot\sec^2x$

(10) $y'=\dfrac{1}{2\sqrt{1+2\cot x}}\cdot(-2\csc^2x)$
$=\dfrac{-\csc^2x}{\sqrt{1+2\cot x}}$

(11) $y'=-\csc^2(\pi-\dfrac{1}{x})\cdot\dfrac{1}{x^2}$

(12) $y=\sin[(x^3-1)^4]$
$y'=\cos[(x^3-1)^4]\cdot 4(x^3-1)^3\cdot 3x^2$
$=12x^2(x^3-1)^3\cos(x^3-1)^4$

EX 2

(1) $y'=-4(1+\cos2t)^{-5}\cdot(-\sin2t)\cdot 2$
$=8(1+\cos2t)^{-5}\cdot\sin2t$

(2) $y=(\sin x)^3$
$y'=3(\sin x)^2\cdot\cos x$
$=3\sin^2x\cos x$

(3) $y'=-12x\cos(x^2)\sin(x^2)$

(4) $y=\tan(x^3)+[\tan x]^3$
$y'=\sec^2(x^3)\cdot 3x^2+3[\tan x]^2\cdot\sec^2x$
$=3x^2\sec^2(x^3)+3\tan^2x\sec^2x$

(5) $y'=2x^{-2/3}\sec^3(x^{1/3})\tan(x^{1/3})$

(6)

$y=[\sin(2x)]^3+[\cos(2x)]^3$

$y'=3[\sin(2x)]^2\cdot\cos(2x)\cdot 2+3[\cos(2x)]^2\cdot(-\sin(2x))\cdot 2$

$=6\sin^2(2x)\cos(2x)-6\cos^2(2x)\sin(2x)$

(7)

$y=\tan^2(\sin^3t)$

$\Rightarrow\dfrac{dy}{dt}=2\tan(\sin^3t)\cdot\sec^2(\sin^3t)\cdot(3\sin^2t\cdot(\cos t))$

$=6\tan(\sin^3t)\sec^2(\sin^3t)\sin^2t\cos t$

EX 3

(1) $y'=\sin x+x\cos x$

(2) $y'=2x\cos\left(\dfrac{1}{x}\right)+x^2\left(\sin\left(\dfrac{1}{x}\right)\right)\cdot\left(-\dfrac{1}{x^2}\right)$
$=2x\cos\left(\dfrac{1}{x}\right)-\sin\left(\dfrac{1}{x}\right)$

(3) $y'=\dfrac{6\csc x-3x\csc x\cot x}{2\sqrt{\csc x}}$

(4) $y'=\dfrac{\cos x(1+\cos x)-\sin x(-\sin x)}{(1+\cos x)^2}$

$=\dfrac{\cos x+\cos^2x+\sin^2x}{(1+\cos x)^2}$

$=\dfrac{\cos x+1}{(1+\cos x)^2}=\dfrac{1}{1+\cos x}$

(5) $y' = 4\cos 4x \cos^2 x - 2\sin 4x \cos x \sin x$

(6) $y' = (\sec\sqrt{x})\left[\dfrac{\tan\sqrt{x}\tan(\frac{1}{x})}{2\sqrt{x}} - \dfrac{\sec^2(\frac{1}{x})}{x^2}\right]$

(7)

$$y = \frac{(\cos x)^2}{\sin x}$$
$$y' = \frac{2(\cos x)(-\sin x)(\sin x) - (\cos x)^2 \cdot \cos x}{\sin^2 x}$$
$$= \frac{-2\cos x \cdot \sin^2 x - \cos^3 x}{\sin^2 x} = \frac{-\cos x(2\sin^2 x + \cos^2 x)}{\sin^2 x}$$

EX 4

(1) $y' = \dfrac{\cos(2x+3y) - 2x\sin(2x+3y) - y\cos x}{\sin x + 3x\sin(2x+3y)}$

(2)

$-\sin(x+y) \cdot (1+y') + \cos(x-y) \cdot (1-y') = 0$

$-\sin(x+y) - y'\sin(x+y) + \cos(x-y) - y'\cos(x-y) = 0$

$-\sin(x+y) + \cos(x-y) = y'\sin(x+y) + y'\cos(x-y)$

$-\sin(x+y) + \cos(x-y) = y'[\sin(x+y) + \cos(x-y)]$

$\therefore \dfrac{dy}{dx} = \dfrac{-\sin(x+y) + \cos(x-y)}{\sin(x+y) + \cos(x-y)}$

(3) $y' = \dfrac{-1 - y\sec^2(xy)}{x\sec^2(xy)}$

(4)

$2x + 2y \cdot y' = \sec(xy)\tan(xy) \cdot (y + xy')$

$2x + 2yy' = y\sec(xy)\tan(xy) + xy'\sec(xy)\tan(xy)$

$y'[2y - x\sec(xy)\tan(xy)] = y\sec(xy)\tan(xy) - 2x$

$\therefore \dfrac{dy}{dx} = \dfrac{y\sec(xy)\tan(xy) - 2x}{2y - x\sec(xy)\tan(xy)}$

EX 5

(1) $y = -\dfrac{\pi}{2}x + \pi$

(2) $f(\pi) = \pi^2\cos\pi = -\pi^2 \Rightarrow \therefore (\pi, -\pi^2)$

$y' = 2x\cos x - x^2\sin x$

$f'(\pi) = 2\pi\cos\pi - \pi^2\sin\pi$

$= -2\pi$

$\therefore y + \pi^2 = -2\pi(x - \pi)$

(3) $y - \dfrac{1}{4} = -\dfrac{\sqrt{3}}{4}\left(x - \dfrac{\pi}{6}\right)$

(4) $f\left(\dfrac{\pi}{4}\right) = 2\left(\cot\left(\dfrac{\pi}{4}\right)\right)^2 - \left(\cos\dfrac{\pi}{4}\right)^2$

$= 2(1)^2 - \left(\dfrac{1}{\sqrt{2}}\right)^2 = \dfrac{3}{2}$

$\Rightarrow \therefore \left(\dfrac{\pi}{4}, \dfrac{3}{2}\right)$

$y = 2(\cot x)^2 - (\cos x)^2$

$y' = 4(\cot x) \cdot (-\csc^2 x) - 2(\cos x)(-\sin x)$

$= -4\cot x \cdot \csc^2 x + 2\cos x\sin x$

$f'\left(\dfrac{\pi}{4}\right) = -4\cot\dfrac{\pi}{4}\csc^2\dfrac{\pi}{4} + 2\cos\dfrac{\pi}{4} \cdot \sin\dfrac{\pi}{4}$

$= -4(1) \cdot (\sqrt{2})^2 + 2 \cdot \dfrac{1}{\sqrt{2}} \cdot \dfrac{1}{\sqrt{2}}$

$= -8 + 1 = -7$

$\therefore y - \dfrac{3}{2} = -7\left(x - \dfrac{\pi}{4}\right)$

(5) $y = 2x$

EX 6

(1) $= 24\sin 2x \cdot \cos^2 2x - 12\sin^3 2x$

(2) $= 54\cos 3x \cdot \sin^2 3x - 27\cos^3 3x$

3.10 Derivatives of Exponential Functions :

EX 1

(1) $y' = e^{5-7x} \cdot (-7)$
$= -7e^{5-7x}$

(2) $y' = e^{\sec x} \cdot \sec x \tan x$

(3) $y' = 2xe^{-x^2}\sin(e^{-x^2})$

(4) $y' = 3^{-x^2} \cdot (-2x) \cdot \ln 3$
$= -2x(3)^{-x^2}\ln 3$

(5) $y' = (1-x)e^{-x} + 3e^{3x}$

(6) $y' = a^{\cos x}(-\sin x) \cdot \ln a$
$= -a^{\cos x}\sin x \ln a$

(7) $y' = 3x^2 \cdot e^{2x} + x^3 \cdot e^{2x} \cdot 2$
$= x^2e^{2x}(3+2x)$

(8) $y' = (27x^4 - 18x^3 + 6x^2 + 18x - 6)e^{x^3}$

(9) $y' = \dfrac{e^{-2x} \cdot (-2) \cdot (x) - e^{-2x} \cdot 1}{x^2}$
$= \dfrac{-e^{-2x}(2x+1)}{x^2}$

(10) $y' = \dfrac{3}{2}\cos\left(\dfrac{x}{2}\right)e^{3\sin\left(\frac{x}{2}\right)}$

(11) $y' = \pi^{x\cos x} \cdot (\cos x + x \cdot (-\sin x))\ln \pi$
$= \pi^{x\cos x}(\cos x - x\sin x)\ln \pi$

(12) $y' = \dfrac{-e^{-x} \cdot (e^x - 1) - (e^{-x}+1) \cdot e^x}{(e^x-1)^2}$
$= \dfrac{-1 + e^{-x} - 1 - e^x}{(e^x-1)^2}$
$= \dfrac{e^{-x} - 2 - e^x}{(e^x-1)^2}$

(13) $e^{x^2y} = 2x + 2y$

$\Rightarrow e^{x^2y}(x^2y' + 2xy) = 2 + 2y'$

$\Rightarrow x^2e^{x^2y}y' + 2xye^{x^2y} = 2 + 2y'$

$\Rightarrow x^2e^{x^2y}y' - 2y' = 2 - 2xye^{x^2y}$

$\Rightarrow y' = \dfrac{2 - 2xye^{x^2y}}{x^2e^{x^2y} - 2}$

(14) $y' = (x^2 + 2x)e^x\cos(x^2e^x)$

(15) $y' = 2e^{-x}(4e^{-x}+2)^{-\frac{3}{2}}$

(16) $e^{x+y} \cdot (1+y') = 6x \cdot y + 3x^2 \cdot y'$

$e^{x+y} + y'e^{x+y} = 6xy + 3x^2y'$

$y'(e^{x+y} - 3x^2) = 6xy - e^{x+y}$

$\therefore \dfrac{dy}{dx} = \dfrac{6xy - e^{x+y}}{(e^{x+y} - 3x^2)}$

(17) $e^{2x} = \sin(x+3y)$

$\Rightarrow 2e^{2x} = (1+3y')\cos(x+3y)$

$\Rightarrow 1 + 3y' = \dfrac{2e^{2x}}{\cos(x+3y)}$

$\Rightarrow 3y' = \dfrac{2e^{2x}}{\cos(x+3y)} - 1$

$\Rightarrow y' = \dfrac{2e^{2x} - \cos(x+3y)}{3\cos(x+3y)}$

EX 2

(1) $f(1)=\dfrac{e^{-1}}{\sqrt{1}}=\dfrac{1}{e} \Rightarrow \therefore \left(1,\dfrac{1}{e}\right)$

$$y'=\dfrac{(-e^{-x})\sqrt{x}-e^{-x}\cdot\dfrac{1}{2\sqrt{x}}}{x}$$

$$f'(1)=\dfrac{(-e^{-1})-e^{-1}\cdot\dfrac{1}{2}}{1}$$

$$=-\dfrac{1}{e}-\dfrac{1}{2e}=-\dfrac{3}{2e}$$

$$\therefore y-\dfrac{1}{e}=-\dfrac{3}{2e}(x-1)$$

(2) $f(1)=e^1= \Rightarrow \therefore (1,e)$

$$y'=2xe^{\frac{1}{x}}+x^2e^{\frac{1}{x}}\left(-\dfrac{1}{x^2}\right)$$

$$f'(1)=2e+e^1\cdot(-1)$$

$$=e$$

$\therefore y-e=e(x-1)$ (or $y=ex$)

EX 3

(1) $y'=e^{3x}\cdot 3=3e^{3x}\cdots\cdots$ ①

Let the tangent point be (x,y) tangent slope at (x,y) = secant slope between $(0,0)$ and (x,y)

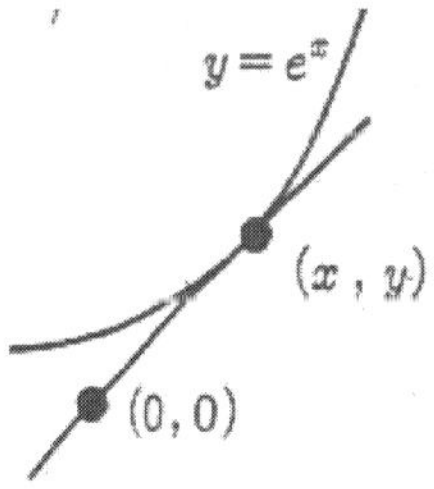

$$y'=\dfrac{y-0}{x-0}$$

$$y'=\dfrac{y}{x}\cdots\cdots ②$$

By ① and ②, $3e^{3x}=\dfrac{e^{3x}}{x}$

$$3=\dfrac{1}{x} \Rightarrow \therefore x=\dfrac{1}{3}$$

$$slope=f'\left(\dfrac{1}{3}\right)=3e$$

$$\therefore y=3ex$$

(2) $y=-\dfrac{1}{e^2}(x+2)$

EX 4

$$f'(x)=e^x\cos x-e^x\sin x$$
$$=e^x(\cos x-\sin x)\cdots\cdots ①$$

$$f''(x)=e^x\cdot(\cos x-\sin x)+e^x(-\sin x-\cos x)$$
$$=e^x(\cos x-\sin x-\sin x-\cos x)$$
$$=2e^x\sin x\cdots\cdots ②$$

Plug ① and ② into original equation

$$[-2e^x\sin x]-p[e^x(\cos x-\sin x)]+q[e^x\cos x]=0$$
$$-2e^x\sin x-pe^x\cos x+pe^x\sin x+qe^x\cos x=0$$
$$-2\sin x-p\cos x+p\sin x+q\cos x=0$$

$(-2+p)\sin x+(-p+q)\cos x=0$ for *all real νmbers x*

$-2+p=0$ and $-p+q=0$

$$\therefore p=2,\ q=2$$

3.11 derivatives of Logarithmic Functions:

EX 1

(1) $y' = \dfrac{3}{3x}$

(2) $y' = \dfrac{1}{\ln x} \cdot \dfrac{1}{x}$

(3) $y' = \dfrac{\cos x}{\sin x}$

(4) $y' = \dfrac{1}{\tan x} \cdot \sec^2 x$
$= \dfrac{\sec^2 x}{\tan x}$

(5) $y' = 5(\ln x)^4 \cdot \dfrac{1}{x}$
$= \dfrac{5(\ln x)^4}{x}$

(6) $y' = \dfrac{1}{\sin x} \cdot \cos x \cdot \dfrac{1}{\ln 3}$
$= \dfrac{\cot x}{\ln 3}$

(7) $y' = \dfrac{1}{x \ln 2}$

(8) $y' = \dfrac{1}{x\cos x}(\cos x - x\sin x)$
$= \dfrac{\cos x - x\sin x}{x\cos x}$

(9) $y' = \dfrac{1}{x}(\log_2 3)3^{\log_2 x}$

(10) $y' = 2x\ln x + x^2 \cdot \dfrac{1}{x}$
$= 2x\ln x + x$

(11) $y' = (\ln x)^{1/2} + \dfrac{1}{2(\ln x)^{\frac{1}{2}}}$

(12) $y' = \dfrac{1}{\sec x + \tan x}(\sec x \tan x + \sec^2 x)$

$= \dfrac{\sec x(\tan x + \sec x)}{\sec x + \tan x}$

$= \sec x$

(13)

$y' = \sec^2[\log(x^2+4)] \cdot \dfrac{1}{x^2+4} \cdot 2x \cdot \dfrac{1}{\ln 10}$
$= \dfrac{2x\sec^2[\log(x^2+4)]}{(x^2+4)\ln 10}$

(14) $y' = 4x^7(\ln x)^3 + 8x^7(\ln x)^4$

(15) $y' = \dfrac{\frac{1}{x} \cdot x^2 - \ln x \cdot 2x}{x^4}$
$= \dfrac{x - 2x\ln x}{x^4} = \dfrac{1-2\ln x}{x^3}$

(16) $\dfrac{1}{xy} \cdot (y + xy') = 1 - y'$

$y + xy' = xy - xyy'$

$y'(x + xy) = xy - y$

$\therefore \dfrac{dy}{dx} = \dfrac{y(x-1)}{x(y+1)}$

(17) $y + 1 = x \cdot [\ln y]^3$

$y' = [\ln y]^3 + x \cdot 3[\ln y]^2 \cdot \dfrac{1}{y} \cdot y'$

$yy' = y\ln^3 y + 3x\ln^2 y \cdot y'$

$y'(y - 3x\ln^2 y) = y\ln^3 y$

$\therefore \dfrac{dy}{dx} = \dfrac{yln^3 y}{(y - 3xln^2 y)}$

(18) $y' = \dfrac{\tan(\ln x)}{x}$

EX 2

(1) $y' = \dfrac{6}{2x+1} + \dfrac{8x}{x^2+1}$

(2) $y' = \dfrac{1}{x} - 1$

(3) $y' = \dfrac{10x}{x^2+1} + \dfrac{1}{2(1-x)}$

(4) $y=\ln\sqrt{\dfrac{(x+1)^5}{(x+2)^{20}}}=\dfrac{1}{2}[5\ln(x+1)-20\ln(x+2)]$

$$\Rightarrow y'=\frac{1}{2}\left(\frac{5}{x+1}-\frac{20}{x+2}\right)=\frac{5}{2}\left[\frac{(x+2)-4(x+1)}{(x+1)(x+2)}\right]$$

$$=-\frac{5}{2}\left[\frac{3x+2}{(x+1)(x+2)}\right]$$

(5) $y'=\dfrac{1}{2}\left[\cot x-\tan x-\dfrac{4}{x(1+2\ln x)}\right]$

(6)

$$y=\log_7\left(\frac{\sin\theta\cos\theta}{e^\theta 2^\theta}\right)$$
$$=\frac{\ln(\sin\theta)+\ln(\cos\theta)-\ln e^\theta-2^\theta}{\ln 7}$$
$$=\frac{\ln(\sin\theta)+\ln(\cos\theta)-\theta-\theta\ln 2}{\ln 7}$$
$$\Rightarrow\frac{dy}{d\theta}=\frac{\cos\theta}{(\sin\theta)(\ln 7)}-\frac{\sin\theta}{(\cos\theta)(\ln 7)}-\frac{1}{\ln 7}-\frac{\ln 2}{\ln 7}$$
$$=\left(\frac{1}{\ln 7}\right)(\cot\theta-\tan\theta-1-\ln 2)$$

EX 3

(1) $\ln y=\ln x^{\ln x}$

$\ln y=\ln x\cdot\ln x$

$\dfrac{1}{y}\cdot y'=\dfrac{1}{x}\ln x+\ln x\cdot\dfrac{1}{x}$

$y'=\dfrac{2y}{x}\ln x$

(2) $y'=(\sin x)^x[\ln(\sin x)+x\cot x]$

(3) $y'=(x+1)^x[\dfrac{x}{x+1}+\ln(x+1)]$

(4) $\ln y=\ln(\sin x)^{\sin x}$

$\ln y=\sin x\ln(\sin x)$

$\dfrac{1}{y}\cdot y'=\cos x\cdot\ln(\sin x)+\sin x\cdot\dfrac{1}{\sin x}\cdot\cos x$

$y'=y\cos x(\ln(\sin x)+1)$

(5) $y'=\sqrt{(x^2+1)(x-1)^2}\left(\dfrac{x}{x^2+1}+\dfrac{1}{x-1}\right)$

(6) $\ln y=\ln(\ln x)^{\ln x}$

$\ln y=\ln x\cdot\ln(\ln x)$

$\dfrac{1}{y}y'=\dfrac{1}{x}\ln(\ln x)+\ln x\cdot\dfrac{1}{\ln x}\cdot\dfrac{1}{x}$

$y'=\dfrac{y}{x}(\ln(\ln x)+1)$

(7) $y=\dfrac{x\sqrt{x^2+1}}{(x+1)^{2/3}}\left[\dfrac{1}{x}+\dfrac{x}{x^2+1}-\dfrac{2}{3(x+1)}\right]$

(8)

$$y=\frac{\theta\sin\theta}{\sqrt{\sec\theta}}\Rightarrow\ln y=\ln\theta+\ln(\sin\theta)-\frac{1}{2}\ln(\sec\theta)$$
$$\Rightarrow\frac{1}{y}\frac{dy}{d\theta}=\left[\frac{1}{\theta}+\frac{\cos\theta}{\sin\theta}-\frac{(\sec\theta)(\tan\theta)}{2\sec\theta}\right]$$
$$\Rightarrow\frac{dy}{d\theta}=\frac{\theta\sin\theta}{\sqrt{\sec\theta}}\left(\frac{1}{\theta}+\cot\theta-\frac{1}{2}\tan\theta\right)$$

(9)

$$\ln y=\ln\left(\frac{e^{-x}\cos x}{\sqrt[3]{x^2-x}}\right)$$

$$\ln y=\ln e^{-x}+\ln(\cos x)-\ln(\sqrt[3]{x^2-x})$$

$$\ln y=-x+\ln(\cos x)-\frac{1}{3}\ln(x^2-x)$$

$$\frac{1}{y}\cdot y'=-1+\frac{1}{\cos x}(-\sin x)-\frac{1}{3}\cdot\frac{1}{x^2-x}\cdot(2x-1)$$

$$y'=y(-1-\tan x-\frac{2x-1}{3(x^2-x)})$$

EX 4

(1) $\ln y = \ln\left[\dfrac{(x+1)^2}{x^3(x-2)}\right]$

$\ln y = 2\ln(x+1) - 3\ln x - \ln(x-2)$

$\dfrac{1}{y}y' = 2 \cdot \dfrac{1}{x+1} - 3 \cdot \dfrac{1}{x} - \dfrac{1}{x-2}$

$y' = y(\dfrac{2}{x+1} - \dfrac{3}{x} - \dfrac{1}{x-2})$

When $x=1$, $y = \dfrac{(1+1)^2}{1(1-2)} = -4$

$f'(1) = -4\left(\dfrac{2}{2} - \dfrac{3}{1} - \dfrac{1}{-1}\right) = -4(1-3+1) = 4$

(2) $= \dfrac{1}{2} \cdot \dfrac{7}{8} = \dfrac{7}{16}$

EX 5

Let

$y = \sqrt{2x + \ln(f(x))}$

$y' = \dfrac{1}{2\sqrt{2x+\ln(f(x))}} \cdot \left(2 + \dfrac{1}{f(x)} \cdot f'(x)\right)$

$y'(2) = \dfrac{1}{2\sqrt{4+\ln(f(2))}} \cdot \left(2 + \dfrac{1}{f(2)} \cdot f'(2)\right)$

$= \dfrac{1}{2\sqrt{4+\ln 1}} \cdot (2+1 \cdot 4) = \dfrac{1}{4}(2+4) = \dfrac{3}{2}$

3.12 Inverse Trigonometric Functions :

EX 1

(1) $-\dfrac{\pi}{6}$ (2) $-\dfrac{\pi}{3}$

(3) $\dfrac{\pi}{3}$ (4) $\dfrac{\pi}{2}$

EX 2

(1) $= \sin^{-1}\left(-\dfrac{1}{2}\right)$
$= -\dfrac{\pi}{6}$

(2) $= \sin\left(-\dfrac{\pi}{6}\right)$
$= -\dfrac{1}{2}$

(3) $= \sin^{-1}\left(\dfrac{\sqrt{3}}{2}\right)$
$= \dfrac{\pi}{3}$

(4) $= \cos\left(-\dfrac{\pi}{2}\right)$
$= 0$

EX 3

(1) $\dfrac{2}{3}\pi$ (2) $\dfrac{5}{6}\pi$

(3) $\dfrac{1}{6}\pi$ (4) $\dfrac{1}{2}\pi$

EX 4

(1) $= \cos^{-1}\left(-\dfrac{\sqrt{3}}{2}\right)$
$= \dfrac{5\pi}{6}$

(2) $= \cos\left(\dfrac{2\pi}{3}\right)$
$= -\dfrac{1}{2}$

(3) $= \cos^{-1}\left(\dfrac{\sqrt{3}}{2}\right)$
$= \dfrac{\pi}{6}$

(4) $= \tan(0)$
$= 0$

EX 5

(1) $-\dfrac{\pi}{4}$ (2) $\dfrac{\pi}{6}$ (3) $\dfrac{\pi}{3}$ (4) 0

EX 6

(1) $=\tan^{-1}(-1)$
$=-\dfrac{\pi}{4}$

(2) $=\tan\left(-\dfrac{\pi}{6}\right)$
$=-\dfrac{1}{\sqrt{3}}$

(3) $=\tan^{-1}\left(-\dfrac{1}{\sqrt{3}}\right)$
$=-\dfrac{\pi}{6}$

(4) $\cos\left(-\dfrac{\pi}{3}\right)$
$=\dfrac{1}{2}$

EX 7

(1) $=\sin^{-1}\left(\dfrac{1}{2}\right)$
$=\dfrac{\pi}{6}$

(2) $\sin^{-1}\left(-\dfrac{\sqrt{3}}{2}\right)$
$=-\dfrac{\pi}{3}$

(3) $=\cos^{-1}\left(-\dfrac{1}{\sqrt{2}}\right)$
$=\dfrac{3\pi}{4}$

(4) $=\sin^{-1}(0)$
$=0$

(5) $=\cos^{-1}\left(\dfrac{1}{2}\right)$
$=\dfrac{\pi}{3}$

(6) $=\cos^{-1}(0)$
$=\dfrac{\pi}{2}$

(7) $=\dfrac{\pi}{2}-\tan^{-1}\left(\dfrac{1}{\sqrt{3}}\right)$
$=\dfrac{\pi}{2}-\dfrac{\pi}{6}=\dfrac{\pi}{3}$

(8) $=\dfrac{\pi}{2}-\tan^{-1}(-\sqrt{3})$
$=\dfrac{\pi}{2}-\left(-\dfrac{\pi}{3}\right)=\dfrac{5\pi}{6}$

EX 8

(1) $\sin\left(\cos^{-1}\dfrac{\sqrt{2}}{2}\right)=\sin\left(\dfrac{\pi}{4}\right)=\dfrac{1}{\sqrt{2}}$

(2) $\sec\left(\cos^{-1}\dfrac{1}{2}\right)=\sec\left(\dfrac{\pi}{3}\right)=2$

(3) $\tan\left(\sin^{-1}\left(-\dfrac{1}{2}\right)\right)=\tan\left(-\dfrac{\pi}{6}\right)=-\dfrac{1}{\sqrt{3}}$

(4) $\cot\left(\sin^{-1}\left(-\dfrac{\sqrt{3}}{2}\right)\right)=\cot\left(-\dfrac{\pi}{3}\right)=-\dfrac{1}{\sqrt{3}}$

3.13 Derivatives of Inverse Trigonometric Functions

EX 1

(1) $y'=\dfrac{-6}{x\sqrt{x^4-9}}$

(2) $y'=-\dfrac{1}{\sqrt{1-(-3x+2)^2}}\cdot(-3)$
$=\dfrac{3}{\sqrt{1-(-3x+2)^2}}$

(3) $y'=\dfrac{-1}{\sqrt{2x-x^2}}$

(4) $y'=-2\cdot\dfrac{1}{\sqrt{1-(\sqrt{x-1})^2}}\cdot\dfrac{1}{2\sqrt{x-1}}$
$=\dfrac{-1}{\sqrt{2-x}\,\sqrt{x-1}}$

(5) $y'=\dfrac{1}{x[1+(\ln x)^2]}$

(6) $y'=-\dfrac{1}{1+\left(\dfrac{1}{x}\right)^2}\cdot\dfrac{-1}{x^2}$
$=\dfrac{1}{x^2+1}$

(7)

$y=\ln(x^2+4)-x\tan^{-1}\left(\dfrac{x}{2}\right)$

$\Rightarrow\dfrac{dy}{dx}=\dfrac{2x}{x+4}-\tan^{-1}\left(\dfrac{x}{2}\right)-x\left[\dfrac{\left(\dfrac{1}{2}\right)}{1+\left(\dfrac{x}{2}\right)^2}\right]$

$=\dfrac{2x}{x^2+4}-\tan^{-1}\left(\dfrac{x}{2}\right)-\dfrac{2x}{4+x^2}=-\tan^{-1}\left(\dfrac{x}{2}\right)$

(8)

$y'=-\dfrac{1}{1+(\csc(x^2))^2}\cdot(-\csc(x^2)\cot(x^2))\cdot 2x$
$=\dfrac{2x\csc(x^2)\cot(x^2)}{1+\csc^2(x^2)}$

(9) $y'=\dfrac{1}{|2x+1|\sqrt{x^2+x}}$

(10) $y' = \dfrac{1}{\sqrt{1-((1-x^2)^2)^2}} \cdot 2(1-x^2) \cdot (-2x)$

$= \dfrac{-4x(1-x^2)}{\sqrt{1-(1-x^2)^4}}$

(11)

$y = \tan^{-1}\sqrt{x^2-1} + \csc^{-1}x = \tan^{-1}(x^2-1)^{1/2} + \csc^{-1}x$

$\Rightarrow \dfrac{dy}{dx} = \dfrac{\left(\frac{1}{2}\right)(x^2-1)^{-1/2}(2x)}{1+\left[(x^2-1)^{1/2}\right]^2} - \dfrac{1}{|x|\sqrt{x^2-1}}$

$= \dfrac{1}{x\sqrt{x^2-1}} - \dfrac{1}{|x|\sqrt{x^2-1}} = 0,\ \text{for } x > 1$

(12) $y' = \dfrac{1}{2\sqrt{x^2-1}} \cdot 2x - \dfrac{1}{|x|\sqrt{x^2-1}},\ x > 1$

$= \dfrac{x}{\sqrt{x^2-1}} - \dfrac{1}{x\sqrt{x^2-1}}$

$= \dfrac{x^2-1}{x\sqrt{x^2-1}}$

$= \dfrac{\sqrt{x^2-1}}{x}$

EX 2

(1) $y' = 4(\cos^{-1}x)^3 \cdot \dfrac{-1}{\sqrt{1-x^2}}$

$y'(0) = 4(\cos^{-1}(0))^3 \cdot \dfrac{-1}{\sqrt{1}}$

$= 4\left(\dfrac{\pi}{2}\right)^3 \cdot (-1)$

$= -\dfrac{\pi^3}{2}$

(2) $y' = \tan^{-1}(x^2-1) + x \cdot \dfrac{1}{1+(x^2-1)^2} \cdot 2x$

$= \tan^{-1}(x^2-1) + \dfrac{2x^2}{1+(x^2-1)^2}$

$y'(-\sqrt{2}) = \tan^{-1}(1) + \dfrac{4}{1+1}$

$= \dfrac{\pi}{4} + 2$

EX 3

(1)

$f(4) = \sec^{-1}(2) = \cos^{-1}\left(\dfrac{1}{2}\right) = \dfrac{\pi}{3} \Rightarrow \therefore \left(4, \dfrac{\pi}{3}\right)$

$y' = \dfrac{1}{|\sqrt{x}|\sqrt{(\sqrt{x})^2-1}} \cdot \dfrac{1}{2\sqrt{x}}$

$= \dfrac{1}{\sqrt{x} \cdot \sqrt{x-1}} \cdot \dfrac{1}{2\sqrt{x}}$

$= \dfrac{1}{2x\sqrt{x-1}}$

$f'(4) = \dfrac{1}{8\sqrt{3}} = \dfrac{\sqrt{3}}{24}$

$\therefore y - \dfrac{\pi}{3} = \dfrac{\sqrt{3}}{24}(x-4)$

(2)

$f(\sqrt{3}) = (6\tan^{-1}(\sqrt{3}))^3 = \left(6 \cdot \dfrac{\pi}{3}\right)^3$

$= (2\pi)^3 = 8\pi^3$

$\Rightarrow \therefore (\sqrt{3}, 8\pi^3)$

$y' = 3(6\tan^{-1}x)^2 \cdot 6 \cdot \dfrac{1}{1+x^2}$

$f'(\sqrt{3}) = 3(6\tan^{-1}(\sqrt{3}))^2 \cdot 6 \cdot \dfrac{1}{1+3}$

$= 3 \cdot \left(6 \cdot \dfrac{\pi}{3}\right)^2 \cdot 6 \cdot \dfrac{1}{4}$

$= 18\pi^2$

$\therefore y - 8\pi^3 = 18\pi^2(x - \sqrt{3})$

3.14 Derivatives of Inverse Functions :

EX 1

(1) To find the slope of g at the point where $x=6$, we must look at the point on f where $y=6$, namely, $(2,6)$. Since $f'(2)=\frac{1}{3}$, $g'(6)=3$.

(2) Since f passes through the point $(3,8)$, f^{-1} must pass through the point $(8,3)$,
Furthermore,
since the graph of f has slope 5 at $(3,8)$, the graph of f^{-1} must have slope $\frac{1}{5}$ at $(8,3)$.

EX 2

(1) $\frac{1}{3}$

(2)

$$\frac{df}{dx}=3x^2-6x \Rightarrow \left.\frac{df^{-1}}{dx}\right|_{x=f(3)}=\frac{1}{\left.\frac{df}{dx}\right|_{x=3}}=\frac{1}{9}$$

(3) $\frac{1}{6}$

(4)

$$\left.\frac{df^{-1}}{dx}\right|_{x=4}=\left.\frac{df^{-1}}{dx}\right|_{x=f(2)}=\frac{1}{\left.\frac{df}{dx}\right|_{x=2}}=\frac{1}{\left(\frac{1}{3}\right)}=3$$

(5) $\frac{1}{2}$

(6) $f(g(x))=x \Rightarrow g(x)=f^{-1}(x)$

$$g'(2)=(f^{-1})'(2)=\frac{1}{f'(f^{-1}(2))}$$

$$=\frac{1}{f'(0)}=\frac{1}{2}$$

$(\therefore f^{-1}(2)=x, \quad 2=f(x)\Rightarrow 2=e^x+x+1$
$e^x+x+1=2\Rightarrow \quad \therefore x=0$

$f'(x)=e^x+1\Rightarrow \quad f'(0)=2\ )$

CHAPTER 04.

Applications of Derivatives

4.1 Critical Points

EX 1

(1) $x=0$ and $x=4$.

(2) $y=x^2+\dfrac{2}{x}$

$$\Rightarrow y'=2x-\frac{2}{x^2}=\frac{2x^3-2}{x^2}=0$$
$$\Rightarrow 2x^3-2=0$$
$$\Rightarrow x=1;\frac{2x^3-2}{x^2}=undefined$$
$$\Rightarrow x^2=0\Rightarrow x=0$$

The domain of the function is $(-\infty,0)\cup(0,\infty)$, thus $x=0$ is not in the domain, so the only critical point is $x=1$.

(3) $x=0, x=1$, and $x=2$.

(4) $y=4x^{\frac{1}{3}}+x^{\frac{4}{3}}$

$$y'=\frac{4}{3}x^{-\frac{2}{3}}+\frac{4}{3}x^{\frac{1}{3}}$$
$$=\frac{4}{3\cdot x^{2/3}}+\frac{4}{3}x^{1/3}$$
$$=\frac{4+4x}{3\cdot x^{2/3}}$$
$$=\frac{4(1+x)}{3x^{2/3}}$$
$$y'=0\,at\;x=-1$$
$$y'=undefined\;at\;x=0$$
$$\therefore x=-1,0$$

(5) $x=e^{-1/2}$

(6) $y'=1+2\cos x$

$$y'=0\Rightarrow 1+2\cos x=0$$
$$\cos x=-\frac{1}{2}$$
$$\therefore x=\frac{2\pi}{3},\frac{4\pi}{3}$$

4.2 Extrema : Maximum & Minimum value

EX 1

(1) neither

(2) absolute minimum, relative minimum

(3) relative maximum

(4) relative minimum

(5) relative maximum

(6) neither

(7) relative minimum

(8) absolute maximum

EX 2

(1) increasing on $(0,1)$ and $(2,\infty)$, decreasing on $(-\infty,0)$ and $(1,2)$.

(2) $Domain : x>0$

$$h'(x)=\frac{1}{x}-\frac{1}{x^2}=\frac{x-1}{x^2}$$
$$h'(x)=0\;at\;x=1$$

x	0		1	
y'		$-$	0	$+$

y is increasing on $x>1$

y is decreasing on $0<x<1$

(3) y is increasing on $0\le x<\dfrac{\pi}{6}$, $\dfrac{5\pi}{6}<x\le\pi$

y is decreasing on $\dfrac{\pi}{6}<x<\dfrac{5\pi}{6}$

(4) $y' = 3x^2e^{-x} - x^3e^{-x}$
$\quad = x^2e^{-x}(3-x)$
$y' = 0$ at $x = 0, 3$

x		0		3	
y'	+	0	+	0	−

y is increasing on $x < 0,\ 0 < x < 3$

y is decreasing on $x > 3$

(5) y is increasing on $-2 < x < 2$

y is decreasing on $x < -2, x > 2$

(6) $f(x) = x^2\ln x$
$\Rightarrow f'(x) = x + 2x\ln x = x(1+2\ln x)$
$\Rightarrow$ a critical point at
$x = e^{-1/2} \Rightarrow f' = [---\ |\ +++,$
$\quad 0 \quad e^{-1/2}$

increasing on $(e^{-1/2}, \infty)$,

decreasing $(0, e^{-1/2})$

EX 3

(1) a local maximum is $k(3) = 162$ at $t = 3$,

a local minimum is $k(-3) = -162$ at $t = -3$.

(2) *Vertical Asymptote* : $x = 0$

$$y' = \frac{e^x \cdot x - e^x \cdot 1}{x^2} = \frac{e^x(x-1)}{x^2}$$

$y' = 0$ at $x = 1$
$y' = undefined$ at $x = 0$

		V.A.			
x		0		1	
y'	−	∅	−	0	+

Relative Minimum $= f(1) = e$

(3) Relative Minimum $= f(0) = 0$

Relative Maximum $= f(2) = \dfrac{4}{e^2}$

(4) $f(x) = x^{1/3}(x+8) = x^{4/3} + 8x^{1/3}$
$\Rightarrow f'(x) = \frac{4}{3}x^{1/3} + \frac{8}{3}x^{-2/3} = \frac{4(x+2)}{3x^{2/3}}$

$\Rightarrow$ critical points at $x = 0, -2$
$\Rightarrow f' = ---\ |\ +++)(+++,$
$\quad -2 \quad 0$

increasing on $(-2, 0) \cup (0, \infty)$,

decreasing on $(-\infty, -2)$,

a local minimum is

$f(-2) = -6\sqrt[3]{2} \approx -7.56$ at $x = -2$.

(5) local maximum is

$g(-2) = 3\sqrt[3]{4} \approx 4.762$ at $x = -2$,

a local minimum is

$g(0) = 0$ at $x = 0$.

EX 4

(1)

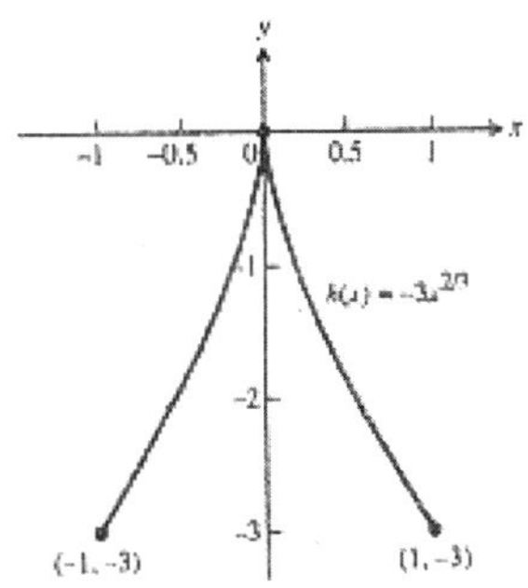

$h(x) = -3x^{2/3} \Rightarrow h'(x) = -2x^{-2/3} \Rightarrow$ a critical point at $x = 0$; $h(-1) = -3$, $h(0) = 0$, $h(1) = -3 \Rightarrow$ the absolute maximum is 0 at $x = 0$ and the absolute minimum is -3 at $x = 1$ and at $x = -1$.

(2) Absolute Maximum $= f(2) = 16$

Absolute Minimum $= f(-2) = f(4) = -16$

(3) Absolute Minimum $= f(1) = 0$

Absolute Maximum $= f(e) = \dfrac{\ln e}{e} = \dfrac{1}{e}$

(4)

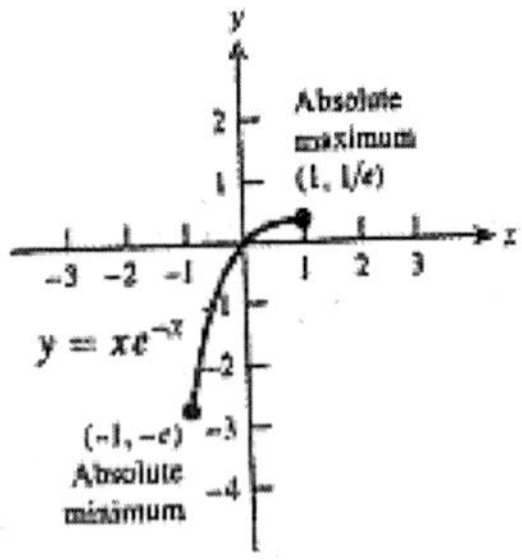

$g(x)=xe^{-x} \Rightarrow g'(x)=e^{-x}-xe^{-x}$

$\Rightarrow$ a critical

point at $x=1$; $g(-1)=-e$, and $g(1)=\frac{1}{e}$,

$\Rightarrow$ the absolute maximum is $\frac{1}{e}$ at $x=1$

and the absolute minimum is $-e$ at $x=-1$.

(5) Absolute Minimum $=f(0)=f(2)=0$

Absolute Maximum $=f(\sqrt{2})=2$

(6) $y=4x-6x^{\frac{2}{3}}$

$y'=4-4x^{-\frac{1}{3}}$

$=4-\frac{4}{x^{1/3}}=\frac{4x^{1/3}-4}{x^{1/3}}=\frac{4(x^{1/3}-1)}{x^{1/3}}$

$y'=0$ *at* $x=1$, $y'=$ *undefined at* $x=0$

x	-1		0		1		8
y'		$+$	$\varnothing$	$-$	0	$+$	
y	-10	↗	0	↘	-2	↗	8

Absolute Maximum $=f(8)=8$

Absolute Minimum $=f(-1)=-10$

(7) $x=0,\pi,2\pi$ $\quad x=\frac{\pi}{3},\frac{5\pi}{3}$

x	0		$\pi/3$		π		$5\pi/3$		2π
y'	0	$+$	0	$-$	0	$+$	0	$-$	0
y	1	↗	$\frac{3}{2}$	↘	-3	↗	$\frac{3}{2}$	↘	1

Absolute Maximum $=f\left(\frac{\pi}{3}\right)=f\left(\frac{5\pi}{3}\right)=\frac{3}{2}$

Absolute Minimum $=f(\pi)=-3$

4.3 Concavity

EX 1

(1)

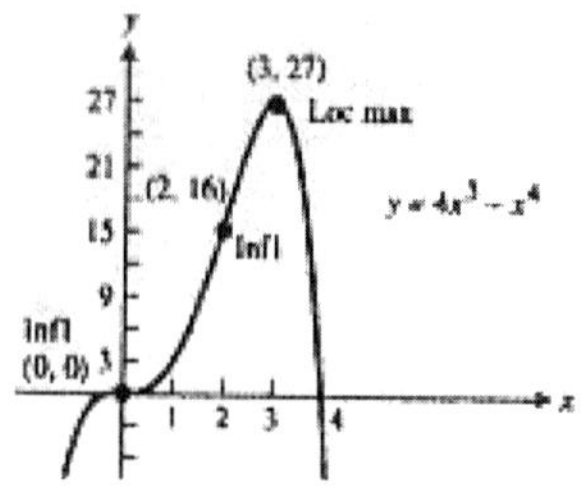

When $y=4x^3-x^4$,

then $y'=12x^2-4x^3=4x^2(3-x)$

and $y''=24x-12x^2=12x(2-x)$. The curve r-ise on $(-\infty,3)$ and falls on $(3,\infty)$. At $x=3$ there is a local maximum, but there is no loc-al minimum, The graph is concave up on $(0, 2)$, and concave down on $(-\infty,0)$ and $(2,\infty)$. There are inflection points at $x=0$ and $x=2$.

(2)

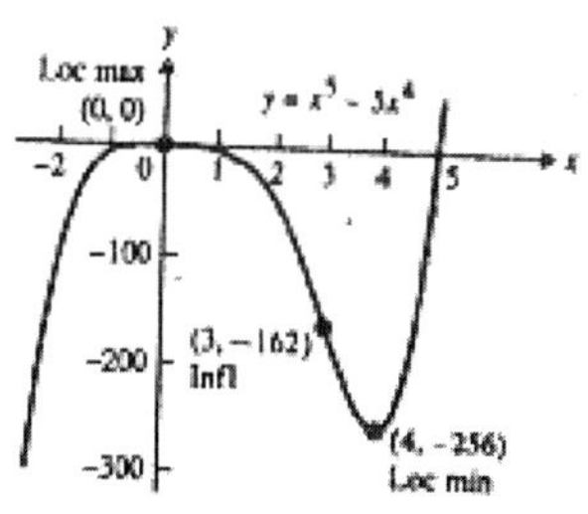

When $y=x^5-5x^4$, then $y'=5x^4-20x^3$ $=5x^3(x-4)$ and $y''=20x^3-60x^2$ $=20x^2(x-3)$. The curve rises on $(-\infty,0)$ and $(4,\infty)$, and falls on $(0,4)$. There is a local m-aximum at $x=0$, and a local minium at $x=4$ the curve is concave down on $(-\infty,3)$ and concave up on $(3,\infty)$. At $x=3$ there is a point of inflection.

(3) y is concave up on $x > e^{-3/2}$

y is concave down on $0 < x < e^{-3/2}$

There are inflection point at $x = e^{-3/2}$

(4)

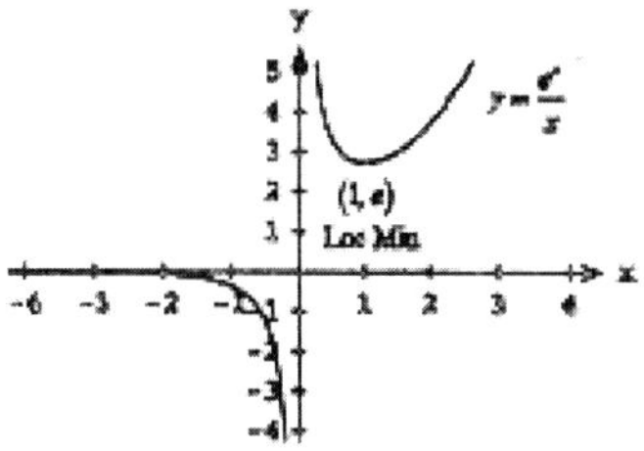

$y = \frac{e^x}{x}$, $\Rightarrow$ $y' = \frac{xe^x - e^x}{x^2}$

$\Rightarrow y' = ---\ |\ ---\ |\ +++$ (sign changes at 0 and 1)

$\Rightarrow$ the graph is rising on $(1, \infty)$, falling on $(-\infty, 0)$ and $(0, 1)$; a local minimum is e at $x = 1$;

$y'' = \frac{x^2(xe^x + e^x - e^x) - (xe^x - e^x)2x}{x^4}$

$= \frac{(x^2 - 2x + 2)e^x}{x^3}$

$\Rightarrow y'' = ---\ |\ +++$ (sign change at 0)

$\Rightarrow$ the graph is concave up on $(0, \infty)$, concave down on $(-\infty, 0)$, but has no inf-lcction points.

EX 2

(1) $a = -6,\ b = 24$

(2) $f(-1) = 3$
$-1 + a - b + c = 3$
$a - b + c = 4 \cdots\cdots ①$

$f'(-1) = 0$
$f'(x) = 3x^2 + 2ax + b$
$f(-1) = 3 - 2a + b = 0 \cdots\cdots ②$

$f''(-3) = 0$
$f''(x) = 6x + 2a$
$f''(-3) = -18 + 2a = 0$
$\therefore a = 9 \cdots\cdots\cdots\cdots ③$

② $3 - 2(9) + b = 0$
$\therefore b = 15$

① $9 - 15 + c = 4$
$\therefore\ c = 10$

(3) $a = -4$, $b = -4$

(4) $f'(1) = -9$
$f'(x) = 3ax^2 + 2bx$
$f'(x) = 3a + 2b = -9 \cdots\cdots ①$

$f''(2) = 0$
$f''(x) = 6ax + 2b$
$f''(2) = 12a + 2b = 0 \cdots\cdots ②$

$f(2) = -12$
$8a + 4b + c = -12 \cdots\cdots ③$

① $3a + 2b = -9$
② $12a + 2b = 0$

① $-$ ② $\ -9a = -9$
$\therefore a = 1$

① $3(1) + 2b = -0$
$2b = -12$
$\therefore b = -6$

③ $8(1) + 4(-6) + c = -12$
$8 - 24 + c = -12$
$\therefore c = 4$

$\therefore a = 1\ b = -6\ c = 4$

EX 3

	$\frac{dy}{dx}$	$\frac{d^2y}{dx^2}$	Behaviour of the graph of f
A	0	+	f has a horizontal tangent slope and the graph of f is concave up. f has a relative minimum.
B	0	0	f has a horizontal tangent slope and f'' changes sign from – to +. f has a point of inflection.
C	+	0	f is increasing and f'' changes sign from + to –. f has a point of inflection
D	0	–	f has a horizontal tangent slope and the graph of f is concave down. f has a relative maximum.
E	–	–	f is decreasing and the graph of f is concave down.

F	undefined	undefined	f' changes sign from − to +. f has a relative minimum. f'' changes sign from − to +. f has a point of inflection
G	undefined	undefined	f has a vertical tangent and f'' changes sign from + to −. f has a point of inflection.

EX 4

(1) $f'(x) = 3x^2 - 6x - 9$
$= 3(x^2 - 2x - 3)$
$= 3(x-3)(x+1)$
$f'(x) = 0$ at $x = 3, -1$
$f''(x) = 6x - 6$

$f''(3) = 18 - 6 = 12 > 0$

$\therefore$ Relative Minimum $= f(3) = -27$

$f''(-1) = -6 - 6 = -12 < 0$

$\therefore$ Relative Maximum $= f(-1) = 5$

(2) $f'(x) = 2e^x + 2xe^x$
$= 2e^x(1+x)$
$f'(x) = 0$ at $x = -1$

$f''(x) = 2e^x + (2e^x + 2xe^x)$
$= 4e^x + 2xe^x$
$= 2e^x(2+x)$

$f''(-1) = 2e^{-1}(2-1) = \frac{2}{e} > 0$

$\therefore$ Relative Minimum $= f(-1) = -2e^{-1}$

EX 5

(1) $6x - 6y^2y' + 12 = 0$
$6x + 12 = 6y^2y'$
$x + 2 = y^2y'$
$y' = \frac{x+2}{y^2}$
$\therefore \frac{dy}{dx} = \frac{x+2}{y^2}$

(2) $\frac{x+2}{y^2} = 0$
$x + 2 = 0$
$\therefore x = -2$

$3(-2)^2 - 2y^3 + 12(-2) + 28 = 0$
$12 - 2y^3 - 24 + 28 = 0$
$16 - 2y^3 = 0$
$y = 2$

$\therefore (-2,2)$

(3) $\frac{d^2y}{dx^2} = \frac{1 \cdot y^2 - (x+2) \cdot 2yy'}{(y^2)^2}$

$\therefore \left.\frac{d^2y}{dx}\right|_{(-2,2)} = \frac{4-0}{16} = \frac{1}{4}$

(4) At P$(-2,2)$, $\frac{dy}{dx} = 0$ and $\frac{d^2y}{dx^2} > 0$.

By the second derivative test, the curve has a relative minimum at the point P.

EX 6

$2x + 2y \cdot y' - 2y' = 0$
$y'(y-1) = -x$
$\frac{dy}{dx} = \frac{-x}{y-1}$, $\frac{dy}{dw} = 0 \Rightarrow x = 0$
$(0)^2 + y^2 - 2y = 0$
$y(y-2) = 0 \Rightarrow y = 0, 2$

$\therefore \frac{dy}{dx} = 0$ at $(0,0)$, $(0,2)$

$\frac{d^2y}{dx^2} = \frac{(-1)(y-1) - (-x)(y')}{(y-1)^2}$
$= \frac{-y + 1 + xy'}{(y-1)^2}$

$\left.\frac{d^2y}{dx^2}\right|_{(0,0)} = \frac{0+1+0}{(0-1)^2} = 1 > 0$

$\therefore$ The curve has a relative minimum at $(0,0)$

$\left.\frac{d^2y}{dx^2}\right|_{(0,2)} = \frac{-2+1+0}{(2-1)^2} = -1 < 0$

$\therefore$ The curve has a relative maximum at $(0,2)$

4.4 Curve Sketching

EX 1

(1)

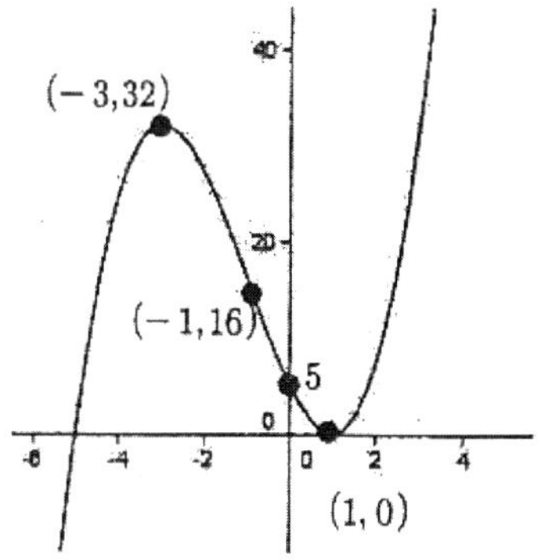

(2) $y' = -6x^2 - 6x + 12$

$\quad = -6(x^2 + x - 2)$

$\quad = -6(x+2)(x-1)$

$y' = 0 \ at \ x = -2, 1$

$y'' = -12x - 6$

$\quad = -6(2x+1)$

$y'' = 0 \ at \ x = -\dfrac{1}{2}$

x		-2		$-1/2$		1	
y'	$-$	0	$+$	$+$	$+$	0	$-$
y''	$+$	$+$	$+$	0	$-$	$-$	$-$
y	↘	-16	↗	$-5/2$	↗	11	↘

$f(-2) = 16 - 12 - 24 + 4 = -16$

$f(-\frac{1}{2}) = \frac{1}{4} - \frac{3}{4} - 6 + 4 = -\frac{5}{2}$

$f(1) = -2 - 3 + 12 + 4 = 11$

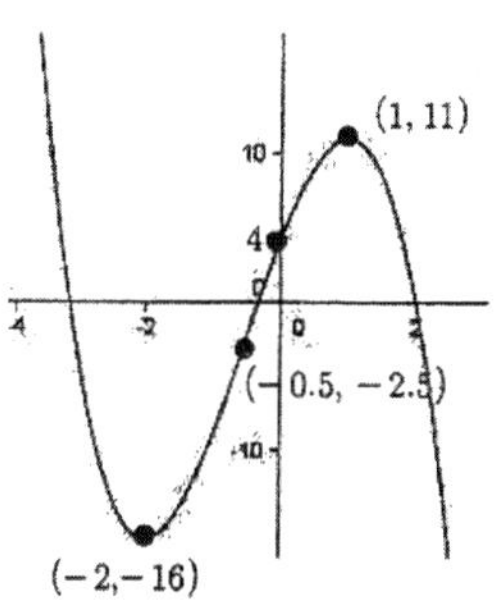

(3)

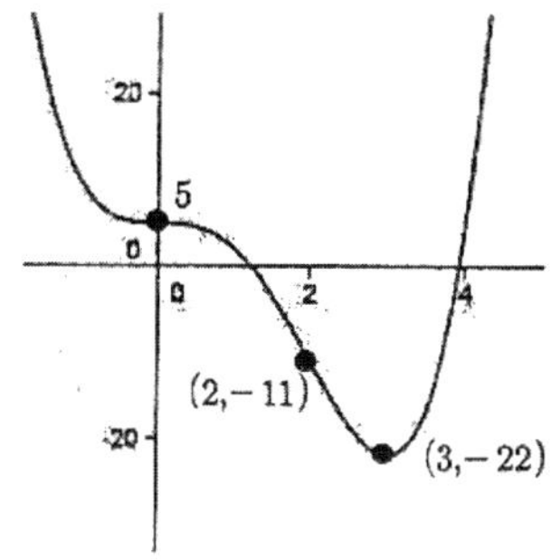

(4) $y' = -4x^3 + 12x$

$\quad = -4x(x^2 - 3)$

$y' = 0 \ at \ x = 0, \pm\sqrt{3}$

$y'' = -12x^2 + 12$

$\quad = -12(x^2 - 1)$

$y'' = 0 \ at \ x = \pm 1$

x		$-\sqrt{3}$		-1		0		1		$\sqrt{3}$	
y'	$+$	0	$-$	$-$	$-$	0	$+$	$+$	$+$	0	$-$
y''	$-$	$-$	$-$	0	$+$	$+$	$+$	0	$-$	$-$	$-$
y	↗	10	↘	6	↘	1	↗	6	↗	10	↘

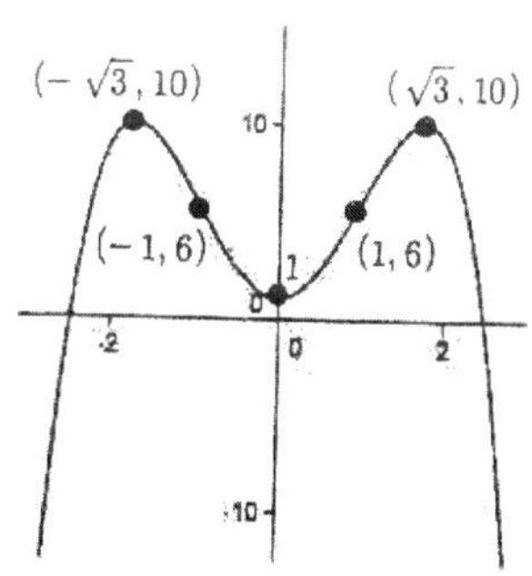

(5)

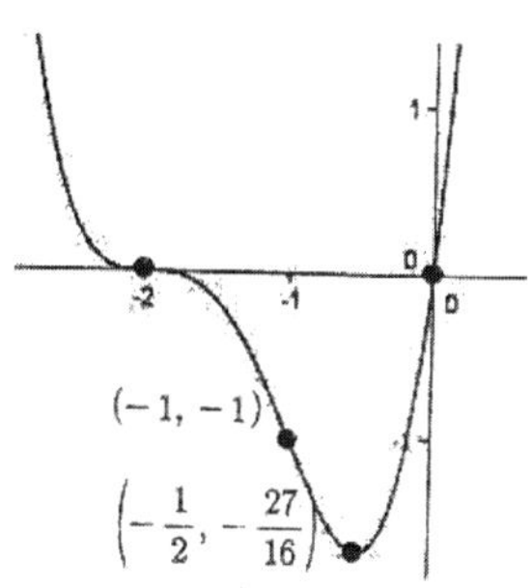

EX 2

(1)

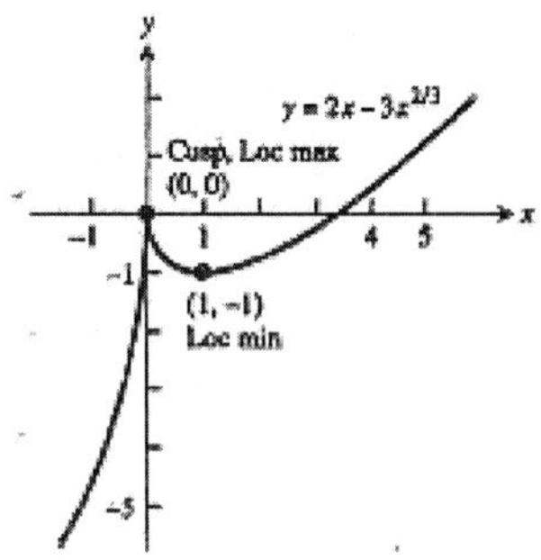

when $y=2x-3x^{2/3}$, then $y'=2-2x^{-1/3}$ and $y''=\frac{2}{3}x^{-4/3}$. The curve is rising on $(-\infty,0)$ and $(1,\infty)$, and falling on $(0,1)$. There is a local maximum at $x=0$ and a local minimum at $x=1$. The curve is concave up on $(-\infty,0)$ and $(0,\infty)$. There are no points of inflection, but a cusp exists at $x=0$.

(2)

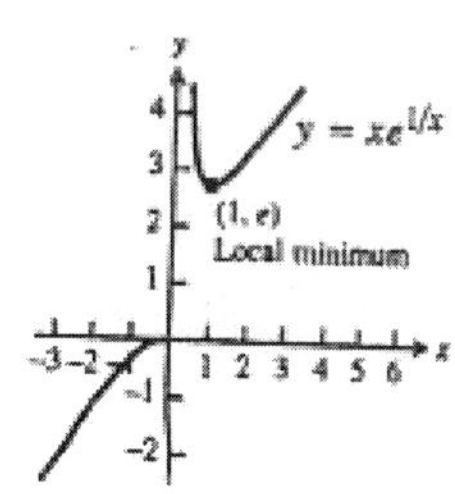

When $y=xe^{1/x}$,then

$$y'=-\frac{xe^{1/x}}{x^2}+e^{1/x}=e^{1/x}(1-\frac{1}{x})\text{ and}$$

$$y''=e^{1/x}(\frac{1}{x^2})+(1-\frac{1}{x})(-\frac{e^{1/x}}{x^2})$$

$$=\frac{e^{1/x}}{x^2}(\frac{1}{x})=\frac{e^{1/x}}{x^3}.$$

The curve is rising on $(1,\infty)$and $(-\infty,0)$ and falling on $(0,1)$. The curve is concave down on $(-\infty,0)$ and concave up on $(0,\infty)$. There is a local minimum of e at $x=1$, but there are no inflection points.

(3)

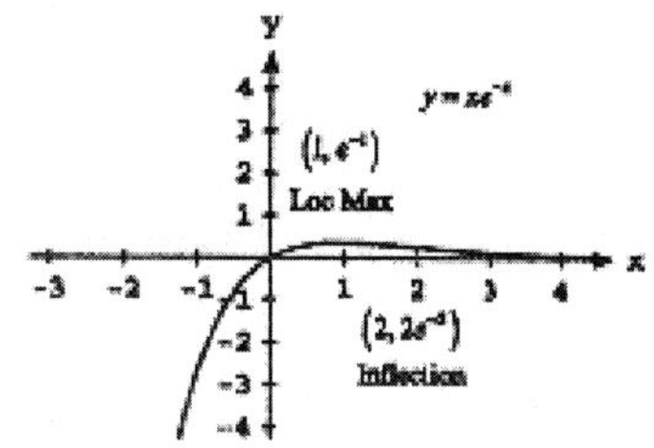

$y=xe^{-x}\Rightarrow y'=e^{-x}-xe^{-x}=(1-x)e^{-x}$

$\Rightarrow y'=+++|---$ $\Rightarrow$ the graph is
1

increasing on $(-\infty,1)$ and decreasing on $(1,\infty)$; a local maximum is e^{-1} at $x=1$;

$y''=(x-1)e^{-x}+(-1)e^{-x}=(x-2)e^{-x}$

$\Rightarrow y''=---|+++\Rightarrow$ Graph is concave
2

up on $(2,\infty)$,concave down on$(-\infty,2)\Rightarrow$ point of inflection at $(2,2e^{-2})$.

(4)

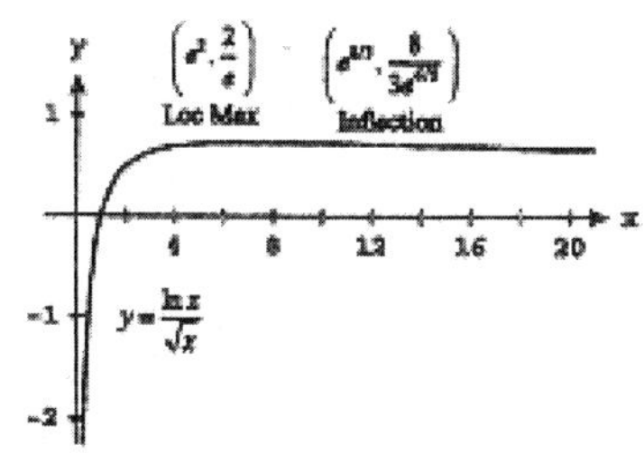

$$y=\frac{\ln x}{\sqrt{x}}$$

$$\Rightarrow y'=\frac{\sqrt{x}\cdot(\frac{1}{x})-\ln x\cdot\frac{1}{2\sqrt{x}}}{(\sqrt{x})^2}=\frac{2-\ln x}{2x^{3/2}}$$

$\Rightarrow y'=($ $+++|---$
0 e^2

$\Rightarrow$ the graph is increasing on $(0,e^2)$. decreasing on (e^2,∞); a local maximum is $\frac{2}{e}$ at $x=e^2$;

$$y''=\frac{2x^{3/2}(-\frac{1}{x})-(2-\ln x)\cdot 2\cdot\frac{2}{3}x^{1/2}}{(2x^{3/2})^2}=\frac{3\ln x-8}{4x^{5/2}}$$

$\Rightarrow y''=($ $---|+++\Rightarrow$ the graph is concave
0 $e^{8/3}$

up on $(e^{8/3},\infty)$, concave down on $(0, e^{8/3}) \Rightarrow$

point of inflection is $(e^{8/3}, \frac{8}{3e^{4/3}})$.

(5)

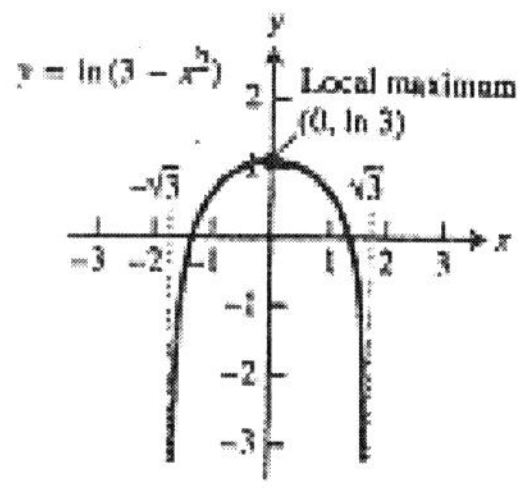

$y = \ln(3 - x^2), \Rightarrow y' = \frac{-2x}{3-x^2} = \frac{2x}{x^2-3}$

$\Rightarrow y' = (\underset{-\sqrt{3}}{}+++\underset{0}{|}---\underset{\sqrt{3}}{)} \Rightarrow$ the graph is

rising on $(-\sqrt{3}, 0)$, falling on $(0, \sqrt{3})$; a local minimum is $\ln 3$ at $x = 0$; $y'' = \frac{(x^2-3)(2)-(2x)(2x)}{(x^2-3)^2} = \frac{-2(x^2+3)}{(x^2-3)^2}$

$\Rightarrow y'' = (\underset{-\sqrt{3}}{}---\underset{\sqrt{3}}{)}$

$\Rightarrow$The graph is concave down on $(-\sqrt{3}, \sqrt{3})$.

(6)

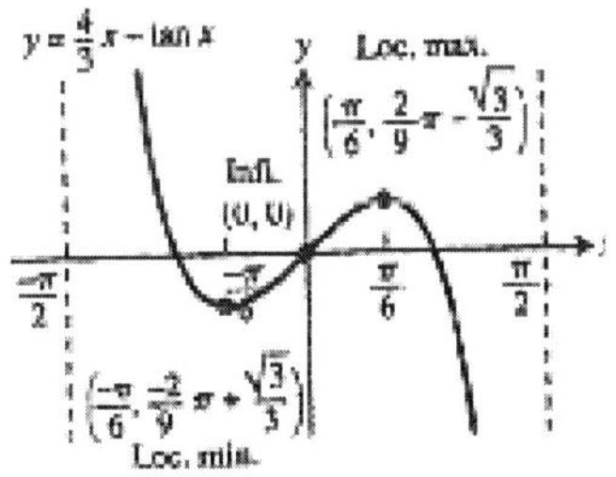

When $y = \frac{4}{3}x - \tan x$, then $y' = \frac{4}{3} - \sec^2 x$ and $y'' = -2\sec^2 x \tan x$. The curve is increasing on $(-\frac{\pi}{6}, \frac{\pi}{6})$, and decreasing on $(-\frac{\pi}{2}, -\frac{\pi}{6})$ and $(\frac{\pi}{6}, \frac{\pi}{2})$.

At $x = -\frac{\pi}{6}$ there is a local minimum, at $x = \frac{\pi}{6}$ there is a local maxi- mum, There are no absolute maxima or absolute minima. The curve is concave up on $(-\frac{\pi}{2}, 0)$, and is concave down on $(0, \frac{\pi}{2})$.

At $x = 0$ there is a pint of inflection.

EX 3

(1)

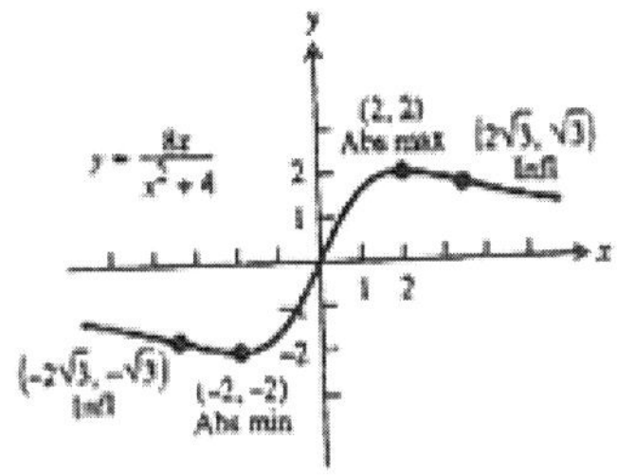

When $y = \frac{8x}{x^2+4}$, then $y' = \frac{-8(x^2-4)}{(x^2+4)^2}$ and $y'' = \frac{16x(x^2-12)}{(x^2+4)^3}$. The curve is falling on $(-\infty, -2)$ and $(2, \infty)$, and is rising on $(-2, 2)$. There is a local and absolute minimum at $x = -2$, and a local and absolute maximum at $x = 2$. The curve is concave down on $(-\infty, -2\sqrt{3})$ and $(0, 2\sqrt{3})$, and concave up on $(-2\sqrt{3}, 0)$ and $(2\sqrt{3}, \infty)$. There are points of inflection at $x = -2\sqrt{3}$, $x = 0$, and $x = 2\sqrt{3}$. $y = 0$ is a horizontal asymptote.

(2)

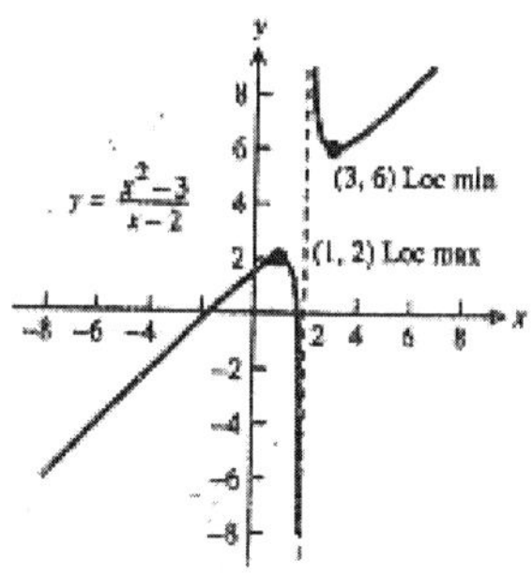

When $y=\dfrac{x^2-3}{x-2}$

then $y'=\dfrac{2x(x-2)-(x^3-3)(1)}{(x-2)^2}$

$=\dfrac{(x-3)(x-1)}{(x-2)^2}$

and

$$y''=\frac{(2x-4)(x-2)^2-(x^2-4x+3)2(x-2)}{(x-2)^4}$$

$$=\frac{2}{(x-2)^3}.$$

The curve is rising on$(-\infty,1)$ and $(3,\infty)$, and falling on $(1,2)$ and $(2,3)$. There is a local maximum at$x=1$ and a local minimum at $x=3$.The curve is concave down on $(-\infty,2)$ and concave up on $(2,\infty)$.There are no points of inflection because $x=2$ is not in the domain.

(3)

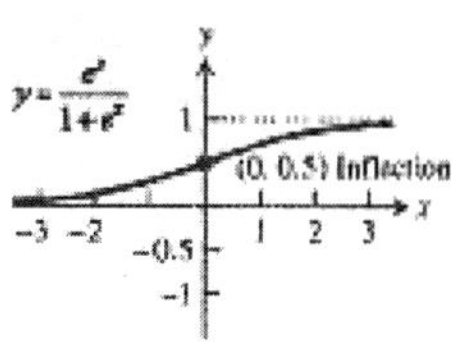

$$y=\frac{e^x}{1+e^x}$$

$\Rightarrow y'=\dfrac{(e^x+1)e^x-e^x\cdot e^x}{(1+e^x)^2}=\dfrac{e^x}{(1+e^x)^2}$

$\Rightarrow y' = +++$

$\Rightarrow$the graph is increasing $(-\infty,\infty)$;

$$y''=\frac{(e^x+1)^2\cdot e^x-e^x\cdot 2(e^x+1)}{(1+e^x)^4}$$

$=\dfrac{e^x(1-e^x)}{(1+e^x)^3}\Rightarrow y'' = +++\underset{0}{|}---$

$\Rightarrow$ the graph is concave up on $(-\infty,0)$, concave down on $(0,\infty)$ $\Rightarrow$ point of inflection is $(0,\frac{1}{2})$.

(4)

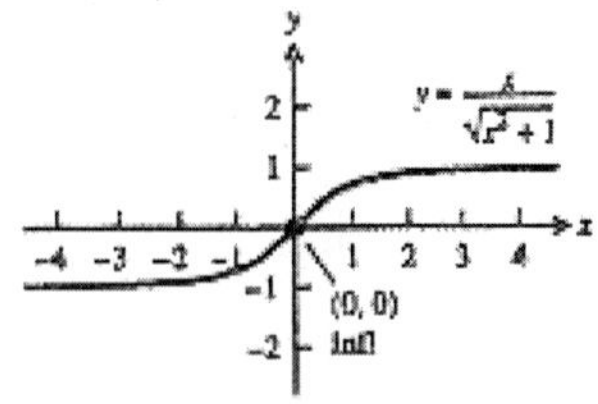

When $y=\dfrac{x}{\sqrt{x^2+1}}$; then $y'=\dfrac{1}{(x^2+1)^{3/2}}$ and $y''=\dfrac{-3x}{(x^2+1)^{3/2}}$. The curve is increasing on $(-\infty,\infty)$. There are no local or absolute extrema. The curve is concave up on $(-\infty,0)$ and concave down on $(0,\infty)$. At $x=0$ there is a point of inflection.

4.5 Relation of $f'(x)$, $f''(x)$ to the Graph of $f(x)$

EX 1

(1)

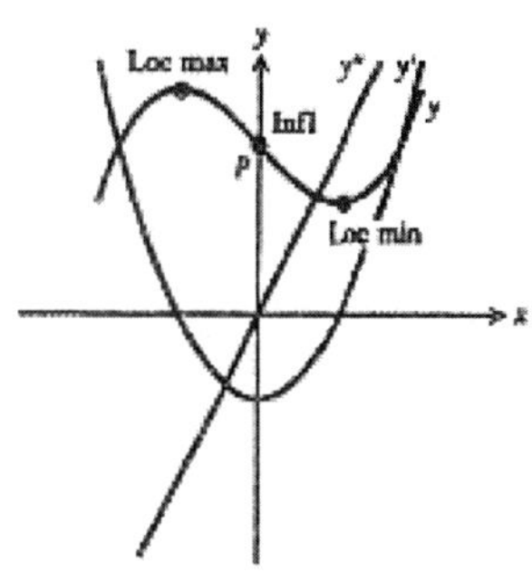

The graph of $y=f''(x)\Rightarrow$ the graph of $y=f(x)$ is concave up on $(0,\infty)$, concave down on $(-\infty,0)$ $\Rightarrow$a point of inflection at $x=0$; the graph of $y=f'(x)$

$\Rightarrow y' = +++\,|\,---\,|\,+++$ $\Rightarrow$the graph $y=f(x)$ has both a local maximum and a local minimum.

(2)

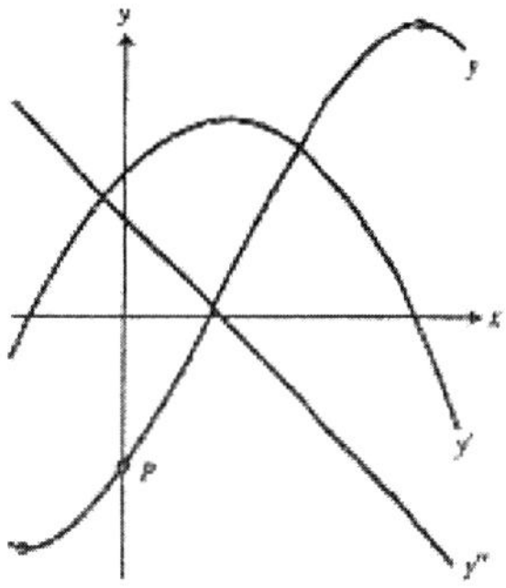

The graph of

$y=f''(x) \Rightarrow y''=$ $+++ \mid ---$

$\Rightarrow$ the graph of $y=f(x)$ has a point of inflection, the graph of

$y=f'(x) \Rightarrow y'=$ $--- \mid +++ \mid ---$

$\Rightarrow$ the graph of $y=f(x)$ has both a local maximum and a local minimum.

(3)

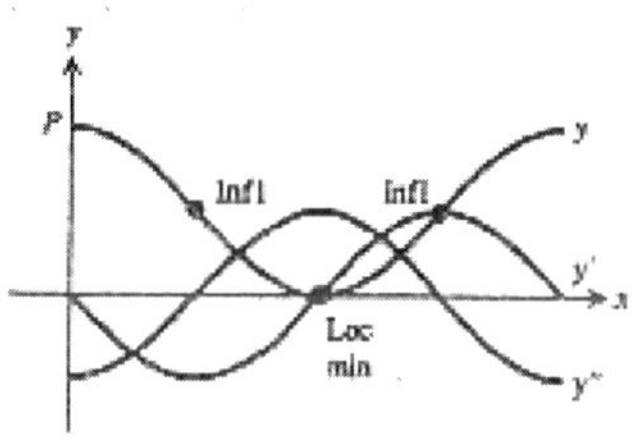

The graph of

$y=f''(x) \Rightarrow y''=$ $+++ \mid --- \Rightarrow$ the graph of $y=f(x)$ has two points of inflection, the graph of

$y=f'(x) \Rightarrow y'=$ $--- \mid +++ \Rightarrow$ the graph of $y=f(x)$ has both a local minimum.

(4)

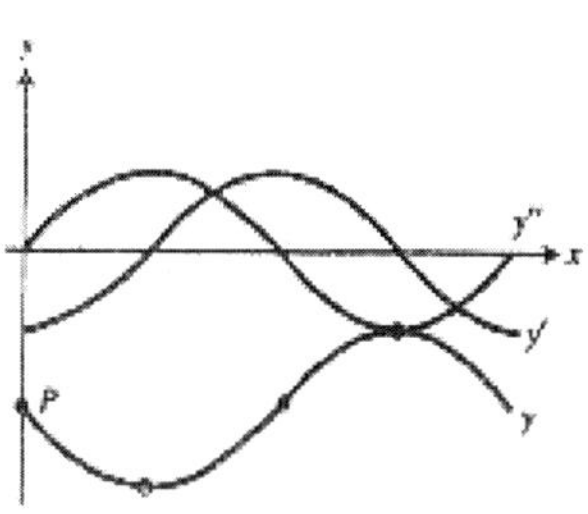

The graph of

$y=f''(x) \Rightarrow y''=$ $--- \mid +++ \mid --- \Rightarrow$ the graph of $y=f(x)$ has a point of inflection;

the graph of

$y=f'(x) \Rightarrow y'=$ $--- \mid +++ \mid --- \Rightarrow$ the graph of $y=f(x)$ has both a local maximum and a local minimum.

EX 2

Point	y'	y''
P	−	+
Q	+	0
R	+	−
S	0	−
T	−	−

EX 3

(1)

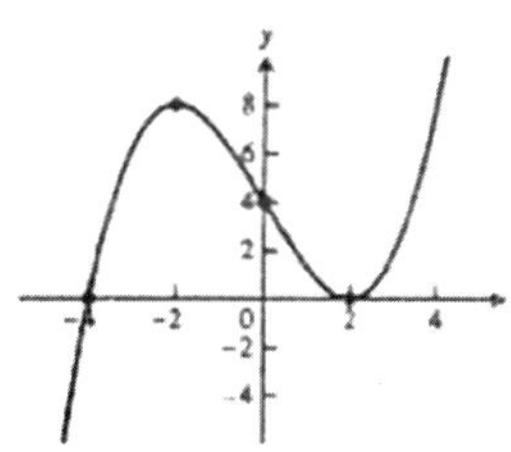

(2)

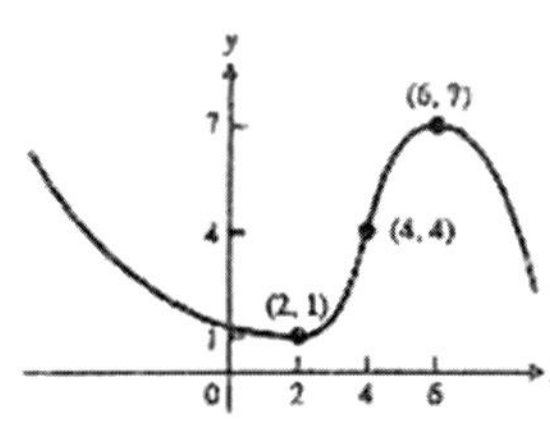

(3)

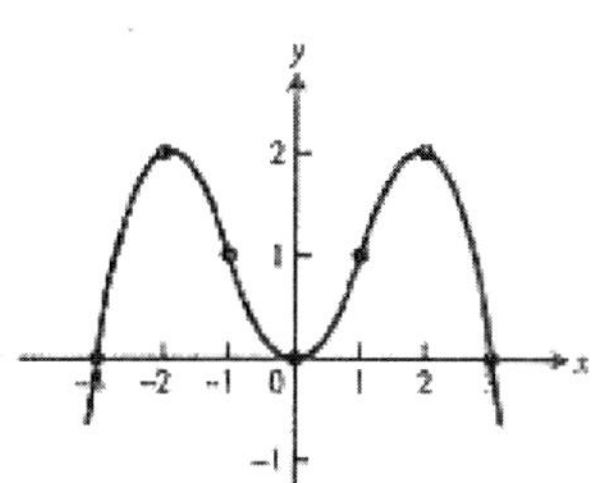

EX 4

(1) $f'(-1)=0,\ f'(1)=-2,\ f'(5)=2,$
$f''(-1)=-1,\ f''(1)=DNE,\ f''(2)=2,$
$f''(3)=DNE$

(2) $\frac{f'(3)-f'(-1)}{3-(-1)}=\frac{2-0}{4}=\frac{1}{2}$

(3) f is increasing when $f'>0$
$\therefore -3\le x<-1,\ 2<x<4,\ 4<x\le 5$

(4) f is concave down when $f''<0$ or f' is decreasing.
$\therefore -2<x<1,\ 3<x<4$

(5) f has a relative minimum when f' changes sign from - to +. This occurs at $x=2$. f has a relative maximum when f' changes sign from + to -. This occur at $x=-1$.

(6) f has a point of inflection when f'' changes sign from + to -, or vice versa. This occurs at $x=1,3,4$

EX 5

(1) $c<x<e$

(2) f' is increasing when f is concave up.
$\therefore b<x<d$

(3) $x=c$

(4) $x=b,\ d$

EX 6

(1) $x=6$ is the only critical point at which f' changes sign from negative to positive. Therefore, f has a local minimum at $x=6$.

(2) From part (1), the absolute minimum occurs either at $x=6$ or at an endpoint.

$$f(0)=f(8)+\int_8^0 f'(x)dx$$
$$=f(8)-\int_0^8 f'(x)dx=4-12=-8$$
$$f(6)=f(8)+\int_8^6 f'(x)dx$$
$$=f(8)-\int_6^8 f'(x)dx=4-7=-3$$
$$f(8)=4$$

The absolute minimum value of f on the closed interval [0,8] is -8.

(3) The graph of f is concave down and increasing on $0<x<1$. and $3<x<4$, because f' is decreasing and positive on these intervals.

(4) $g'(x)=3[f(x)]^2\cdot f'(x)$
$g'(3)=3[f(3)]^2\cdot f'(3)=3(-\frac{5}{2})^2\cdot 4=75$

EX 7

(1) f has a point of inflection when f'' changes sign. This occurs at $x=b,d$.

(2) f is concave up when $f''>0$.
$\therefore b<x<d$

(3) f' is increasing when $f''>0$.
$\therefore b<x<d$

(4) f' has a relative maximum when f'' changes sign from + to - .
$\therefore x=d.$

(5) f' is concave down when f'' is decreasing.
$\therefore c<x<e,\ f<x<g.$

4.6 Motion A long a Line

EX 1

(1) $v(t)=s'(t)=3t^2-12$
$a(t)=v'(t)=6t$

(2) The particle changes direction when v changes sign.

$v(t)=0$ *when* $t=2$
$\therefore t=2$

(3) v is increasing when $a=v'>0$.

$6t>0$

$\therefore t>0$

(4) The speed of the particle is decreasing when the signs of v and a are opposite.

$\therefore 0<t<2$

EX 2

(1) $s(2)-s(1)=-1-1=-2m$

(2) $\dfrac{s(2)-s(1)}{2-1}=\dfrac{-2}{1}=-2m/\sec$

(3) $v(t)=s'(t)=3t^2-12t+9$

$a(t)=v'(t)=6t-12$

(4) The particle is at rest when $v=0$

$v(t)=3t^2-12t+9$
$=3(t^2-4t+3)$
$=(t-1)(t-3)$
$\therefore t=1,3$ sec

(5) $s(4)=1m$
$v(4)=9m/\sec$
speed $=9m/\sec$
$a(4)=12m/\sec^2$

(6) The particle is moving away from the origin when he signs of $s(t)$ and $v(t)$ are the same. $s(4)=1>0$ and $v(4)=9>0$

$\therefore$ The particle is moving away from the origin at $t=4$

(7) $\dfrac{v(2)-v(0)}{2-0}=\dfrac{-3-9}{2}=-6m/\sec^2$

(8) The particle is moving in the positive direction when $v>0$.

$v(t)=3(t-1)(t-3)$
$\therefore 0\le t<1,\ t>3$

(9) v is decreasing when $a=v'<0$

$a(t)=6(t-2)$
$\therefore 0\le t<2$

(10) The speed of the particle is increasing when the signs of $v(t)$ and $a(t)$ are the same.

$\therefore 1<t<2, t>3$

(11) $\therefore 4+4+4=12m$

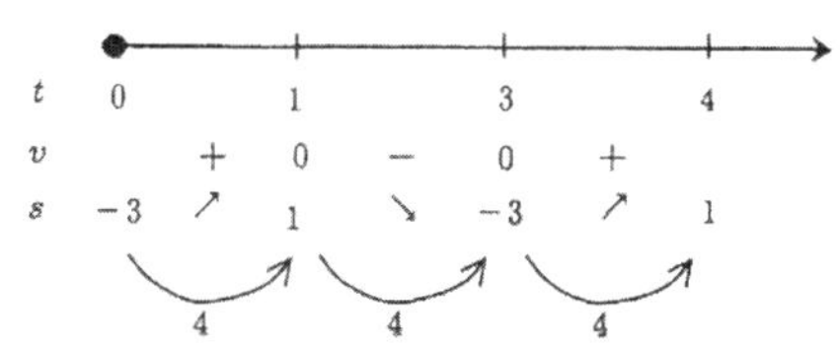

EX 3

(1) The particle is at rest when $v=0$.

$\therefore t=0,3,5,7$ sec

(2) The particle changes direction when v changes sign.

$\therefore t=5,7$ sec

(3) *speed* $=|v(6)|=2m/\sec$

(4) $a(5)=v'(5)=-2m/\sec^2$

(5) $\dfrac{v(7)-v(1)}{7-1}=\dfrac{0-2}{6}=-\dfrac{1}{3}m/\sec^2$

(6) The particle is moving in the negative direction when $v<0$.

$\therefore 5<t<7$

(7) The speed is decreasing when

$v>0, a<0$ or $v<0, a>0$.

$\therefore 2<t<3,\ 4<t<5,\ 6<t<7$

(8) $4+2+2+1=9m$

(9) $t=5\,\text{sec}$

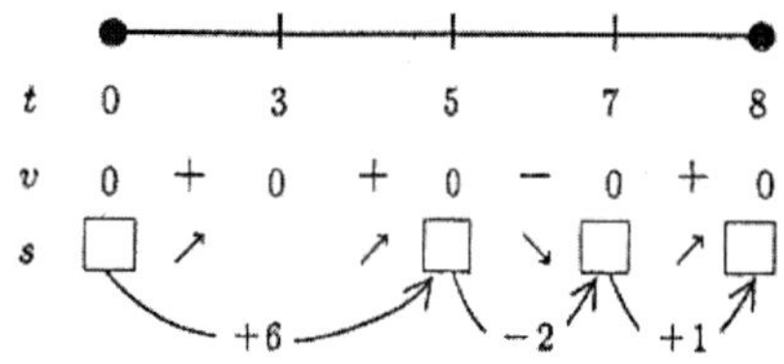

EX 4

(1) Since $v(t)<0$ for $0<t<3$ and $5<t<6$, and $v(t)>0$ for $3<t<5$, we consider $t=3$ and $t=6$.

$$x(3)=-2+\int_0^3 v(t)dt = -2-8=-10$$

$$x(6)=-2+\int_0^6 v(t)dt =-2-8+6-2=-9$$

Therefore, the particle is farthest left at time $t=3$ when its position is $x(3)=-10$.

(2) The particle moves continuously and monotonically from $x(0)=-2$ to $x(3)=-10$. Similarly, the particle moves continuously and monotonically from $x(3)=-10$ to $x(5)=-7$ and also from $x(5)=-7$ to $x(6)=-9$.

By the Intermediate Value Theorem, there are three values of t for which the particle is at $x(t)=-8$.

(3) The speed is decreasing on the interval $2<t<3$ since on this interval $v<0$ and v is increasing.

(4) The acceleration is negative on the intervals $0<t<1$ and $4<t<6$ since velocity is decreasing on these intervals.

4.7 Motion Along a Curve

EX 1

(1) (E) Since $x=3\cos\frac{\pi}{3}t$ and $y=2\sin\frac{\pi}{3}t$,

we note that $\left(\frac{x}{3}\right)^2+\left(\frac{y}{2}\right)^2=1$.

(2) (A) Note that $y=-\pi\sin\frac{\pi}{3}ti+\frac{2\pi}{3}\cos\frac{\pi}{3}tj$.

At $t=3$.

$$|v|=\sqrt{(-\pi\cdot 0)^2+\left(\frac{2\pi}{3}\cdot(-1)\right)^2}=\frac{2}{3}\pi$$

(3) (B) $a=-\frac{\pi^2}{3}\cos\frac{\pi}{3}ti-\frac{2\pi^2}{3}tj$. At $t=3$,

$$|a|=\sqrt{\left(-\frac{\pi^2}{3}\cdot -1\right)^2+\left(-\frac{2\pi^2}{9}\cdot 0\right)^2}=\frac{\pi^2}{3}$$

(4) (D) he slope of the curve is the slope of v, namely, $\dfrac{dy/dt}{dx/dt}$.

At $t=\frac{1}{2}$, the slope is equal to

$$\frac{\frac{2\pi}{3}\cdot\cos\frac{\pi}{6}}{-\pi\cdot\sin\frac{\pi}{6}}=-\frac{2}{3}\cot\frac{\pi}{6}=-\frac{2\sqrt{3}}{3}$$

EX 2

(1) Eliminating t yields the equation

$y=-\frac{1}{4}x^2+2x$. (E)

(2) $|v|=\sqrt{\left(\frac{dx}{dt}\right)^2+\left(\frac{dy}{dt}\right)^2}=\sqrt{2^2+(4-2t)^2}$. (B)

(3) Since $|v|=2\sqrt{t^2-4t+5}$.

$$\frac{d|v|}{dt}=\frac{2t-4}{\sqrt{t^2-4t+5}}=0$$

if $t=2$. We note that as t increases through 2, the signs of $|v|'$ are -,0,+, assuring a minimum of $|v|$ at $t=2$.

Evaluate $|v|$ at $t=2$. (A)

(4) The direction of a is $\tan^{-1}\dfrac{\frac{d^2y}{dt^2}}{\frac{d^2x}{dt^2}}$. Since

$\dfrac{d^2x}{dt^2}=0$ and $\dfrac{d^2y}{dt^2}=-2$, the acceleration

is always directed downward. Its

magnitude, $\sqrt{0^2+(-2)^2}$, is 2 for all t.

(C)

EX 3

(1) $\left.\dfrac{dx}{dt}\right|_{t=2}=\dfrac{2}{e^2}$

Because $\left.\dfrac{dx}{dt}\right|_{t=2}>0$, the particle is

moving to the right at time $t=2$.

$\left.\dfrac{dy}{dx}\right|_{t=2}=\dfrac{dy/dt|_{t=2}}{dx/dt|_{t=2}}=3.055$(or 3.054)

(2) $x=1+\int_2^4 \sqrt{(t+2)}/e^t dt$

(3) Speed = $\sqrt{(x'(4))^2+(y'(4))^2}=0.575\ (0.574)$

Acceleration $=\langle x''(4), y''(4)\rangle$

$=\langle -0.041, 0.989\rangle$

EX 4

(1) Speed $=\sqrt{x'(4)^2+y'(4)^2}=2.912$

(2) The slope is 2, so $\dfrac{\frac{dy}{dt}}{\frac{dx}{dt}}=2$, or

$\ln(t^2+1)-2\arctan\left(\dfrac{t}{1+t}\right)$. Since $t>0$,

$t=1.35766$. At this time, the acceleration

is $\langle x''(t), y''(t)\rangle|_{t=1.35766}=\langle 0.135, 0.955\rangle$.

4.8 Modeling and Optimization

EX 1

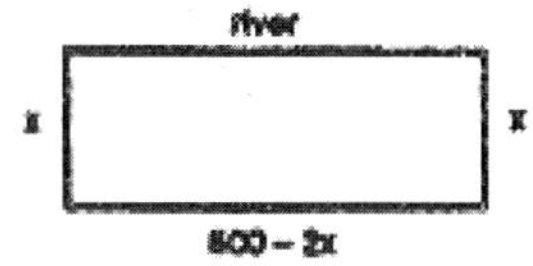

The area is $A(x)=x(800-2x)$, Where $0\le x\le 400$. Solving $A'(x)=800-4x=0$ $\Rightarrow x=200$. With $A(0)=A(400)=0$, the

maximum area is $A(200)=80{,}000m^2$. The

dimensions are $200m$ by $400m$.

EX 2

(1) -5 or 5

(2) Let the two numbers be x, $18-x$.

$$\begin{aligned}P&=x^2\cdot(18-x)\\&=18x^2-x^3\\P'&=36x-3x^2\\&=3x(12-x), x>0\\P'&=0\ at\ x=12\end{aligned}$$

$\therefore 12$ or 6

EX 3

(1) $V(\frac{5}{3})=\dfrac{2450}{27}\approx 91in^3$ must be the

maximum volume of the box with

dimensions $\dfrac{14}{3}\times\dfrac{35}{3}\times\dfrac{5}{3}$ inches.

(2)

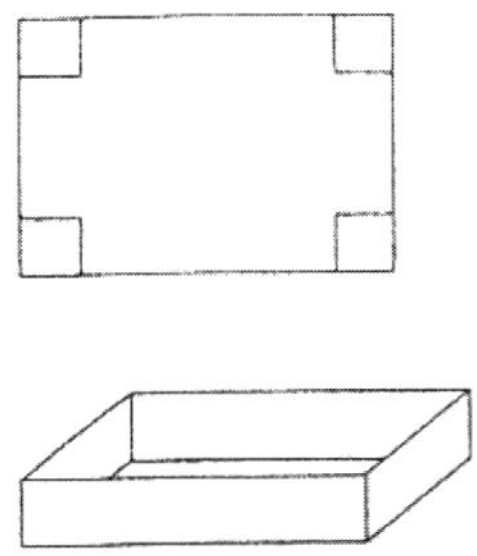

$$V = x(10-2x)(16-2x)$$
$$= 4x^3 + 52x^2 + 160x$$
$$V' = 12x^2 - 104x + 160$$
$$= 4(3x^2 - 26x + 40)$$
$$= 4(x-2)(3x-20), 0 < x < 5$$
$$V' = 0 \text{ at } x = 2$$

$\therefore 12cm$ *by* $6cm$ *by* $2cm$

EX 4

(1) $300cm^2$

(2)

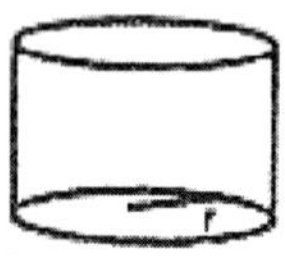

A volume $V = \pi r^2 h = 1000 \Rightarrow h = \frac{1000}{\pi r^2}$. The amount of material is the surface area given by the sides and bottom of the can $\Rightarrow S = 2\pi rh + \pi r^2 = \frac{2000}{r} + \pi r^2$, $0 < r$. Then $\frac{dS}{dr} = -\frac{2000}{r^2} + 2\pi r = 0 \Rightarrow \frac{\pi r^3 - 1000}{r^2} = 0$. The critical points are 0 and $\frac{10}{\sqrt[3]{\pi}}$, but 0 is not in the domain. Since $\frac{d^2S}{dr^2} = \frac{4000}{r^3} + 2\pi > 0$, we have a minimum surface area when $r = \frac{10}{\sqrt[3]{\pi}}$ cm and $h = \frac{1000}{\pi r^2} = \frac{10}{\sqrt[3]{\pi}}$ cm.

(3) $4ft$ *by* $4ft$ *by* $16ft$

(4)

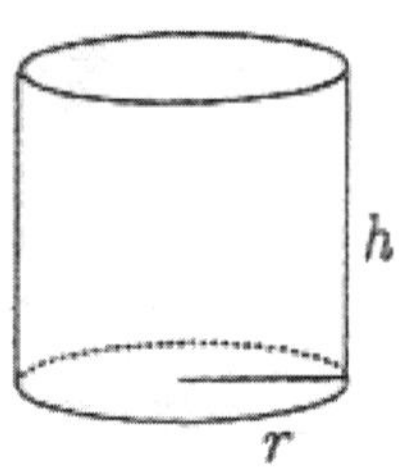

$$V = \pi r^2 h = 1000$$

$$h = \frac{1000}{\pi r^2} \cdots\cdots\cdots ①$$

$$A = 2\pi r^2 + 2\pi rh \cdots\cdots\cdots ②$$

① → ②

$$A = 2\pi r^2 + 2\pi r\left(\frac{1000}{\pi r^2}\right)$$
$$= 2\pi r^2 + \frac{2000}{r}$$
$$A' = 4\pi r - \frac{2000}{r^2}$$
$$= \frac{4\pi r^3 - 2000}{r^2}$$
$$= \frac{4(\pi r^3 - 500)}{r^2}$$

$$A' = 0 \Rightarrow \pi r^3 - 500 = 0$$

$$r^3 = \frac{500}{\pi}$$
$$r = \sqrt[3]{\frac{500}{\pi}}$$
$$= 5.41926$$

$$h = \frac{1000}{\pi(5.41926)^2}$$
$$= 10.8385$$

$\therefore radius = 5.419cm$
$height = 10.839cm$

EX 5

(1) $\frac{24\pi}{4+\pi}$

(2)

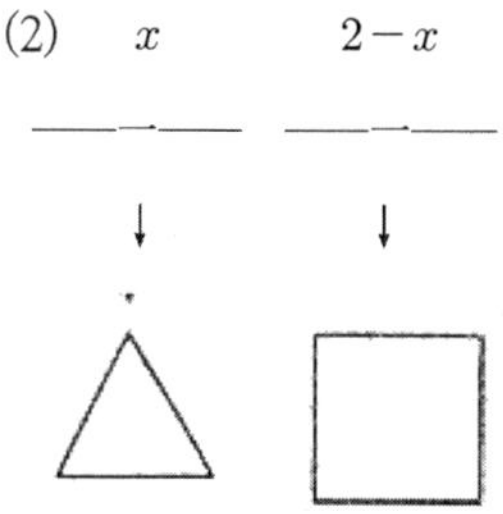

$l = \frac{x}{3}$ $\qquad s = \frac{20-x}{4}$

$$A=\frac{\sqrt{3}}{4}l^2+s^2$$
$$=\frac{\sqrt{3}}{4}\left(\frac{x}{3}\right)^2+\left(\frac{20-x}{4}\right)^2$$
$$=\frac{\sqrt{3}x^2}{36}+\frac{(20-x)^2}{16}$$
$$A'=\frac{2\sqrt{3}x}{36}+\frac{2(20-x)\cdot(-1)}{16}$$
$$=\frac{\sqrt{3}x}{18}-\frac{20-x}{8}$$
$$=\frac{4\sqrt{3}x-9(20-x)}{72}$$
$$=\frac{4\sqrt{3}x+9x-180}{72}$$
$$=\frac{(4\sqrt{3}+9)x-180}{72}$$
$$A'=0\ at\ x=\frac{180}{4\sqrt{3}+9}$$
$$=11.301$$

EX 6

(1) $10\sqrt{2}$ *by* $10\sqrt{2}$ *cm*

(2) Let $(x,y)=(x,\sqrt{9-x^2})$ be the coordinates of the corner that intersects the semicircle. Then base $=2x$ and height $=y=\sqrt{9-x^2}$, thus the area of the inscribed rectangle is given by

$A=(2x)\sqrt{9-x^2}, 0\le x\le 3$.Then

$$A'=2\sqrt{9-x^2}+(2x)\frac{-x}{\sqrt{9-x^2}}$$
$$=\frac{2(9-x^2)-2x^2}{\sqrt{9-x^2}}$$
$$=\frac{18-4x^2}{\sqrt{9-x^2}},$$

$A'=0\Rightarrow 18-4x^2=0\Rightarrow x=\pm\frac{3\sqrt{2}}{2}$,

only $x=\frac{3\sqrt{2}}{2}$ lies in $0\le x\le 3$. A is continuous on the closed interval $0\le x\le 3$

$\Rightarrow$ A has an absolute maxima and absolute minima. $A(0)=0$, $A(3)=0$,and

$A(\frac{3\sqrt{2}}{2})=(3\sqrt{2})(\frac{3\sqrt{2}}{2})=9\Rightarrow$absolute maxima. Base of rectangle is $3\sqrt{2}$ and height is $\frac{3\sqrt{2}}{2}$.

(3) $4000\pi cm^3$

(4)

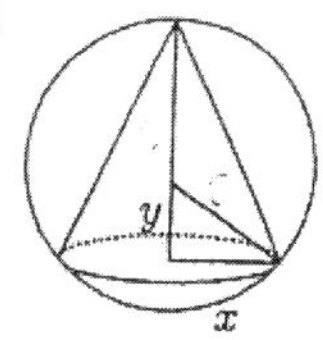

The volume of the cone is $V=\frac{1}{3}\pi r^2h$,

where $r=x=\sqrt{9-y^2}$ and $h=y+3$

Thus,

$$V(y)=\frac{\pi}{3}(9-y^2)(y+3)$$
$$=\frac{\pi}{3}(27+9y-3y^2-y^3)$$

$$V'(y)=\frac{\pi}{3}(9-6y-3y^2)=\pi(1-y)(3+y).$$

The critical points are -3 and 1, but -3 is not in the domain. Thus

$V''(1)=\frac{\pi}{3}(-6-6(1))<0\Rightarrow$ at $y=1$ we have a maximum volume of

$V(1)=\frac{\pi}{3}(8)(4)=\frac{32\pi}{3}$ cubic units.

EX 7

(1)

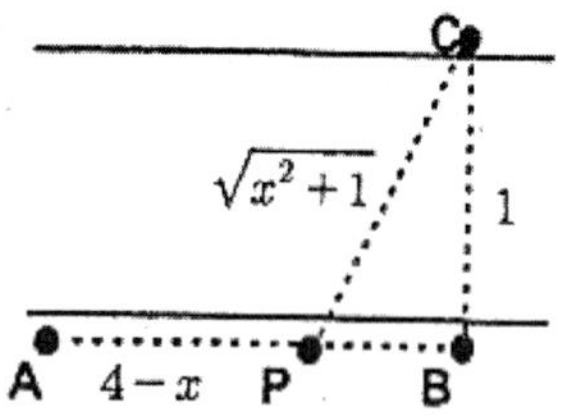

$\$50/m = \$50000/km$
$\$30/m = \$30000/km$

$$C = 50000\sqrt{x^2+1} + 30000(4-x)$$
$$C' = 50000 \cdot \frac{1}{2\sqrt{x^2+1}} \cdot 2x + 30000(-1)$$
$$= \frac{50000x - 30000\sqrt{x^2+1}}{\sqrt{x^2+1}}$$
$$= \frac{10000(5x - 3\sqrt{x^2+1})}{\sqrt{x^2+1}}$$
$$C' = 0 \Rightarrow 5x - 3\sqrt{x^2+1} = 0$$
$$5x = 3\sqrt{x^{2+1}}$$
$$25x^2 = 9(x^2+1)$$
$$25x^2 = 9x^2 + 9$$
$$16x^2 = 9$$
$$x^2 = \frac{9}{16},\ x > 0 \Rightarrow \therefore x = \frac{3}{4}$$

(2)

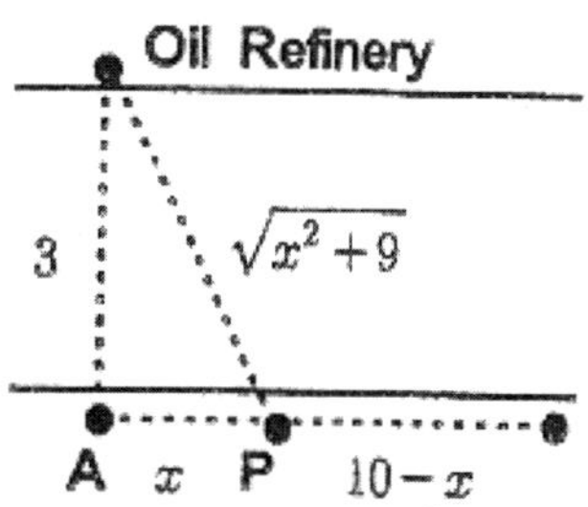

$\$20/m = \$20000/km$
$\$25/m = \$25000/km$

$$C = 25000\sqrt{x^2+9} + 20000(10-x)$$
$$C' = 25000 \cdot \frac{1}{2\sqrt{x^2+9}}(2x) + 20000(-1)$$
$$= \frac{25000x - 20000\sqrt{x^2+9}}{\sqrt{x^2+9}}$$
$$\frac{5000(5x - 4\sqrt{x^2+9})}{\sqrt{x^2+9}}$$
$$C' = 0 \Rightarrow 5x - 4\sqrt{x^2+9} = 0$$
$$5x = 4\sqrt{x^2+9}$$
$$25x^2 = 16(x^2+9)$$
$$25x^2 = 16x^2 + 144$$
$$25x^2 = 16(x^2+9)$$
$$9x^2 = 144$$
$$x^2 = 16, x > 0 \Rightarrow x = 4$$

(3)

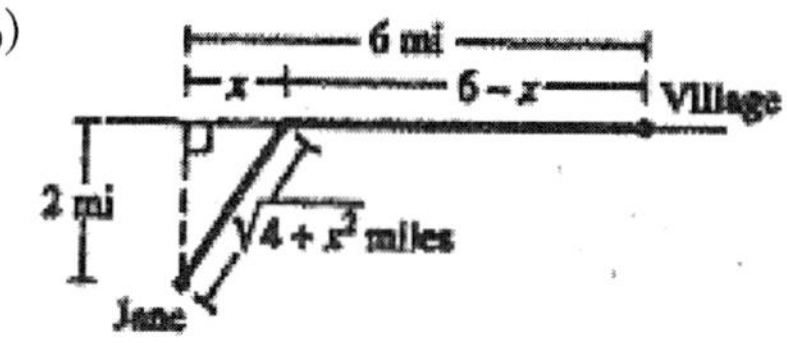

Let x be this distance from the point on the shoreline nearest Jane's boat to the point where she lands her boat. Then she needs to row $\sqrt{4+x^2}$ mi at 2 mph and walk $6-x$ at 5 mph. The total amount of time to reach the village is

$$f(x) = \frac{\sqrt{4+x^2}}{2} + \frac{6-x}{5} \text{ hours } (0 \le x \le 6).$$

Then $f'(x) = \frac{1}{2}\frac{1}{2\sqrt{4+x^2}}(2x) - \frac{1}{5}$
$$= \frac{x}{2\sqrt{4+x^2}} - \frac{1}{5}.$$

Solving $f'(x) = 0$, We have :

$$\frac{x}{\sqrt{4+x^2}} = \frac{1}{5}$$
$$\Rightarrow 5x = 2\sqrt{4+x^2}$$
$$\Rightarrow 25x^2 = 4(4+x^2)$$
$$\Rightarrow 21x^2 = 16 \Rightarrow x \pm \frac{4}{\sqrt{21}}.$$

We discard the negative value of x because it is not in the domain. Checking the endpoints and critical point, we have

$f(0) = 2.2 f(\frac{4}{\sqrt{21}}) \approx 2.12$, and $f(6) \approx 3.16$.

Jane should land her boat $\frac{4}{\sqrt{21}} \approx 0.87$ miles down the shoreline from the point nearest her boat.

EX 8

(1) (4,4)

(2)

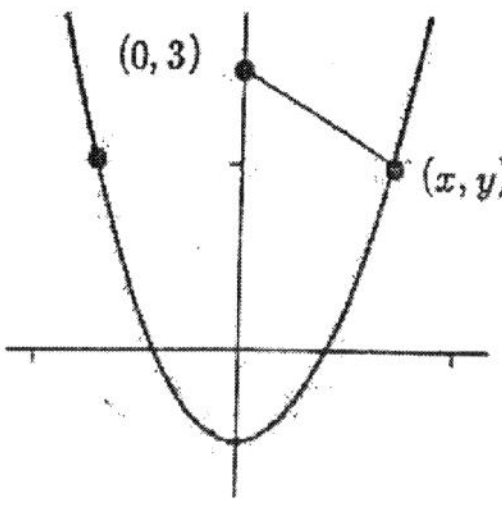

$y = \frac{1}{4}x^2 - 1$

$4y = x^2 - 4$

$x^2 = 4y + 4 \cdots\cdots ①$

Distance Formula

$l = \sqrt{(x-0)^2 + (y-3)^2}$

$= \sqrt{x^2 + y^2 - 6y + 9} \cdots\cdots ②$

① → ②

$l = \sqrt{(4y+4) + y^2 - 6y + 9}$

$= \sqrt{y^2 - 2y + 13}$

$l' = \frac{1}{2\sqrt{y^2 - 2y + 13}}(2y-2)$

$l' = 0 \ at \ y = 1$

$\therefore (\pm 2\sqrt{2}, 1)$

(3) $2\sqrt{3}$

(4) $x^2 - y^2 = -1$

$y^2 = x^2 + 1 \cdots\cdots ①$

$l = \sqrt{(x-4)^2 + y^2} \cdots\cdots ②$

① → ②

$l = \sqrt{x^2 - 8x + 16 + x^2 + 1}$

$= \sqrt{2x^2 - 8x + 17}$

$l' = \frac{1}{\sqrt{2x^2 - 8x + 17]}}(4x - 8)$

$l' = 0 \ at \ x = 2$

$\therefore (2, \pm \sqrt{5})$

(5) (2,1)

EX 9

(1) $6\sqrt{3}$

(2)

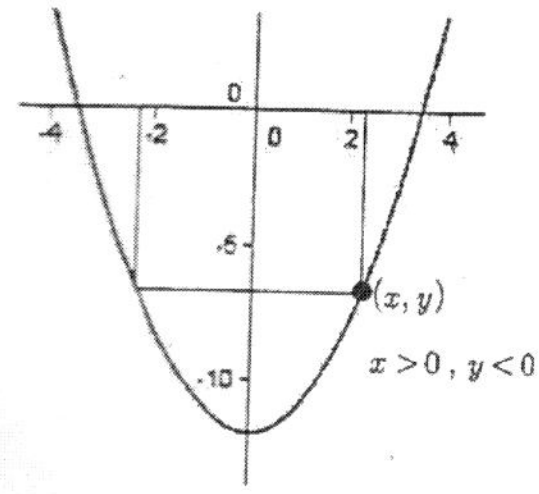

$A = 2x \cdot (-y)$

$= -2xy$

$= -2x(x^2 - 12)$

$= -2x^3 + 24x$

$A' = -6x^2 + 24$

$= -6(x^2 - 4), \ x > 0$

$A' = 0 \ at \ x = 2$

$A = 32$

(3)

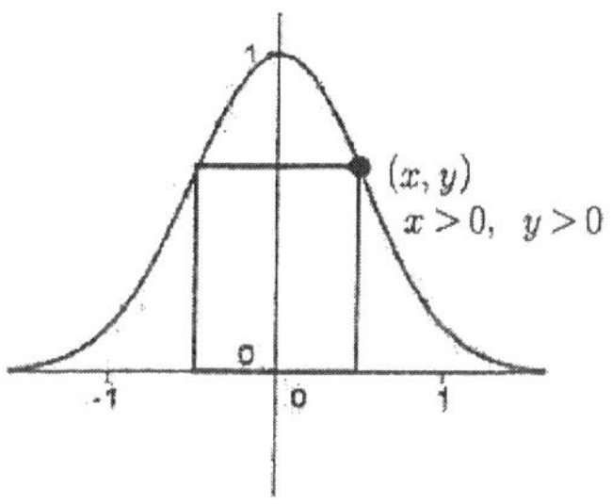

$A = 2x \cdot y$

$- 2xe^{-2x^2}$

$A' = 2e^{-2x^2} + 2xe^{-2x^2}(-4x)$

$= 2e^{-2x^2} - 8x^2e^{-2x^2}$

$= 2e^{-2x^2}(1 - 4x^2), \ x > 0$

$A' = 0 \ at \ x = \frac{1}{2}$

$A(\frac{1}{2}) = \frac{1}{\sqrt{e}}$

4.9 Related Rates

EX 1

(1)

(a) $\dfrac{dV}{dt}=1\ volt/\text{sec}$

(b) $\dfrac{dI}{dt}=-\dfrac{1}{3}\ amp/\text{sec}$

(c) $\dfrac{dV}{dt}=R\left(\dfrac{dI}{dt}\right)+I\left(\dfrac{dR}{dt}\right)\Rightarrow\dfrac{dR}{dt}=\dfrac{1}{I}\left(\dfrac{dV}{dt}-\dfrac{V}{I}\dfrac{dI}{dt}\right)$

(d) $\dfrac{dR}{dt}=\dfrac{1}{2}\left[1-\dfrac{12}{2}(-\dfrac{1}{3})\right]=\dfrac{3}{2}\ \Omega s/\text{sec},$

R is increasing

EX 2

(1)

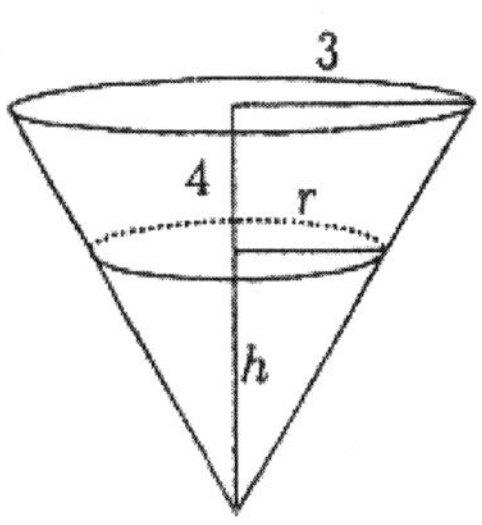

$$\frac{dV}{dt}=18\pi\ m^3/\text{min}$$

$$\left.\frac{dh}{dt}\right|_{h=2}=?$$

$$\frac{3}{4}=\frac{r}{h},\ 4r=\frac{3}{4}h\cdots\cdots ①$$

$$V=\frac{1}{3}\pi r^2h\cdots\cdots ②$$

$$① \rightarrow ②$$

$$V=\frac{\pi}{3}\left(\frac{3}{4}h\right)^2\bullet h$$

$$=\frac{3\pi}{16}h^3$$

$$\frac{dV}{dt}=\frac{9\pi}{16}h^2\bullet\frac{dh}{dt}$$

$$18\pi=\frac{9\pi}{16}(2)^2\bullet\frac{dh}{dt}$$

$$\therefore\ \left.\frac{dh}{dt}\right|_{h=2}=8\ m/\text{min}$$

(2)

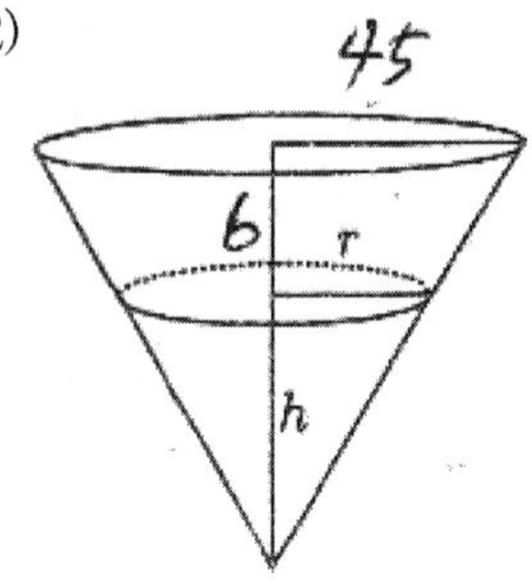

$V=\dfrac{1}{3}\pi r^2h$ and

$$r=\frac{15h}{2}\Rightarrow V=\frac{1}{3}\pi\left(\frac{15h}{2}\right)^2h=\frac{75\pi h^3}{4}$$

$$\Rightarrow\frac{dV}{dt}=\frac{225\pi h^2}{4}\frac{dh}{dt}$$

$$\Rightarrow\left.\frac{dh}{dt}\right|_{h=2}=\frac{4(-50)}{225\pi(5)^2}=\frac{-8}{225\pi}$$

$$\approx-0.0113m/\text{min}$$

$$r=\frac{15h}{2}\Rightarrow\frac{dr}{dt}=\frac{15}{2}\frac{dh}{dt}$$

$$\Rightarrow\left.\frac{dr}{dt}\right|_{h=5}=(\frac{15}{2})(\frac{-8}{225\pi})=\frac{-4}{15\pi}\approx-0.0849\,\text{m/sec}$$

(3) If $V=\dfrac{4}{3}\pi r^3, r=5,$ and $\dfrac{dV}{dt}=100\pi\ ft^3/\text{min},$

then $\dfrac{dV}{dt}=4\pi r^2\dfrac{dr}{dt}\Rightarrow\dfrac{dr}{dt}=1\ ft/\text{min}.$ Then

$S=4\pi r^2\Rightarrow\dfrac{dS}{dt}=8\pi r\dfrac{dr}{dt}$

$=8\pi(5)(1)=40\pi\ ft^2/\text{min}.$ the rate at which the surface area is increasing.

EX 3

Given: $\dfrac{dx}{dt}=5ft/\text{sec}$, the ladder is 13 ft long,

and $x=12, y=5$ at the instant of time

(a) Since $x^2+y^2=169$

$\Rightarrow \frac{dy}{dt}=-\frac{x}{y}\frac{dx}{dt}=-(\frac{12}{5})(5)=-12ft/\text{sec}$,

the ladder is sliding down the wall

(b) The area of the triangle formed by the ladder and walls is

$A=\frac{1}{2}xy \Rightarrow \frac{dA}{dt}=\left(\frac{1}{2}\right)\left(x\frac{dy}{dt}+y\frac{dx}{dt}\right)$.

The area is changing at

$\frac{1}{2}[12(-12)+5(5)]=-\frac{119}{2}=-59.5ft^2/\text{sec}$.

(c) $\cos\theta=\frac{x}{13} \Rightarrow -\sin\theta\frac{d\theta}{dt}=\frac{1}{13}\bullet\frac{dx}{dt}$

$\Rightarrow \frac{d\theta}{dt}=-\frac{1}{13\sin\theta}\bullet\frac{dx}{dt}$

$=-(\frac{1}{5})(5)$

$=-1\,rad/\text{sec}$

EX 4

(1) Let s represent the length of the rope and x the horizontal distance of the boat from the dock.

(a) We have

$s^2=x^2+36 \Rightarrow \frac{dx}{dt}=\frac{s}{x}\frac{ds}{dx}=\frac{s}{\sqrt{s^2-36}}\frac{ds}{dt}$.

Therefore, the boat is approaching the dock at

$\left.\frac{dx}{dt}\right|_{x=10}=\frac{10}{\sqrt{10^2-36}}(-2)=-2.5ft/\text{sec}$.

(b) $\cos\theta=\frac{6}{s}$

$\Rightarrow -\sin\theta\frac{d\theta}{dt}=-\frac{6}{s^2}\frac{ds}{dt} \Rightarrow \frac{d\theta}{dt}=\frac{6}{s^2\sin\theta}\frac{ds}{dt}$.

Thus, $s=10, x=8$, and

$\sin\theta=\frac{8}{10} \Rightarrow \frac{d\theta}{dt}=\frac{6}{10^2(\frac{8}{10})}\bullet(-2)$

$=-\frac{3}{20}rad/\text{sec}$

(2) Let s =distance of car from foot of perpendicular in the textbook diagram

$\Rightarrow \tan\theta=\frac{s}{1320} \Rightarrow \sec^2\theta\frac{d\theta}{dt}=\frac{1}{132}\frac{ds}{dt}$

$\Rightarrow \frac{d\theta}{dt}=\frac{\cos^2\theta}{132}\frac{ds}{dt}; \frac{ds}{dt}=-264$ and

$\theta=0 \Rightarrow \frac{d\theta}{dt}=-2rad/\text{sec}$. A half second later the car has traveled 132 ft right of the perpendicular

$\Rightarrow |\theta|=\frac{\pi}{4}, \cos^2\theta=\frac{1}{2}$, and $\frac{ds}{dt}=264$

(since s is increases)

$\Rightarrow \frac{d\theta}{dt}=\frac{\left(\frac{1}{2}\right)}{132}(264)=1\,rad/\text{sec}$.

EX 5

(1)

(a)

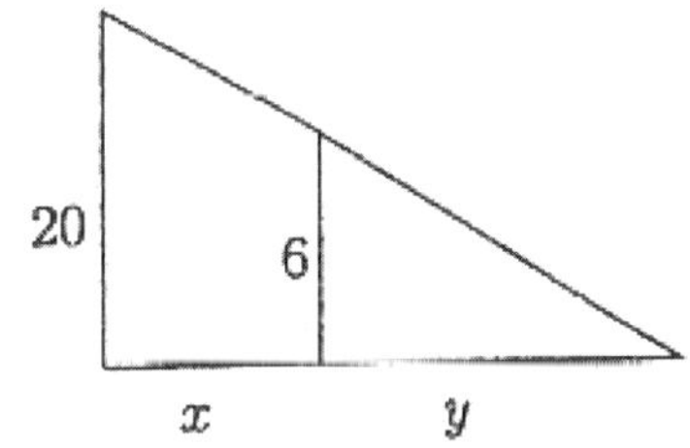

$\frac{dx}{dt}=-3ft/\text{sec}, \frac{dy}{dt}=?$

$\frac{6}{20}=\frac{y}{x+y}$

$6x+6y=20y$

$6x=14y$

$3x=7y$

$3\frac{dx}{dt}=7\frac{dy}{dt}$

$\frac{dy}{dt}=\frac{3}{7}\frac{dx}{dt}=\frac{3}{7}(-3)$

$=-\frac{9}{7}ft/\text{sec}$

(b)

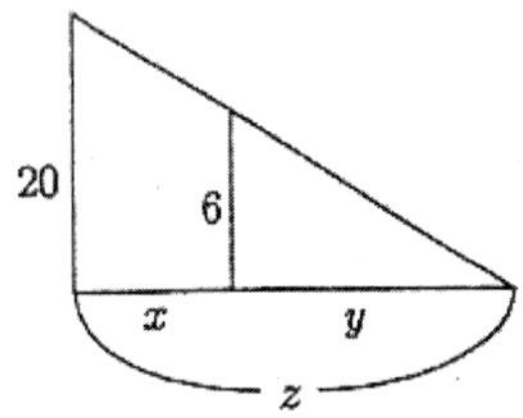

$$\frac{dz}{dt}=?$$

$$z=x+y$$
$$\frac{dz}{dt}=\frac{dx}{dt}+\frac{dy}{dt}$$
$$=-3-\frac{9}{7}$$
$$=-\frac{30}{7}ft/\text{sec}$$

(2) Let s represent the distance between the bicycle and balloon, h the height of the balloon and x the horizontal distance between the galloon and the bicycle. The relationship between the variables is

$$s^2=h^2+x^2$$

$$\Rightarrow\frac{dx}{dt}=\frac{1}{s}(h\frac{dh}{dt}+x\frac{dx}{dt})$$

$$\Rightarrow\frac{ds}{dt}=\frac{1}{85}[68(1)+51(17)]=11ft/\text{sec}.$$

(3)

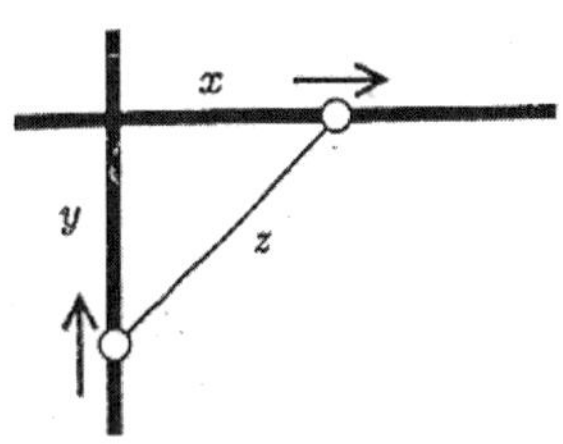

$$\frac{dy}{dt}=-30mph,\ \frac{dx}{dt}=50mph$$
$$\left.\frac{dz}{dt}\right|_{t=1}=?$$

$$x^2+y^2=z^2$$
$$2x\frac{dx}{dt}=2y\frac{dy}{dt}=2z\frac{dz}{dt}$$

$$\frac{dz}{dt}=\frac{x\frac{dx}{dt}+y\frac{dy}{dt}}{z}$$

When $t=1$,
$$x=120,\ y=50$$
$$z=\sqrt{(120)^2+(50)^2}=130$$
$$\therefore\left.\frac{dz}{dt}\right|_{t=1}=\frac{(120)(50)+(50)(-30)}{130}$$
$$=\frac{450}{13}mph$$

EX 6

(1)

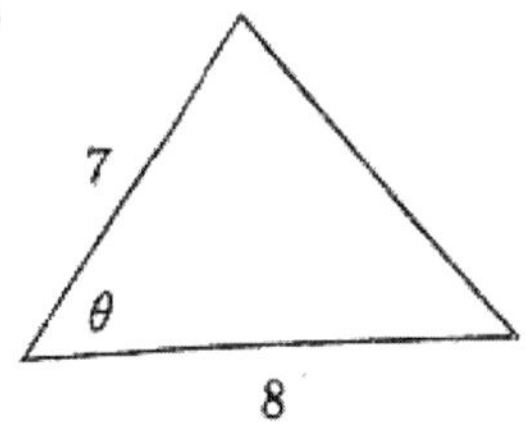

$$\frac{d\theta}{dt}=-\frac{\pi}{14}\frac{rad}{\text{sec}}$$
$$\left.\frac{dA}{dt}\right|_{\theta=\frac{\pi}{3}}=?$$

$$A=\frac{1}{2}\cdot7\cdot8\cdot\sin\theta$$
$$=28\sin\theta$$
$$\left.\frac{dA}{dt}\right|_{\theta=\frac{\pi}{3}}=28\cos\theta\cdot\frac{d\theta}{dt}$$
$$=28\cos\left(\frac{\pi}{3}\right)\cdot\left(-\frac{\pi}{14}\right)$$
$$=-\pi cm^2/\text{sec}$$

(2)

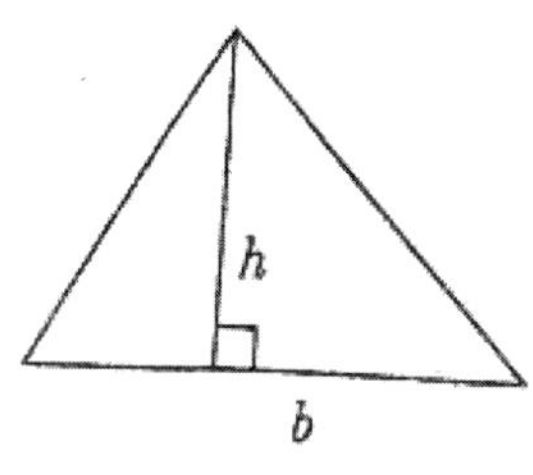

$$\frac{dh}{dt}=2cm/\text{sec}$$
$$\frac{db}{dt}=-3cm/\text{sec}$$
$$\frac{dA}{dt}=?$$

$A=\frac{1}{2}bh$

$\frac{dA}{dt}=\frac{1}{2}\frac{db}{dt}\cdot h+\frac{1}{2}b\cdot\frac{dh}{dt}$

When $b=8$ and $A=40, h=10$

$\frac{dA}{dt}=\frac{1}{2}(-3)(10)+\frac{1}{2}(8)(2)$

$=-7cm^2/\sec$

(3)

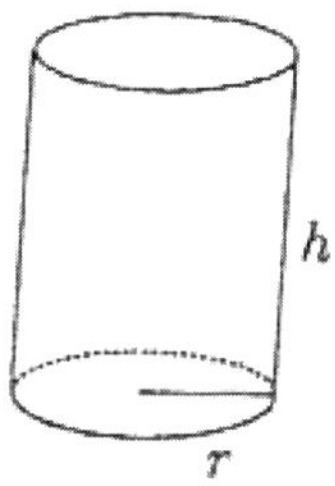

$\frac{dh}{dt}=6cm/\min, \frac{dV}{dt}=0, \left.\frac{dr}{dt}\right|_{h=15}=?$

$V=\pi r^2 h$

$\frac{dV}{dt}=2\pi r\cdot\frac{dr}{dt}\cdot h+\pi r^2\cdot\frac{dh}{dt}$

When $h=15$,

$240\pi=\pi r^2(15)$

$16=r^2, r>0$

$\therefore r=4$

$0=2\pi(4)\frac{dr}{dt}(15)+\pi(4)^2\cdot(6)$

$0=120\pi\frac{dr}{dt}+96\pi$

$\therefore \left.\frac{dr}{dt}\right|_{h=15}=-\frac{4}{5}cm/\min$

EX 7

(1) $y=\sqrt{x}$

$\left.\frac{dy}{dt}\right|_{x=4}=\frac{1}{2\sqrt{x}}\cdot\frac{dx}{dt}$

$=\frac{1}{2\sqrt{4}}\cdot 4$

$=1unit/\sec$

(2) $z^2=x^2+y^2$

$2z\frac{dz}{dt}=2x\frac{dx}{dt}+2y\frac{dy}{dt}$

$\frac{dz}{dt}=\frac{x\frac{dx}{dt}+y\frac{dy}{dt}}{z}$

When $x=4,\ y=\sqrt{4}=2$

$\left.\frac{dz}{dt}\right|_{x=4}=\frac{(4)(4)+(2)(1)}{\sqrt{16+4}}$

$=\frac{18}{2\sqrt{5}}=\frac{9}{\sqrt{5}}\ unit/\sec$

(3) $\tan\theta=\frac{y}{x}=\frac{\sqrt{x}}{x}=x^{-\frac{1}{2}}$

$\sec^2\theta\cdot\frac{d\theta}{dt}=-\frac{1}{2}x^{-\frac{3}{2}}\cdot\frac{dx}{dt}$

When $x=4$and $y=2,\ z=2\sqrt{5}$

$\therefore \cos\theta=\frac{4}{2\sqrt{5}}=\frac{2}{\sqrt{5}}$

$\left(\frac{\sqrt{5}}{2}\right)^2\cdot\frac{d\theta}{dt}=-\frac{1}{2}(4)^{-\frac{3}{2}}\cdot 4$

$\left.\frac{d\theta}{dt}\right|_{x=4}=-\frac{5}{4}\cdot\frac{1}{2}\cdot\frac{1}{8}\cdot 4$

$=-\frac{1}{5}rad/\sec$

EX 8

Let a represent the distance between point O and ship A, b the distance between point O and ship B, and D the distance between the ships, By the Law of Cosines,

$D^2=a^2+b^2-2ab\cos 120°$

$\Rightarrow\frac{dD}{dt}=\frac{1}{2D}\left[2a\frac{da}{dt}+2b\frac{db}{dt}+a\frac{db}{dt}+b\frac{da}{dt}\right]$

When $a=5, \frac{da}{dt}=14, b=3,$ and $\frac{db}{dt}=21$, then

$\frac{dD}{dt}=\frac{413}{2D}$ where $D=7$. The ships are

moving $\frac{dD}{dt}=29.5$ knots apart.

4.10 Mean Value Theorem (M.V.T)

EX 1

(1) $c=\frac{1}{2}$

(2) When $f(x)=x^{2/3}$ for $0 \le x \le 1$, then

$\frac{f(1)-f(0)}{1-0}=f'(c) \Rightarrow 1=\left(\frac{2}{3}\right)c^{-1/3} \Rightarrow c=\frac{8}{27}$.

(3) $c=1$

(4) When $f(x)=\sqrt{x-1}$ for $1 \le x \le 3$, then

$\frac{f(3)-f(1)}{3-1}=f'(c) \Rightarrow \frac{\sqrt{2}}{2}=\frac{1}{2\sqrt{c-1}} \Rightarrow c=\frac{3}{2}$

(5) $c=\pm\sqrt{1-\frac{4}{\pi^2}} \approx \pm 0.771$

(6) When $f(x)=\ln(x-1)$ for $2 \le x \le 4$, then

$f'(c)=\frac{f(4)-f(2)}{4-2} \Rightarrow \frac{1}{c-1}=\frac{\ln 3-\ln 1}{2}$

$\Rightarrow c-1=\frac{2}{\ln 3}$

$\Rightarrow c=1+\frac{2}{\ln 3} \approx 2.820$

(7) $\frac{1+\sqrt{7}}{3} \approx 1.22$ and $\frac{1-\sqrt{7}}{3} \approx -0.549$

are both in the interval $-1 \le x \le 2$.

(8) When $g(x)=\begin{cases} x^3 & -2 \le x \le 0 \\ x^2 & 0 < x \le 2 \end{cases}$, then

$\frac{g(2)-g(-2)}{2-(-2)}=g'(c)$. If $-2 \le x < 0$, then

$g'(x)=3x^2 \Rightarrow 3=g'(c) \Rightarrow 3c^2=3 \Rightarrow c=\pm 1$.

Only $c=-1$ is in the interval, If $0 < x \le 2$, then

$g'(x)=2x \Rightarrow 3=g'(c) \Rightarrow 2c=3 \Rightarrow c=\frac{3}{2}$.

EX 2

(1)

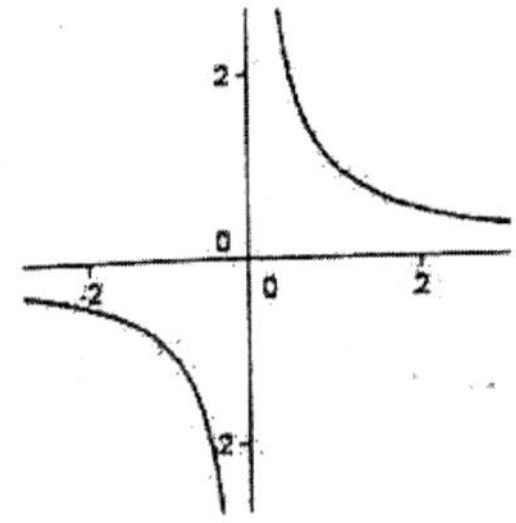

$f(x)$ is discontinuous at $x=0$.

(2)

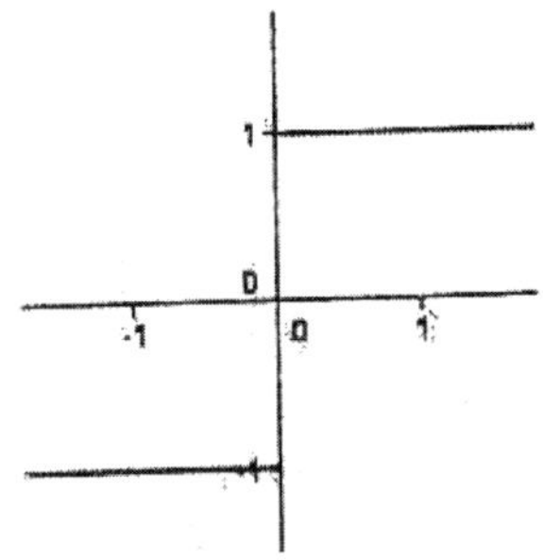

$f(x)$ is discontinuous at $x=0$.

(3)

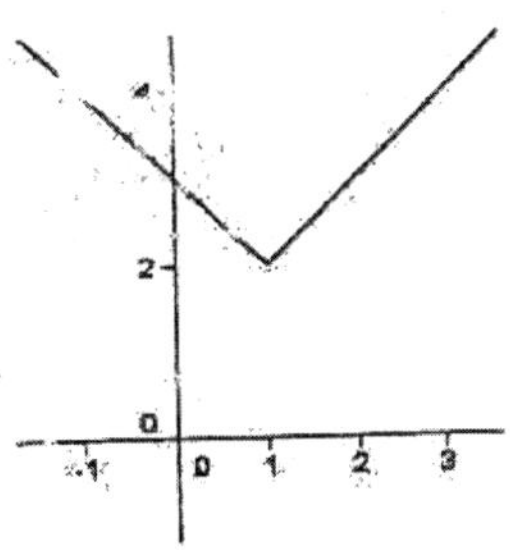

$f(x)$ is not differentiable at $x=1$

(4)

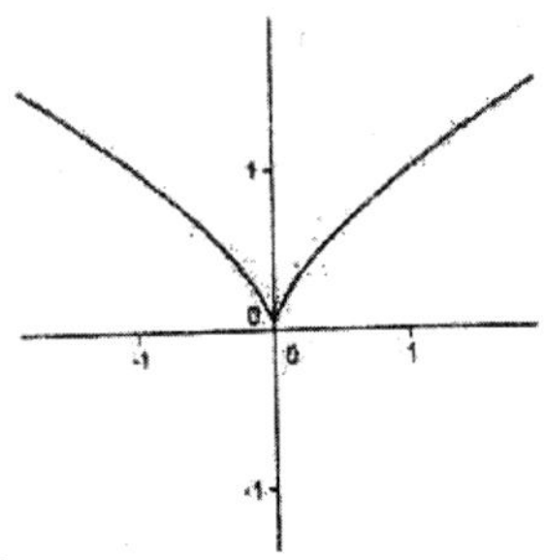

$f(x)$ is not differentiable at $x=0$.

EX 3

(1) $f'(x)=2^x\ln2-\cos x$

$$f'(c)=\frac{f(\pi)-0}{\pi-0}$$

$$2^c\ln2-\cos c=\frac{(2^\pi-0)-(2^0-0)}{\pi}$$

$$2^c\ln2-\cos c=\frac{2^\pi-1}{\pi},0<c<\pi$$

$\therefore c=1.743$

(2) $f'(x)=3\cos^{-1}x+3x\left(\frac{-1}{\sqrt{1-x^2}}\right)$

$$f'(c)=\frac{f(\frac{1}{2})-f(0)}{\frac{1}{2}-0}$$

$$3\cos^{-1}c-\frac{3c}{\sqrt{1-c^2}}=\frac{\frac{3}{2}\bullet\frac{\pi}{3}-0}{\frac{1}{2}}$$

$$3\cos^{-1}c-\frac{3c}{\sqrt{1-c^2}}=\pi,\ 0<c<\frac{1}{2}$$

$\therefore c=0.256$

(3) Note $f(0)=f(\sqrt{3})=0$ and $f'(x)$ exists on the given interval. By the MVT, there is a number, c,in the interval such that $f'(c)=0$. If $c=1$,then $6c^2-6=0$.(-1 is not in the interval.) $c=\pm1$

$\therefore c=1$(-1 is not in the interval.)

(4) Since $f(x)=3^x-x^3$then $f'(x)=3^x\ln3-3x^2$. Furthermore, f is continuous on [0,3] and f' is differentiable on (0,3), so MVT applies. We therefore seek c such that

$f'(c)=\frac{f(3)-f(0)}{3}=-\frac{1}{3}$.Solving

$3^c\ln3-3c^2=-\frac{1}{3}$with calculator, we find that c may be either 1.244 or 2.727. These values are the x-coordinates of points on the graph of $f(x)$ at which the tangents are parallel to the secand through points (0,1) and (3,0) on the curve.

4.11 Local Linear Approximation

EX 1

(1) Let $f(x)=\sqrt{x}$

$f(16)=\sqrt{16}=4\Rightarrow(16,4)$

$$f'(x)=\frac{1}{2\sqrt{x}}$$

$$f'(16)=\frac{1}{2\sqrt{16}}=\frac{1}{8}$$

$$y-4=\frac{1}{8}(x-16)$$

$$y=\frac{1}{8}(x-16)+4$$

$\sqrt{x}\approx\frac{1}{8}(x-16)+4$ *at* $x=16$

$\sqrt{16.08}\approx\frac{1}{8}(0.08)+4=4.01$

(2) Let $f(x)=\sqrt[3]{x}$

$f(8)=\sqrt[3]{8}=2\Rightarrow(8,2)$

$$f'(x)=\frac{1}{3}x^{-\frac{2}{3}}$$

$$f'(8)=\frac{1}{3}(8)^{-\frac{2}{3}}=\frac{1}{3}\bullet\frac{1}{4}=\frac{1}{12}$$

$$y-2=\frac{1}{12}(x-8)+2$$

$\sqrt[3]{x}\approx\frac{1}{12}(x-8)+2$ *at* $x=8$

$\sqrt[3]{7.4}\approx\frac{1}{12}(-0.6)+2=1.95$

(3) Let $f(x)=\sqrt{x}$

$f(64)=\sqrt{64}=8\Rightarrow(64,8)$

$$f'(x)=\frac{1}{2\sqrt{x}}$$

$$f'(64)=\frac{1}{16}$$

$$y-8=\frac{1}{16}(x-64)$$

$$y=\frac{1}{16}(x-64)+8$$

$\sqrt{x}\approx\frac{1}{16}(x-64)+8$ *near* $x=64$

$\sqrt{63}\approx\frac{1}{16}(-1)+8=\frac{127}{16}$(or7.938)

(4) Let $f(x)=x^{10}$

$f(1)=1 \Rightarrow (1,1)$
$f'(x)=10x^9$
$f'(1)=10$
$y-1=10(x-1)$
$y=10(x-1)+1$

$x^{10} \approx 10(x-1)+1$ $near$ $x=1$
$(1.01)^{10} \approx 10(0.01)+1=1.1$

(5) Let $f(x)=2^x$

$f(4)=2^4=16 \Rightarrow (4,16)$
$f'(x)=2^x \cdot \ln 2$
$f'(4)=16\ln 2$
$y-16=16\ln 2(x-4)$
$y=16\ln 2(x-4)+16$

$2^x \approx 16\ln 2 \cdot (x-4)+16$ $near$ $x=4$
$2^{3.99} \approx 16\ln 2 \cdot (-0.01)+16$
$\approx 16-\dfrac{4}{25}\ln 2$ (or 15.889)

(6) Let $f(x)=\tan^{-1}x$

$f(1)=\tan^{-1}(1)=\dfrac{\pi}{4} \Rightarrow \left(1,\dfrac{\pi}{4}\right)$

$f'(x)=\dfrac{1}{1+x^2}$

$f'(1)=\dfrac{1}{2}$

$y-\dfrac{\pi}{4}=\dfrac{1}{2}(x-1)$

$y=\dfrac{1}{2}(x-1)+\dfrac{\pi}{4}$

$\tan^{-1}x \approx \dfrac{1}{2}(x-1)+\dfrac{\pi}{4}$ $near$ $x=1$

$\tan^{-1}(1.01) \approx \dfrac{1}{2}(0.01)+\dfrac{\pi}{4}$

$\approx \dfrac{\pi}{4}+0.005$ (or 0.790)

EX 2

(1) $f(x)=x^3-2x+3$
$\Rightarrow f'(x)=3x^2-2$
$\Rightarrow L(x)=f'(2)(x-2)+f(2)=10(x-2)+7$
$\Rightarrow L(x)=10x-13$ at $x=2$

(2) $f(x)=\sqrt{x^2+9}=(x^2+9)^{\frac{1}{2}}$
$\Rightarrow f'(x)=\left(\dfrac{1}{2}\right)(x^2+9)^{-1/2}(2x)$
$=\dfrac{x}{\sqrt{x^2+9}}$
$\Rightarrow L(x)=f'(-4)(x+4)+f(-4)$
$=-\dfrac{4}{5}(x+4)+5$
$\Rightarrow L(x)=-\dfrac{4}{5}x+\dfrac{9}{5}$ at $x=-4$

(3) $f(x)=x+\dfrac{1}{x}$
$\Rightarrow f'(x)=1-x^{-2}$
$\Rightarrow L(x)=f(1)+f'(1)(x-1)$
$=2+0(x-1)=2$

(4) $f(x)=x^{1/3}$
$\Rightarrow f'(x)=\dfrac{1}{3x^{2/3}}$
$\Rightarrow L(x)=f'(-8)(x-(-8))+f(-8)$
$=\dfrac{1}{12}(x+8)-2$
$\Rightarrow L(x)=\dfrac{1}{12}x-\dfrac{4}{3}$

(5) $f(x)=\tan x$
$\Rightarrow f'(x)=\sec^2 x$
$\Rightarrow L(x)=f(\pi)+f'(\pi)(x-\pi)$
$=0+1(x-\pi)=x-\pi$

EX 3

(1) $r(5.4) = r(5) + r'(5)\Delta t$
$= 30 + 2(0.4) = 30.8 ft$

Since the graph of r is concave down on the interval $5 < t < 5.4$, this estimate is greater than $r(5.4)$.

(2) $\frac{dV}{dt} = 3\left(\frac{4}{3}\right)\pi r^2 \frac{dr}{dt}$.

$$\left.\frac{dV}{dt}\right|_{t=5} = 4\pi(30)^2\, 2 = 7200\pi\ ft^3/\text{min}$$

EX 4

(1) $f(2) = 4 \Rightarrow (2,4)$
$f'(2) = 3$
$y - 4 = 3(x-2)$
$y = 3(x-2) + 4$

$f(x) \approx 3(x-2) + 4$ *near* $x = 2$
$f(2.10) \approx 3(0.1) + 4 = 4.3$

(2) $f'' > 0$ *on* $0 < x < 3 \Rightarrow$ The graph of f is concave up.

$\therefore$ The approximate value is less than the actual value.

4.12 Newton's Method

EX 1

(1) $x_1 = 2$

$x_2 = 1.636363$

$x_3 = 1.530392$

$x_4 = 1.521441$

$x_5 = 1.521380$

(2) $x_1 = 1$

$x_2 = 1.285714$

$x_3 = 1.216457$

$x_4 = 1.208996$

$x_5 = 1.208918$

(3) $x_1 = 2$

$x_2 = 0.995140$

$x_3 = 1.030107$

$x_4 = 1.029867$

(4) $x_1 = 0$

$x_2 = 0.5$

$x_3 = 0.510958$

$x_4 = 0.510973$

EX 2

$y = x^2 + x - 1 \Rightarrow y' = 2x + 1$

$$\Rightarrow x_{n+1} = x_n - \frac{x_n^2 + x_n - 1}{2x_n + 1};\ x_0 = 1$$

$$\Rightarrow x_1 = 1 - \frac{1+1-1}{2+1} = \frac{2}{3}$$

$$\Rightarrow x_2 = \frac{2}{3} - \frac{\frac{4}{9} + \frac{2}{3} - 1}{\frac{4}{3} + 1}$$

$$= \frac{2}{3} - \frac{4+6-9}{12+9}$$

$$= \frac{2}{3} - \frac{1}{21}$$

$$= \frac{13}{21} \approx 0.61905;\ x_0 = -1$$

$$\Rightarrow x_1 = 1 - \frac{1-1-1}{-2+1} = -2$$

$$\Rightarrow x_2 = -2 - \frac{4-2-1}{-4+1} = -\frac{5}{3} \approx -1.66667$$

EX 3

$y=x^3+3x+1 \Rightarrow y'=3x^2+3$

$$\Rightarrow x_{n+1}=x_n-\frac{x_n^3+3x_n+1}{3x_n^2+3};\ x_0=0$$
$$\Rightarrow x_1=0-\frac{1}{3}=-\frac{1}{3}$$
$$\Rightarrow x_2=-\frac{1}{3}-\frac{-\frac{1}{27}-1+1}{\frac{1}{3}+3}$$
$$=-\frac{1}{3}+\frac{1}{90}=-\frac{29}{90}\approx -0.32222$$

EX 4

$y=x^4+x-3 \Rightarrow y'=4x^3+1$

$$\Rightarrow x_{n+1}=x_n-\frac{x_n^4+x_n-3}{4x_n^3+1};\ x_0=1$$
$$\Rightarrow x_1=1-\frac{1+1-3}{4+1}=\frac{6}{5}$$
$$\Rightarrow x_2=\frac{6}{5}-\frac{\frac{1296}{625}+\frac{6}{5}-3}{\frac{864}{125}+1}$$
$$=\frac{6}{5}-\frac{1296+750-1875}{4320+625}$$
$$=\frac{6}{5}-\frac{171}{4945}=\frac{5763}{4945}\approx 1.16542;\ x_0=-1$$
$$\Rightarrow x_1=-1-\frac{1-1-3}{-4+1}=-2$$
$$\Rightarrow x_2=-2-\frac{16-2-3}{-32+1}$$
$$=-2+\frac{11}{31}=-\frac{51}{31}\approx -1.64516$$

4.13 Approximating Derivatives Numerically

EX 1

The best approximation to $f'(0.10)$ is

$\frac{f(0.20)-f(0.10)}{0.20-0.10}$.

EX 2

Use the symmetric difference quotient; then

$$f'(1.5)=\frac{f(1.6)-f(1.4)}{1.6-1.4}=\frac{8}{0.2}$$

EX 3

$$\frac{5^{2.3}-5^{1.97}}{2.03-1.97}=40.25158$$

EX 4

$$f'(2)=\frac{f(2)-f(1.98)}{2-1.98}=\frac{4.00-4.10}{0.02}=-5$$

EX 5

Since an estimate of the answer for Question [EX4] is $f'(2)=-5$, them

$$(f^{-1})'(4)=\frac{1}{f'(2)}=\frac{1}{-5}=-0.2$$

EX 6

(1) $\frac{f(12)-f(0)}{12-0}=\frac{45-9}{12}=\frac{36}{12}=3$

(2) $f(0)\approx\frac{f(2)-f(0)}{2-10}=\frac{13-9}{2}=\frac{4}{2}=2$

$f'(4)\approx\frac{f(6)-f(2)}{6-2}=\frac{28-13}{4}=\frac{15}{4}$

$f'(9)\approx\frac{f(10)-f(8)}{10-8}=\frac{43-37}{2}=6$

$f'(12)\approx\frac{f(12)-f(10)}{12-10}=\frac{45-43}{2}=1$

(3)

$$(f^{-1})'(19)=\frac{1}{f'(f^{-1}(19))}=\frac{1}{f'(4)}\approx\frac{1}{\frac{15}{4}}=\frac{4}{15}$$

EX 7

(1) $C'(3.5)\approx\dfrac{C(4)-C(3)}{4-3}$

$=\dfrac{12.8-11.2}{1}=1.6$ ounces/min

(2) C is differentiable $\Rightarrow$ C is continuous(on the close interval)

$\dfrac{C(4)-C(2)}{4-2}=\dfrac{12.8-8.8}{2}=2$

Therefore, by the Mean Value Theorem. there is at least one time t, $2<t<4$ for which $C'(t)=2$.

(3) $B'(t)=-16(-0.4)e^{-0.4t}=6.4e^{-0.4t}$

$B'(5)=6.4e^{-0.4(5)}=\dfrac{6.4}{e^2}$ ounces/min

4.14 L'Hopital's Rule

(1) $\frac{1}{4}$

(2) $$\lim_{x\to\infty}\frac{5x^3-2x}{7x^3+3}=\lim_{x\to\infty}\frac{15x^2-2}{21x^2}=\lim_{x\to\infty}\frac{30x}{42x}=\lim_{x\to\infty}\frac{30}{42}=\frac{5}{7}$$

(3) -16

(4) $$\lim_{\theta\to\pi/2}\frac{1-\sin\theta}{1+\cos 2\theta}=\lim_{\theta\to\pi/2}\frac{-\cos\theta}{-2\sin 2\theta}=\lim_{\theta\to\pi/2}\frac{\sin\theta}{-4\cos 2\theta}=\frac{1}{(-4)(-1)}=\frac{1}{4}$$

(5) 3

(6) $$\lim_{\theta\to 0}\frac{3^{\sin\theta}-1}{\theta}=\lim_{\theta\to 0}\frac{3^{\sin\theta}((\ln 3)(\cos\theta))}{1}=\frac{(3^0)(\ln 3)(1)}{1}=\ln 3$$

(7) $\ln 2$

(8) $$\lim_{y\to 0}\frac{\sqrt{5y+25}-5}{y}=\lim_{y\to 0}\frac{(5y+25)^{1/2}-5}{y}=\lim_{y\to 0}\frac{\left(\frac{1}{2}\right)(5y+25)^{-1/2}(5)}{1}=\lim_{y\to 0}\frac{5}{2\sqrt{5y+25}}=\frac{1}{2}$$

(9) $\ln 2$

(10) $\lim_{x\to1^+}\left(\frac{1}{x-1}-\frac{1}{\ln x}\right)=\lim_{x\to1^+}\left(\frac{\ln x-(x-1)}{(x-1)(\ln x)}\right)$

$$=\lim_{x\to1^+}\left(\frac{\frac{1}{x}-1}{(\ln x)+(x-1)\left(\frac{1}{x}\right)}\right)$$

$$=\lim_{x\to1^+}\left(\frac{1-x}{(x\ln x)+x-1}\right)$$

$$=\lim_{x\to1^1}\left(\frac{-1}{(\ln x+1)+1}\right)$$

$$=\frac{-1}{(0+1)+1}=-\frac{1}{2}$$

(11) 0

(12) $\lim_{\theta\to0}\frac{\theta-\sin\theta\cos\theta}{\tan\theta-\theta}=\lim_{\theta\to0}\frac{1+\sin^2\theta-\cos^2\theta}{\sec^2\theta-1}$

$$=\lim_{\theta\to0}\frac{2\sin^2\theta}{\tan^2\theta}$$

$$=\lim_{\theta\to0}2\cos^2\theta$$

$$=2$$

(13) e

(14) The limit leads to the indeterminate form 1^∞

Let $f(x)=(\ln x)^{1/(x-e)}\Rightarrow \ln f(x)=\frac{\ln(\ln x)}{x-e}$

$$=\lim_{x\to e^+}\ln f(x)=\lim_{x\to e^+}\frac{\ln(\ln x)}{x-e}=\lim_{x\to e^+}\frac{\left(\frac{1}{x\ln x}\right)}{1}$$

$$=\frac{1}{e}.$$

Therefore $(\ln x)^{1/(x-e)}=\lim_{x\to e^+}f(x)$

$$=\lim_{x\to e^+}e^{\ln f(x)}=e^{1/e}$$

(15) e

(16) The limit lead to the indeterminate form 1^∞.

Let $f(x)=(e^x+x)^{1/x}\Rightarrow \ln f(x)=\frac{\ln(e^x+x)}{x}\Rightarrow$

$$\lim_{x\to0}\ln f(x)=\lim_{x\to0}\frac{\ln(e^x+x)}{x}\ \lim_{x\to0}\frac{e^x+1}{e^x+x}=2$$

Therefore

$$\lim_{x\to0}(e^x+x)^{1/x}=\lim_{x\to0}f(x)=\lim_{x\to0}e^{\ln f(x)}=e^2$$

(17) 1

(18)

$$\lim_{x\to0^+}x(\ln x)^2=\lim_{x\to0^+}\left(\frac{(\ln x)^2}{\frac{1}{x}}\right)=\lim_{x\to0^+}\left(\frac{2(\ln x)\frac{1}{x}}{-\frac{1}{x^2}}\right)=\lim_{x\to0^+}\left(\frac{2\ln x}{-\frac{1}{x}}\right)$$

$$=\lim_{x\to0^+}\left(\frac{\frac{2}{x}}{\frac{1}{x^2}}\right)=\lim_{x\to0^+}\left(\frac{2x^2}{x}\right)=\lim_{x\to0^+}(2x)=0$$

(19) 0

CHAPTER 05. Integration

5.1. indefinite integrals

EX 1

(1) $= 3x - 2x^2 + \frac{1}{2}x^4 + C$

(2) $= \int 6x^{\frac{7}{3}}dx$

$= \frac{9}{5}x^{\frac{10}{3}} + C$

(3) $= \frac{1}{5}x^5 + \frac{3}{5}x^{\frac{5}{3}} + \frac{2}{x} - \frac{1}{2}x^{\frac{2}{3}} + C$

(4) $= 2\ln|x| + \frac{1}{x} + \sqrt{x} + C$

(5) $= \int \frac{1}{\ln 2} \cdot \frac{1}{x}dx$

$= \frac{\ln|x|}{\ln 2} + C$

(6) $= \int (x^3 - 27)dx$

$= \frac{1}{4}x^4 - 27x + C$

(7) $= \int (x^2 - 1)dx = \frac{1}{3}x^3 - x + C$

(8) $= 5x + C$

(9) $= \frac{1}{4}t^4 + \frac{3}{2}t^2 + \frac{2}{7}t^{\frac{7}{2}} + 2t^{\frac{3}{2}} + C$

EX 2

(1) $= \frac{1}{2}x^2 - 6x + 12\ln|x| + \frac{8}{x} + C$

(2) $= \int (x^{\frac{3}{2}} - 2x^{\frac{1}{2}})dx$

$= \frac{2}{5}x^{\frac{5}{2}} - 2 \cdot \frac{2}{3}x^{\frac{3}{2}} + C$

$= \frac{2}{5}x^{\frac{5}{2}} - \frac{4}{3}x^{\frac{3}{2}} + C$

(3) $\frac{1}{2}x^2 - \ln|x| + \frac{1}{x} + C$

(4) $\frac{1}{3}x^3 - x - \frac{1}{4x} + C$

(5) $= \int (x^{\frac{1}{2}} + x)x^{\frac{2}{3}}dx$

$= \int (x^{\frac{7}{6}} + x^{\frac{5}{3}})dx$

$= \frac{6}{13}x^{\frac{13}{6}} + \frac{3}{8}x^{\frac{8}{3}} + C$

(6) $= 2y^{\frac{1}{2}} - \frac{2}{3}y^{3/2} + \frac{1}{10}y^{5/2} + C$

EX 3

(1) $= e^x + 4x + C$

(2) $= \frac{5^x}{\ln 5} - \frac{7^x}{\ln 7} + C$

(3) $= \frac{10^x}{\ln 10} + 10\ln|x| + \frac{7}{2}x^{-2} + C$

EX 4

(1) $= 5\sin x - \cos x + C$

(2) $= 2\ln|\sec x| + 3\ln|\sin x| + C$

(3) $= \int \csc x\, dx$
$= \ln|\csc x - \cot x| + C$

(4) $= \ln|\sec x| - 3\cos x + C$

(5) $= \sec x + 5\sin x + C$

(6) $= \tan x - 4\ln|\csc x - \cot x| + C$

(7) $- \cot x + C$

(8) $= \int \frac{1-\cos^2 x}{1-\cos x} dx$

$= \int \frac{(1-\cos x)(1+\cos x)}{1-\cos x} dx$

$= \int (1+\cos x) dx$

$= x + \sin x + C$

(9) $= \sec x + C$

(10) $= \int \frac{1}{\sin x} \cdot \frac{\cos x}{\sin x} dx$

$= \int \csc x \cot x\, dx$

$= -\csc x + C$

(11) $= \tan x - x + C$

(12) $= \int (\csc^2 x - 1) dx$

$= -\cot x - x + C$

(13) $= e^x - \cot x + C$

(14) $= \int (10^x - \frac{1}{\cot x}) dx$

$= \int (10^x - \tan x) dx$

$= \frac{10^x}{\ln 10} - \ln|\sec x| + C$

EX 5

(1) $= 5\sin^{-1} x + C$

(2) $= 3\tan^{-1} x + C$

(3) $= -7\sec^{-1}|x| + C$

(4) $= 4\tan^{-1} x + C$

EX 6

(1) $= \frac{1}{8}(2x+1)^4 + C$

(2) $= \frac{1}{27}(3x-2)^9 + C$

(3) $= \frac{1}{3}(\frac{1}{2}x+3)^6 + C$

(4) $= -\frac{2}{9}(1-3x)^{\frac{3}{2}} + C$

(5) $= 7\sqrt{2x+1} + C$

(6) $= (1-x)^{-1} + C = \frac{1}{1-x} + C$

EX 7

(1) $= \frac{1}{7}e^{7x} + \frac{1}{3}e^{3x} + C$

(2) $= -\frac{1}{2}e^{-2x} + \frac{1}{2}\sin 2x + C$

(3) $= -\frac{1}{2}e^{-2x} - 3e^{-\frac{1}{3}x} + C$

(4) $= -4e^{-x} + \frac{3^{x-1}}{\ln 3} + C$

(5) $= \frac{10^{5x}}{5\ln 10} - \frac{1}{2}\cos 2x + C$

(6) $= \int (4^{3x} + 4^x) dx$
$= \frac{4^{3x}}{3\ln 4} + \frac{4^x}{\ln 4} + C$

(7) $= \int (3^{x+1})^{\frac{1}{2}} dx$

$= \int 3^{\frac{1}{2}x+\frac{1}{2}} dx$

$= \frac{2(3)^{\frac{1}{2}x+\frac{1}{2}}}{\ln 3} + C$

(8) $= \frac{1}{3}\sin 3x - 2\cos(\frac{x}{2}) + C$

(9) $= \frac{1}{3}\sin(3x+5) + C$

(10) $= \frac{1}{5}\sec 5x + C$

(11) $= \frac{1}{2}e^{2x} + 2x - \frac{1}{2}e^{-2x} + C$

(12) $= \frac{1}{2}\tan 2x + \frac{1}{2}\ln|\sec 2x| + C$

EX 8

(1) $= -\frac{3}{4}\cos 2x + C$

(2) $= \int \frac{1}{2}\sin\theta d\theta$

$= -\frac{1}{2}\cos\theta + C$

(3) $= -\cos 4x + C$

(4) $- \int 6\sin 6x dx$

$= -\cos 6x + C$

(5) $= \frac{1}{2}x - \frac{1}{2}\sin x + C$

(6) $= \int \frac{1-\cos 6x}{2} dx$

$= \int \left(\frac{1}{2} - \frac{1}{2}\cos 6x\right) dx$

$= \frac{1}{2}x - \frac{1}{12}\sin 6x + C$

(7) $\frac{1}{2}x + \frac{1}{4}\sin 2x + C$

(8) $= \int \frac{1+\cos 8x}{2} dx$

$= \int \left(\frac{1}{2} + \frac{1}{2}\cos 8x\right) dx$

$= \frac{1}{2}x + \frac{1}{16}\sin 8x + C$

EX 9

(1) $= 5\ln|\sin x| + C$

(2) $= \ln|x^3 + x| + C$

(3) $= 7\ln|x+1| + C$

(4) $= \int \frac{1}{2} \cdot \frac{2x}{x^2+1} dx$

$= \frac{1}{2}\ln|x^2+1| + C$

$= \frac{1}{2}\ln(x^2+1) + C$

(5) $= 3\ln|x^2-4| + C$

(6) $\int \frac{2}{-3} \cdot \frac{-3}{1-3x} dx$

$= -\frac{2}{3}\ln|1-3x| + C$

(7) $= -5\ln|1+\cos x| + C$

(8) $= \int \frac{1}{4} \cdot \frac{4x}{2x^2+1} dx$

$= \frac{1}{4}\ln(2x^2+1) + C$

(9) $= \ln(e^x + e^{-x}) + C$

EX 10

(1) $f(x) = x + \ln x + 2$

(2) $\dfrac{dy}{dx} = 2^x - x^{\frac{5}{2}} + \sin x$

$y = \dfrac{2^x}{\ln 2} - \dfrac{2}{7}x^{\frac{7}{2}} - \cos x + C$

$y(0) = \dfrac{1}{\ln 2}$ $: C = 1$

$\therefore y = \dfrac{2^x}{\ln 2} - \dfrac{2}{7}x^{\frac{7}{2}} - \cos x + 1$

(3) $f(x) = 8\tan^{-1}x + \tan x + C$

$f(0) = 1; C = 1$

$\therefore f(x) = 8\tan^{-1}x + \tan x + 1$

(4) $y = 2\ln|2x-1| + \dfrac{1}{2} \cdot \dfrac{8}{4}(2x+1)^4 + C$

$y = 2\ln|2x-1| + (2x+1)^4 + C$

$y(0) = 3 :$

$C = 2$

$\therefore y = 2\ln|2x-1| + (2x+1)^4 + 2$

(5) $f(x) = \ln(x^2+4) + \dfrac{1}{2}x + \dfrac{1}{4}\sin 2x - \ln 4$

(6) $f'(x) = 2e^{\frac{1}{2}x} - 6x^2 + 2x + C$

$f'(0) = -1 :$

$C = -3$

$\therefore f'(x) = 2e^{\frac{1}{2}x} - 6x^2 + 2x - 3$

$f(x) = 4e^{\frac{1}{2}x} - 2x^3 + x^2 - 3x + D$

$f(0) = 1 :$

$D = -3$

$\therefore f(x) = 4e^{\frac{1}{2}x} - 2x^3 + x^2 - 3x - 3$

(7) $y = \dfrac{1}{x} + 2x - 2$

(8) $\dfrac{dy}{dx} = -e^{-x} + 3x^2 + C$

$y = e^{-x} + x^3 + Cx + D$

$y(0) = 2$

$D = 1$

$y(1) = \dfrac{1}{e} - 1 :$

$C = -3$

$\therefore y = e^{-x} + x^3 - 3x + 1$

5.2 Intergration by substitution

EX 1

(1) $= -\dfrac{2}{5}(1-x^2)^{5/4} + C$

(2) Let $u = x^2 - 2x + 3$ $\quad = \displaystyle\int u^{10} \cdot \dfrac{1}{2}du$

$\dfrac{du}{dx} = 2x - 2$ $\quad = \dfrac{1}{2} \cdot \dfrac{1}{11}u^{11} + C$

$\dfrac{1}{2}du = (x-1)dx$ $\quad = \dfrac{1}{22}(x^2-2x+3)^{11} + C$

(3) $= \dfrac{-1}{x^4+1} + C$

(4) Let $u = x^2 - 6x$ $\quad = \displaystyle\int \dfrac{1}{u^{\frac{1}{4}}} \cdot \dfrac{1}{2}du$

$\dfrac{du}{dx} = 2x - 6$ $\quad = \displaystyle\int \dfrac{1}{2}u^{-\frac{1}{4}}du$

$\dfrac{1}{2}du = (x-3)dx$ $\quad = \dfrac{1}{2} \cdot \dfrac{4}{3}u^{\frac{3}{4}} + C$

$= \dfrac{2}{3}(x^2-6x)^{\frac{3}{4}} + C$

(5) $= -6(1-x^3)^{1/2} + C$

(6) Let $u = 4x - x^2$ $\quad = \displaystyle\int \dfrac{1}{\sqrt{u}} \cdot -\dfrac{1}{2}du$

$\dfrac{du}{dx} = 4 - 2x$ $\quad = \displaystyle\int -\dfrac{1}{2}u^{-\frac{1}{2}}du$

$\dfrac{du}{dx} = -2(x-2)$ $\quad = -\dfrac{1}{2} \cdot 2u^{\frac{1}{2}} + C$

$-\dfrac{1}{2}du = (x-2)dx$ $\quad = -\sqrt{4x-x^2} + C$

(7) $= -\dfrac{1}{3}(7-3x^2)^{3/2} + C$

(8) Let $u = 1 + \sqrt{x}$ $\quad = \displaystyle\int u^5 \cdot 2du$

$\dfrac{du}{dx} = \dfrac{1}{2\sqrt{x}}$ $\quad = 2 \cdot \dfrac{1}{6}u^6 + C$

$2du = \dfrac{1}{\sqrt{x}}dx$ $\quad = \dfrac{1}{3}(1+\sqrt{x})^6 + C$

(9) $= \dfrac{2^{\sqrt{x}+1}}{\ln^2} + C$

(10) Let $tu = \sqrt{x}-1$

$\frac{du}{dx} = \frac{1}{2\sqrt{x}}$

$2du = \frac{1}{\sqrt{x}}dx$

$= \int \frac{1}{u} \cdot 2du$

$= 2 \cdot \ln|u| + C$

$= 2\ln|\sqrt{x}-1| + C$

EX 2

(1) $= \frac{1}{8}(\ln x)^4 + C$

(2) Let $u = \ln x$

$\frac{du}{dx} = \frac{1}{x}$

$du = \frac{1}{x}dx$

$= \int \frac{1}{u} \cdot du$

$= \ln|u| + C$

$= \ln|\ln x| + C$

(3) $= \sqrt{\ln^2 x + 1} + C$

(4) Let $u = \ln(\ln x)$

$\frac{du}{dx} = \frac{1}{\ln x} \cdot \frac{1}{x}$

$du = \frac{1}{x\ln x}dx$

$= \int \frac{1}{u} \cdot du$

$= \ln|u| + C$

$= \ln|\ln(\ln x)| + C$

(5) $\frac{1}{2}\sin^6\left(\frac{x}{3}\right) + C$

(6) Let $u = \cos 3x$

$\frac{du}{dx} = -3\sin 3x$

$-\frac{1}{3}du = \sin 3x dx$

$= \int u^3 \cdot -\frac{1}{3}du$

$= -\frac{1}{3} \cdot \frac{1}{4}u^4 + C$

$= -\frac{1}{12}\cos^4 3x + C$

(7) $= \frac{e^{\sec \pi x}}{\pi} + C$

(8) Let $u = \tan\theta$

$\frac{du}{d\theta} = \sec^2\theta$

$du = \sec^2\theta d\theta$

$= \int u^2 \cdot du$

$= \frac{1}{3}u^3 + C$

$= \frac{1}{3}\tan^3\theta + C$

(9) $= e^{\sin^{-1}x} + C$

(10) $= \int \sec^2\theta \cdot \sec^2\theta d\theta$

Let $u = \tan\theta$

$\frac{du}{dx} = \sec^2\theta$

$du = \sec^2\theta d\theta$

$= \int (1+\tan^2\theta) \cdot \sec^2\theta d\theta$

$= \int (1+u^2)du$

$= u + \frac{1}{3}u^3 + C$

$= \tan\theta + \frac{1}{3}\tan^3\theta + C$

(11) $= \ln|\tan^{-1}x| + C$

(12) Let $u = 1+\tan 2\theta$

$\frac{du}{d\theta} = 2 \cdot \sec^2 2\theta$

$\frac{1}{2}du = \sec^2 2\theta d\theta$

$= \int \frac{1}{u} \cdot \frac{1}{2}du$

$= \frac{1}{2}\ln|u| + C$

$= \frac{1}{2}\ln|1+\tan 2\theta| + C$

(13) $= \frac{1}{2\cos(2x+1)} + C$

(14) Let $u = 1+\sin 2x$

$\frac{du}{dx} = 2\cos 2x$

$\frac{1}{2}du = \cos 2x dx$

$= \int \frac{1}{\sqrt{u}} \cdot \frac{1}{2}du$

$= \int \frac{1}{2}u^{-\frac{1}{2}}du$

$= \frac{1}{2} \cdot 2u^{\frac{1}{2}} + C$

$= \sqrt{1+\sin 2x} + C$

(15) $= e^{\sin^2 x} + C$

(16) Let $u = \cos x$

$\frac{du}{dx} = -\sin x$

$-du = \sin x dx$

$= \int \frac{1}{1+u^2} \cdot -du$

$= -\tan^{-1}u + C$

$= -\tan^{-1}(\cos x) + C$

EX 3

(1) $= \ln|1+e^x| + C$

(2)

$\int \frac{f'(x)}{f(x)}dx = \ln|f(x)| + C$

$\int \frac{e^{2x}}{e^{2x}+1}dx = \int \frac{1}{2} \cdot \frac{2e^{2x}}{e^{2x}+1}dx = \frac{1}{2}\ln(e^{2x}+1) + C$

(3) $= -e^{-x^2} + C$

(4) Let $u = x^2+2$

$\frac{du}{dx} = 2x$

$\frac{1}{2}du = xdx$

$= \int e^{-u}\frac{1}{2}du$

$= -\frac{1}{2}e^{-u} + C$

$= -\frac{1}{2}e^{-x^2-2} + C$

(5) $= \frac{2}{3}\sqrt{(\tan^{-1}x)^3} + C$

(6) Let $u=\sin^{-1}(2x)$

$\dfrac{du}{dx}=\dfrac{1}{\sqrt{1-(2x)^2}}\cdot 2=\dfrac{2}{\sqrt{1-4x^2}}$

$\dfrac{1}{2}du=\dfrac{1}{\sqrt{1-4x^2}}dx$

$=\int u\cdot\dfrac{1}{2}du=\dfrac{1}{4}u^2+C=\dfrac{1}{4}(\sin^{-1}(2x))^2+C$

(7) $=\tan^{-1}(e^x)+C$

(8) Let $u=e^{2x}$

$\dfrac{du}{dx}=e^{2x}\cdot 2$

$\dfrac{1}{2}du=e^{2x}dx$

$=\int\dfrac{1}{1+u^2}\cdot\dfrac{1}{2}du=\tan^{-1}u+C=\dfrac{1}{2}\tan^{-1}(e^{2x})+C$

(9) $=\dfrac{1}{2}x^2-x+4\ln|x+1|+C$

(10) Let $u=e^x$

$\dfrac{du}{dx}=e^x$

$du=e^x dx$

$=\int\dfrac{e^x\cdot e^x}{e^x+1}dx=\int\dfrac{u}{u+1}du=\int\left(1+\dfrac{-1}{u+1}\right)du=u-\ln|u+1|+C=e^x-\ln(e^x+1)+C$

EX 4

(1) $=\dfrac{1}{2}\tan^{-1}(2x)+C$

(2) $=\sin^{-1}\left(\dfrac{x}{4}\right)+C$

(3) $=\tan^{-1}(x+1)+C$

(4) $=3\cdot\dfrac{1}{3}\tan^{-1}\left(\dfrac{x}{3}\right)+C=\tan^{-1}\left(\dfrac{x}{3}\right)+C$

(5) $=\sin^{-1}\left(\dfrac{x}{3}\right)+C$

(6) $=\int\dfrac{1}{4}\cdot\dfrac{1}{\left(\dfrac{3}{2}\right)^2+x^2}dx$ where $a=\dfrac{3}{2}$

$=\dfrac{1}{4}\cdot\dfrac{1}{3/2}\tan^{-1}\left(\dfrac{x}{3/2}\right)$

$=\dfrac{1}{6}\tan^{-1}\left(\dfrac{2x}{3}\right)+C$

(7) $=-\dfrac{1}{2}\sqrt{4x-x^2}+C$

(8) $=\dfrac{1}{3}\int\dfrac{e^x dx}{1+\left(\dfrac{e^x}{3}\right)^2}=3\cdot\dfrac{1}{3}\int\dfrac{\dfrac{e^x}{3}dx}{1+\left(\dfrac{e^x}{3}\right)^2}$

$=\tan^{-1}\dfrac{e^x}{3}+C$

(9) $=\ln(x^2+4)+\dfrac{1}{2}\tan^{-1}\dfrac{x}{2}+C$

(10) Let $u=4-x^2$

$\dfrac{du}{dx}=-2x$

$-\dfrac{1}{2}du=xdx$

$=\int\dfrac{x}{\sqrt{4-x^2}}dx+\int\dfrac{1}{\sqrt{4-x^2}}dx$

$=\int\dfrac{1}{\sqrt{u}}\cdot-\dfrac{1}{2}du+\sin^{-1}\left(\dfrac{x}{2}\right)+C$

$=-\sqrt{u}+\sin^{-1}\left(\dfrac{x}{2}\right)+C$

$=-\sqrt{4-x^2}+\sin^{-1}\left(\dfrac{x}{2}\right)+C$

(11) $=\int\dfrac{1}{x^2-2x+1+4}dx=\int\dfrac{1}{(x-1)^2+4}dx=\dfrac{1}{2}\tan^{-1}\left(\dfrac{x-1}{2}\right)+C$

EX 5

$f(x)=\int\dfrac{1}{\sqrt{9-4x^2}}dx$

$=\int\dfrac{1}{2}\cdot\dfrac{1}{\sqrt{9/4-x^2}}dx$

$=\dfrac{1}{2}\sin^{-1}\left(\dfrac{x}{3/2}\right)+C$

$=\dfrac{1}{2}\sin^{-1}\left(\dfrac{2x}{3}\right)+C$

$f\left(-\dfrac{3}{2}\right)=\dfrac{\pi}{12}$: $\dfrac{\pi}{12}=\dfrac{1}{2}\sin^{-1}(-1)+C$

$\dfrac{\pi}{12}=\dfrac{1}{2}\cdot\left(-\dfrac{\pi}{2}\right)+C$

$\dfrac{\pi}{12}+\dfrac{\pi}{4}=C$

$C=\dfrac{\pi}{3}$

$\therefore f(x)=\dfrac{1}{2}\sin^{-1}\left(\dfrac{2x}{3}\right)+\dfrac{\pi}{3}$

5.3 Integration by Parts :

EX 1

(1) $= xsinx + \cos x + C$

(2) $= -2xe^{-\frac{1}{2}x} - \int -2e^{-\frac{1}{2}x}dx$

$= -2xe^{-\frac{1}{2}} - 4e^{-\frac{1}{2}x} + C$

(3) $= \frac{1}{3}xsin3x - \int \frac{1}{3}\sin 3x\,dx$

$= \frac{1}{3}xsin3x + \frac{1}{9}\cos 3x + C$

(4) $= -\frac{1}{2}x\cos 2x - \int -\frac{1}{2}\cos 2x dx$

$= -\frac{1}{2}x\cos 2x + \frac{1}{4}\sin 2x + C$

(5) $= e^x(x^2 - 2x + 2) + C$

(6) $= xlnx - \int 1\,dx$

$= xlnx - x + C$

(7) $= \frac{x^4}{16}(4\ln x - 1) + C$

(8) $= -\frac{1}{x}\ln x - \int -\frac{1}{x^2}dx$

$= -\frac{\ln x}{x} - \frac{1}{x} + C$

(9) $= 2\sqrt{x}\ln x - \int \frac{1}{x} \cdot 2\sqrt{x}\,dx$

$= 2\sqrt{x}\ln x - \int \frac{2}{\sqrt{x}}dx$

$= 2\sqrt{x}\ln x - 4\sqrt{x} + C$

(10) $= xtan^{-1}x - \int \frac{x}{1+x^2}dx$

$= xtan^{-1}x - \frac{1}{2}\ln(1+x^2) + C$

(11) $= \frac{1}{2}x^2\sec^{-1}x - \int \frac{x}{2\sqrt{x^2-1}}dx$

$= \frac{1}{2}x^2 \cdot \sec^{-1}x - \int \frac{1}{2\sqrt{t}} \cdot \frac{1}{2}dt$

$= \frac{1}{2}x^2\sec^{-1}x - \frac{1}{2}\sqrt{t} + C$

$= \frac{1}{2}x^2\sec^{-1}x - \frac{1}{2}\sqrt{x^2-1} + C$

EX 2

(1) $= \frac{1}{2}x^2(\ln x)^2 - \int x\ln x dx$

$= \frac{1}{2}x^2(\ln x)^2 - \left[\frac{1}{2}x^2\ln x - \int \frac{1}{2}xdx\right]$

$= \frac{1}{2}x^2(\ln x)^2 - \frac{1}{2}x^2\ln x + \frac{1}{4}x^2 + C$

(2) $= -\frac{1}{x}(\ln x)^2 - \int -\frac{2(\ln x)}{x^2}dx$

$= -\frac{1}{x}(\ln x)^2 + 2\int \frac{\ln x}{x^2}dx$

$= -\frac{1}{x}(\ln x)^2 + 2\left[-\frac{1}{x}\ln x - \int -\frac{1}{x^2}dx\right]$

$= -\frac{1}{x}(\ln x)^2 + 2\left[-\frac{1}{x}\ln x - \frac{1}{x}\right]$

$= -\frac{1}{x}(\ln x)^2 - \frac{2}{x}\ln x - \frac{2}{x} + C$

EX 3

(1) $= -e^{2x}\cos x + 2\int e^{2x}\cos x dx$

$= -e^{2x}\cos x + 2\left[e^{2x}\sin x - \int 2e^{2x}\sin x dx\right]$

$= -e^{2x}\cos x + 2e^{2x}\sin x - 4\int e^{2x}\sin x dx$

Let $\int e^{2x}\sin x dx = I$

$I = -e^{2x}\cos x + 2e^{2x}\sin x - 4I$

$5I = -e^{2x}\cos x + 2e^{2x}\sin x$

$\therefore I = \frac{e^{2x}(-\cos x + 2\sin x)}{5} + C$

(2) $= \frac{1}{2}e^x\sin 2x - \frac{1}{2}\int e^x\sin 2x dx$

$= \frac{1}{2}e^x\sin 2x - \frac{1}{2}\left[-\frac{1}{2}e^x\cos 2x - \int -\frac{1}{2}e^x\cos 2x dx\right]$

$= \frac{1}{2}e^x\sin 2x + \frac{1}{4}e^x\cos 2x - \frac{1}{4}\int e^x\cos 2x dx$

Let $\int e^x\cos 2x = I$

$I = \frac{1}{2}e^x\sin 2x + \frac{1}{4}e^x\cos 2x - \frac{1}{4}I$

$4I = 2e^x\sin 2x + e^x\cos 2x - I$

$5I = e^x(2\sin 2x + \cos 2x)$

$\therefore I = \frac{e^x(2\sin 2x + \cos 2x)}{5} + C$

(3) $= \dfrac{e^{2x}}{13}(3\sin 3x + 2\cos 3x) + C$

(4) $= -\dfrac{e^{-2x}}{4}(\sin 2x + \cos 2x) + C$

EX 4

(1) $= x^2e^x - 2xe^x + 2e^x + C$

(2) $= e^{-x}(-x^3 - 3x^2 - 6x - 6) + C$

(3) $= e^{-\frac{x}{2}}(-2x^3 - 12x^2 - 48x - 96) + C$

(4) $= \dfrac{1}{3}x^2\sin 3x + \dfrac{2}{9}x\cos 3x - \dfrac{2}{27}\sin 3x + C$

(5)

$= -2x^3\cos\dfrac{1}{2}x + 12x^2\sin\dfrac{1}{2}x + 48x\cos\dfrac{1}{2}x - 96\sin\dfrac{1}{2}x + C$

EX 5

(1) Let $u = \sqrt{x}$, $u^2 = x$, $2u\dfrac{du}{dx} = 1$, $2u\,du = dx$

$$= \int e^u \cdot 2u\,du = \int 2ue^u du = 2u(e^u) - 2(e^u) + C = 2\sqrt{x}e^{\sqrt{x}} - 2e^{\sqrt{x}} + C$$

(2) Let $u = \sqrt{x}$, $u^2 = x$, $2u\,du = dx$

$$= \int \cos u \cdot 2u\,du = \int 2u \cdot \cos u\,du = 2u(\sin u) - 2(-\cos u) + C = 2\sqrt{x}\sin\sqrt{x} + 2\cos\sqrt{x} + C$$

(3) Let $u = x^2$, $du = 2x\,dx$, $\dfrac{1}{2}du = x\,dx$

$$= \int x^2 \cdot e^{-x^2} \cdot x\,dx = \int u \cdot e^{-u} \cdot \frac{1}{2}du = \int \frac{1}{2}ue^{-u}du = \frac{1}{2}u(-e^{-u}) - \frac{1}{2}(e^{-u}) + C = -\frac{1}{2}x^2e^{-x^2} - \frac{1}{2}e^{-x^2} + C$$

(4) Let $u = x^2$, $du = 2x\,dx$, $\dfrac{1}{2}du = x\,dx$

$$= \int x^2 \cdot \sin(x^2) \cdot x\,dx = \int u \cdot \sin u \cdot \frac{1}{2}du = \int \frac{1}{2}u\sin u\,du = \frac{1}{2}u(-\cos u) - \frac{1}{2}(-\sin u) + C = -\frac{1}{2}x^2\cos(x^2) + \frac{1}{2}\sin(x^2) + C$$

5.4 Rectangular Approximation Method (RAM) :

EX 1

(a) $L(4) = \dfrac{1}{2}\left[f(0) + f\left(\dfrac{1}{2}\right) + f(1) + f\left(\dfrac{3}{2}\right)\right] = \dfrac{1}{2}\left[4 + \dfrac{15}{4} + 3 + \dfrac{7}{4}\right] = \dfrac{25}{4}$ (or 6.25)

(b) $R(4) = \dfrac{1}{2}\left[f\left(\dfrac{1}{2}\right) + f(1) + f\left(\dfrac{3}{2}\right) + f(2)\right] = \dfrac{1}{2}\left[\dfrac{15}{4} + 3 + \dfrac{7}{4} + 0\right] = \dfrac{17}{4}$ (or 4.25)

(c) $M(4) = \dfrac{1}{2}\left[f\left(\dfrac{1}{4}\right) + f\left(\dfrac{3}{4}\right) + f\left(\dfrac{5}{4}\right) + f\left(\dfrac{7}{4}\right)\right] = \dfrac{1}{2}\left[\dfrac{63}{16} + \dfrac{55}{16} + \dfrac{39}{16} + \dfrac{15}{16}\right] = \dfrac{43}{8}$ (or 5.375)

5.5 Trapezoidal Approximation :

EX 1

make a table of values for $f(x) = \dfrac{120}{x}$:

x	1	2	3	4	5	6
$f(x)$	120	60	40	30	24	20

Then:

$L(5) = (120+60+40+30+24) \cdot 1 = 274;$

$R(5) = (60+40+30+24+20) \cdot 1 = 174;$

$T(5) = \dfrac{1}{2}(120+2 \cdot 60+2 \cdot 40+2 \cdot 30+2 \cdot 24+20) = 224.$

The calculator finds that $\int_1^6 \dfrac{120}{x}dx$ is approximately 215.011.

EX 2

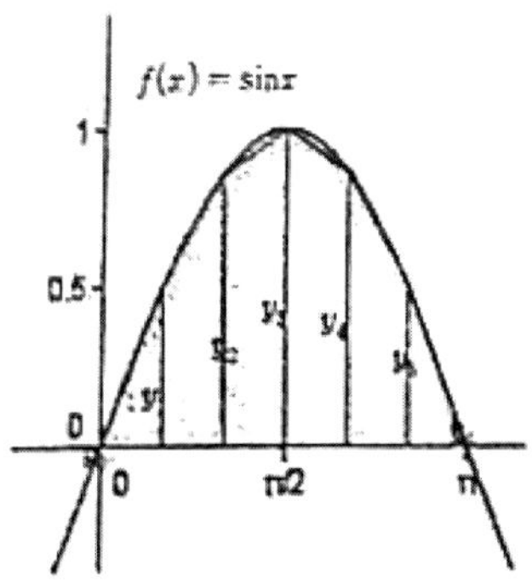

$T(6) = \dfrac{h}{2}[y_0 + 2y_1 + 2y_2 + 2y_3 + 2y_4 + 2y_5 + y_6]$

$= \dfrac{\frac{\pi}{6}}{2}\left[0+2 \cdot \dfrac{1}{2}+2 \cdot \dfrac{\sqrt{3}}{2}+2 \cdot 1+2\dfrac{\sqrt{3}}{2}+2 \cdot \dfrac{1}{2}+0\right]$

$= \dfrac{\pi}{12}[1+\sqrt{3}+2+\sqrt{3}+1]$

$= \dfrac{(2+\sqrt{3})\pi}{6}$

EX 3

(1) $L(4) = 1 \cdot [f(-1)+f(0)+f(1)+f(2)]$

$= 0+1+2+9 = 12$

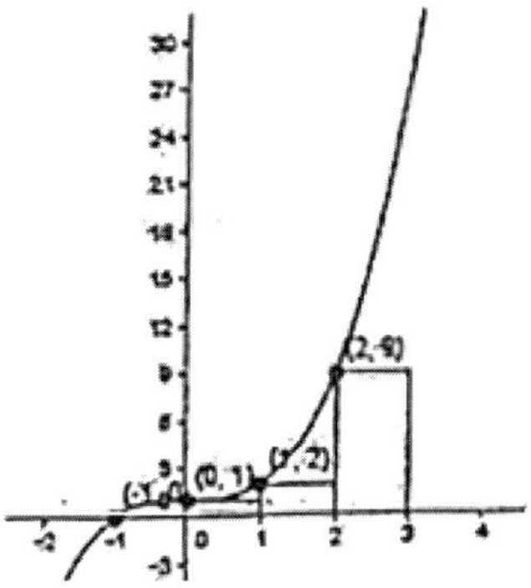

(2) $R(4) = 1 \cdot [f(0)+f(1)+f(2)+f(3)]$

$= 1+2+9+28 = 40$

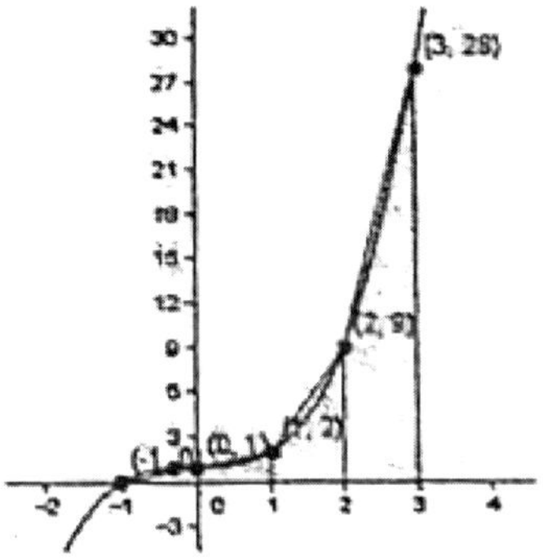

(3) $T(4)$

$= \dfrac{1}{2}[f(-1)+2f(0)+2f(1)+2f(2)+f(3)]$

$= \dfrac{1}{2}[0+2 \cdot 1+2 \cdot 2+2 \cdot 9+28]$

$= 26$

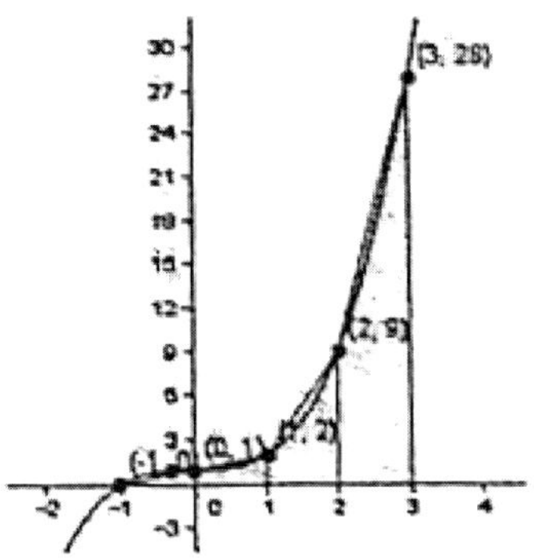

EX 4

(1) Any number of curves will do. The graph has only to pass through the points given in the table of speeds, as does the graph in Figure N6-12a.

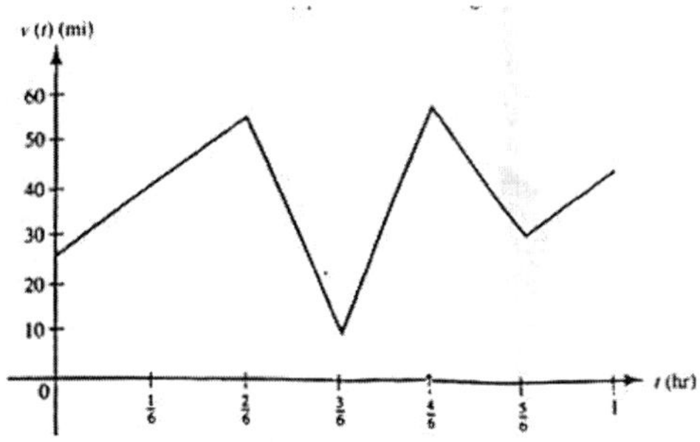

FIGURE N6-12a

(2) $L(6) = (26 + 40 + 55 + 10 + 60 + 32) \cdot \frac{1}{6} = 37\frac{1}{6}\, mi;$

$R(6) = (40 + 55 + 10 + 60 + 32 + 45) \cdot \frac{1}{6} = 40\frac{1}{3}\, mi;$

$T(6)$
$= \frac{1/6}{2}(26 + 2 \cdot 40 + 2 \cdot 55 + 2 \cdot 10 + 2 \cdot 60 + 2 \cdot 32 + 45)$
$= 38\frac{3}{4}\, mi.$

EX 5

(1) $L(5) = 1 \cdot 4 + 2 \cdot 6 + 3 \cdot 8 + 3 \cdot 9 + 1 \cdot 7$
$= 74\,L$

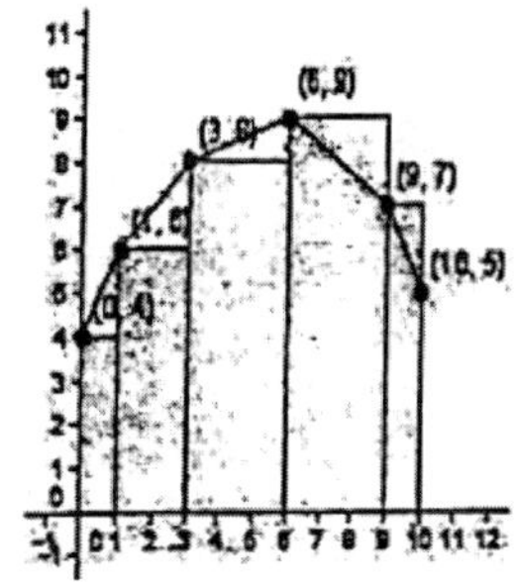

(2) $R(5) = 1 \cdot 6 + 2 \cdot 8 + 3 \cdot 9 + 3 \cdot 7 + 1 \cdot 5$
$= 75\,L$

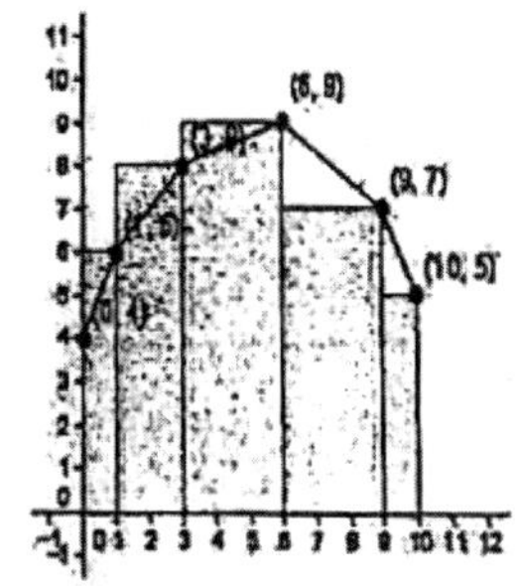

(3) $T(5)$

$= \frac{(4+6) \cdot 1}{2} + \frac{(6+8) \cdot 2}{2} + \frac{(8+9) \cdot 3}{2} + \frac{(9+7) \cdot 3}{2} + \frac{(7+5) \cdot 1}{2}$

$= 5 + 14 + \frac{51}{2} + 24 + 6$

$= 74.5\,L$

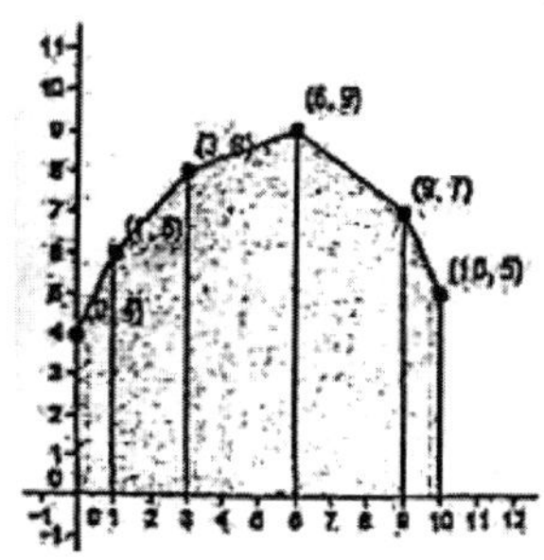

5.6 Definition of Area as a Limit

EX 1

$$y = -2x+3,\ [0,1]\left(note : \Delta x = \frac{1-0}{n} = \frac{1}{n}\right)$$

$$s(n) = \sum_{i=1}^{n} f\left(\frac{i}{n}\right)\left(\frac{1}{n}\right) = \sum_{i=1}^{n}\left[-2\left(\frac{i}{n}\right)+3\right]\left(\frac{1}{n}\right)$$

$$= 3 - \frac{2}{n^2}\sum_{i=1}^{n} i = 3 - \frac{2(n+1)n}{2n^2} = 3 - \frac{n+1}{n}$$

$$Area = \lim_{n\to\infty} s(n) = 2$$

EX 2

$$y = x^2+1,\ [0,3]\left(note : \Delta x = \frac{3-0}{n} = \frac{3}{n}\right)$$

$$S(n) = \sum_{i=1}^{n} f\left(\frac{3i}{n}\right)\left(\frac{3}{n}\right) = \sum_{i=1}^{n}\left[\left(\frac{3i}{n}\right)^2+1\right]\left(\frac{3}{n}\right)$$

$$= \frac{27}{n^3}\sum_{i=1}^{n} i^2 + \frac{3}{n}\sum_{i=1}^{n} 1$$

$$= \frac{27}{n^3}\,\frac{n(n+1)(2n+1)}{6} + \frac{3}{n}(n)$$

$$= \frac{9}{2}\,\frac{2n^2+3n+1}{n^2} + 3$$

$$Area = \lim_{n\to\infty} S(n) = \frac{9}{2}(2) + 3 = 12$$

EX 3

$$\Delta x = \frac{2-0}{n} = \frac{2}{n}$$

$$x_k = 0 + k\Delta x = \frac{2k}{n}$$

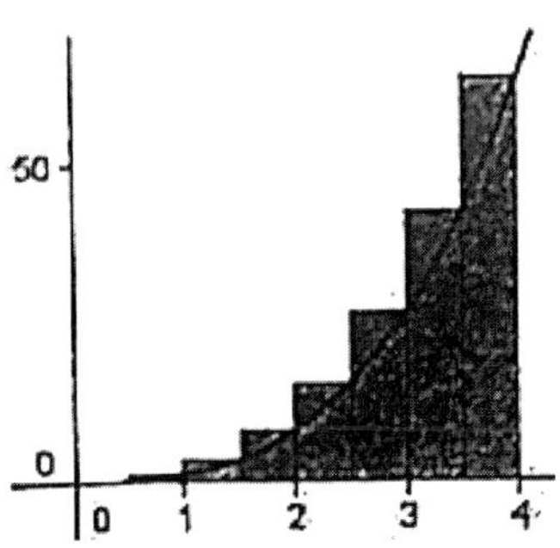

$Area$

$$= \lim_{n\to\infty}\sum_{k=1}^{n} f(x_k)\Delta x$$

$$= \lim_{n\to\infty}\sum_{k=1}^{n} f\left(\frac{2k}{n}\right)\frac{2}{n}$$

$$= \lim_{n\to\infty}\sum_{k=1}^{n} \frac{8k^3}{n^3}\cdot\frac{2}{n}$$

$$= \lim_{n\to\infty}\sum_{k=1}^{n} \frac{16k^3}{n^4}$$

$$= \lim_{n\to\infty}\left[\frac{16}{n^4}\sum_{k=1}^{n} k^3\right]$$

$$= \lim_{n\to\infty}\frac{16}{n^4}\cdot\left[\frac{n(n+1)}{2}\right]^2$$

$$= \lim_{n\to\infty}\frac{16}{n^4}\cdot\frac{n^2(n+1)^2}{4}$$

$$= \lim_{n\to\infty}\frac{4(n+1)^2}{n^2} = 4$$

EX 4

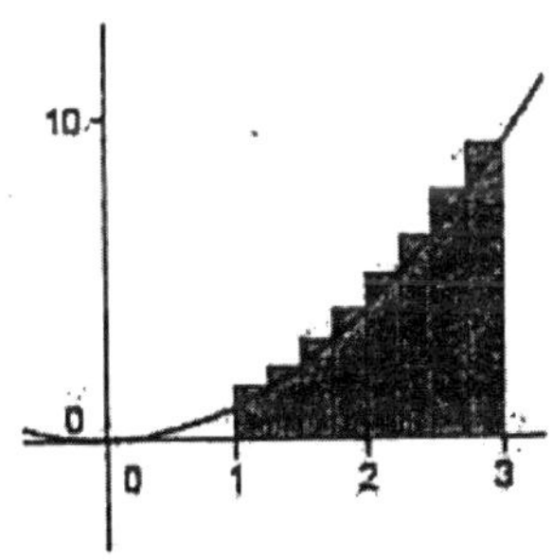

$$\Delta x = \frac{3-1}{n} = \frac{2}{n}$$

$$x_k = 1 + k\Delta x = 1 + \frac{2k}{n}$$

$Area$

$$= \lim_{n\to\infty}\sum_{k=1}^{n} f(x_k)\Delta x$$

$$= \lim_{n\to\infty}\sum_{k=1}^{n} f\left(1+\frac{2k}{n}\right)\frac{2}{n}$$

$$= \lim_{n\to\infty}\sum_{k=1}^{n}\left(1+\frac{2k}{n}\right)^2\frac{2}{n}$$

$$= \lim_{n\to\infty}\sum_{k=1}^{n}\left(1+\frac{4k}{n}+\frac{4k^2}{n^2}\right)\frac{2}{n}$$

$$= \lim_{n\to\infty}\sum_{k=1}^{n}\left(\frac{2}{n}+\frac{8k}{n^2}+\frac{8k^2}{n^3}\right)$$

$$= \lim_{n\to\infty}\left[\frac{2}{n}\sum_{k=1}^{n}1 + \frac{8}{n^2}\sum_{k=1}^{n}k + \frac{8}{n^3}\sum_{k=1}^{n}k^2\right]$$

$$= \lim_{n\to\infty}\left[\frac{2}{n}\cdot n + \frac{8}{n^2}\cdot\frac{n(n+1)}{2} + \frac{8}{n^3}\cdot\frac{n(n+1)(2n+1)}{6}\right]$$

$$= \lim_{n\to\infty}\left[2+\frac{4(n+1)}{n}+\frac{4(n+1)(2n+1)}{3n^2}\right]$$
$$= 2+4+\frac{8}{3}=\frac{26}{3}$$

EX 5

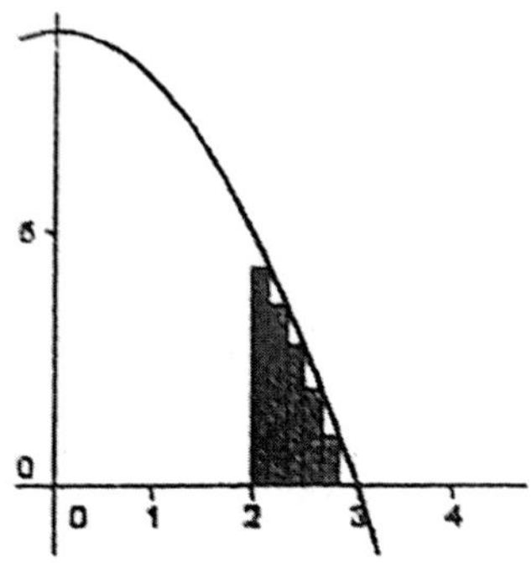

$$\Delta x=\frac{3-2}{n}=\frac{1}{n}$$
$$x_k=2+k\Delta x=2+\frac{k}{n}$$

Area

$$=\lim_{n\to\infty}\sum_{k=1}^{n} f(x_k)\Delta x$$
$$=\lim_{n\to\infty}\sum_{k=1}^{n} f\left(2+\frac{k}{n}\right)\frac{1}{n}$$
$$=\lim_{n\to\infty}\sum_{k=1}^{n}\left[9-\left(2+\frac{k}{n}\right)^2\right]\frac{1}{n}$$
$$=\lim_{n\to\infty}\sum_{k=1}^{n}\left(9-4-\frac{4k}{n}-\frac{k^2}{n^2}\right)\frac{1}{n}$$
$$=\lim_{n\to\infty}\sum_{k=1}^{n}\left(\frac{5}{n}-\frac{4k}{n^2}-\frac{k^2}{n^3}\right)$$

$$=\lim_{n\to\infty}\left[\frac{5}{n}\sum_{k=1}^{n}1-\frac{4}{n^2}\sum_{k=1}^{n}k-\frac{1}{n^3}\sum_{k=1}^{n}k^2\right]$$
$$=\lim_{n\to\infty}\left[\frac{5}{n}\cdot n-\frac{4}{n^2}\cdot\frac{n(n+1)}{2}-\frac{1}{n^3}\cdot\frac{n(n+1)(2n+1)}{6}\right]$$
$$=\lim_{n\to\infty}\left[5-\frac{2(n+1)}{n}-\frac{(n+1)(2n+1)}{6n^2}\right]$$
$$=5-2-\frac{1}{3}=\frac{8}{3}$$

5.7 Definite Integrals

EX 1

(1) $\int_1^9 -2f(x)dx = -2\int_1^9 f(x)dx$

$= -2(-1)=2$

(2) $\int_7^9 [f(x)+h(x)]dx$

$=\int_7^9 f(x)\,dx+\int_7^9 h(x)\,dx = 5+4 = 9$

(3) $\int_7^9 [2f(x)-3h(x)]dx$

$=2\int_7^9 f(x)dx-3\int_7^9 h(x)\,dx = 2(5)-3(4)$

$=-2$

(4) $\int_9^1 f(x)\,dx = -\int_1^9 f(x)\,dx = -(-1)=1$

(5) $\int_1^7 f(x)\,dx = \int_1^9 f(x)\,dx - \int_7^9 f(x)\,dx$

$= -1-5 = -6$

(6) $\int_9^7 [h(x)-f(x)]\,dx = \int_7^9 [f(x)-h(x)]dx$

$=\int_7^9 f(x)\,dx - \int_7^9 h(x)\,dx = 5-4=1$

EX 2

(1) $=\int_{-1}^2 f(x)dx$

$=2$

(2) $= -\int_2^4 f(x)dx$

$=-3$

(3) $= -\left[\int_{-1}^2 f(x)dx+\int_2^4 f(x)dx\right]$

$= -(2+3)$

$= -5$

(4) 0

(5) $= 3\int_{-1}^{2} f(x)dx + 2\int_{-1}^{2} g(x)dx$
$= 3(2) + 2(-1)$
$= 4$

(6)

$= 2\int_{-1}^{4} f(x)dx - 3\int_{-1}^{4} g(x)dx$
$= 2\left[\int_{-1}^{2} f(x)dx + \int_{2}^{4} f(x)dx\right] - 3\left[\int_{-1}^{2} g(x)dx + \int_{2}^{4} g(x)dx\right]$
$= 2(2+3) - 3[-1+(-4)]$
$= 10 + 15 = 25$

EX 3

(1) 0

(2) $= -\int_{a}^{c} e^x \cos x\, dx$
$= -\left[\int_{a}^{b} e^x \cos x\, dx + \int_{b}^{c} e^x \cos x\, dx\right]$
$= -(r+s) = -r-s$

(3) $= -\int_{c}^{a} x^2 \ln(x+1)dx$
$= -p$

(4) $= 2\int_{a}^{c} e^x \cos x dx - 3\int_{a}^{c} x^2 \ln(x+1)dx$
$-2(r+s) - 3(-p)$
$= 2r + 2s + 3p$

EX 4

(1) $\int_{-2}^{5} f(x)dx$
$= \int_{-2}^{4} f(x)dx + \int_{4}^{2} f(x)dx + \int_{2}^{5} f(x)dx$
$= 3 + (-2) + 6$
$= 7$

(2) $\int_{-1}^{3} f(x)dx$
$= \int_{-1}^{6} f(x)dx + \int_{6}^{-3} f(x)dx + \int_{-3}^{3} f(x)dx$
$= 1 + 4 + 2$
$= 7$

5.8 Fundamental Theorem of Calculus, Part Ⅰ :

EX 1

(1) $-\frac{4}{3}$

(2) $= \left[4x + 4x^2 - 2x^3\right]_0^2$
$= 8$

(3) $= \frac{25}{3}$

(4) $= \int_{-1}^{8} x^{-\frac{1}{3}} dx$
$= \left[\frac{3}{2}x^{\frac{2}{3}}\right]_{-1}^{8} = \frac{9}{2}$

(5) $= 1 - \frac{1}{3}\ln 2$

(6) $= \int_{2}^{1} (9x^2 + 12 + \frac{4}{x^2})dx$
$= \left[3x^3 + 12x - \frac{4}{x}\right]_2^1 = -35$

(7) $= 1$

(8) $= \left[\ln|\sec\theta| - \sec\theta\right]_0^{\frac{\pi}{3}}$
$= \ln 2 - 1$

EX 2

(1) $= 2$

(2) $= \left[2\ln(x^2+1)\right]_0^1$
$= \ln 4$

(3) $= \frac{8\pi}{3}$

(4) $= \left[3\tan^{-1}x\right]_{-1}^{\sqrt{3}}$
$= 3\tan^{-1}(\sqrt{3}) - 3\tan^{-1}(-1)$
$= 3 \cdot \frac{\pi}{3} - 3 \cdot \left(-\frac{\pi}{4}\right)$
$= \frac{7}{4}\pi$

(5) $= \frac{\pi}{6}$

(6) $= \left[\frac{1}{3}\tan^{-1}\left(\frac{x}{3}\right)\right]_0^{\sqrt{3}}$
$= \frac{1}{3}\tan^{-1}\left(\frac{\sqrt{3}}{3}\right) - \frac{1}{3}\tan^{-1}(0)$
$= \frac{\pi}{18}$

(7) $= \frac{\pi}{8} - \frac{1}{4}$

(8) $= \int_{\frac{\pi}{12}}^{\frac{\pi}{6}} \frac{1+\cos 4\theta}{2} d\theta$
$= \int_{\frac{\pi}{12}}^{\frac{\pi}{6}} \left(\frac{1}{2} + \frac{1}{2}\cos 4\theta\right) d\theta$
$= \left[\frac{1}{2}\theta + \frac{1}{8}\sin 4\theta\right]_{\frac{\pi}{12}}^{\frac{\pi}{6}}$
$= \frac{\pi}{24}$

EX 3

(1) 5.333
(2) 6.714
(3) 8.524
(4) 8.085

5.9 Fundamental Theorem of Calculus, Part Ⅱ :

EX 1

(1) $= e^{-x}\sec^2 x$
(2) $= \sqrt{1+x^2}$
(3) $= x^4 + \frac{3}{\sqrt{1-x^2}}$
(4) $= (x-3)\ln|x+1|$

EX 2

(1) $\sqrt{2+\cos^3 x}$

(2) $= \frac{d}{dx}\int_e^x \frac{\ln(u)}{u} du$
$= \frac{\ln x}{x}$

(3) $= \frac{6}{3+x^4}$

(4) $= \frac{d}{dx}\int_o^x u\cos u du$
$= x\cos x$

EX 3

(1) $= -\frac{\sec^2 x}{1+\tan^2 x}$

(2) $\frac{d}{dx}\int_{2x}^{x^2} f(t)dt$
$= f(x^2) \cdot 2x - f(2x) \cdot 2$
$= 2xf(x^2) - 2f(2x)$

(3) $= \frac{1}{2\sqrt{x}} \cos \sqrt{x}$

(4) $\frac{d}{dx}\int_{-x}^{x^2} \sqrt{t^2+1}\,dt$
$= \sqrt{(x^2)^2+1} \cdot 2x - \sqrt{(-x)^2+1} \cdot (-1)$
$= 2x\sqrt{x^4+1} + \sqrt{x^2+1}$

(5) $= \dfrac{\cos x}{1-\sin^2 x} + \dfrac{\sin x}{1-\cos^2 x}$

(6) $\dfrac{d}{dx}\displaystyle\int_{\frac{1}{x}}^{3x} \cos(2t)\,dt$

$= \cos(2(3x)) \cdot 3 - \cos\left(2\left(\dfrac{1}{x}\right)\right) \cdot \left(-\dfrac{1}{x^2}\right)$

$= 3\cos 6x + \dfrac{1}{x^2}\cos\left(\dfrac{2}{x}\right)$

EX 4

(1) $= \dfrac{2e^{x^2}}{x} - \dfrac{e^{\sqrt{x}}}{2x}$

(2) $\dfrac{dy}{dx} = \ln(\ln x) \cdot \dfrac{1}{x} - \ln(x+1) \cdot 1$

$= \dfrac{\ln(\ln x)}{x} - \ln(x+1)$

(3) $= \dfrac{\sec^2 x}{1-\tan^2 x} - \dfrac{3}{1-9x^2}$

(4) $\sin^2 2x \Rightarrow (\sin 2x)^2, \quad \cos^2 x \Rightarrow (\cos x)^2$

$\dfrac{dy}{dx} = \cos^2(\sin^2 2x) \cdot 2(\sin 2x) \cdot \cos 2x$

$\cdot\, 2 - \cos^2(\cos^2 x) \cdot 2(\cos x) \cdot (-\sin x)$

$= 4\cos^2(\sin^2 2x)\sin 2x \cos 2x + 2\cos^2(\cos^2 x)\cos x \sin x$

EX 5

(1) Differentiate each side with respect to x.

$\dfrac{d}{dx}\displaystyle\int_2^x f(t)\,dt = \dfrac{d}{dx}(x^2 \sin 3x)$

$f(x) = 2x \cdot \sin 3x + x^2 \cdot \cos 3x \cdot 3$

$f(\pi) = 2\pi \cdot \sin 3\pi + \pi^2 \cdot \cos 3\pi \cdot 3$

$= -3\pi^2$

(2) $A(2) = \displaystyle\int_4^4 \dfrac{1}{1+t^2}\,dt$ Differentiate each side

$\therefore A(2) = 0$

with respect to x.

$A'(x) = \dfrac{d}{dx}\displaystyle\int_4^{x^2} \dfrac{1}{1+t^2}\,dt$

$= \dfrac{1}{1+(x^2)^2} \cdot 2x = \dfrac{2x}{1+x^4}$

$\therefore A'(2) = \dfrac{4}{17}$

EX 6

x		0	
y''	+	0	−

$\therefore f$ is concave up on $x < 0$

$f'(x) = \dfrac{d}{dx}\displaystyle\int_1^x \dfrac{1}{t^4+1}\,dt = \dfrac{1}{x^4+1} = (x^4+1)^{-1}$

$f''(x) = -(x^4+1)^{-2} \cdot 4x^3$

$= -\dfrac{4x^3}{(x^4+1)^2}$

$f''(x) = 0$ at $x = 0$

5.10 Area between a Curve and the x -axis :

EX 1

(1) $Area = 3$

(2)

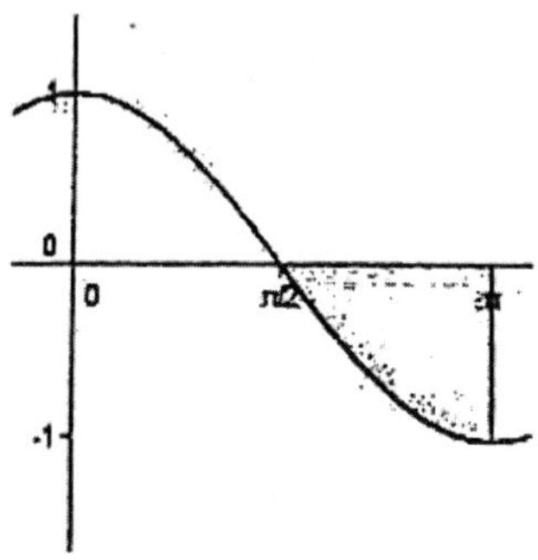

$$Area = -\int_{\frac{\pi}{2}}^{\pi} \cos x dx$$
$$= -[\sin x]_{\frac{\pi}{2}}^{\pi}$$
$$= -\left(\sin\pi - \sin\frac{\pi}{2}\right)$$
$$= -(0-1) = 1$$

(3) $Area = \frac{1}{2}\ln 2$

(4)

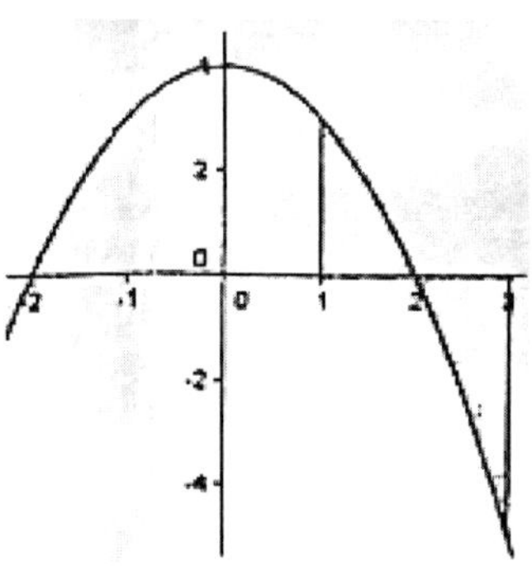

$Area$
$$= \int_1^2 (4-x^2)dx - \int_2^3 (4-x^2)dx$$
$$= \left[4x - \frac{1}{3}x^3\right]_1^2 - \left[4x - \frac{1}{3}x^3\right]_2^3$$
$$= \frac{5}{3} + \frac{7}{3} = 4$$

(5)

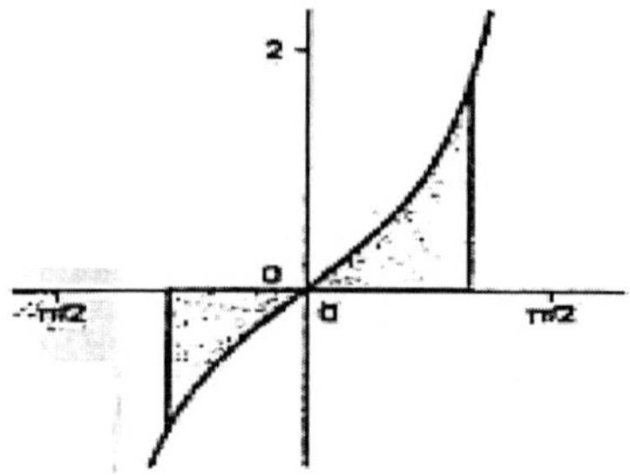

$Area$
$$= -\int_{-\frac{\pi}{4}}^{0} \tan x dx + \int_0^{\frac{\pi}{3}} \tan x dx$$
$$= -[\ln|\sec x|]_{-\frac{\pi}{4}}^{0} + [\ln|\sec x|]_0^{\frac{\pi}{3}}$$
$$= \ln\sqrt{2} + \ln 2 = \ln(2\sqrt{2})$$

(6)

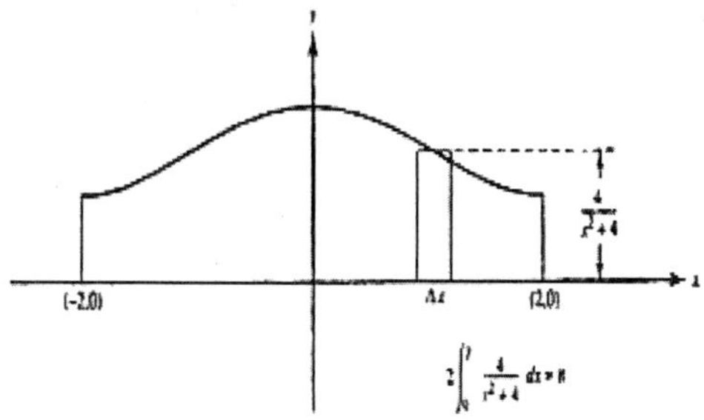

$$Area = 2\int_0^2 \frac{4}{x^2+4}dx = \pi$$

(7)

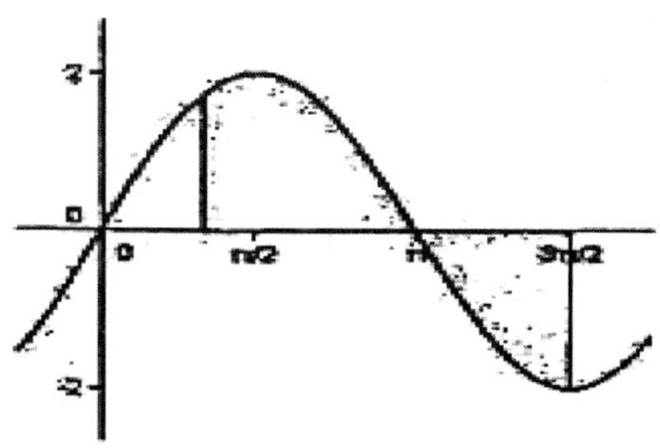

$$Area = \int_{\frac{\pi}{3}}^{\pi} 2\sin x dx - \int_{\pi}^{\frac{3}{2}\pi} 2\sin x dx$$
$$= -2[\cos x]_{\frac{\pi}{3}}^{\pi} + 2[\cos x]_{\pi}^{\frac{3}{2}\pi}$$
$$= -2\left[-1-\left(\frac{1}{2}\right)\right] + 2[0-(-1)]$$
$$= 3 + 2 = 5$$

(8)

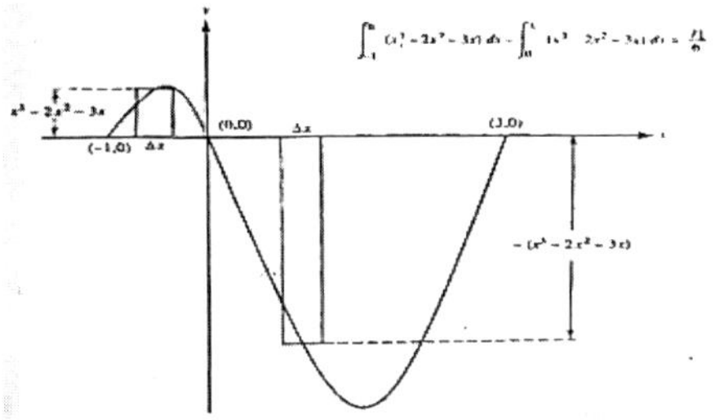

$$Area = \int_{-1}^{0}(x^3-2x^2-3x)dx$$

$$-\int_{0}^{3}(x^3-2x^2-3x)dx$$

$$=\frac{71}{6}$$

EX 2

$$=\int_{-1}^{1}f(x)dx+\int_{1}^{3}f(x)dx$$

$$=\int_{-1}^{1}(2x+1)dx+\int_{1}^{3}(x^2+2)dx$$

$$=[x^2+x]_{-1}^{1}+\left[\frac{1}{3}x^3+2x\right]_{1}^{3}=\frac{44}{3}$$

EX 3

(1) $\frac{23}{3}$

(2)

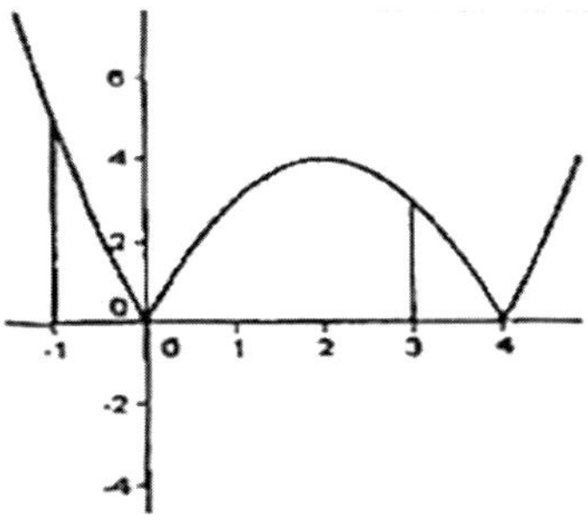

$$=\int_{-1}^{0}(x^2-4x)dx+\int_{0}^{3}(-x^2+4x)dx$$

$$=\left[\frac{1}{3}x^3-2x^2\right]_{-1}^{0}+\left[-\frac{1}{3}x^3+2x^2\right]_{0}^{3}$$

$$=\frac{7}{3}+9=\frac{34}{3}$$

(3) 4

(4)

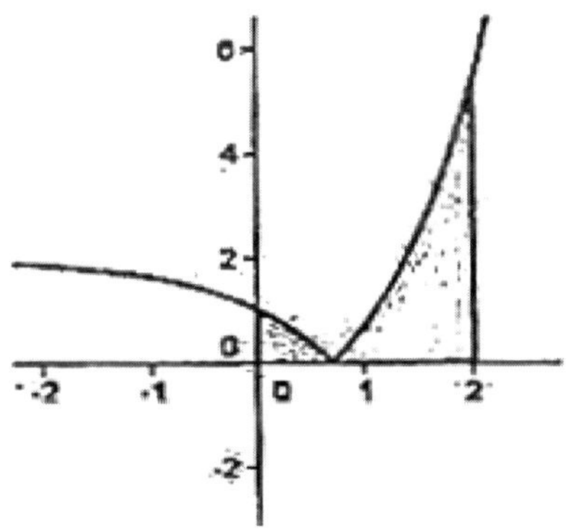

$e^x-2=0$
$e^x=2,\ x=\ln 2$

$$=\int_{0}^{\ln 2}(-e^x+2)dx+\int_{\ln 2}^{2}(e^x-2)dx$$

$$=[-e^x+2x]_{0}^{\ln 2}+[e^x-2x]_{\ln 2}^{2}$$

$$=e^2+4\ln 2-7$$

EX 4

(1) $= -10$

(2) $= 10$

(3) $= -10+10-5$
$= -5$

(4) $= 10+10+5$
$= 25$

(5) $\int_b^a f(x)dx = -\int_a^b f(x)dx = -10$

(6) $\int_d^b f(x)dx = -\int_b^d f(x)dx$

$= -\left[\int_b^c f(x)dx + \int_c^d f(x)dx\right]$

$= -[10-5] = -5$

EX 5

(1) π

(2)

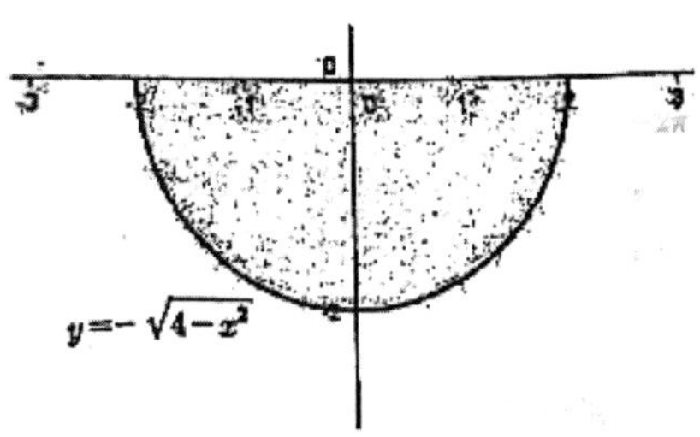

$= -\frac{1}{2}\pi(2)^2$

$= -2\pi$

(3) $30-\frac{9}{2}\pi$

(4)

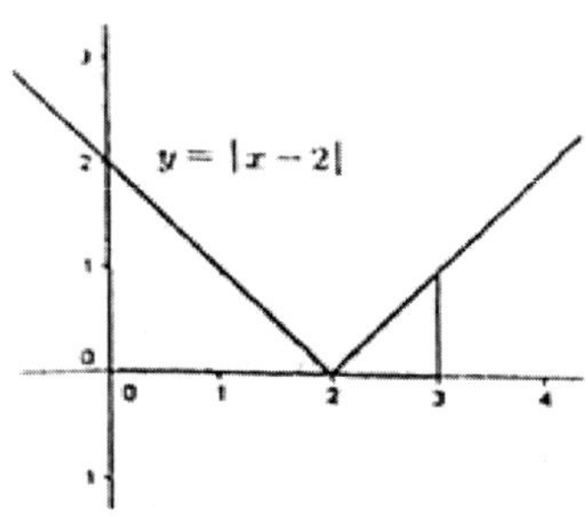

$= \frac{2 \cdot 2}{2} + \frac{1 \cdot 1}{2}$

$= \frac{5}{2}$

(5) 7

(6)

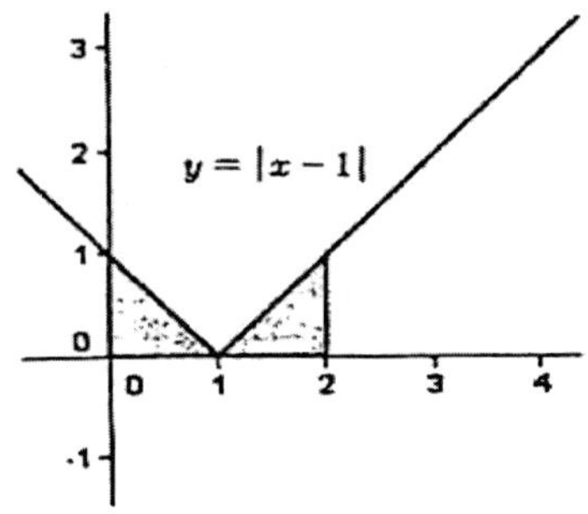

$= \int_0^2 |x-1|dx + \int_0^2 3x^2 dx$

$= 1 + [x^3]_0^2$

$= 9$

EX 6

(1) 0

(2) $= \int_1^3 f(x)dx = \int_3^5 f(x)dx$

$= 2 - \frac{1}{2}\pi(1)^2 = 2 - \frac{\pi}{2}$

(3) $\frac{\pi}{2} - 3$

(4) $= \int_{-3}^{-1} f(x)dx + \int_{-1}^{0} f(x)dx$

$= -(4-\pi) + (-1) = \pi - 5$

(5) $2 + \frac{\pi}{2}$

(6) Let $u = x-3$

$\frac{du}{dx} = 1$

$du = dx$

$if\ x = 0,\ u = -3$

$if\ x = 8,\ u = 5$

$= \int_{-3}^{5} f(u)du = \int_{-3}^{5} f(x)dx$

$= \int_{-3}^{0} f(x)dx + \int_{0}^{5} f(x)dx$

$= \pi - 5 + \left(3 - \frac{\pi}{2}\right)$

$= \frac{\pi}{2} - 2$

(7) 4

(8) Let $u = 2x$

$\frac{du}{dx} = 2$

$\frac{1}{2}du = dx$

$if\ x = 1,\ u = 2$

$if\ x = 2,\ u = 4$

$$= \int_2^4 f(u)\frac{1}{2}du = \frac{1}{2}\int_2^4 f(x)dx = \frac{1}{2}\left[\int_2^3 f(x)dx + \int_3^4 f(x)dx\right] = \frac{1}{2}\left[\frac{1}{2} - \frac{\pi}{4}\right] = \frac{1}{4} - \frac{\pi}{8}$$

EX 7

(1) $g(3) = \int_1^3 f(t)dt = 2$

$$g(-3) = \int_1^{-3} f(x)dx = -\int_{-3}^1 f(x)dx$$

$$= -\left[\int_{-3}^{-1} f(x)dx + \int_{-1}^0 f(x)dx + \int_0^1 f(x)dx\right]$$

$$= -\left[-\frac{1}{4}\pi(2)^2 - 1 + 1\right] = \pi$$

$$g(x) = \int_1^x f(t)dt \Rightarrow g'(x) = f(x)$$

$$\Rightarrow g''(x) = f'(x)$$

$g'(3) = f(3) = 0,$

$g''(3) = f'(3) = -1$

(2) $\frac{g(3) - g(-3)}{3 - (-3)} = \frac{2 - \pi}{6}$

(3) $\frac{g'(3)\quad g'(-1)}{3 - (-1)} = \frac{f(3) - f(-1)}{4}$

$= \frac{0 - (-2)}{4} = \frac{1}{2}$

(4) $g(-1) = \int_1^{-1} f(t)dt = 0 \Rightarrow (-1, 0)$

$g'(-1) = f(-1) = -2$

$\therefore y - 0 = -2(x + 1)$

(5) $g(x) \approx -2(x+1)\ near\ x = -1$

$g(-1.1) \simeq -2(-0.1) = 0.2$

(6) $g\ is\ decreasing\ when\ g' = f < 0.$

$\therefore\ -4 < x < 0,\ 3 < x < 5$

(7) $g\ is\ concave\ up\ when\ g'' = f' > 0.$

$\therefore -1 < x < 1,\ 4 < x < 5$

(8) g has a relative maximum when

$g' = f$ changes sign from + to −

$\therefore x = 3$

(9) g has a point of inflection when

$g'' = f'$ changes sign.

$\therefore x = -1, 1, 4$

(10)

x	-3		0		3		5
$g' = f$	0	−	0	+	0	−	0
g		↘		↗		↘	

$g(0)$ and $g(5)$ are the candidates for the absolute minimum.

$$g(0) = \int_1^0 f(t)dt = -\int_0^1 f(t)dt = -1$$

$$g(5) = \int_1^5 f(t)dt = \int_1^3 f(t)dt + \int_3^5 f(t)dt = 2 - 1 = 1$$

$g(0) < g(5)$

$\therefore$ absolute minimum $= g(0) = -1$

5.11 Substitution for Definite Intergrals :

EX 1

(1) 4

(2) $u=1+t^4$

$\Rightarrow du = 4t^3\,dt \Rightarrow \frac{1}{4}du = t^3\,dt; t = 0$

$\Rightarrow u = 1, t = 1 \Rightarrow u = 2$

$$\int_0^1 t^3(1+t^4)\,dt = \int_1^2 \frac{1}{4}u^3 du$$
$$= \left[\frac{u^4}{16}\right]_1^2 = \frac{2^4}{16} - \frac{1^4}{16} = \frac{15}{16}$$

(3) $4\tan^{-1}\frac{\pi}{4}$

(4) Let $u = 4+r^2$

$\Rightarrow du = 2r\,dr \Rightarrow \frac{1}{2}du = r\,dr; r = -1$

$\Rightarrow u = 5, r = 1 \Rightarrow u = 5$

$$\int_{-1}^1 \frac{5r}{(4+r^2)}dr = 5\int_5^5 \frac{1}{2}u^{-2}du = 0$$

(5) $\frac{\pi}{3}$

(6) Let $u = \ln x \Rightarrow du = \frac{1}{x}dx; x = 2$

$\Rightarrow u = \ln 2$ and $x = 4 \Rightarrow u = \ln 4;$

$$\int_2^4 \frac{dx}{x(\ln x)^2} = \int_{\ln 2}^{\ln 4} u^{-2}du = \left[-\frac{1}{u}\right]_{\ln 2}^{\ln 4}$$
$$= -\frac{1}{\ln 4} + \frac{1}{\ln 2} = -\frac{1}{2\ln 2} + \frac{1}{\ln 2}$$
$$= \frac{1}{2\ln 2} = \frac{1}{\ln 4}$$

(7) $3^{5/2}-1$

(8) Let $u = \tan\theta \Rightarrow du = \sec^2\theta\,d\theta; \theta = 0$

$\Rightarrow u = 0, \theta = \frac{\pi}{4} \Rightarrow u = 1;$

$$\int_0^{\pi/4} (1+e^{\tan\theta})\sec^2\theta\,d\theta$$
$$= \int_0^{\pi/4} \sec^2\theta\,d\theta + \int_0^1 e^u du$$
$$= [\tan\theta]_0^{\pi/4} + [e^u]_0^1$$
$$= \left[\tan\left(\frac{\pi}{4}\right) - \tan(0)\right] + (e^1 - e^0)$$
$$= (1-0)+(e-1) = e$$

(9) 0

(10) Let $u = \cos 2\theta \Rightarrow du = -2\sin 2\theta\,d\theta$

$\Rightarrow -\frac{1}{2}du = \sin 2\theta\,d\theta; \theta = 0$

$\Rightarrow u = 1, \theta = \frac{\pi}{6} \Rightarrow u = \cos 2\left(\frac{\pi}{6}\right) = \frac{1}{2}$

$$\int_0^{\pi/6} \cos^{-3}2\theta \sin 2\theta\,d\theta = \int_1^{1/2} u^{-3}\left(-\frac{1}{2}du\right)$$
$$= -\frac{1}{2}\int_1^{1/2} u^{-3}du = \left[-\frac{1}{2}\left(\frac{u^{-2}}{-2}\right)\right]_1^{1/2}$$
$$= \frac{1}{4\left(\frac{1}{2}\right)^2} - \frac{1}{4(1)^2} = \frac{3}{4}$$

(11) π

(12) $\displaystyle\int_{\sqrt{2}}^2 \frac{\sec^2(\sec^{-1}x)}{x\sqrt{x^2-1}}dx$

$$= \int_{\pi/4}^{\pi/3} \sec^2 u\,du, \text{ where } u$$
$$= \sec^{-1}x \text{ and } du = \frac{dx}{x\sqrt{x^2-1}}; x$$
$$= \sqrt{2} \Rightarrow u = \frac{\pi}{4}, x = 2 \Rightarrow u = \frac{\pi}{3}$$
$$= [\tan u]_{\pi/4}^{\pi/3} = \tan\frac{\pi}{3} - \tan\frac{\pi}{4} = \sqrt{3} - 1$$

(13) $\sqrt{\ln 2}$

(14) Let $u = \tan x \Rightarrow du = \sec^2 x\,dx; x = 0$

$\Rightarrow u = 0, x = \frac{\pi}{4} \Rightarrow u = 1$

$$\int_0^{\pi/4} \tan x \sec^2 x\,dx$$
$$= \int_0^1 u\,du = \left[\frac{u^2}{2}\right]^1 = \frac{1^2}{2} - 0 = \frac{1}{2}$$

EX 2

(1) $f'(x) = g(x) \Rightarrow \int g(x)dx = f(x)+C$

$= [f(x)+x^3]_{-1}^{1}$

$= [f(1)+1]-[f(-1)-1]$

$= 6-(-1) = 7$

(2)

Let $u = 2x$

$du = 2dx$

$\frac{1}{2}du = dx$

if $x = -2, u = -4$

if $x = 2, \quad u = 4$

$= \int_{-4}^{4} g(u) \cdot \frac{1}{2}du = \int_{-4}^{4} \frac{1}{2}g(x)dx$

$= \left[\frac{1}{2}f(x)\right]_{-4}^{4}$

$= \frac{1}{2}(f(4)-f(-4))$

$= \frac{1}{2}(2-(-3)) = \frac{5}{2}$

(3) Let $u = \frac{x}{2}$

$du = \frac{1}{2}dx$

if $x = -2, u = -1$

if $x = 2, \quad u = 1$

$= \int_{-1}^{1} 3g(u) \cdot 2du$

$= \int_{-1}^{1} 6g(x)dx$

$= [6f(x)]_{-1}^{1}$

$= 6[f(1)-f(-1)]$

$= 6(5-0)$

$= 30$

(4) Let $u = x^2$

$du = 2xdx$

$\frac{1}{2}du = x\,dx$

if $x = -2, u = 4$

if $x = 0, \quad u = 0$

$= \int_{4}^{0} g(u) \cdot \frac{1}{2}du$

$= \int_{4}^{0} \frac{1}{2}g(x)dx$

$= \left[\frac{1}{2}f(x)\right]_{4}^{0}$

$= \frac{1}{2}[f(0)-f(4)]$

$= \frac{1}{2}[-2-2] = -2$

5.12 Integration by Parts for Definite Intergrals :

EX 1

(1) $= \frac{2}{e}$

(2) $= [x(-\cos x)-1(-\sin x)]_{\frac{\pi}{3}}^{\frac{\pi}{2}}$

$= [-x\cos x+\sin x]_{\frac{\pi}{3}}^{\frac{\pi}{2}}$

$= \frac{\pi}{6}-\frac{\sqrt{3}}{2}+1$

(3) $= \frac{\pi^2-4}{8}$

(4) Let $u = \ln x, \; v' = x$

$u' = \frac{1}{x}, \; v = \frac{1}{2}x^2$

$= \left[\frac{1}{2}x^2\ln x\right]_1^e - \int_1^e \frac{1}{2}xdx$

$= \left[\frac{1}{2}x^2\ln x\right]_1^e - \left[\frac{1}{4}x^2\right]_1^e$

$= \frac{1}{4}e^2+\frac{1}{4}$

(5) 1

(6) Let $u = \tan^{-1}x, \; v' = 1$

$u' = \frac{1}{1+x^2}, \; v = x$

$= [x\tan^{-1}x]_0^1 - \int_0^1 \frac{x}{1+x^2}dx$

$= [x\tan^{-1}x]_0^1 - \left[\frac{1}{2}\ln(1+x^2)\right]_0^1$

$= \frac{\pi}{4}-\ln\sqrt{2}$

(7) $= \frac{5}{12}\pi-\frac{\sqrt{3}}{2}$

(8) Let $u = \sqrt{x}$

$u^2 = x$

$2udu = dx$

if $x = 0, u = o$

if $x = 4, u = 2$

$= \int_0^2 e^u \cdot 2udu$

$= \int_0^2 2ue^u du$

$= \left[2u(e^u) - 2(e^u)\right]_0^2$

$= \left[e^u(2u-2)\right]_0^2$

$= 2e^2 + 2$

EX 2

(1) $f'(e) = e^2$

An equation for the line tangent to the graph of f at the point (e,2) is y-2= $e^2(x-e)$.

(2) $f''(x) = x + 2xln\,x$.

For $1 < x < 3, x > 0$ and ln $x > 0$, so $f''(x) > 0$. Thus, the graph of f is concave up on (1,3).

(3) Since $f(x) = \int (x^2\ln x)\,dx$, we consider integration by parts.

$u = \ln x \qquad dv = x^2\,dx$

$du = \frac{1}{x}dx \qquad v = \int (x^2)\,dx = \frac{1}{3}x^3$

Therefore,

$f(x) = \int (x^2 \ln x)\,dx$

$= \frac{1}{3}x^3 \ln x - \int \left(\frac{1}{3}x^3 \cdot \frac{1}{x}\right) dx$

$= \frac{1}{3}x^3 \ln x - \frac{1}{9}x^3 + C.$

Since $f(e) = 2$, $2 = \frac{e^3}{3} - \frac{e^3}{9} + C$ and

$C = 2 - \frac{2}{9}e^3$.

Thus, $f(x) = \frac{x^3}{3}\ln x - \frac{1}{9}x^3 + 2 - \frac{2}{9}e^3$.

EX 3

(1)

$f'(x) < 0$ for $0 < x < 3$ and $f'(x) > 0$ for $x > 3$

Therefore, f has a relative minimum at $x = 3$.

(2) $f''(x) = e^x + (x-3)e^2 = (x-2)e^x$

$f''(x) > 0$ for $x > 2$

$f'(x) < 0$ for $0 < x < 3$

Therefore, the graph of f is both decreasing and concave up on the interval $2 < x < 3$.

(3) $f(3) = f(1) + \int_1^3 f'(x)\,dx$

$= 7 + \int_1^3 (x-3)e^x\,dx$

$u = x - 3 \qquad dv = e^x\,dx$

$du = dx \qquad v = e^x$

$f(3) = 7 + (x-3)e^x|_1^3 - \int_1^3 e^x\,dx$

$= 7 + |(x-3)e^x - e^x|_1^3$

$= 7 + 3e - e^3$

5.13 Even and Odd Functions :

EX 1

(1) $= \int_{-1}^{1} (x^4+3x^2+5)dx$

$= 2\int_{0}^{1} (x^4+3x^2+5)dx$

$= 2\left[\frac{1}{5}x^5+x^3+5x\right]_0^1$

$= \frac{62}{5}$

(2) 0

(3) $= \int_{-2}^{2} (3x^2+2)dx$

$= 2\int_{0}^{2} (3x^2+2)dx$

$= 2[x^3+2x]_0^2$

$= 24$

(4) $= \int_{-\pi}^{\pi} 1\,dx$

$= 2\int_{0}^{\pi} 1\,dx$

$= 2[x]_0^{\pi} = 2\pi$

5. 14 Integration by Partial fraction

(1) $\ln\frac{27}{4}$

(2) $\frac{1}{x^2+2x} = \frac{A}{x}+\frac{B}{x+2}$

$\Rightarrow 1 = A(x+2)+Bx;\ x=0$

$\Rightarrow A = \frac{1}{2};x = -2 \Rightarrow B = -\frac{1}{2};$

$\int \frac{dx}{x^2+2x} = \frac{1}{2}\int\frac{dx}{x} - \frac{1}{2}\int\frac{dx}{x+2}$

$= \frac{1}{2}[\ln|x| - \ln|x+2|] + C$

(3) $y - 2\ln|y+1| + C$

(4) $\frac{x+4}{x^2+5x-6} = \frac{A}{x+6}+\frac{B}{x-1}$

$\Rightarrow x+4 = A(x-1)+B(x+6);x=1$

$\Rightarrow B = \frac{5}{7};x = -6$

$\Rightarrow A = \frac{2}{-7} = \frac{2}{7};$

$\int \frac{x+4}{x^2+5x-6}dx$

$= \frac{2}{7}\int\frac{dx}{x+6}+\frac{5}{7}\int\frac{dx}{x-1}$

$= \frac{2}{7}\ln|x+6| + \frac{5}{7}\ln|x-1| + C$

$= \frac{1}{7}\ln|(x+6)^2(x-1)^5| + C$

(5)

$\frac{1}{t^3+t^2-2t} = \frac{A}{t}+\frac{B}{t+2}+\frac{C}{t-1}$

$\Rightarrow 1 = A(t+2)(t-1) + Bt(t-1) + Ct(t+2);t=0$

$\Rightarrow A = -\frac{1}{2};t = -2$

$\Rightarrow B = \frac{1}{6};t = 1 \Rightarrow C = \frac{1}{3};\int\frac{dt}{t^3+t^2-2t}$

$= -\frac{1}{2}\int\frac{dt}{t}+\frac{1}{6}\int\frac{dt}{t+2}+\frac{1}{3}\int\frac{dt}{t-1}$

$= -\frac{1}{2}\ln|t|+\frac{1}{6}\ln|t+2| + \frac{1}{3}\ln|t-1| + C$

(6) $\dfrac{1}{(x+1)(x^2+1)} = \dfrac{A}{x+1} + \dfrac{Bx+C}{x^2+1}$

$\Rightarrow 1 = A(x^2+1) + (Bx + C)(x+1); x = -1$

$\Rightarrow A = \dfrac{1}{2};$

coefficient of x^2

$= A+B \Rightarrow A+B = 0 \Rightarrow B =-\dfrac{1}{2}$;constant

$= A + C \Rightarrow A + C = 1$

$\Rightarrow C = \dfrac{1}{2}; \int_0^1 \dfrac{dx}{(x+1)(x^2+1)}$

$$\frac{1}{2}\int_0^1 \frac{dx}{x+1} + \frac{1}{2}\int_0^1 \frac{(-x+1)}{x^2+1}dx$$
$$= \left[\frac{1}{2}\ln|x+1| - \frac{1}{4}\ln(x^2+1) + \frac{1}{2}\tan^{-1}x\right]_0^1$$
$$= \left(\frac{1}{2}\ln 2 - \frac{1}{4}\ln 2 + \frac{1}{2}\tan^{-1}1\right) - \left(\frac{1}{2}\ln 1 - \frac{1}{4}\ln 1 + \frac{1}{2}\tan^{-1}0\right)$$
$$= \frac{1}{4}\ln 2 + \frac{1}{2}\left(\frac{\pi}{4}\right) = \frac{(\pi + 2\ln 2)}{8}$$

(7) $\displaystyle\int \frac{1}{x-1} + \frac{2}{(x-1)^2} - \frac{1}{x+1}dx$

$= \ln|x-1| - \dfrac{2}{x-1} - \ln|x+1| + c$

(8) $\displaystyle\int \frac{1}{x} + \frac{x}{x^2+4} - \frac{1}{x^2+4}dx$

$= \ln|x| + \dfrac{1}{2}\ln(x^2+4) - \dfrac{1}{2}\tan^{-1}\left(\dfrac{x}{2}\right) + c$

CHAPTER 06.

Applications of Definite Integrals

6.1 Average Value of a Function

EX 1

(1) y_{avg}

$$=\left(\frac{1}{\sqrt{3}-0}\right)\int_0^{\sqrt{3}}(x^2-1)dx$$

$$=\frac{1}{\sqrt{3}}\int_0^{\sqrt{3}}x^2dx-\frac{1}{\sqrt{3}}\int_0^{\sqrt{3}}1dx$$

$$=\frac{1}{\sqrt{3}}\left(\frac{(\sqrt{3})^3}{3}\right)-\frac{1}{\sqrt{3}}(\sqrt{3}-0)$$

$$=1-1=0.$$

(2) $y_{avg}=\frac{1}{2-(-1)}\int_{-1}^{2}y\,dx$

$$=\frac{1}{3}\int_{-1}^{2}(x-3)^2\,dx$$

$$=\frac{1}{3}\left[\frac{1}{3}(x-3)^3\right]_{-1}^{2}$$

$$=\frac{1}{9}[(-1)^3-(-4)^3]=\frac{1}{9}[-1+64]=7$$

(3) y_{avg}

$$=\left(\frac{1}{3-(-1)}\right)\int_{-1}^{3}(|x|-1)dx$$

$$=\frac{1}{4}\int_{-1}^{1}(|x|-1)dx+\frac{1}{4}\int_{1}^{3}(|x|-1)dx$$

$$=\frac{1}{4}(-1+2)=\frac{1}{4}$$

(4) $y_{avg}=\frac{1}{2}\int_{-1}^{1}\frac{20}{x^2+1}dx$

$$=\frac{1}{2}[20\tan^{-1}x]_{-1}^{1}$$

$$=10[\tan^{-1}x]_{-1}^{1}$$

$$=10\left[\frac{\pi}{4}-\left(-\frac{\pi}{4}\right)\right]=5\pi$$

(5) $y_{avg}=\frac{1}{\ln 3}\int_0^{\ln 3}(e^x-1)dx$

$$=\frac{1}{\ln 3}[e^x-x]_0^{\ln 3}$$

$$=\frac{1}{\ln 3}[(e^{\ln 3}-\ln 3)-1]$$

$$=\frac{1}{\ln 3}[3-\ln 3-1]=\frac{2-\ln 3}{\ln 3}$$

(6) $y_{avg}=\frac{1}{\pi}\int_0^{\pi}y\,dx$

$$=\frac{1}{\pi}\int_0^{\pi}\sin x\,dx$$

$$=\frac{1}{\pi}[-\cos x]_0^{\pi}$$

$$=\frac{1}{\pi}(1-(-1))=\frac{2}{\pi}$$

EX 2

R_{avg}

$$= \frac{1}{18-0}\int_0^{18} R(t)\,dt$$

$$= \frac{1}{18}\left[\frac{(8+13)\cdot 6}{2} + \frac{(13+12)\cdot 6}{2} + \frac{(12+8)\cdot 6}{2}\right]$$

$$= \frac{1}{18}[63+75+60] = 11\,(L/\text{min})$$

EX 3

(1) Average velocity =

$$\frac{1}{4-0}\int_0^4 (5t-t^2+100)\,dt = 104\frac{2}{3} \text{ mph.}$$

(2) (a) Average velocity = v_{avg}

$$= \frac{1}{3-1}\int_1^3 v(t)dt = \frac{1}{2}\int_1^3 (12t^2+1)dt$$

$$= \frac{1}{2}[4t^3+t]_1^3 = \frac{1}{2}[111-5] = 53$$

(b) Average accerleration =

$$a_{avg} = \frac{v(3)-v(1)}{3-1} = \frac{109-13}{2} = 48$$

EX 4

(C) Average area

$$= \frac{1}{5-2}\int_2^5 \pi x^2 dx = \frac{\pi}{3}\left(\frac{125-8}{3}\right) = 13\pi \text{in}^2.$$

EX 5

(A) Average volume

$$= \frac{1}{4-0}\int_0^4 \frac{30}{(t+1)^2}dt \simeq 6 \text{ mg.}$$

EX 6

(D) $\frac{1}{25-0}\int_0^{25} P(t)dt = \frac{1}{25}\int_0^{25} 6e^{0.024t} \simeq 8.2$

(billion people).

EX 7

(1) $V_{avg} = \frac{1}{10}\int_0^{10} V(t)\,dt$

$$= \frac{1}{10}\int_0^{10} 35{,}000e^{-0.223t}dt = \int_0^{10} 3{,}500e^{-0.223t}dt$$

$$= \$14{,}007.40$$

(2) $P_{avg} = \frac{1}{30-10}\int_{10}^{30} P(t)dt$

$$= \frac{1}{20}\int_{10}^{30}\frac{20{,}000}{1+299e^{-0.2t}}dt = \int_{10}^{30}\frac{1{,}000}{1+299e^{-0.2t}}dt$$

$$= 4148.442 \simeq 4148$$

6.2 Area Between Curves

EX 1

(1) 48/5

(2)

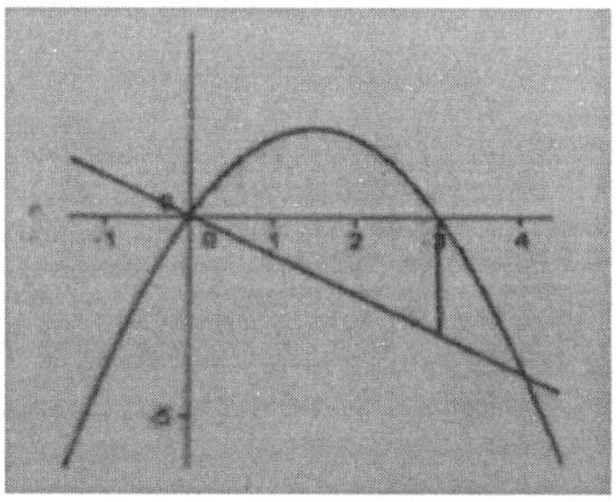

$$Area = \int_0^3 [-x^2+3x-(-x)]\,dx$$

$$= \int_0^2 (-x^2+4x)dx$$

$$= \left[-\frac{1}{3}x^3+2x^2\right]_0^3$$

$$= (-9+18) = 9$$

(3)

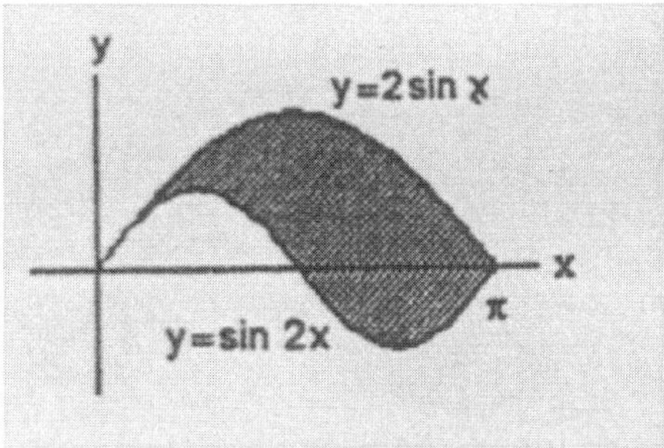

$$\Rightarrow A = \int_0^{\pi} (2\sin x - \sin 2x)\,dx$$

$$= \left[-2\cos x + \frac{\cos 2x}{2}\right]_0^{\pi}$$

$$= \left[-2(-1) + \frac{1}{2}\right] - \left(-2 \cdot 1 + \frac{1}{2}\right) = 4$$

(4)

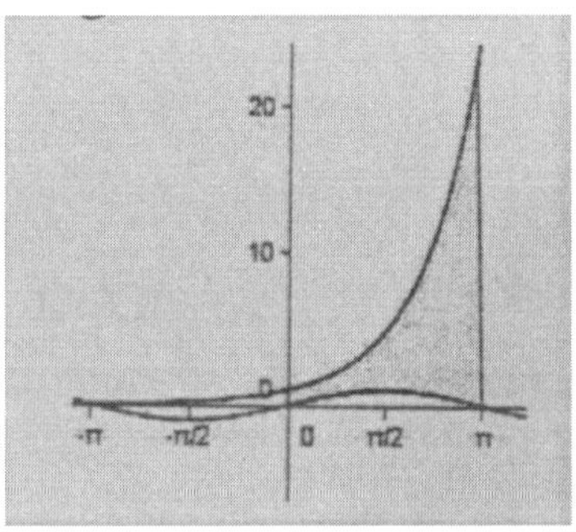

$$Area = \int_0^{\pi} (e^x - \sin x)dx$$

$$= \left[e^x + \cos x\right]_0^{\pi}$$

$$= (e^{\pi} - 1) - (1 + 1)$$

$$= e^{\pi} - 3$$

(5) $\pi/2$

(6)

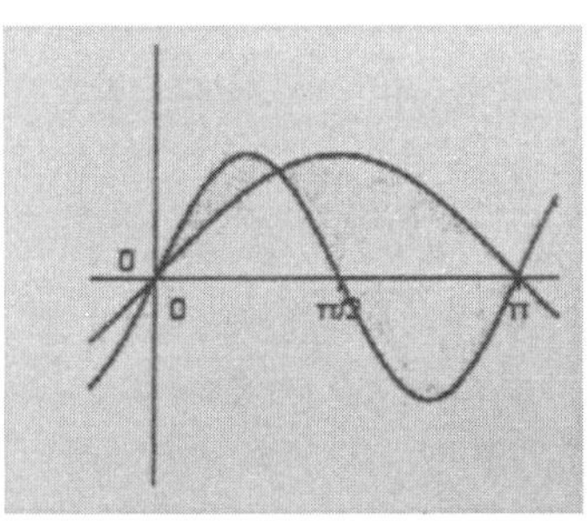

$$Area = \int_0^{\frac{\pi}{3}} (\sin 2x - \sin x)dx + \int_{\frac{\pi}{3}}^{\pi} (\sin x - \sin 2x)dx$$

$$= \left[-\frac{1}{2}\cos 2x + \cos x\right]_0^{\frac{\pi}{3}} + \left[-\cos x + \frac{1}{2}\cos 2x\right]_{\frac{\pi}{3}}^{\pi}$$

$$= \left(\frac{1}{4} + \frac{1}{2}\right) - \left(-\frac{1}{2} + 1\right) + \left(1 + \frac{1}{2}\right) - \left(-\frac{1}{2} - \frac{1}{4}\right)$$

$$= \frac{1}{4} + \frac{9}{4} = \frac{5}{2}\}$$

EX 2

(1) $9/2$

(2)

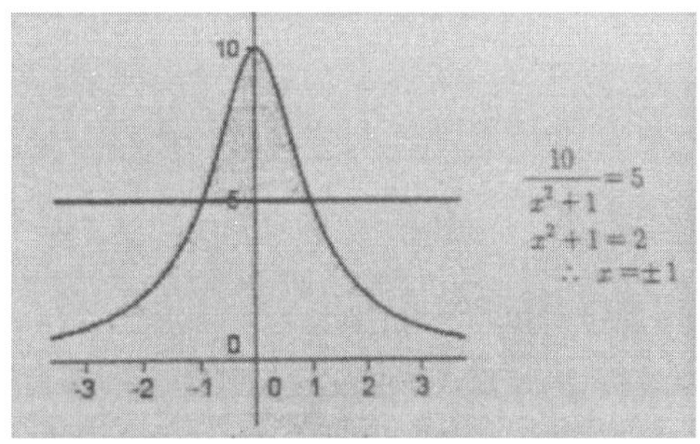

$$Area = \int_{-1}^{1} \left(\frac{10}{x^2+1} - 5\right)dx$$

$$= 2\int_0^{1} \left(\frac{10}{x^2+1} - 5\right)dx$$

$$= 2\left[10\tan^{-1}x - 5x\right]_0^1$$

$$= 2\left[\left(\frac{10\pi}{4} - 5\right) - 0\right] = 5\pi - 10$$

(3) $\frac{3}{2} - \ln 4$

(4)

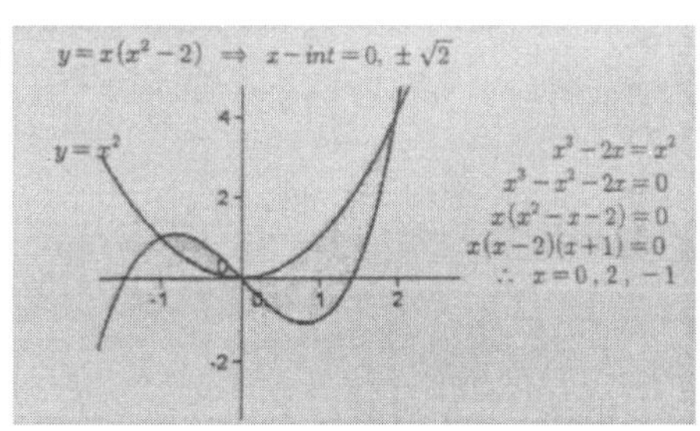

$$Area = \int_{-1}^{0} (x^3 - 2x - x^2)dx + \int_{0}^{2} (x^2 - x^3 + 2x)\, dx$$

$$= \left[\frac{1}{4}x^4 - x^2 - \frac{1}{3}x^3\right]_{-1}^{0} + \left[\frac{1}{3}x^3 - \frac{1}{4}x^4 + x^2\right]_{0}^{2}$$

$$= 0 - \left(\frac{1}{4} - 1 + \frac{1}{3}\right) + \left(\frac{8}{3} - 4 + 4\right) - 0$$

$$= \frac{5}{12} + \frac{8}{3} = \frac{37}{12}$$

(5) 131/4

(6) $y = |x^2 - 4| = \begin{cases} x^2 - 4, x \le -2 \text{ or } x \ge 2 \\ 4 - x^2, -2 \le x \le 2 \end{cases}$

for $x \le -2$ and $x \ge 2$: $x^2 - 4 = \frac{x^2}{2} + 4$

$\Rightarrow 2x^2 - 8 = x^2 + 8 \Rightarrow x^2 = 16 \Rightarrow x = \pm 4;$

for $-2 \le x \le 2$: $4 - x^2 = \frac{x^2}{2} + 4$

$\Rightarrow 8 - 2x^2 = x^2 + 8$

$\Rightarrow x^2 = 0 \Rightarrow x = 0$; by symmetry of the graph,

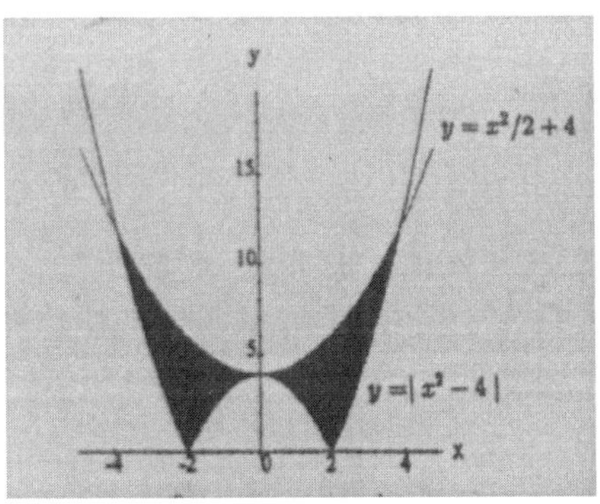

$$A = 2\int_{0}^{2}\left[\left(\frac{x^2}{2} + 4\right) - (4 - x^2)\right]dx$$

$$+ 2\int_{2}^{4}\left[\left(\frac{x^2}{2} + 4\right) - (x^2 - 4)\right]dx$$

$$= 2\left[\frac{x^3}{2}\right]_0^2 + 2\left[8x - \frac{x^3}{6}\right]_2^4$$

$$= 2\left(\frac{8}{2} - 0\right) + 2\left(32 - \frac{64}{6} - 16 + \frac{8}{6}\right)$$

$$= 40 - \frac{56}{3} = \frac{64}{3}$$

EX 3

(1) $Area = 10.225$

$[-10, 10] \times [-10, 10]$

$x_1 = -1.746749$
$x_2 = 4.8478775$

(2) Area= 6.2347

$[-10, 10] \times [-10, 10]$

$x_1 = -0.585$
$x_2 = 2.131$

(3) $Area = 92.3188504 \approx 92.319$

$[-10, 10] \times [-10, 30]$

$x_1 = -4.470857$
$x_2 = 2.5879522$

(4) $Area = 72.5308651 \approx 72.531$

$[-10, 10] \times [-50, 10]$

$x_1 = -6.076252$
$x_2 = 0$
$x_3 = 2.7429189$

EX 4

(1)18

(2)

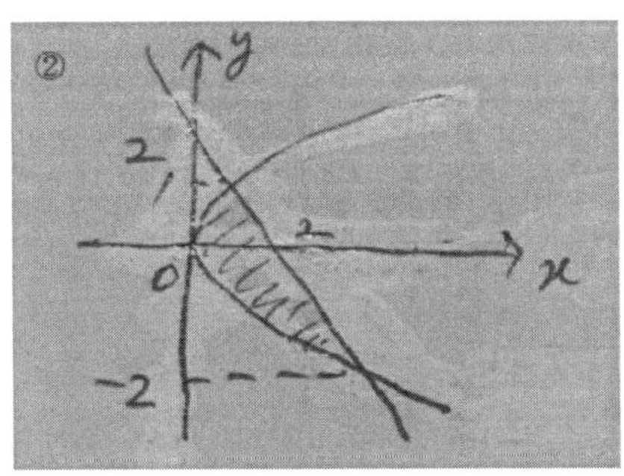

$$Area = \int_{-2}^{1} (x_1 - x_2)dy$$

$$= \int_{-2}^{1} (2 - y - y^2)dy$$

$$= \left[2y - \frac{1}{2}y^2 - \frac{1}{3}y^3\right]_{-2}^{1}$$

$$= \left(2 - \frac{1}{2} - \frac{1}{3}\right) - \left(-4 - 2 + \frac{8}{3}\right) = \frac{9}{2}$$

(3) 4

(4)

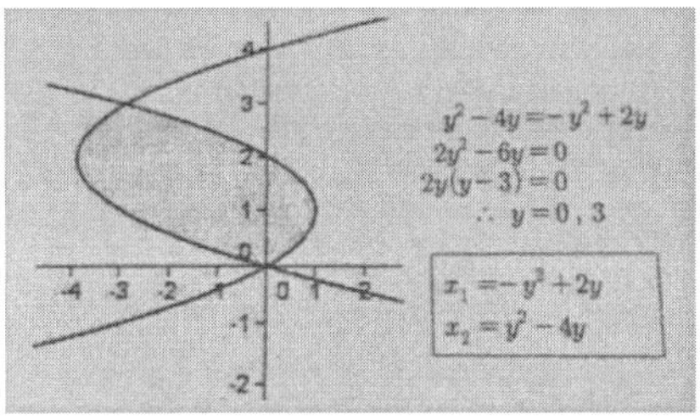

$$
\begin{aligned}
Area &= \int_0^3 (x_1-x_2)dy \\
&= \int_0^3 [(-y^2+2y-(y^2-4y)]dy \\
&= \int_0^3 (-2y^2+6y)dy \\
&= \left[-\frac{2}{3}y^3+3y^2\right]_0^3 = -18+27=9
\end{aligned}
$$

(5)

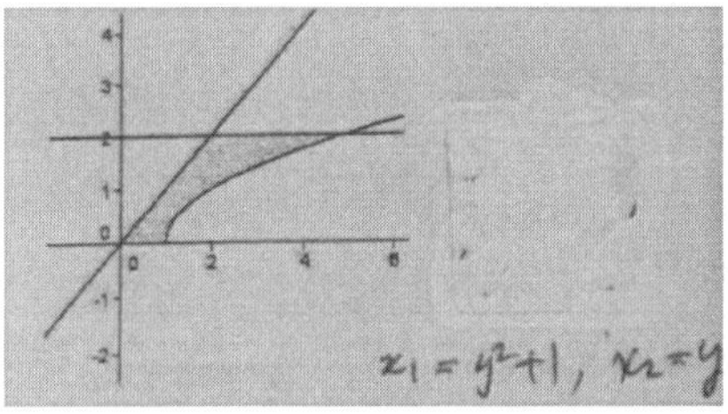

$$
\begin{aligned}
Area &= \int_0^2 (x_1-x_2)dy \\
&= \int_0^2 (y^2+1-y)dy \\
&= \left[\frac{1}{3}y^3+y-\frac{1}{2}y^2\right]_0^2 \\
&= \frac{8}{3}+2-2=\frac{8}{3}
\end{aligned}
$$

(6)

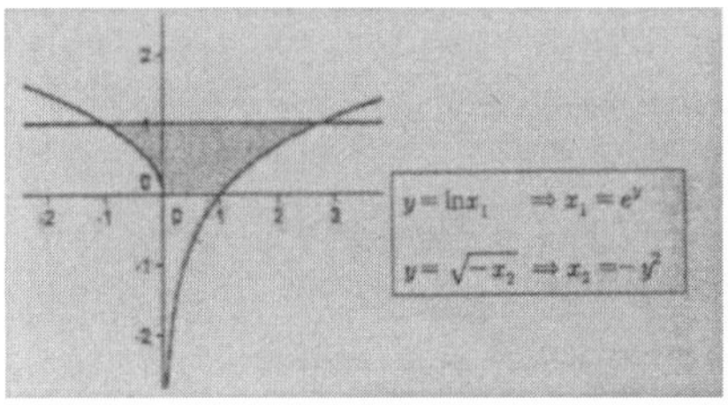

$$
\begin{aligned}
Area &= \int_0^1 (x_1-x_2)dy \\
&= \int_0^1 (e^y+y^2)dy \\
&= \left[e^y+\frac{1}{3}y^3\right]_0^1 \\
&= \left(e+\frac{1}{3}\right)-(1-0)=e-\frac{2}{3}
\end{aligned}
$$

(7)

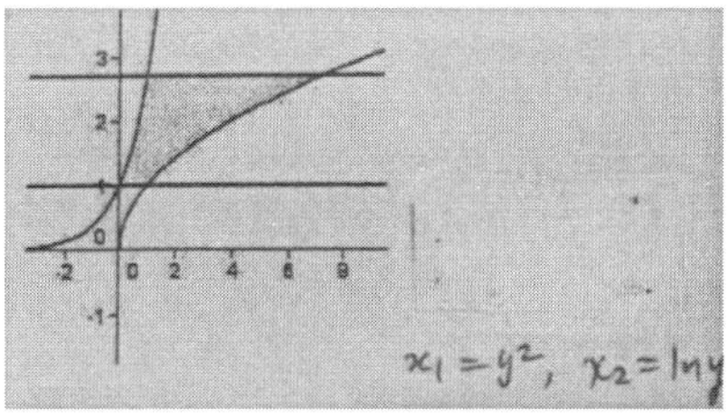

$$
\begin{aligned}
Area &= \int_1^e (x_1-x_2)dy \\
&= \int_1^e (y^2-\ln y)dy \\
&= \left[\frac{1}{3}y^3+(y\ln y+y)\right]_1^e = \left[\frac{1}{3}y^3-y\ln y+y\right]_1^e \\
&= \left(\frac{1}{3}e^3-e\ln e+e\right)-\left(\frac{1}{3}-\ln 1+1\right) \\
&= \frac{1}{3}e^3-\frac{4}{3}
\end{aligned}
$$

(8)

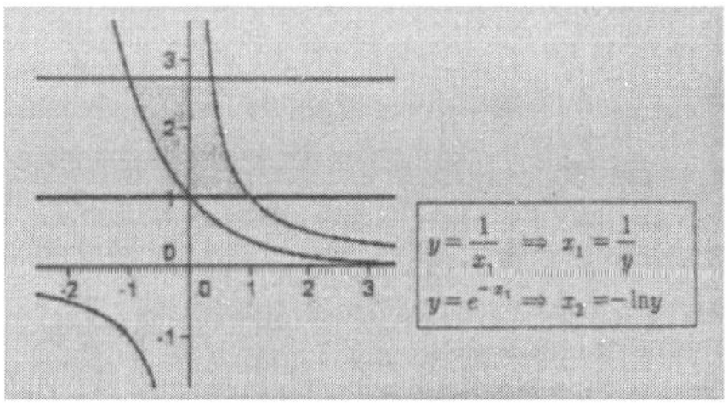

$$
\begin{aligned}
Area &= \int_1^e (x_1-x_2)dy \\
&= \int_1^e \left(\frac{1}{y}+\ln y\right)dy \\
&= [\ln|y|+y\ln y-y]_1^e \\
&= (\ln e+e\ln e-e)-(\ln 1+\ln 1-1) \\
&= 1-(-1)=2
\end{aligned}
$$

6.3 Volume of Solids with Known Cross Sections:

EX 1

$Area = (2y)^2 = 4y^2$

$V = \int_0^3 4y^2 dx = \int_0^3 4x\,dx$

$= [2x^2]_0^3$

$= 18$

EX 2

$$A(x) = \frac{\pi\,(diameter)^2}{4}$$
$$= \frac{\pi[(2-x^2)-x^2]^2}{4} = \frac{\pi[2(1-x^2)]^2}{4}$$
$$= \pi(1-2x^2+x^4); a = -1, b = 1;$$

$$V = \int_a^b A(x)\,dx = \int_{-1}^1 \pi(1-2x^2+x^4)dx$$

$$= \pi\left[x - \frac{2}{3}x^3 + \frac{x^5}{5}\right]_{-1}^1 = 2\pi\left(1-\frac{2}{3}+\frac{1}{5}\right) = \frac{16\pi}{15}$$

EX 3

$2\sqrt{3}$

EX 4

$$A(x) = \frac{(diagonal)^2}{2} = \frac{[\sqrt{1-x^2}-(-\sqrt{1-x^2})]^2}{2}$$
$$= \frac{(2\sqrt{1-x^2})^2}{2} = 2(1-x^2);\ a = -1, b = 1;$$

$$V = \int_a^b A(x)\,dx = 2\int_{-1}^1 (1-x^2)dx$$

$$= 2\left[x - \frac{x^3}{3}\right]_{-1}^1 = 4\left(1-\frac{1}{3}\right)$$

EX 5

(1) *Area*

$$= \int_0^4 \sqrt{x}\,dx + \frac{1}{2}\cdot 2\cdot 2$$

$$= \frac{2}{3}x^{3/2}\Big|_{x=0}^{x=4} + 2 = \frac{22}{3}$$

(2) $y = \sqrt{x} \Rightarrow x = y^2$
$y = 6 - x \Rightarrow x = 6 - y$

Width $= (6-y) - y^2$

Volume $= \int_0^2 2y(6-y-y^2)\,dy$

(c) $g'(x) = -1$

Thus a line perpendicular to the graph of g has slope 1

$$f'(x) = \frac{1}{2\sqrt{x}}$$
$$\frac{1}{2\sqrt{x}} = 1 \Rightarrow x = \frac{1}{4}$$

The point P has coordinates $\left(\frac{1}{4}, \frac{1}{2}\right)$

EX 6

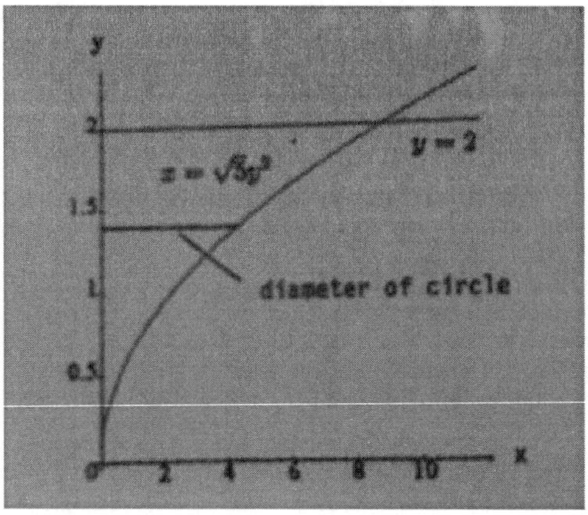

$$A(y) = \frac{\pi}{4}(diameter)^2 = \frac{\pi}{4}(\sqrt{5}y^2 - 0)^2 = \frac{5\pi}{4}y^4;$$

$c = 0,\ d = 2;\ V = \int_c^d \frac{5\pi}{4}y^4 dy$

$$= \left[\left(\frac{5\pi}{4}\right)\left(\frac{y^5}{5}\right)\right]_0^2 = \frac{\pi}{4}(2^5 - 0) = 8\pi$$

EX 7

$Area = (2x)(8x) = 16x^2$

$V = \int_0^1 16x^2 dy = \int_0^1 16y\,dy$

$= [8y^2]_0^1$

$= 8$

EX 8

$A(y) = \frac{1}{2}(\leq g)(\leq g) = \frac{1}{2}[\sqrt{1-y^2} - (-\sqrt{1-y^2})]^2$

$= \frac{1}{2}(2\sqrt{1-y^2})^2 = 2(1-y^2); c = -1, d = 1;$

$V = \int_c^d A(y)dy = \int_{-1}^1 2(1-y^2)\,dy = 2\left[y - \frac{y^3}{3}\right]_{-1}^2$

$= 4\left(1 - \frac{1}{3}\right) = \frac{8}{3}$

EX 9

$2\pi/15$

6.4 Volume of Solids of Revolution : Disks, Washers and Cylindrical Shells:

EX 1

(1)$32\pi/5$

(2)

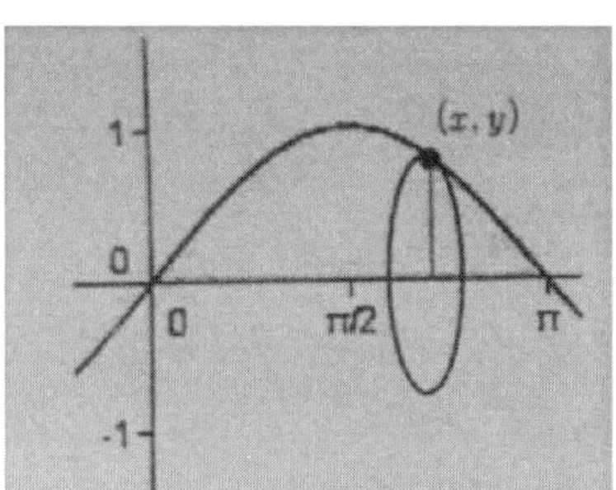

$V = \pi \int_0^\pi y^2\,dx$

$= \pi \int_0^\pi \sin^2 x\,dx$

$= \pi \int_0^\pi \left[\frac{1}{2} - \frac{1}{2}\cos 2x\right] dx$

$= \pi \left[\frac{1}{2}x - \frac{1}{4}\sin 2x\right]_0^\pi$

$= \pi \left[\frac{\pi}{2}\right] = \frac{\pi^2}{2}$

(3) $\frac{\pi}{2}(1 - \frac{1}{e^2})$

(4) $\frac{\pi}{30}$

EX 2

(1) 2π

(2)

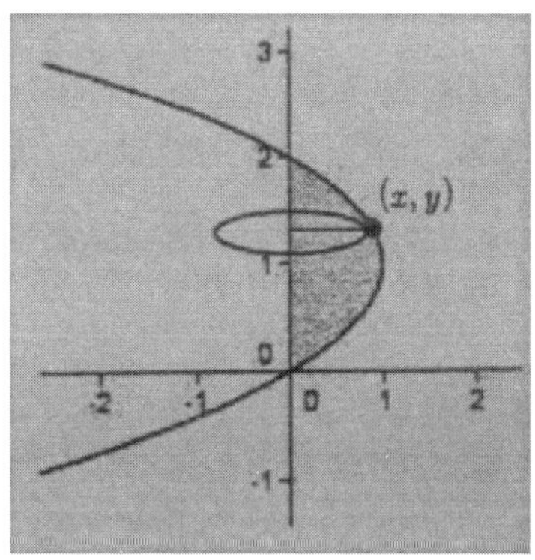

$Area = \pi x^2$

$V = \pi \int_0^2 (-y^2 + 2y)^2\,dy$

$= \pi \int_0^2 [\,y^4 - 4y^3 + 4y^2\,]\,dy$

$= \pi \left[\frac{1}{5}y^5 - y^4 + \frac{4}{3}y^3\right]_0^2$

$= \pi \left[\frac{32}{5} - 16 + \frac{32}{5}\right]$

$= \frac{16\pi}{15}$

(3)2π

(4) 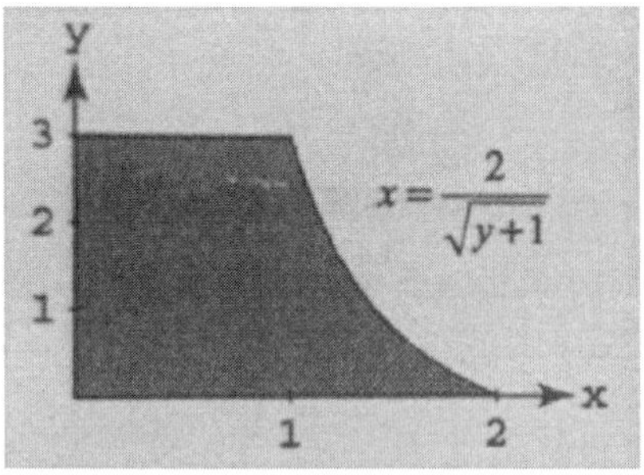

$$R(y)=\frac{2}{\sqrt{y+1}}$$

$$\Rightarrow V=\int_0^3 \pi[R(y)]^2 dy = 4\pi\int_0^3 \frac{1}{y+1}dy$$

$$=4\pi[\ln|y+1|]_0^3$$
$$=4\pi[\ln 4-\ln 1]$$
$$=4\pi\ln 4$$

EX 3

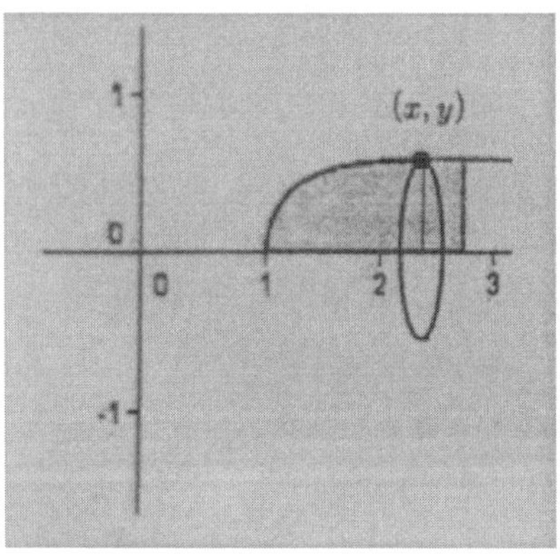

$$Area=\pi y^2$$

$$V=\pi\int_1^e y^2 dx$$
$$=\pi\int_1^e \frac{\ln x}{x}dx$$
$$=\pi\int_0^1 u\,du$$
$$=\pi\left[\frac{1}{2}u^2\right]_0^1$$
$$=\frac{\pi}{2}$$

EX 4

(1) 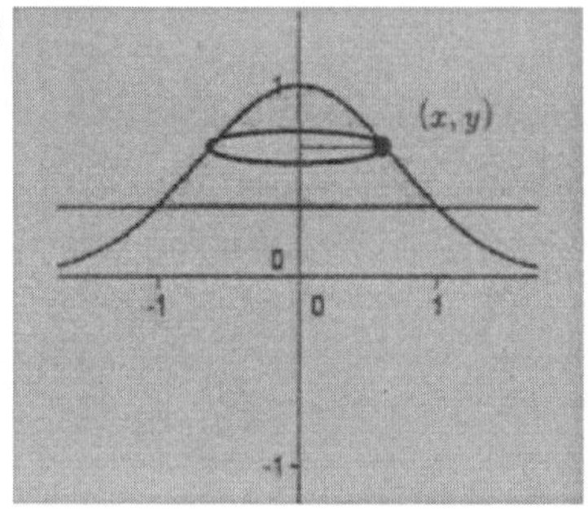

$$Area=\pi x^2$$

$$y=e^{-x^2}\Rightarrow \ln y=-x^2$$

$$x^2=-\ln y$$

$$V=\pi\int_{\frac{1}{e}}^1 x^2 dy$$
$$=\pi\int_{\frac{1}{e}}^1 -\ln y\, dy$$
$$=\pi[-y\ln y+y]_{\frac{1}{e}}^1$$
$$=\pi\left[(0+1)-\left(\frac{1}{e}+\frac{1}{e}\right)\right]$$
$$=\pi\left(1-\frac{2}{e}\right)$$

(2) 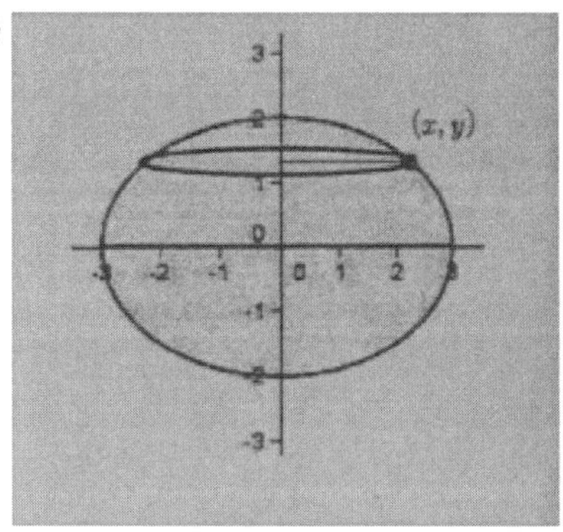

$$Area=\pi x^2$$

$$\frac{x^2}{9}+\frac{y^2}{4}=1\Rightarrow x^2=9-\frac{9}{4}y^2$$

$$
\begin{aligned}
V &= \pi\int_{-2}^{2} x^2 dy \\
&= \pi\int_{-2}^{2}\left(9-\frac{9}{4}y^2\right)dy \\
&= 2\pi\int_{0}^{2}\left(9-\frac{9}{4}y^2\right)dy \\
&= 2\pi\left[9y-\frac{3}{4}y^3\right]_0^2 \\
&= 2\pi(18-6)=24\pi
\end{aligned}
$$

EX 5

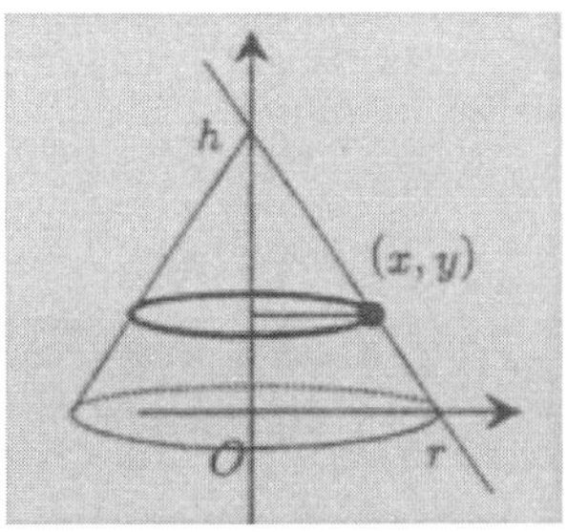

$Area=\pi x^2$

$$y=-\frac{h}{r}x+h$$

$$\Rightarrow x=r\left(1-\frac{y}{h}\right)$$

$$
\begin{aligned}
V &= \pi\int_0^h x^2 dy=\pi\int_0^h r^2\left(1-\frac{y}{h}\right)^2 dy \\
&= \pi r^2\left[y-\frac{y^2}{h}+\frac{y^3}{3h^2}\right]_0^h=\pi r^2\left[h-h+\frac{h}{3}\right] \\
&= \frac{1}{3}\pi r^2 h
\end{aligned}
$$

EX 6

(1)$117\pi/5$

(2)

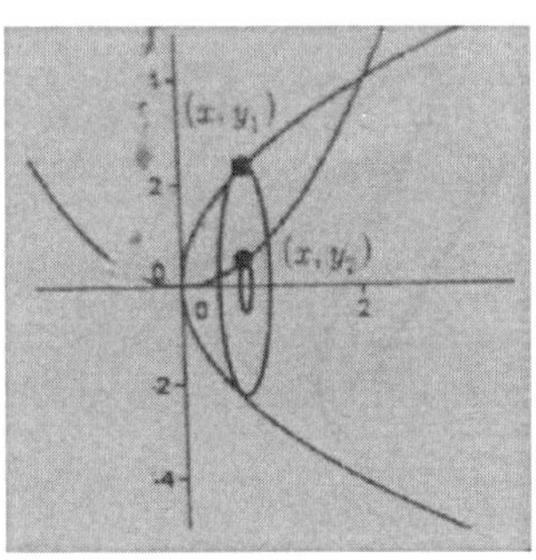

$$
\begin{aligned}
(x^2)^2 &= 8x \\
x^4 &= 8x \\
x(x^3-8) &= 0 \\
x &= 0,\ 2
\end{aligned}
$$

$$
\begin{aligned}
y_1{}^2 &= 8x \\
y_2 &= x^2
\end{aligned}
$$

$$
\begin{aligned}
V &= \pi\int_0^2 (y_1{}^2-y_2{}^2)dx \\
&= \pi\int_0^2 (8x-x^4)dx \\
&= \pi\left[4x^2-\frac{1}{5}x^5\right]_0^2 \\
&= \pi\left(16-\frac{32}{5}\right)=\frac{48\pi}{5}
\end{aligned}
$$

(3) $256\pi/5$

(4)

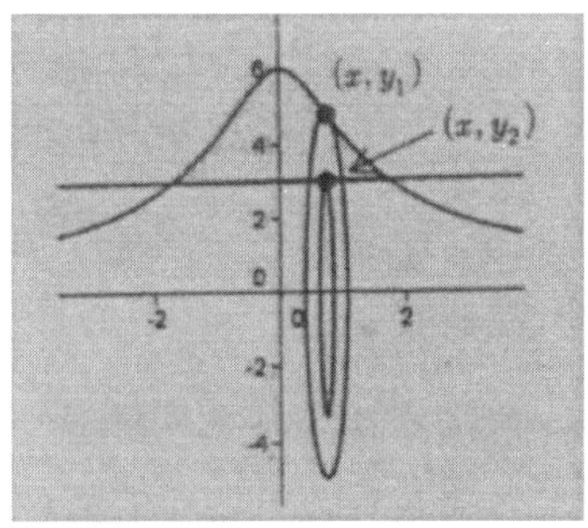

$$
\begin{aligned}
\frac{6}{\sqrt{x^2+1}} &= 3 \\
\sqrt{x^2+1} &= 2 \\
x^2+1 &= 4 \\
x &= \pm\sqrt{3}
\end{aligned}
$$

$$
\begin{aligned}
V &= \pi\int_{-\sqrt{3}}^{\sqrt{3}} (y_1{}^?-y_2{}^?)dx \\
&= \pi\int_{-\sqrt{3}}^{\sqrt{3}}\left(\frac{36}{x^2+1}-9\right)dx \\
&= 2\pi\int_0^{\sqrt{3}}\left(\frac{36}{x^2+1}-9\right)dx \\
&= 2\pi[36\tan^{-1}x-9x]_0^{\sqrt{3}} \\
&= 2\pi\left(36\cdot\frac{\pi}{3}-9\sqrt{3}\right) \\
&= 2\pi(12\pi-9\sqrt{3}) \\
&= 6\pi(4\pi-3\sqrt{3})
\end{aligned}
$$

(5) $\pi(\pi-2)$

(6)

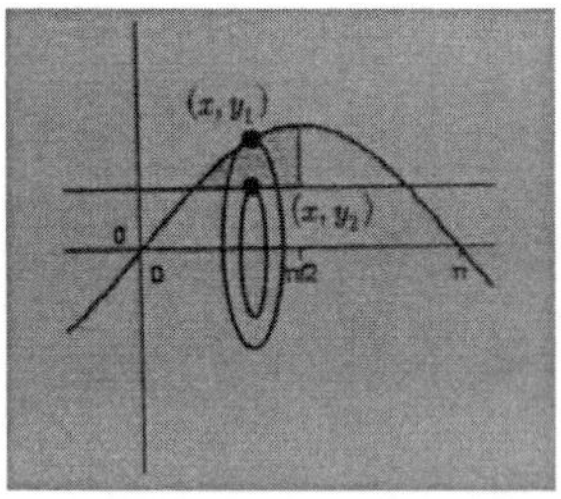

$$V=\pi\int_{\frac{\pi}{6}}^{\frac{\pi}{2}}({y_1}^2-{y_2}^2)dx$$
$$=\pi\int_{\frac{\pi}{6}}^{\frac{\pi}{2}}\left[\sin^2 x-\frac{1}{4}\right]dx$$
$$=\pi\int_{\frac{\pi}{6}}^{\frac{\pi}{2}}\left[\frac{1}{2}-\frac{1}{2}\cos 2x-\frac{1}{4}\right]dx$$
$$=\pi\left[\frac{1}{4}x-\frac{1}{4}\sin 2x\right]_{\frac{\pi}{6}}^{\frac{\pi}{2}}$$
$$=\pi\left[\left(\frac{\pi}{8}-0\right)-\left(\frac{\pi}{24}-\frac{\sqrt{3}}{8}\right)\right]$$
$$=\pi\left(\frac{\pi}{12}+\frac{\sqrt{3}}{8}\right)$$

EX 7

(1) 8π

(2)

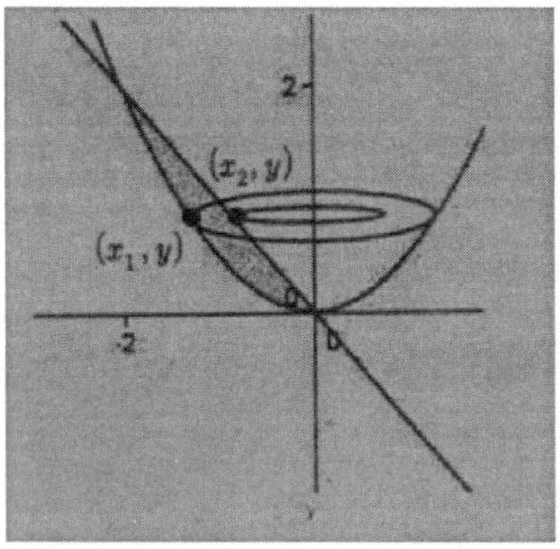

$$\frac{1}{2}x^2=-x$$
$$x^2=-2x$$
$$x(x+2)=0$$
$$x=0,\ -2\Rightarrow y=0,\ 2$$
$$y=\frac{1}{2}{x_1}^2\Rightarrow {x_1}^2=2y$$
$$y=-x_2\Rightarrow {x_2}^2=y^2$$

$$V=\pi\int_0^2({x_1}^2-{x_2}^2)dy$$
$$=\pi\int_0^2(2y-y^2)\,dy$$
$$=\pi\left[y^2-\frac{1}{3}y^3\right]_0^2$$
$$=\pi\left(4-\frac{8}{3}\right)$$
$$=\frac{4\pi}{3}$$

(3)

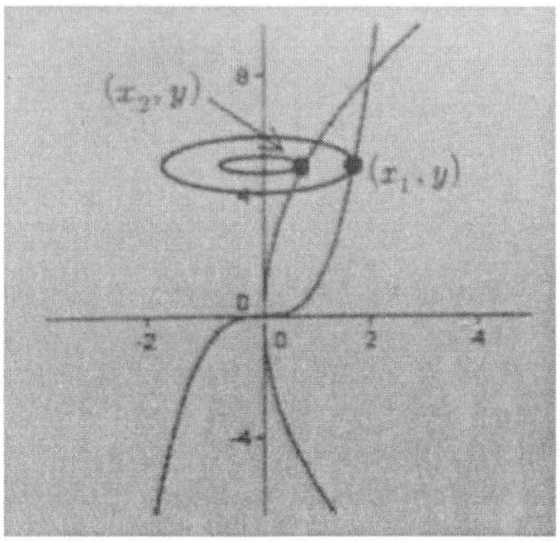

$$(x^3)^2=32x$$
$$x^6=32x$$
$$x(x^5-32)=0$$
$$x=0,\ 2\Rightarrow y=0,\ 8$$
$$y={x_1}^3\Rightarrow x_1=y^{\frac{1}{3}}$$
$$y^2=32x_2\Rightarrow x_2=\frac{1}{32}y^2$$

$$V=\pi\int_0^8({x_1}^2-{x_2}^2)dy$$
$$=\pi\int_0^8\left[y^{\frac{2}{3}}-\frac{y^4}{1024}\right]dy$$
$$=\pi\left[\frac{3}{5}y^{\frac{5}{3}}-\frac{y^5}{5\cdot 1024}\right]_0^8$$
$$=\pi\left[\frac{3}{5}(8)^{\frac{5}{3}}-\frac{(8)^5}{5\cdot 1024}\right]$$
$$=\pi\left(\frac{3}{5}(2)^5-\frac{2^{15}}{5\cdot 1024}\right)$$
$$=\pi\left(\frac{96}{5}-\frac{2^5}{5}\right)$$
$$=\frac{64\pi}{5}$$

(4)

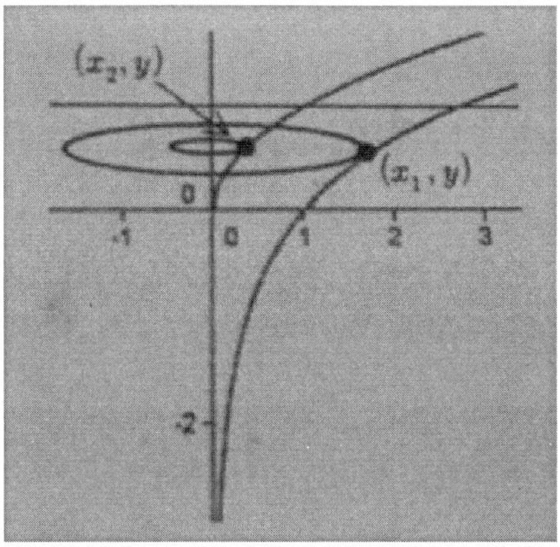

$y = \ln x_1 \Rightarrow x_1 = e^y$
$y = \sqrt{x_2} \Rightarrow x_2 = y^2$

$$
\begin{aligned}
V &= \pi \int_0^1 ({x_1}^2 - {x_2}^2)\,dy \\
&= \pi \int_0^1 (e^{2y} - y^4)\,dy \\
&= \pi \left[\frac{1}{2} e^{2y} - \frac{1}{5} y^5 \right]_0^1 \\
&= \pi \left[\left(\frac{1}{2} e^2 - \frac{1}{5} \right) - \frac{1}{2} \right] \\
&= \pi \left(\frac{e^2}{2} - \frac{7}{10} \right)
\end{aligned}
$$

EX 8

(1) $r(x) = 0$ and $R(x) = 1 - x^2$

$$
\begin{aligned}
&\Rightarrow V = \int_{-1}^{1} \pi([R(x)]^2 - [r(x)]^2)\,dx \\
&= \pi \int_{-1}^{1} (1 - x^2)^2 dx = \pi \int_{-1}^{1} (1 - 2x^2 + x^4)\,dx \\
&= \pi \left[x - \frac{2x^3}{3} + \frac{x^5}{5} \right]_{-1}^{1} = 2\pi \left(1 - \frac{2}{3} + \frac{1}{5} \right) \\
&= 2\pi \left(\frac{15 - 10 + 3}{15} \right) = \frac{16\pi}{15}
\end{aligned}
$$

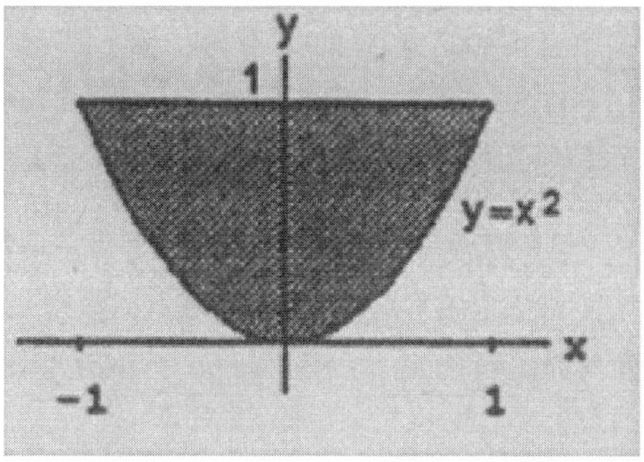

(2) $r(x) = 1$ and $R(x) = 2 - x^2$

$$
\begin{aligned}
&\Rightarrow V = \int_{-1}^{1} \pi([R(x)]^2 - [r(x)]^2)\,dx \\
&= \pi \int_{-1}^{1} [(2 - x^2)^2 - 1]\,dx \\
&= \pi \int_{-1}^{1} (4 - 4x^2 + x^4 - 1)\,dx \\
&= \pi \int_{-1}^{1} (3 - 4x^2 + x^4)\,dx \\
&= \pi \left[3x - \frac{4}{3} x^3 + \frac{x^5}{5} \right]_{-1}^{1} = 2\pi \left(3 - \frac{4}{3} + \frac{1}{5} \right) \\
&= \frac{2\pi}{15} (45 - 20 + 3) = \frac{56\pi}{15}
\end{aligned}
$$

EX 9

(1) $r(y) = 0$ and $R(y) = 1 - \frac{y}{2}$

$$
\begin{aligned}
&\Rightarrow V = \int_0^2 \pi([R(y)]^2 - [r(y)]^2)\,dy \\
&= \pi \int_0^2 \left(1 - \frac{y}{2} \right)^2 dy = \pi \int_0^2 \left(1 - y + \frac{y^2}{4} \right) dy \\
&= \pi \left[y - \frac{y^2}{2} + \frac{y^3}{12} \right]_0^2 = \pi \left(2 - \frac{4}{2} + \frac{8}{12} \right) = \frac{2\pi}{3}
\end{aligned}
$$

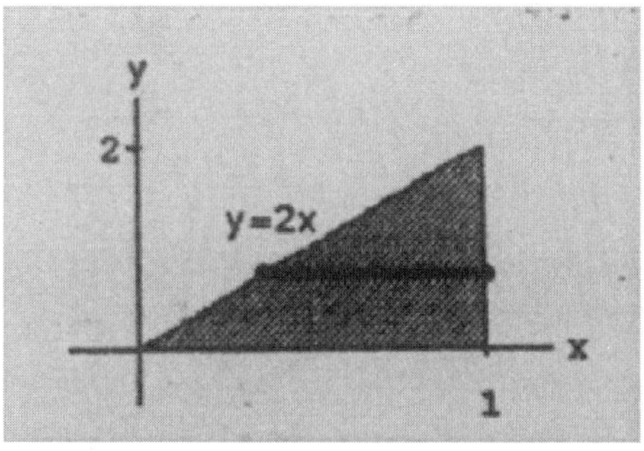

(2) $r(y)=1$ and $R(y)=2-\dfrac{y}{2}$

$$\Rightarrow V=\int_0^2 \pi([R(y)]^2-[r(y)]^2)dy$$
$$=\pi\int_0^2\left[\left(2-\frac{y}{2}\right)^2-1\right]dy$$
$$=\pi\int_0^2\left(4-2y+\frac{y^2}{4}-1\right)dy$$
$$=\pi\int_0^2\left(3-2y+\frac{y^2}{4}\right)dy$$
$$=\pi\left[3y-y^2+\frac{y^3}{12}\right]_0^2$$
$$=\pi\left(6-4+\frac{8}{12}\right)=\pi\left(2+\frac{2}{3}\right)=\frac{8\pi}{3}$$

EX 10

$e^{2x-x^2}=2$ when $x=0.446057,\ 1.553943$

Let $P=0.446057$ and $Q=1.553943$

(1) Area of $R=\int_P^Q (e^{2x-x^2}-2)dx=0.514$

(2) $e^{2x-x^2}=1$ when $x=0,\ 2$

Area of $S=\int_0^2 (e^{2x-x^2}-1)dx$ − Area of R

$=2.06016$ − Area of R = 1.546

OR

$$\int_0^P (e^{2x-x^2}-1)dx+(Q-P)\cdot 1+\int_Q^2 (e^{2x-x^2}-1)dx$$

$=0.219064+1.107886+0.219064=1.546$

(3) Volume $=\pi\int_P^Q((e^{2x-x^2}-1)^2-(2-1)^2)dx$

EX 11

(1) Area

$$=\int_0^9 (6-2\sqrt{x})dx=\left(6x-\frac{4}{3}x^{3/2}\right)\Bigg|_{x=0}^{x=9}=18$$

(2) Volume $=\pi\int_0^9((7-2\sqrt{x})^2-(7-6)^2)dx$

(3) Solving $y=2\sqrt{x}$ for x yields $x=\dfrac{y^2}{4}$

Each rectangular cross section has area

$$\left(3\frac{y^2}{4}\right)\left(\frac{y^2}{4}\right)=\frac{3}{16}y^4$$

Volume$=\int_0^6 \frac{3}{16}y^4dy$

EX 12

(1) $5\pi/6$

(2)

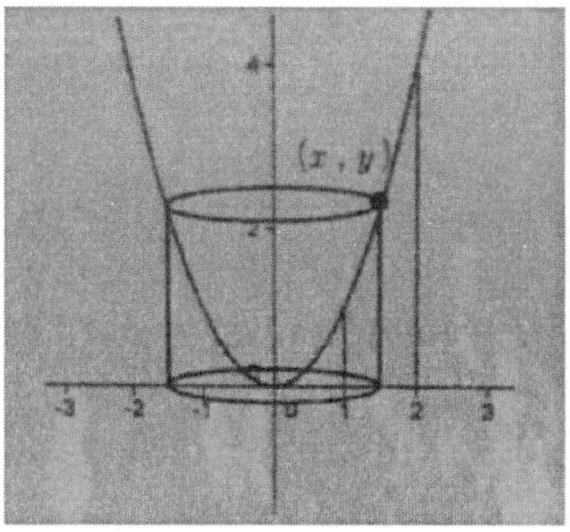

$$V=2\pi\int_1^2 xy\,dx$$
$$=2\pi\int_1^2 x\cdot x^2dx$$
$$=2\pi\int_1^2 x^3dx$$
$$=2\pi\left(\frac{1}{4}x^4\right)_1^2$$
$$=2\pi\left(4-\frac{1}{4}\right)=\frac{15\pi}{2}$$

(3) $7\pi/15$

(4)

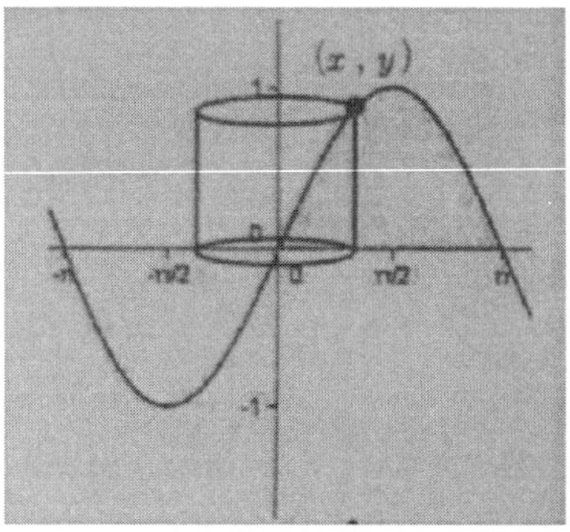

$$V = 2\pi \int_0^{\pi} xy\,dx$$
$$= 2\pi \int_0^{\pi} x \sin x\,dx$$

(Use repeated integration by parts)

$$= 2\pi [x(-\cos x) - 1(-\sin x)]_0^{\pi}$$
$$= 2\pi[-x\cos x + \sin x]_0^{\pi}$$
$$= 2\pi[\pi + 0 - (0+0)]$$
$$= 2\pi^2$$

(5) $8\pi/3$

EX 13

(1) $5\pi/6$

(2) $c=0, d=1$

$$V = \int_c^d 2\pi \begin{pmatrix} shell \\ radius \end{pmatrix} \begin{pmatrix} shell \\ height \end{pmatrix} dy$$
$$= \int_0^1 2\pi y(2y - y^2 - y)dy$$
$$= 2\pi \int_0^1 y(y - y^2)dy = 2\pi \int_0^1 (y^2 - y^3)dy$$
$$= 2\pi \left[\frac{y^3}{3} - \frac{y^4}{4} \right]_0^1 = 2\pi \left(\frac{1}{3} - \frac{1}{4} \right) = \frac{\pi}{6}$$

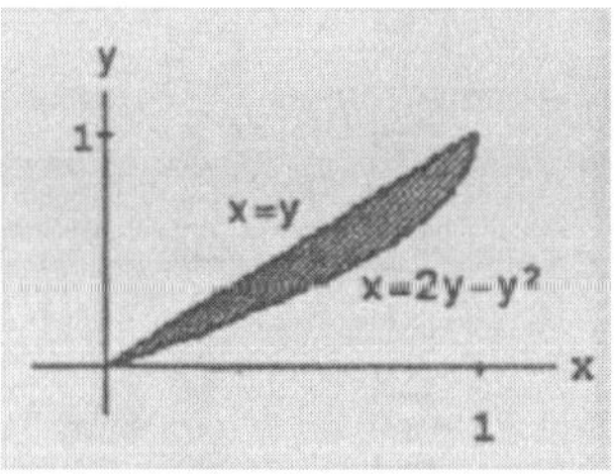

(3) $8\pi/3$

(4) $c=0, d=2$;

$$V = \int_c^d 2\pi \begin{pmatrix} shell \\ radius \end{pmatrix} \begin{pmatrix} shell \\ height \end{pmatrix} dy$$
$$= \int_0^2 2\pi y(2y - y^2)dy$$
$$= 2\pi \int_0^2 (2y^2 - y^3)dy = 2\pi \left[\frac{2y^3}{3} - \frac{y^4}{4} \right]_0^2$$
$$= 2\pi \left(\frac{16}{3} - \frac{16}{4} \right)$$
$$= 32\pi \left(\frac{1}{3} - \frac{1}{4} \right) = \frac{32\pi}{12} = \frac{8\pi}{3}$$

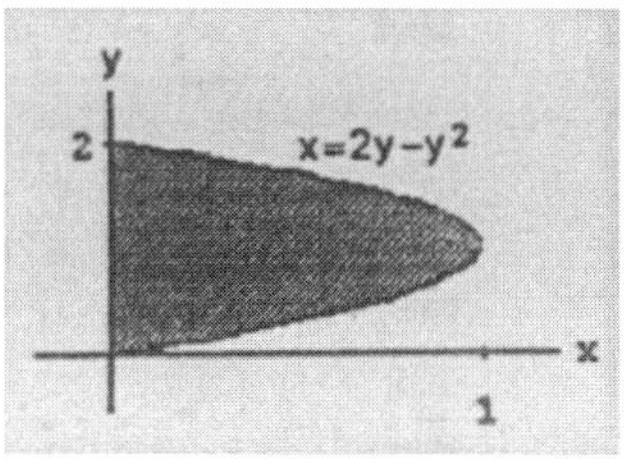

EX 14

(1) $$V = \int_a^b 2\pi \begin{pmatrix} shell \\ radius \end{pmatrix} \begin{pmatrix} shell \\ height \end{pmatrix} dx$$
$$= \int_0^2 2\pi(3-x)(8-x^3)dx$$
$$= 2\pi \int_0^2 (24 - 8x - 3x^3 + x^4)dx$$
$$= 2\pi \left[24x - 4x^2 - \frac{3}{4}x^4 + \frac{1}{5}x^5 \right]_0^2$$
$$= 2\pi \left(48 - 16 - 12 + \frac{32}{5} \right) = \frac{264\pi}{5}$$

(2) $$V = \int_a^b 2\pi \begin{pmatrix} shell \\ radius \end{pmatrix} \begin{pmatrix} shell \\ height \end{pmatrix} dx$$
$$= \int_0^2 2\pi(x+2)(8-x^3)dx$$
$$= 2\pi \int_0^2 (16 + 8x - 2x^3 - x^4)dx$$
$$= 2\pi \left[16x + 4x^2 - \frac{1}{2}x^4 - \frac{1}{5}x^5 \right]_0^2$$
$$= 2\pi \left(32 + 16 - 8 - \frac{32}{5} \right) = \frac{336\pi}{5}$$

(3) $$V = \int_c^d 2\pi \begin{pmatrix} shell \\ radius \end{pmatrix} \begin{pmatrix} shell \\ height \end{pmatrix} dy$$
$$= \int_0^8 2\pi(8-y)y^{1/3}dy$$
$$= 2\pi \int_0^8 (8y^{1/3} - y^{4/3})dy$$
$$= 2\pi \left[6y^{4/3} - \frac{3}{7}y^{7/3} \right]_0^8$$
$$= 2\pi \left(96 - \frac{384}{7} \right) = \frac{576\pi}{7}$$

(4) $V=\int_c^d 2\pi\begin{pmatrix}shell\\radius\end{pmatrix}\begin{pmatrix}shell\\height\end{pmatrix}dy$

$=\int_0^8 2\pi(y+1)y^{1/3}dy=2\pi\int_0^8(y^{4/3}+y^{1/3})dy$

$=2\pi\left[\frac{3}{7}y^{7/3}+\frac{3}{4}y^{4/3}\right]_0^8$

$=2\pi\left(\frac{384}{7}+12\right)=\frac{936\pi}{7}$

EX 15

$V=\int_c^d 2\pi\begin{pmatrix}shell\\radius\end{pmatrix}\begin{pmatrix}shell\\height\end{pmatrix}dy$

$=\int_0^2 2\pi(5-y)\left[\frac{y^2}{2}-\left(\frac{y^4}{4}-\frac{y^2}{2}\right)\right]dy$

$=\int_0^2 2\pi(5-y)\left(y^2-\frac{y^4}{4}\right)dy$

$=2\pi\int_0^2\left(5y^2-\frac{5}{4}y^4-y^3+\frac{y^5}{4}\right)dy$

$=2\pi\left[\frac{5y^3}{3}-\frac{5y^5}{20}-\frac{y^4}{4}+\frac{y^6}{24}\right]_0^2$

$=2\pi\left(\frac{40}{3}-\frac{160}{20}-\frac{16}{4}+\frac{64}{24}\right)=8\pi$

6.5 Definite Integral as Net Change

EX 1

(1) $\int_0^h(10+40t)dt=100$. This occurs at $h=2$ hours after 8 A.M. ∴10 A.M.

(2) Total gallons

$=\lim_{n\to\infty}\sum_{k=1}^{n}1000e^{-0.3t_k}\Delta t=\int_0^8 1000e^{-0.3t}dt\approx 3031.$

(3) The total amount dumped after 7 weeks is

$\int_0^{49}\frac{\sqrt{t}}{180}dt\simeq 1.27$ tons.

(4) $\int_0^T te^{-0.3t}dt=9$. The answer is about 10.2 weeks.

(5) The amount by which your investment of $1000 today will increase in 6 yr is

$\int_0^6 80e^{0.08t}dt$ or $616.07, so it will be worth $(1000+616.07).

Note: $1000e^{0.08(6)}=1616.07$

(6) $\int_0^3(-0.1t^2-0.3t+2)dt\simeq 3.75$ gal.

EX 2

(1) $s(t)=s(0)+\int_0^t v(t)dt$

$=2+\int_0^t(3t^2+2t)dt$

$=2+\left[t^3+t^2\right]_0^t$

$=t^3+t^2+2$

(2) $s(t)=s(\pi)+\int_\pi^t 4\cos 2t\,dt$

$=-3+[2\sin 2t]_\pi^t$

$=-3+2\sin 2t$

(3) $s(t)=s(3)+\int_3^t v(t)dt$
$=1+\int_3^t \frac{1}{\sqrt{1+t}}dt$
$=1+[2\sqrt{1+t}\,]_3^t$
$=1+2\sqrt{1+t}-4$
$=2\sqrt{1+t}-3$

(4) $s(t)=s(1)+\int_1^t v(t)dt$
$=2+\int_1^t \frac{\ln t}{t}dt$
$=2+\left[\frac{1}{2}(\ln t)^2\right]_1^t$
$=2+\frac{1}{2}(\ln t)^2$

(5) $v(t)=v(1)+\int_1^t (6y-3)dt$
$=2+[3t^2-3t]_1^t$
$=3t^2-3t+2$

$s(t)=s(0)+\int_0^t (3t^2-3t+2)dt$
$=2+\left[t^3-\frac{3}{2}t^2+2t\right]_0^t$
$=t^3-\frac{3}{2}t^2+2t+2$

(6) $v(t)=v(0)+\int_0^t 2\sin 2t\,dt$
$=3+[-\cos 2t]_0^t$
$=3-\cos 2t-(-1)$
$=4-\cos 2t$

$s(t)=s(0)+\int_0^t (4-\cos 2t)\,dt$
$=-2+\left[4t-\frac{1}{2}\sin 2t\right]_0^t$
$=-2+4t-\frac{1}{2}\sin 2t$

EX 3

(1) $f(3)=f(1)+\int_1^3 f'(x)\,dx$
$=3+\int_1^3 3\sin(x^2)\,dx=4.390$

(2) $f(2)=f(-1)+\int_{-1}^2 f'(x)\,dx$
$=2+\int_{-1}^2 e^{-x^2}dx=3.629$

(3) $s(4)=s(0)+\int_0^4 v(t)dt$
$=2+\int_0^4 (e^{\sin t}-1)dt=4.796$

(4) $v(5)=v(1)+\int_1^3 a(t)dt$
$=3+\int_1^3 2\cos(t+\ln t)dt=0.673$

(5) Net change
$=\int_{10}^{30} 500e^{-0.02t}dt=6747.977\approx 6748$

EX 4

(1) $\int_0^{12} P(t)dt \approx 46\cdot 4+57\cdot 4+62\cdot 4$
$=660\,ft^3$

(2) $\int_0^{12} R(t)dt=225.594\,ft^3$

(3) $1000+\int_0^{12} P(t)dt-\int_0^{12} R(t)dt=1434.406$

At time $t=12$ hours, the volome of water in the pool is approximately 1434 ft^3

(4) $V'(t)=P(t)-R(t)$
$V'(8)=P(8)-R(8)=60-25e^{-0.4}$
$=43.241$ or $43.242\,ft^3/hr$
$V=\pi(12)^2h$
$\frac{dV}{dt}=144\pi\frac{dh}{dt}$
$\left.\frac{dh}{dt}\right|_{t=8}=\frac{1}{144\pi}\cdot\left.\frac{dV}{dt}\right|_{t=8}$
$=0.095$ or $0.096\,ft/hr$

EX 5

(1) $H'(3.5) \approx \dfrac{H(5)-H(2)}{5-2}$

$= \dfrac{52-60}{3} = -2.666 \text{ or } -2.667$

degrees Celsius per minute

(2) $\dfrac{1}{10}\displaystyle\int_0^{10} H(t)\,dt$ is the average temperature of the tea, in degrees Celsus, over the 10 minutes.

$= 52.95$

(3) $\displaystyle\int_0^{10} H'(t)dt = H(10)-H(0) = 43-66 = -23$

The temperature of the tea drops 23 degrees Celsius from time $t=0$ to time $t=10$ minutes.

(4) $B(10) = 100 + \displaystyle\int_0^{10}$

$B'(t)dt = 34.18275;\quad H(10)-B(10) = 8.817$

The biscuits are 8.817 degrees Celsius cooler than the tea.

6.6 Motion Along a Straight Line (1-Dimension)

EX 1

(1) Displacement $= \dfrac{5\pi}{6} - \dfrac{1}{2}$

Distance$= \dfrac{5\pi}{6} - \dfrac{1}{2}$

(2)

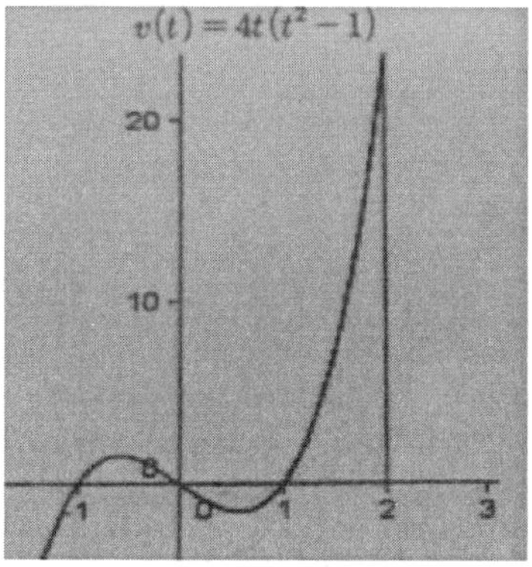

Displacement $= \displaystyle\int_0^2 v(t)dt = \int_0^2 (4t^3-4t)dt$
$= [t^4-2t^2]_0^2 = 16-8 = 8$

Distance

$= \displaystyle\int_0^2 |v(t)|dt$

$= -\displaystyle\int_0^1 (4t^3-4t)\,dt + \int_1^2 (4t^3-4t)dt$

$= -\left[t^4-2t^2\right]_0^1 + \left[t^4-2t^2\right]_1^2$

$= -[1-2]+(16-8)-(1-2)$

$= 1+9 = 10$

(3) Displacement $=0$

Distance$=8$

(4)

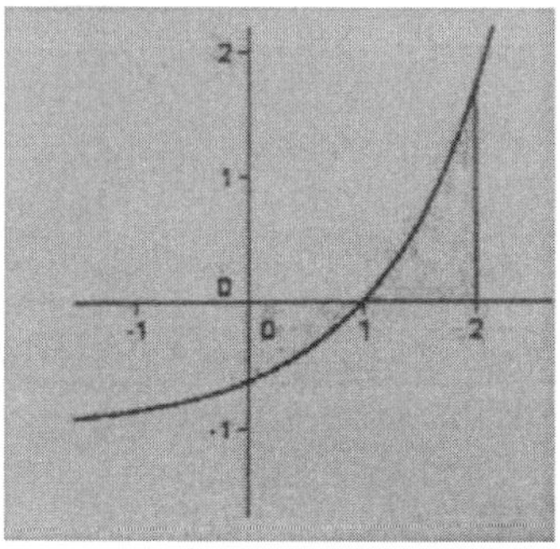

$$v(t)=v(1)+\int_1^t e^{t-1}dt$$
$$=0+[e^{t-1}]_1^t$$
$$=e^{t-1}-1$$

Displacement

$$=\int_0^2 v(t)dt=\int_0^2 (e^{t-1}-1)dt=[e^{t-1}-t]_0^2$$
$$=(e^1-2)-(e^{-1})=e-2-\frac{1}{e}$$

Distance

$$=\int_0^2 |v(t)|dt$$
$$=-\int_0^1 (e^{t-1}-1)\,dt+\int_1^2 (e^{t-1}-1)dt$$
$$=-[e^{t-1}-t]_0^1+[e^{t-1}-t]_1^2$$
$$=-[(1-1)-e^{-1}]+[e-2-(1-1)]$$
$$=e^{-1}+e-2=\frac{1}{e}+e-2$$

EX 2

(1) The squirrel changes direction whenever its velocity changes sign.

This occurs at $t=9$ and $t=15$.

(2) Velocity is 0 at $t=0$, $t=9$, and $t=15$.

t	position at time t
0	0
9	$\frac{9+5}{2}\cdot 20=140$
15	$140-\frac{6+4}{2}\cdot 10=90$
18	$90+\frac{3+2}{2}\cdot 10=115$

The squirrel is farthest from building A at time $t=9$;

its greatest distance from the building is 140

(3) The total distance traveled is

$$\int_0^{18} |v(t)|dt=140+50+25=215$$

(4) For $7<t<10$, $a(t)=\frac{20-(-10)}{7-10}=-10$

$$v(t)=20-10(t-7)=-10t+90$$
$$x(7)=\frac{7+5}{2}\cdot 20=120$$
$$x(t)=x(7)+\int_7^t(-10u+90)du$$
$$=120+(-5u^2+90u)\Big|_{u=7}^{u=t}$$
$$=-5t^2+90t-265$$

EX 3

(1) $a(7.5)=v'(7.5)=\frac{v(8)-v(7)}{8-7}$

$=-0.1$miles/minute^2

(2) $\int_0^{12}|v(t)|dt$ is the distance, in miles, that Caren rode during the 12 minutes from $t=0$ to $t=12$.

$$\int_0^{12}|v(t)|dt=\int_0^2 v(t)\,dt-\int_2^4 v(t)dt+\int_4^{12} v(t)dt$$
$=0.2+0.2+1.4=1.8$ miles

(3) Caren turns around to go back home at time $t=2$ minutes. This is the time at which her velocity changes from positive to negative.

(4) $\int_0^{12} w(t)dt=1.6$; Larry lives 1.6 miles from school.

$\int_0^{12} v(t)\,dt=1.4$; Caren lives 1.4 miles from school.

Therefore, Caren lives closer to school.

EX 4

(1) $a(5)\approx\frac{v(10)-v(0)}{10-0}=\frac{0.3}{10}=0.03$ meters/sec^2

(2) $\int_0^{60}|v(t)|dt$ is the total distance, in meters, Ben rides over the 60-second interval $t=0$

to $t = 60$.

$$\int_0^{60} |v(t)|dt \approx 2.0 \cdot 10 + 2.3(40-10) + 2.5(60-40)$$

= 139meters.

(3) Because $\dfrac{B(60)-B(40)}{60-40} = \dfrac{49-9}{20} = 2$,

The Mean Value Theorem implies there is a time t, $40 < t < 60$, such that $v(t) = 2$

(4) $2L(t)L'(t) = 2B(t)B'(t)$

$$L'(40) = \frac{B(40)B'(40)}{L(40)} = \frac{9 \cdot 2.5}{\sqrt{144+81}}$$

$$= \frac{3}{2}\text{ meters/sec}$$

EX 5

(1) $\int_0^6 f(t)dt = 142.274$ or 142.275 cubic feet

(2) Rate of change is $f(8) - g(8) = -59.582$ or -59.583 cubic feet per hour.

(3) $h(0) = 0$

For $0 < t \le 6$, $h(t)$

$$= h(0) + \int_0^t g(s)ds = 0 + \int_0^t 0\,ds = 0.$$

For $6 < t \le 7$, $h(t) = h(6) + \int_6^t g(s)ds$

$$= 0 + \int_6^t 125\,ds = 125(t-6).$$

For $7 < t \le 9$, $h(t) = h(7) + \int_7^t g(s)ds$

$$= 125 + \int_7^t 108ds = 125 + 108(t-7).$$

$$\text{Thus, } h(t) = \begin{cases} 0 & \text{for } 0 \le t \le 6 \\ 125(t-6) & \text{for } 6 < t \le 7 \\ 125 + 108(t-7) & \text{for } 7 < t \le 9 \end{cases}$$

(4) Amount of snow is

$$\int_0^9 f(t)dt - h(9) = 26.334 \text{ or } 26.335$$

cubic feet.

6.7 Motion Along a Plane Curve

EX 1

(1) $v = (-2\sin 2t + c_1)i + (2\cos t + c_2)j$, and since $v = 2j$ when $t = 0$, it follows that $c_1 = c_2 = 0$. So $v = -2\sin 2ti + 2\cos tj$.
Also $R = (\cos 2t + c_3)i + (2\sin t + c_4)j$; and since $R = i$ when $t = 0$, we see that $c_3 = c_4 = 0$. Finally, then,

$R = \cos 2ti + 2\sin tj$.

(2) From (1) the parametric equations of motion are

$x = \cos 2t, \quad y = 2\sin t.$

By a trigonometric identity,

$$x = 1 - 2\sin^2 t = 1 - \frac{y^2}{2}.$$

P travels in a counterclockwise direction along part of a parabola that has its ver-tex at (1,0) and opens to the left. The Path of the particle is sketched in Figure ; note that $-1 \le x \le 1$, $-2 \le y \le 2$.

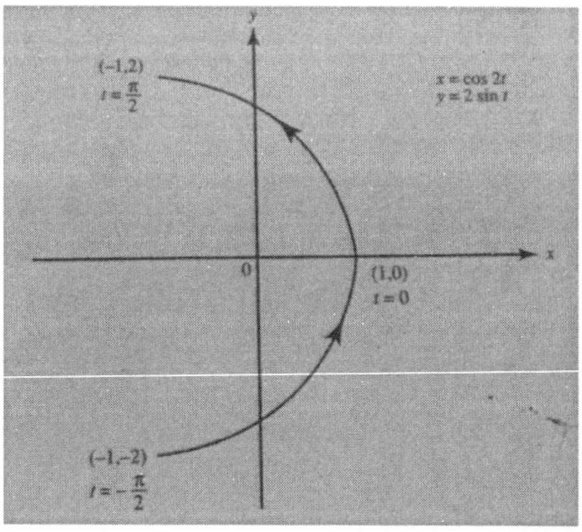

EX 2

(1) $\frac{dy}{dt}=0$ only when $t=0.36735$.

Let $b=0.36735$.

The maximum vertical distance from the water surface to the diver's shoulders is

$y(b)=11.4+\int_0^b \frac{dy}{dt}dt=12.061$ meters.

Alternatively, $y(t)=11.4+3.6t-4.9t^2$, so $y(b)=12.061$ meters.

(2) $y(A)=11.4+\int_0^A \frac{dy}{dt}dt$ when

$=11.4+3.6A-4.9A^2=0$

$A=1.936$ seconds.

(3) $\int_0^A \sqrt{\left(\frac{dx}{dt}\right)^2+\left(\frac{dy}{dt}\right)^2}dt=12.946$ meters

(4) At time A, $\frac{dy}{dx}=\frac{dy/dt}{dx/dt}\bigg|_{t=A}=-19.21913$.

The angle between the path of the diver and the water is

$\tan^{-1}(19.21913)=1.518$ or 1.519.

EX 3

(1) Speed= $\sqrt{(x'(3))^2+(y'(3))^2}$

$=2.828$ meters per second

(2) $x'(t)=2t-4$

Distance

$=\int_0^4 \sqrt{(2t-4)^2+(te^{t-3}-1)^2}\,dt$

$=11.587$ or 11.588 meters

(3) $\frac{dy}{dx}=\frac{dy/dt}{dx/dt}=0$ when $te^{t-3}-1=0$ and $2t-4\neq 0$

This occurs at $t=2.20794$.

Since $x'(2.20794)>0$, the particle is moving toward the right at time $t=2.207$ or 2.208.

(4) $x(t)=5$ at $t=1$ and $t=3$

At time $t=1$, the slope is

$\frac{dy}{dx}\bigg|_{t=1}=\frac{dy/dt}{dx/dt}\bigg|_{t=1}=0.432$.

At time $t=3$, the slope is

$\frac{dy}{dx}\bigg|_{t=3}=\frac{dy/dt}{dx/dt}\bigg|_{t=3}=1$.

$y(1)=y(3)=3+\frac{1}{e}+\int_2^3 \frac{dy}{dt}dt=4$

EX 4

(1) Speed

$=\sqrt{(x'(3))^2+(y'(3))^2}=13.006$ or 13.007

Acceleration

$=\langle x''(3), y''(3)\rangle$

$=\langle 4, -5.566\rangle$ or $\langle 4, -5.467\rangle$

(2) Slope $=\frac{y'(3)}{x'(3)}=0.031$ or 0.032

(3) $x(3)=0+\int_0^3 \frac{dx}{dt}dt=21$

$y(3)=-4+\int_0^3 \frac{dy}{dt}dt=-3.226$

At time $t=3$, the particle is at position $(21, -3.226)$.

(4) Distance $=\int_0^3 \sqrt{\left(\frac{dx}{dt}\right)^2+\left(\frac{dy}{dt}\right)^2}dt=21.091$

6.8 Arc Length

EX 1

Since $x-2=2y^{3/2}$ and $\dfrac{dx}{dy}=3y^{1/2}$,

Equation above yields

$$s=\int_0^1 \sqrt{1+9y}\,dy \approx 2.268,$$

again using the calculator to evaluate the definite integral.

EX 2

$$\int_0^4 \sqrt{1+\frac{9}{4}x}\,dx=\frac{8}{27}(10^{\frac{3}{2}}-1)$$

EX 3

Here $\dfrac{dy}{dx}=\dfrac{\sec x\tan x}{\sec x}$,

$$s=\int_0^{\pi/3}\sqrt{1+\tan^2 x}\,dx=\int_0^{\pi/3}\sec x\,dx$$
$$=\ln(\sec x+\tan x)\,|_0^{\pi/3}=\ln(2+\sqrt{3}).$$

EX 4

$$s=\int_1^2\sqrt{(t^2+1)^2}\,dt=\int_1^2(t^2+1)dt$$
$$=\frac{t^3}{3}+t\Big|_1^2=\frac{10}{3}.$$

EX 5

$$\int_0^{2\pi}\sqrt{(1-\cos t)^2+(\sin t)^2}\,dt=\int_0^{2\pi}\sqrt{2-2\cos t}\,dt$$
$$=\int_0^{2\pi}\sqrt{2-2(1-2\sin^2\frac{t}{2})}\,dt=\int_0^{2\pi}2\sin\frac{t}{2}dt=8$$

EX 6

$$\int_2^3\sqrt{(-e^t\sin t+e^t\cos t)^2+(e^t\cos t+e^t\sin t)^2}\,dt$$
$$=\sqrt{2}e^t\,|_2^3=\sqrt{2}(e^3-e^2)$$

EX 7

(1) $f'(x)=2e^{2x}$

Perimeter $=1+k+e^{2k}+\displaystyle\int_0^k\sqrt{1+(2e^{2x})^2}\,dx$

(2) Volume

$$=\pi\int_0^k(e^{2x})^2dx=\pi\int_0^k e^{4x}dx=\frac{\pi}{4}e^{4x}\Big|_{x=0}^{x=k}$$
$$=\frac{\pi}{4}e^{4k}-\frac{\pi}{4}$$

(3) $\dfrac{dV}{dt}=\pi e^{4k}\dfrac{dk}{dt}$

When $k=\dfrac{1}{2}$, $\dfrac{dV}{dt}=\pi e^2\cdot\dfrac{1}{3}$

EX 8

(1) *Area* $=30\cdot 20-\displaystyle\int_0^{30}f(x)dx=218.028\text{ cm}^2$

(2) volume $=\displaystyle\int_0^{30}\frac{\pi}{2}\left(\frac{f(x)}{2}\right)^2dx=2356.194\text{ cm}^3$

Therefore, the baker needs $2356.194\times 0.05=117.809$ or 117.810 grams of chocolate.

(3) Perimeter

$$=30+\int_0^{30}\sqrt{1+(f'(x))^2}\,dx=81.803 \text{ or } 81.804$$

6.9 Improper Integral

EX 1

(1) $\int_1^{\infty}\frac{dx}{\sqrt{x}}=\lim_{b\to\infty}\int_1^b x^{-1/2}dx=\lim_{b\to\infty}2\sqrt{x}\Big|_1^b$

$=\lim_{b\to\infty}2(\sqrt{b}-1)=+\infty$

Then $\int_1^{\infty}\frac{dx}{\sqrt{x}}$ diverges.

(2) $\int_0^{\infty}\frac{dx}{x^2+9}=\lim_{b\to\infty}\int_0^b\frac{dx}{x^2+9}=\lim_{b\to\infty}\frac{1}{3}\tan^{-1}\frac{x}{3}\Big|_0^b$

$=\lim_{b\to\infty}\frac{1}{3}\cdot\frac{\pi}{2}=\frac{\pi}{6}.$

(3) $\int_0^{\infty}\frac{dy}{e^y}=\lim_{b\to\infty}\int_0^b e^{-y}dy=\lim_{b\to\infty}-(e^{-b}-1)=1.$

(4) $\int_{-\infty}^0\frac{dz}{(z-1)^2}=\lim_{b\to-\infty}\int_b^0(z-1)^{-2}dz$

$=\lim_{b\to-\infty}-\frac{1}{z-1}\Big|_b^0$

$=\lim_{b\to-\infty}-\left(-1-\frac{1}{b-1}\right)=1.$

(5) $\int_0^{\infty}\cos x\,dx=\lim_{b\to\infty}\sin x\Big|_b^0=\lim_{b\to\infty}\sin b.$ Since this limit does not exist (sin b takes on values between -1 and 1 as $b\to\infty$), if follows that the given integral diverges. Note, however, that it does not become infinite; rather, it diverges by oscillation.

(6) $\int_0^1\frac{dx}{x^3}=\lim_{k\to0^+}\int_k^1 x^{-3}dx=\lim_{k\to0^+}-\frac{1}{2x^2}\Big|_k^1$

$=\lim_{k\to0^+}-\frac{1}{2}\left(1-\frac{1}{k^2}\right)=\infty.$

Therefore, this integral diverges.

(7) $\int_0^2\frac{dy}{\sqrt{4-y^2}}=\lim_{k\to2^-}\int_0^k\frac{dy}{\sqrt{4-y^2}}$

$=\lim_{k\to2^-}\sin^{-1}\frac{y}{2}\Big|_0^k=\sin^{-1}1-\sin^{-1}0=\frac{\pi}{2}.$

(8)

$\int_0^2\frac{dx}{(x-1)^{2/3}}$

$=\lim_{k\to1^-}\int_0^k(x-1)^{-2/3}dx+\lim_{m\to1^+}\int_m^2(x-1)^{-2/3}dx$

$=\lim_{k\to1^-}3(x-1)^{1/3}\Big|_0^k+\lim_{m\to1^+}3(x-1)^{1/3}\Big|_m^2$

$=3(0+1)+3(1-0)=6.$

(9) $\int_{-2}^2\frac{dx}{x^2}=\lim_{k\to0^-}\int_{-2}^k x^{-2}dx+\lim_{m\to0^+}\int_m^2 x^{-2}dx=$

$\lim_{k\to0^-}-\frac{1}{x}\Big|_{-2}^k+\lim_{m\to0^+}-\frac{1}{x}\Big|_m^2$ Neither limit exists; the integral diverges.

(10) $\int_0^{\pi/2}\frac{\sin x}{\sqrt{1-\cos x}}dx$

$=\lim_{k\to(\pi/2)^-}2\sqrt{1-\cos x}\,\Big|_0^k=2$

EX 2

(1)

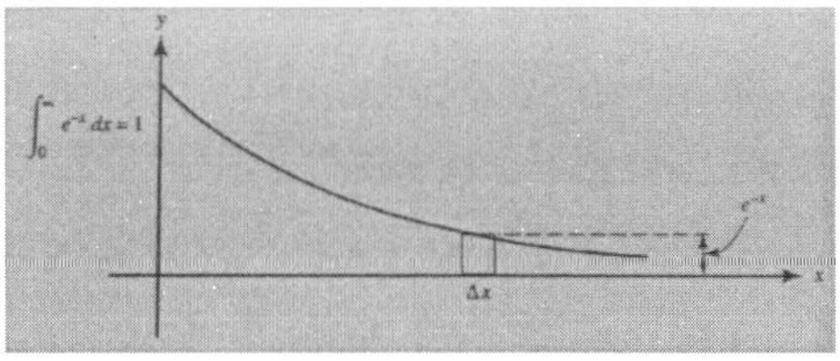

$\int_0^{\infty}e^{-x}\,dx=1$

(2)

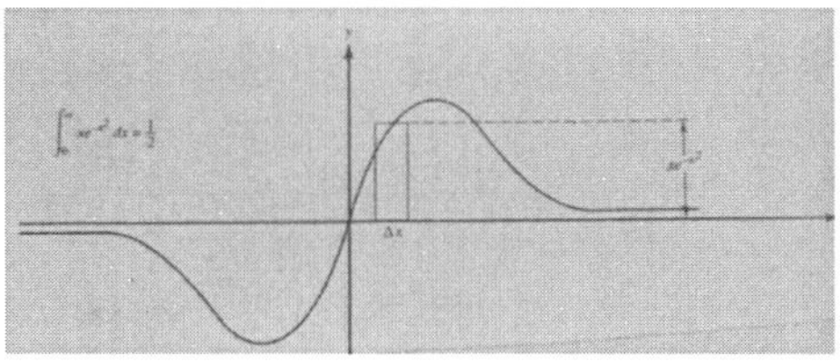

$\int_0^{\infty}xe^{-x^2}dx=\frac{1}{2}$

(3)

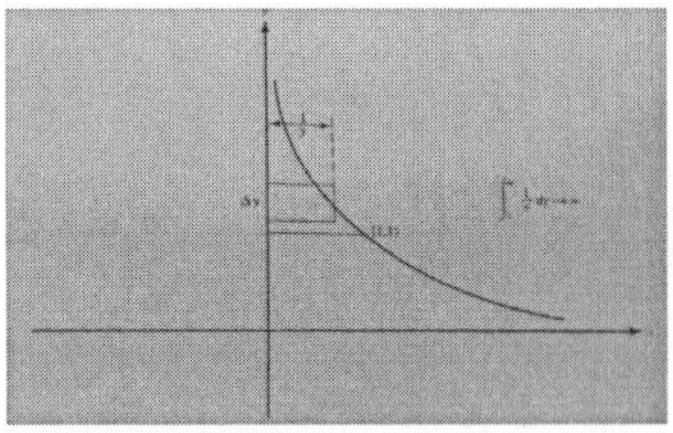

$$\int_1^{\infty} \frac{1}{y} dx = \infty$$

(4)

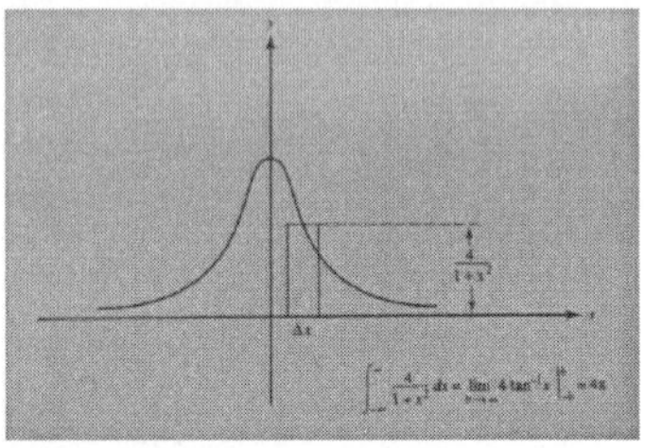

$$\int_{-\infty}^{\infty} \frac{4}{1+x^2} dx = \lim_{b\to\infty} 4\tan^{-1}x|_{-b}^{b} = 4\pi$$

(5)

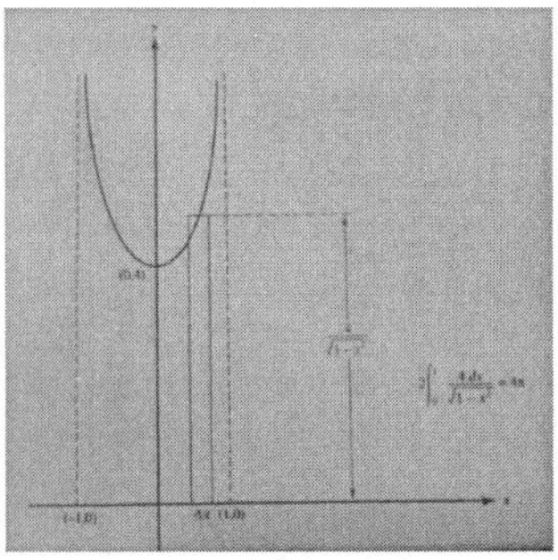

$$2\int_0^1 \frac{4}{\sqrt{1-x^2}} dx = 4\pi$$

EX 3

(1)

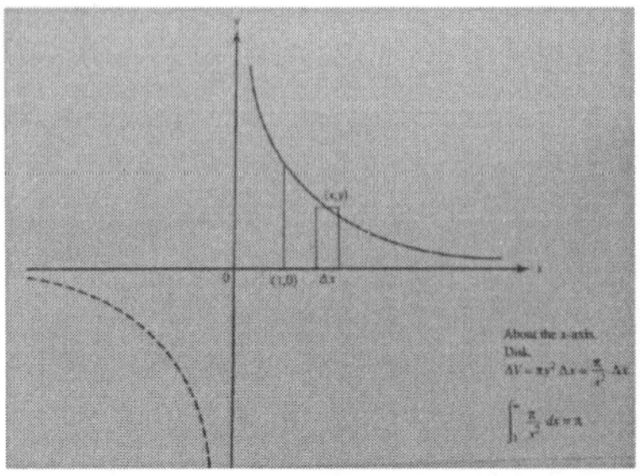

About the $x-$*axis, disk*

$$\triangle V = \pi y^2 \triangle x = \frac{\pi}{x^2}\triangle x, \int_1^{\infty} \frac{\pi}{x^2} dx = \pi$$

(2)

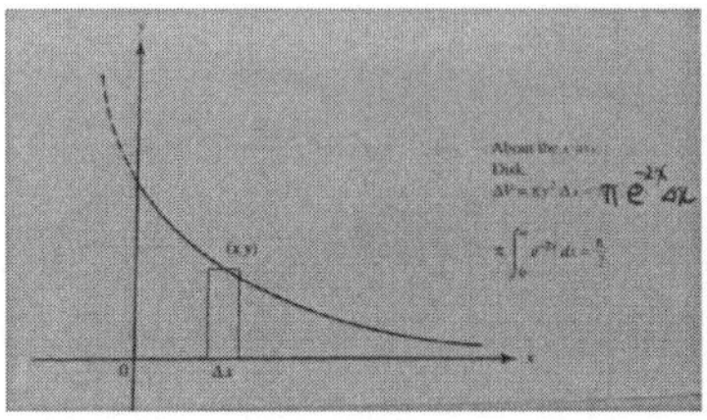

About the $x-$*axis, disk*

$$\triangle V = \pi y^2 \triangle x = \pi e^{-2x}\triangle x, \pi\int_1^{\infty} e^{-2x} dx = \frac{\pi}{2}$$

EX 4

(1) $g'(x) = \dfrac{4(1+4x^2)-4x(8x)}{(1+4x^2)^2} = \dfrac{4(1-4x^2)}{(1+4x^2)^2}$

For $x>0$, $g'(x)=0$ for $x=\frac{1}{2}$.

$g'(x)>0$ for $0<x<\frac{1}{2}$

$g'(x)<0$ for $x>\frac{1}{2}$

$g\left(\frac{1}{2}\right)=1$

Therefore g has a maximum value of 1 at $x=\frac{1}{2}$, and g has no minimum value on the open interval $(0, \infty)$

(2) $\displaystyle\int_1^{\infty}(f(x)-g(x))dx = \lim_{b\to\infty}\int_1^{b} f(x)-g(x))\,dx$

$$= \lim_{b\to\infty}\left(\ln(x)-\frac{1}{2}\ln(1+4x^2)\right)\Bigg|_{x=1}^{x=b}$$

$$= \lim_{b\to\infty}\ln\left(\frac{b\sqrt{5}}{\sqrt{1+4b^2}}\right)$$

$$= \lim_{b\to\infty}\ln\left(\frac{\sqrt{5b^2}}{\sqrt{1+4b^2}}\right)$$

$$= \frac{1}{2}\lim_{b\to\infty}\ln\left(\frac{5b^2}{1+4b^2}\right)$$

$$= \frac{1}{2}\ln\frac{5}{4}$$

CHAPTER 07. Differential Equations

7.1 Slope Fields (Direction Fields) :

EX 1

(1)

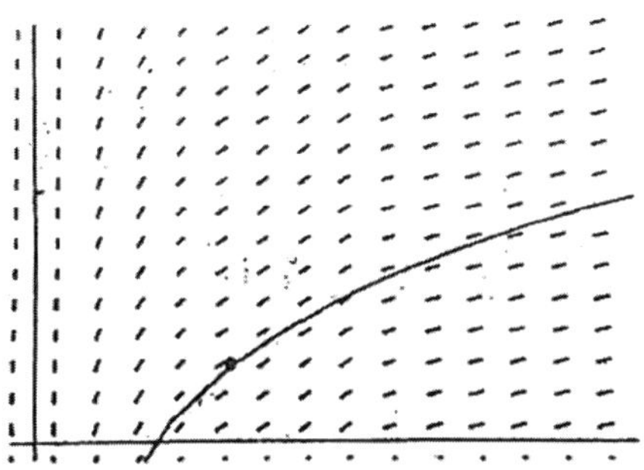

(2) Since we already know that, if

$$\frac{dy}{dx} = \frac{1}{x},\ then\ \ y = \int \frac{1}{x}dx = \ln x + C$$

EX 2

(1)

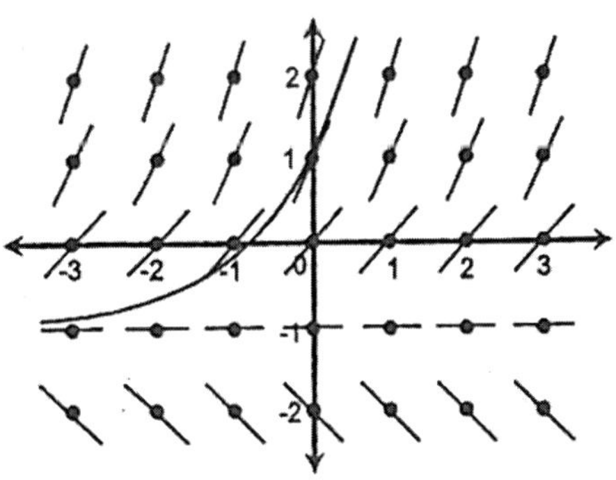

(2)

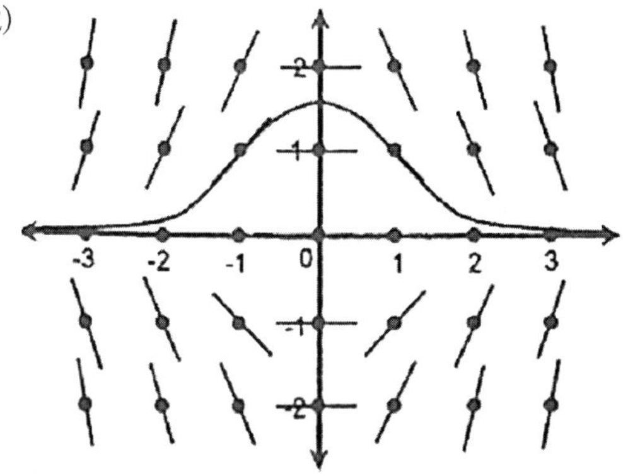

(3)

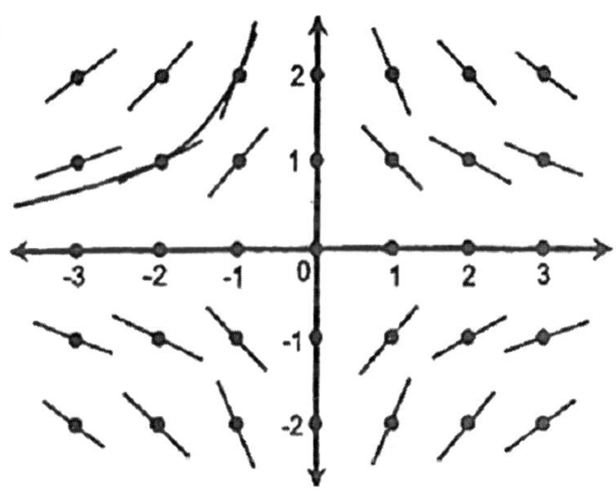

(4)

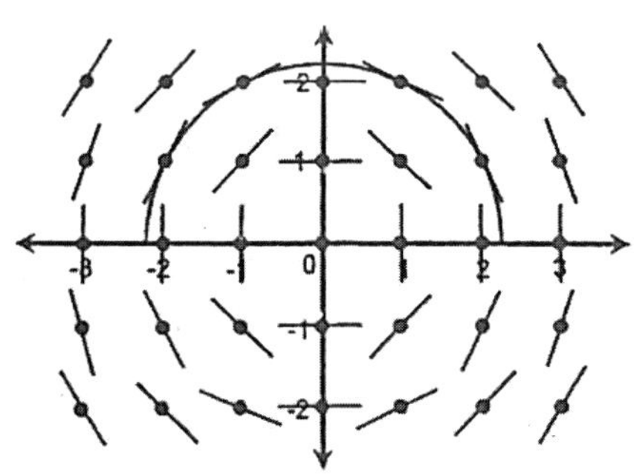

EX 3

(1) (b) (2) (c) (3) (d) (4) (a)

EX 4

(1) (e) (2) (c) (3) (b) (4) (a) (5) (d) (6) (f)

EX 5

(1) (c) (2) (a) (3) (b) (4) (d)

7.2 First Order Separable Differential Equations :

EX 1

(1) $y=\left(\frac{1}{3}x^{3/2}+1\right)^2$

(2)
$$\frac{dy}{y^3}=-2x\,dx$$
$$\int\frac{dy}{y^3}dy=\int -2x\,dx$$
$$-\frac{1}{2y^2}=-x^2+C$$
$$f(0)=1\ :\quad -\frac{1}{2}=0+C,\ \ C=-\frac{1}{2}$$
$$-\frac{1}{2y^2}=-x^2-\frac{1}{2}$$
$$\frac{1}{y^2}=2x^2+1$$
$$y^2=\frac{1}{2x^2+1}\ \Rightarrow\ y=\pm\sqrt{\frac{1}{2x^2+1}}$$

Since the curve passes through (0,1),

$$y=\sqrt{\frac{1}{2x^2+1}}$$

(3) $y=x$

(4) $2x\frac{dy}{dx}=3y$
$$\int\frac{2dy}{y}=\int\frac{3dx}{x}$$
$$2\ln|y|=3\ln|x|+C$$
$$\ln|y|=\frac{3}{2}\ln|x|+\frac{1}{2}C$$
$$|y|=e^{\frac{3}{2}\ln|x|+\frac{1}{2}C}$$
$$y=Ae^{\frac{3}{2}\ln|x|}=A|x|^{\frac{3}{2}\ln e}=A|x|^{\frac{3}{2}}$$
$$y(1)=100\ :\quad 100=A\cdot 1$$
$$A=100$$
$$\therefore y=100|x|^{3/2}$$

EX 2

(1) $y = e^{\sqrt{x}-2}$

(2) $\frac{dy}{dx} = y^2\cos 2x$

$\int \frac{dy}{y^2} = \int \cos 2x\, dx$

$-\frac{1}{y} = \frac{1}{2}\sin 2x + C$

$f\left(\frac{\pi}{4}\right) = -1 : 1 = \frac{1}{2}\sin\left(\frac{\pi}{2}\right) + C,\ C = \frac{1}{2}$

$-\frac{1}{y} = \frac{1}{2}\sin 2x + \frac{1}{2}$

$\frac{1}{y} = -\frac{1}{2}\sin 2x - \frac{1}{2}$

$y = \frac{1}{-\frac{1}{2}\sin 2x - \frac{1}{2}}$

$\therefore y = \frac{-2}{\sin 2x + 1}$

(3) $\frac{1}{2}\ln^2 y = \ln|x| + \frac{1}{2}$

(4) $\frac{dy}{dx} = \frac{y+2}{x}$

$\int \frac{dy}{y+2} = \int \frac{dx}{x}$

$\ln|y+2| = \ln|x| + C$

$|y+2| = e^{\ln|x|+C}$

$y+2 = Ae^{\ln|x|}$

$y+2 = A|x|$

$y(-1) = 3 : 5 = A \cdot 1$

$A = 5$

$y+2 = 5|x|$

$\therefore y = 5|x| - 2$

(5) $y = -\ln(-x+2)$

(6) $\int \frac{dy}{\sqrt{y}} = \int \frac{2}{x^2+1}dx$

$2\sqrt{y} = 2\tan^{-1}x + C$

$y(0) = 4 : 4 = 2\tan^{-1}(0) + C,\ C = 4$

$2\sqrt{y} = 2\tan^{-1}x + 4$

$\sqrt{y} = \tan^{-1}x + 2$

$y = (\tan^{-1}x + 2)^2$

EX 3

(1) $f(x) = \pm\sqrt{x^2+1}$

(2) $\frac{dy}{dx} = y^2 + 1$

$\int \frac{dy}{y^2+1} = \int dx$

$\tan^{-1}y = x + C$

$f(0) = \sqrt{3} : \tan^{-1}(\sqrt{3}) = 0 + C$

$\therefore C = \frac{\pi}{3}$

$\tan^{-1}(y) = x + \frac{\pi}{3}$

$\therefore y = \tan\left(x + \frac{\pi}{3}\right)$

$f\left(\frac{\pi}{2}\right) = \tan\left(\frac{\pi}{2} + \frac{\pi}{3}\right)$

$= \tan\left(\frac{5\pi}{6}\right)$

$= -\frac{1}{\sqrt{3}}$

7.3 Exponential Growth and Decay

EX 1

(1) 9 (2) 7.6 hr

EX 2

$$\frac{dy}{dt}=ky$$
$$\int\frac{dy}{y}=\int kdt$$
$$\ln|y|=kt+C$$
$$\therefore y=Ae^{kt}$$

Let the initial amount be y_0.

Then $y=y_0$ when $t=0$.

$$\therefore y=y_0e^{kt}$$

When $t=45, y=0.8y_0$.

$$0.8y_0=y_0e^{45k}$$
$$0.8=e^{45k},\ 45k=\ln(0.8)$$
$$k=\frac{\ln(0.8)}{45}$$
$$\therefore y=y_0e^{\frac{t}{45}\ln(0.8)}$$

Find t when $y=0.5y_0$.

$$0.5y_0=y_0e^{\frac{t}{45}\ln(0.8)}$$
$$0.5=e^{\frac{t}{45}\ln(0.8)}$$
$$\ln(0.5)=\frac{t}{45}\ln(0.8)$$
$$t=\frac{45\ln(0.5)}{\ln(0.8)}$$
$$=139.783\ hours$$

EX 3

$17.33\,yr$

EX 4

$$\frac{dy}{dt}=ky$$
$$\int\frac{dy}{y}=\int kdt$$
$$\ln|y|=kt+C$$
$$\therefore y=Ae^{kt}$$

When $t=0, y=4$.

$$4=Ae^0$$
$$A=4$$
$$\therefore y=4e^{kt}$$

When $t=2, y=10$

$$10=4e^{2k}$$
$$\frac{5}{2}=e^{2k},\ 2k=\ln\left(\frac{5}{2}\right)$$
$$k=\frac{1}{2}\ln\left(\frac{5}{2}\right)$$
$$\therefore y=4e^{\frac{1}{2}\ln\left(\frac{5}{2}\right)t}$$

Find y when $t=5$.

$$y=4e^{\frac{5}{2}\ln\left(\frac{5}{2}\right)}$$
$$=4\left(\frac{5}{2}\right)^{\frac{5}{2}}$$
$$=39.528$$

EX 5

This is an example of Case (the temperature is decreasing toward the limiting temperature 20℉).

(1) If R(t) is the temperature of the roast at time t, then

$\frac{dR(t)}{dt}=-k[R(t)-20]$ and $R(t)=20+ce^{-kt}$

Since $R(0)=68$℉, we have

$$68=20+c,$$
$$c=48,$$
$$R(t)=20+48e^{-kt},$$

Also, $R(2)=40$℉, so $40=20+48e^{-k\cdot 2}$

and $\quad e^{-k}=0.65,$

yielding $\quad R(t)=20+48(0.65)^t \quad (*)$

and, finally, $\quad R(5)=20+48(0.65)^5=26°F$

(2) Equation (*) in part (1) gives the roast's temperature at time t. We must find t when R=21;

$$21=20+48(0.65)^t,$$
$$\frac{1}{48}=(0.65)^t,$$
$$-\ln 48=t\ln(0.65),$$
$$t=9\,hr.$$

EX 6

$$\frac{dT}{dt}=-k(T-41)$$
$$\int\frac{dT}{T-41}=\int -k\,dt$$
$$\ln|T-41|=-kt+C$$
$$\therefore T-41=Ae^{-kt}$$

When $t=0,\ T=77$

$$77-41=Ae^0$$
$$A=36$$
$$\therefore T-41=36e^{-kt}$$

When $t=10,\ T=68$

$$27=36e^{-10k}$$
$$\frac{3}{4}=e^{-10k},\quad -10k=\ln\left(\frac{3}{4}\right)$$
$$k=-\frac{1}{10}\ln\left(\frac{3}{4}\right)$$
$$\therefore T-41=36e^{\frac{t}{10}\ln\left(\frac{3}{4}\right)}$$

Find t when $T=45$

$$4=36e^{\frac{t}{10}\ln\left(\frac{3}{4}\right)}$$
$$\frac{1}{9}=e^{\frac{t}{10}\ln\left(\frac{3}{4}\right)}$$
$$\ln\left(\frac{1}{9}\right)=\frac{t}{10}\ln\left(\frac{3}{4}\right)$$
$$t=\frac{10\ln(1/9)}{\ln(3/4)}$$
$$=76.337\text{ min}$$

EX 7

Let N(t) be the nimber of people aware of the product at time t.

$N'(t)=k[15{,}000-N(t)]$

The solution of this d.e is $N(t)=15{,}000-ce^{-kt}$

Since $N(0)=0,\ c=15{,}000$ and

$$N(t)=15{,}000(1-e^{-kt}).$$

when $t=6,\ N=1500$

$$1500=15{,}000(1-e^{-6k}),$$
$$e^{-6k}=0.9,$$
$$k=\frac{\ln 0.9}{-6}=0.018.$$

Find t when $N=2700$

$$2700=15{,}000(1-e^{-0.0018t})$$
$$0.18=1-e^{-0.018t},$$
$$e^{-0.018t}=0.82,$$
$$t\simeq 11\,days.$$

EX 8

$$\frac{dT}{dt}=k(250-T)$$
$$\frac{dT}{dt}=-k(T-250)$$
$$\int\frac{dt}{T-250}=\int -k\,dt$$
$$\ln|T-250|=-kt+C$$
$$\therefore T-250=Ae^{-kt}$$

When $t=0,\ T=-5.$

$$-5-250=Ae^0$$
$$A=-255$$
$$\therefore T-250=-255e^{-kt}$$

When $t=7,\ T=100$

$$-150=-255e^{-7k}$$
$$\frac{10}{17}=e^{-7k},\ -7k=\ln\left(\frac{10}{17}\right)$$
$$k=-\frac{1}{7}\ln\left(\frac{10}{17}\right)$$
$$\therefore T-250=-255e^{\frac{t}{7}\ln\left(\frac{10}{17}\right)}$$

Find t when $T=200.$

$$-50=-255e^{\frac{t}{7}\ln\left(\frac{10}{17}\right)}$$
$$\frac{10}{51}=e^{\frac{t}{7}\ln\left(\frac{10}{17}\right)}$$
$$\ln\left(\frac{10}{51}\right)=\frac{t}{7}\ln\left(\frac{10}{17}\right)$$
$$t=\frac{7\ln(10/51)}{\ln(10/17)}$$
$$=21.493 \text{ min}$$

EX 9

$$\frac{dT}{dt}=-k(T-20)$$
$$\int\frac{dT}{T-20}=\int-kdt$$
$$\ln|T-20|=-kt+C$$
$$\therefore T-20=Ae^{-kt}$$

When $t=0,\ T=25.$

$$25-20=Ae^0$$
$$A=5$$
$$\therefore T-20=5e^{-kt}$$

When $t=2,\ T=23.$

$$3=5e^{-2k}$$
$$\frac{3}{5}=e^{-2k},\quad -2k=\ln\left(\frac{3}{5}\right)$$
$$k=-\frac{1}{2}\ln\left(\frac{3}{5}\right)$$
$$\therefore T-20=5e^{\frac{t}{2}\ln\left(\frac{3}{5}\right)}$$

Find t *when* $T=37.$

$$17=5e^{\frac{t}{2}\ln\left(\frac{3}{5}\right)}$$
$$\frac{17}{5}=e^{\frac{t}{2}\ln\left(\frac{3}{5}\right)}$$
$$\ln\left(\frac{17}{5}\right)=\frac{t}{2}\ln\left(\frac{3}{5}\right)$$
$$t=\frac{2\ln(17/5)}{\ln(3/5)}$$
$$=-4.791\ hours$$
$$=4hrs\ 47\min\ ago$$

∴ The murder

occurred

at 7:13 PM.

7.4 Euler's Method

EX 1

(1)

$$y_1=y_0+\left(1-\frac{y_0}{x_0}\right)dx=-1+\left(1-\frac{-1}{2}\right)(.5)=-0.25,$$
$$y_2=y_1+\left(1-\frac{y_1}{x_1}\right)dx=-0.25+\left(1-\frac{-0.25}{2.5}\right)(.5)=0.3,$$
$$y_3=y_2+\left(1-\frac{y_2}{x_2}\right)dx=0.3+\left(1-\frac{0.3}{3}\right)(.5)=0.75;$$

(2)

$$y_1=y_0+x_0(1-y_0)dx=0+1(1-0)(.2)=.2,$$
$$y_2=y_1+x_1(1-y_1)dx=.2+1.2(1-.2)(.2)=.392,$$
$$y_3=y_2+x_2(1-y_2)dx=.392+1.4(1-.392)(.2)=.5622;$$

(3)

$$y_1=y_0+(2x_0y_0+2y_0)dx=3+[2(0)(3)+2(3)](.2)=4.2,$$
$$y_2=y_1+(2x_1y_1+2y_1)dx=4.2+[2(.2)(4.2)+2(4.2)](.2)$$
$$=6.216,$$
$$y_3=y_2+(2x_2y_2+2y_2)dx$$
$$=6.216+[2(.4)(6.216)+2(6.216)](.2)$$
$$=9.6969;$$

(4)

$$y_1=y_0+y_0^2(1+2x_0)dx=1+1^2[1+2(-1)](.5)=.5,$$
$$y_2=y_1+y_1^2(1+2x_1)dx=.5+(.5)^2[1+2(-.5)](.5)=.5,$$
$$y_3=y_2+y_2^2(1+2x_2)dx=.5+(.5)^2[1+2(0)](.5)=.625;$$

EX 2

(1)

$\lim_{x\to0}(f(x)+1)=-1+1=0$ and $\lim_{x\to0}\sin x=0$

Using L' Hospital's Rule,

$$\lim_{x\to0}\frac{f(x)+1}{\sin x}=\lim_{x\to0}\frac{f'(x)}{\cos x}=\frac{f'(0)}{\cos 0}=\frac{(-1)^2\cdot 2}{1}=2$$

(3) $\frac{dy}{dx}=y^2(2x+2)$

$\frac{dy}{y^2}=(2x+2)dx$

$\int\frac{dy}{y^2}=\int(2x+2)dx$

$-\frac{1}{y}=x^2+2x+C$

$-\frac{1}{-1}=0^2+2\cdot 0+C\Rightarrow C=1$

$-\frac{1}{y}=x^2+2x+1$

$y=-\frac{1}{x^2+2x+1}=-\frac{1}{(x+1)^2}$

EX 3

(1) $f\left(\frac{1}{2}\right)=f(1)+\left(\frac{dy}{dx}\Big|_{(1,0)}\right)\cdot\Delta x$

$=0+1\cdot\left(-\frac{1}{2}\right)=-\frac{1}{2}$

$f(0)=f\left(\frac{1}{2}\right)+\left(\frac{dy}{dx}\Big|_{\left(\frac{1}{2},-\frac{1}{2}\right)}\right)\cdot\Delta x$

$=-\frac{1}{2}+\frac{3}{2}\cdot\left(-\frac{1}{2}\right)=-\frac{5}{4}$

(2) Since f is differentiable at $x=1, f$ is continuous at $x=1$, So,

$\lim_{x\to1}f(x)=0=\lim_{x\to1}(x^3-1)$ and we may apply L' Hospital's Rule.

$\lim_{x\to1}\frac{f(x)}{x^3-1}=\lim_{x\to1}\frac{f'(x)}{3x^2}=\frac{\lim_{x\to1}f'(x)}{\lim_{x\to1}3x^2}=\frac{1}{3}$

(3) $\frac{dy}{dx}=1-y$

$\int\frac{1}{1-y}dy=\int 1dx$

$-\ln|1-y|=x+C$

$-\ln1=1+C\Rightarrow -1$

$\ln|1-y|=1-x$

$|1-y|=e^{1-x}$

$f(x)=1-e^{1-x}$

EX 4

(1) $\frac{d^2y}{dx^2}=3+2\frac{dy}{dx}=3+2(3x+2y+1)$

$=6x+4y+5$

(2) if $y=mx+b+e^{rx}$ is a solution, then

$m+re^{rx}=3x+2(mx+b+e^{rx})+1.$

if $r\neq0$: $m=2b+1,\ r=2,\ 0=3+2m,$

$so\ m=-\frac{3}{2},\ r=2,$ and $b=-\frac{5}{4}$

OR

if $r=0$: $m=2b+3,\ r=0,\ 0=3+2m,$

$so\ m=-\frac{3}{2},\ r=0,\ b=-\frac{9}{4}.$

(3)

$f\left(\frac{1}{2}\right)\approx f(0)+f'(0)\cdot\frac{1}{2}=-2+(-3)\cdot\frac{1}{2}=-\frac{7}{2}$

$f'\left(\frac{1}{2}\right)\approx 3\left(\frac{1}{2}\right)+2\left(-\frac{7}{2}\right)+1=-\frac{9}{2}$

$f(1)\approx f\left(\frac{1}{2}\right)+f'\left(\frac{1}{2}\right)\cdot\frac{1}{2}=-\frac{7}{2}+\left(-\frac{9}{2}\right)\cdot\frac{1}{2}=-\frac{23}{4}$

(4)

$g'(0)=3\cdot0+2\cdot k+1=2k+1$

$g(1)\approx g(0)+g'(0)\cdot1=k+(2k+1)=3k+1=0$

$k=-\frac{1}{3}$

7.5 Logistic Growth

EX 1

(1) $k=0.0015$, and $A=150$, and $y=\dfrac{150}{1+ce^{-0.225t}}$

initial condition :

$y(0)=6, \Rightarrow 6=\dfrac{150}{1+c} \Rightarrow 1+c=25 \Rightarrow c=24$

Formula : $y=\dfrac{150}{1+24e^{-0.225t}}$

(2) $100=\dfrac{150}{1+24e^{-0.225}} \Rightarrow 1+24e^{-0.225}=\dfrac{3}{2}$

$\Rightarrow 24e^{-0.225}=\dfrac{1}{2} \Rightarrow e^{-0.225}=\dfrac{1}{48}$

$\Rightarrow -0.225t=-\ln 48$

$\Rightarrow t=\dfrac{\ln 48}{0.225} \approx 17.21\,weeks$

$125=\dfrac{150}{1+24e^{-0.225t}} \Rightarrow 1+24e^{-0.225t}=\dfrac{6}{5}$

$\Rightarrow 24e^{-0.225t}=\dfrac{1}{5} \Rightarrow e^{-0.225t}=\dfrac{1}{120}$

$\Rightarrow -0.225t=-\ln 120$

$\Rightarrow t=\dfrac{\ln 120}{0.225} \approx 21.28$

It will take about 17.21 weeks to reach 100 guppies, and about 21.28 weeks to reach 125 guppies.

EX 2

(1)

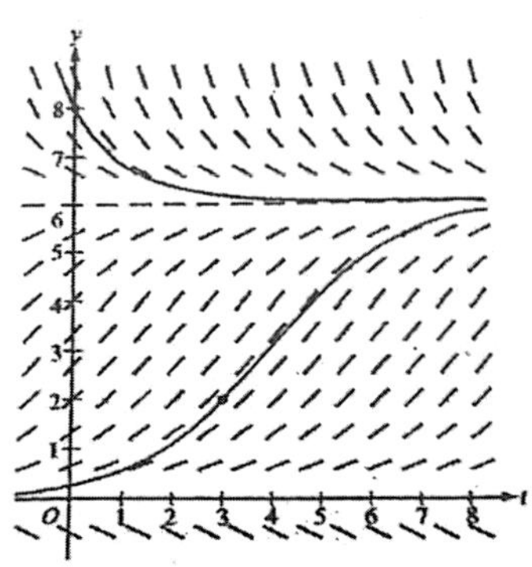

(2) $f\left(\dfrac{1}{2}\right) \approx 8+(-2)\left(\dfrac{1}{2}\right)=7$

$f(1) \approx 7+\left(-\dfrac{7}{8}\right)\left(\dfrac{1}{2}\right)=\dfrac{105}{16}$

(3) $\dfrac{d^2y}{dt^2}=\dfrac{1}{8}\dfrac{dy}{dt}(6-y)+\dfrac{y}{8}\left(-\dfrac{dy}{dt}\right)$

$\left.f(0)=8;\ f'(0)=\dfrac{dy}{dt}\right\}$

The second-degree Taylor polynomial for f about

$t=0$ is $P_2(t)=8-2t+\dfrac{5}{4}t^2$.

$f(1) \approx P_2(1)=\dfrac{29}{4}$

(4) The range of f for $t \geq 0$ is $6 < y \leq 8$.

EX 3

(1) We are told that

$N'(t)=k \cdot N \cdot (50{,}000-N)$,

that $N(0)=100$, and that

$N(1)=130$. The d.e. describing logistic growth leads to $N(t)=\dfrac{50{,}000}{1+ce^{-50{,}000kt}}$.

From $N(0)=100$, *we get*

$100=\dfrac{50{,}000}{1+c}$,

which yields $c=499$, From $N(1)=130$, we get $130=\dfrac{50{,}000}{1+499e^{-50{,}000k}}$,

$130(1+499\,e^{-50{,}000k})=50{,}000,$

$e^{-50{,}000k}=0.77.$

Then $N(t)=\dfrac{50{,}000}{1+499(0.77)^t}$.

(2) We must find $N(8)$, Since $t=0$ represents yesterday:

$N(8)=\dfrac{50{,}000}{1+499(0.77)^8}=798\,people.$

CHAPTER 08. Sequence and Series

8.1 Sequence

EX 1

(1) $\lim_{n\to\infty} 2+(0.1)^n = 2$

(2) ∞

(3) $\lim_{n\to\infty}\left(\frac{n+1}{2n}\right)\left(1-\frac{1}{n}\right)$

$= \lim_{n\to\infty}\left(\frac{1}{2}+\frac{1}{2n}\right)\left(1-\frac{1}{n}\right)=\frac{1}{2}$

(4) $\sqrt{2}$

(5) $\lim_{n\to\infty}\frac{\sin n}{n}=0$

(6) 0

(7) $\lim_{n\to\infty}\left(\frac{3}{n}\right)^{\frac{1}{n}} = \lim_{n\to\infty}\frac{3^{\frac{1}{n}}}{n^{\frac{1}{n}}}=\frac{1}{1}=1$

(8) ∞

(9) $\lim_{n\to\infty}\ln\left(1+\frac{1}{n}\right)^n = \ln\left(\lim_{n\to\infty}\left(1+\frac{1}{n}\right)^n\right)=\ln e = 1$

(10) 5

(11) $\lim_{n\to\infty} n\sin\frac{1}{n}= \lim_{n\to\infty}\frac{\sin\frac{1}{n}}{\frac{1}{n}}=1$

(12) 0

(13) $\lim_{n\to\infty}\frac{4\cdot 4^n+3^n}{4^n}= \lim_{n\to\infty}\frac{4+\left(\frac{3}{4}\right)^n}{1}=4$

(14) 0

EX 2

(1) $\lim_{n\to\infty} a_n = L,\ \ L=\frac{72}{1+L} \Rightarrow L=8$

(2) 2

(3) $\lim_{n\to\infty} a_n = L,\ \ L=\sqrt{8+2L} \Rightarrow L=4$

(4) 9

8.2 Infinite Series

EX 1

(1) 1

(2) $\lim_{n\to\infty}\sum_{k=1}^{n} 3\left(\frac{1}{2k-1}-\frac{1}{2k+1}\right)$

$= \lim_{n\to\infty} 3\left(1-\frac{1}{2n+1}\right)=3$

(3) 1

(4) $\lim_{n\to\infty}\sum_{n=1}^{n}\left(\frac{1}{n^2}-\frac{1}{(n+1)^2}\right)$

$=\left(1-\frac{1}{(n+1)^2}\right)=1$

(5) $-\frac{1}{\ln 2}$

(6) $\lim_{n\to\infty}\sum_{k=1}^{n}(\tan^{-1}k-\tan^{-1}(k+1))$

$= \lim_{n\to\infty}\tan^{-1}(1)-\tan^{-1}(n+1)$

$=\tan^{-1}1-\tan^{-1}\infty$

$=\frac{\pi}{4}-\frac{\pi}{2}=-\pi/4$

EX 2

(1) $2+\sqrt{2}$

(2) $\dfrac{3/2}{1-\left(-\frac{1}{2}\right)}=1$

(3) diverges

(4) $\dfrac{1}{1-\frac{1}{e^2}}=\dfrac{e^2}{e^2-1}$

(5) $\dfrac{2}{9}$

(6) $\dfrac{1}{1-\frac{2}{3}}-\dfrac{1}{1-\frac{1}{3}}=\dfrac{3}{2}$

(7) 4

(8) $\dfrac{1}{1-\frac{e}{\pi}}=\dfrac{\pi}{\pi-e}$

EX 3

(1) $|x|<\dfrac{1}{2}$, $\dfrac{1}{1-2x}$

(2) $|x+1|<1$ or $-2<x<0$, $\dfrac{1}{1+(x+1)}$
$=\dfrac{1}{2+x}$

(3) $x \neq (2k+1)\dfrac{\pi}{2}$, k an integer, $\dfrac{1}{1-\sin x}$

(4) $|\ln x|<1$ or $e^{-1}<x<e$, $\dfrac{1}{1-\ln x}$

EX 4

(1) $\lim\limits_{n\to\infty}\dfrac{n}{n+10}=1 = 1 \neq 0$

$\Rightarrow$diverges

(2) $\lim\limits_{n\to\infty}\dfrac{n(n+1)}{(n+2)(n+3)}$
$=\lim\limits_{n\to\infty}\dfrac{n^2+n}{n^2+5n+6}$
$=\lim\limits_{n\to\infty}\dfrac{2n+1}{2n+5}$
$=\lim\limits_{n\to\infty}\dfrac{2}{2}=1 \neq 0$

$\Rightarrow$ diverges

(3) $\lim\limits_{n\to\infty}\dfrac{1}{n+4}=0$

$\Rightarrow$ test inconclusive

(4) $\lim\limits_{n\to\infty}\dfrac{n}{n^2+3}-\lim\limits_{n\to\infty}\dfrac{1}{2n}=0$

$\Rightarrow$ test inconclusive

(5) $\lim\limits_{n\to\infty}\cos\dfrac{1}{n}=\cos 0=1 \neq 0$

$\Rightarrow$ diverses

(6) $\lim\limits_{n\to\infty}\dfrac{e^n}{e^n+n}$
$=\lim\limits_{n\to\infty}\dfrac{e^n}{e^n+1}$
$=\lim\limits_{n\to\infty}\dfrac{e^n}{e^n}$
$=\lim\limits_{n\to\infty}\dfrac{1}{1}=1 \neq 0$

$\Rightarrow$ diverges

(7) $\lim\limits_{n\to\infty}\ln\dfrac{1}{n}=-\infty \neq 0$

$\Rightarrow$ diverges

(8) $\lim\limits_{n\to\infty}\cos n\pi=$ does not exist

$\Rightarrow$ diverges

EX 5

(1) converges

(2) $\int_1^\infty \frac{1}{x^{0.2}}dx = \lim_{b\to\infty}\int_1^b \frac{1}{x^{0.2}}dx$
$= \lim_{b\to\infty}\left[\frac{5}{4}x^{0.8}\right]_1^b$
$= \infty$

(3) converges

(4) $\int_1^\infty \frac{1}{x+4}dx = \lim_{b\to\infty}\int_1^b \frac{1}{x+4}dx$
$= \lim_{n\to\infty}[\ln(x+4)]_1^b$
$= \infty$

(5) converges

(6) $\int_2^\infty \frac{1}{x(\ln x)^2}dx = \lim_{b\to\infty}\int_2^b \frac{1}{x(\ln x)^2}dx$
$= \lim_{b\to\infty}\left[-\frac{1}{\ln x}\right]_2^b$
$= \frac{1}{\ln 2}$, converges

(7) diverges

(8) $\int_2^\infty \frac{\ln x^2}{x}dx = \lim_{b\to\infty}\int_2^b \frac{2\ln x}{x}dx$
$= \lim_{b\to\infty}[(\ln x)^2]_2^b$
$= \infty$, diverges

EX 6

(1) converges

(2) Compare with $\frac{1}{\sqrt{n}-1} \geq \frac{1}{\sqrt{n}}$, $\sum \frac{1}{\sqrt{n}}$ is

divergent $\sum_{n=2}^\infty \frac{1}{\sqrt{n}-1}$ diverges

(3) converges

(4) Compare with $\sum_{n=1}^\infty \frac{1}{n^3}$, which is a

convergent p-series, since $p=3>1$.
Both series has nonnegative terms for $n \geq 1$. For $n \geq 1$, we have

$n^4 \leq n^4+2$
$\Rightarrow \frac{1}{n^4} \leq \frac{1}{n^4+2}$
$\Rightarrow \frac{n}{n^4} \leq \frac{n}{n^4+2}$
$\Rightarrow \frac{1}{n^3} > \frac{n}{n^4+2} > \frac{n-1}{n^4+2}$.

Then by Comparison Test,

$\sum_{n=1}^\infty \frac{n-1}{n^4+2}$ converges

(5) diverges

(6) Compare with $\sum_{n=1}^\infty \frac{1}{3^n}$, which is a

convergent geometric series, since

$|r| < \left|\frac{1}{3}\right| < 1$, Both series have

nonnegative terms for $n \geq 1$.

For $n \geq 1$, we have $n \cdot 3^n \geq 3^n$

$\Rightarrow \frac{1}{n \cdot 3^n} \leq \frac{1}{3^n}$.

Then by Comparison Test,

$\sum_{n=1}^\infty \frac{1}{n \cdot 3^n}$ converges.

EX 7

(1) converges

(2) Compare with $\sum_{n=1}^\infty \frac{1}{\sqrt{n}}$, which is a

divergent p-series, since $p=\frac{1}{2} \leq 1$.

$$\lim_{n\to\infty}\frac{a_n}{b_n} = \lim_{n\to\infty}\frac{\sqrt{\frac{n+1}{n^2+2}}}{1/\sqrt{n}} = \lim_{n\to\infty}\sqrt{\frac{n^2+n}{n^2+2}} = 1$$

Then by Limit Comparison Test,

$\sum_{n=1}^\infty \sqrt{\frac{n+1}{n^2+2}}$ diverges.

(3) diverges

(4) Compare with $\sum_{n=1}^{\infty}\frac{1}{2^n}$, which is a convergent geometric series, since $|r|=\left|\frac{1}{2}\right|<1$.

$$\lim_{n\to\infty}\frac{a_n}{b_n}=\lim_{n\to\infty}\frac{2^n}{3+4^n}/1/2^n$$

$$=\lim_{n\to\infty}\frac{4^n}{3+4^n}=1>0$$

Then by Limit comparison Test $\sum 2^n/3+4^n$ converges

(5) diverges

(6) Compare with $\sum_{n=1}^{\infty}\frac{1}{n^2}$, which a convergent p-series, since $p=2>1$.

$$\lim_{n\to\infty}\frac{a_n}{b_n}=\lim_{n\to\infty}\frac{\ln\left(1+\frac{1}{n^2}\right)}{1/n^2}$$

$=1>0$. Then by Limit comparision Test.

$\sum_{n=1}^{\infty}\ln(1+\frac{1}{n^2})$ converges

EX 8

(1) converges

(2) $$\lim_{n\to\infty}\frac{\frac{(n+1)+2}{3^{n+1}}}{\frac{n+2}{3^n}}=\lim_{n\to\infty}\left(\frac{n+3}{3^n\cdot 3}\cdot\frac{3^n}{n+2}\right)$$

$$=\lim_{n\to\infty}\left(\frac{n+3}{3n+6}\right)$$

$$=\lim_{n\to\infty}\left(\frac{1}{3}\right)=\frac{1}{3}<1$$

$$\Rightarrow\sum_{n=1}^{\infty}\frac{n+2}{3^n}\text{ converges}$$

(3) diverges

(4) $$\lim_{n\to\infty}\frac{\frac{2^{(n+1)+1}}{(n+1)\cdot 3^{(n+1)-1}}}{\frac{2^{n+1}}{n\cdot 3^{n-1}}}$$

$$=\lim_{n\to\infty}\left(\frac{2^{n+1}\cdot 2}{(n+1)\cdot 3^{n-1}\cdot 3}\cdot\frac{n\cdot 3^{n-1}}{2^{n+1}}\right)$$

$$=\lim_{n\to\infty}\frac{2n}{3n+3}=\frac{2}{3}$$

$$\Rightarrow\sum_{n=1}^{\infty}\frac{2^{n+1}}{n\cdot 3^{n-1}}\text{ converge}$$

(5) converges

(6) $$\lim_{n\to\infty}\left(\frac{\frac{3^{(n+1)+2}}{\ln(n+1)}}{\frac{3^{n+2}}{\ln n}}\right)=\lim_{n\to\infty}\left(\frac{3^{n+1}\cdot 3^2}{\ln(n+1)}\cdot\frac{\ln n}{3^{n+2}}\right)$$

$$=\lim_{n\to\infty}\left(\frac{3\ln n}{\ln(n+1)}\right)$$

$$=\lim_{n\to\infty}\frac{\frac{3}{n}}{\frac{1}{n+1}}$$

$$=\lim_{n\to\infty}\left(\frac{3n+3}{n}\right)$$

$$=\lim_{n\to\infty}\left(\frac{3}{1}\right)$$

$$=3>1$$

$$\Rightarrow\sum_{n=2}^{\infty}\frac{3^{n+2}}{\ln n}\text{ diverges}$$

(7) converges

EX 9

(1) converges

(2) $\frac{4^n}{(3n)^n}\geq 0$ for $n\geq 1$;

$$\lim_{n\to\infty}\sqrt[n]{\frac{4^n}{(3n)^n}}$$

$$=\lim_{n\to\infty}\left(\frac{4}{3n}\right)$$

$$=0<1$$

$$\Rightarrow\sum_{n=1}^{\infty}\frac{4^n}{(3n)^n}\text{ converges}$$

(3) diverges

(4) $\dfrac{8}{\left(3+\dfrac{1}{n}\right)^{2n}} \geq 0$ for all $n \geq 1$;

$$\lim_{n\to\infty}\sqrt[n]{\frac{8}{\left(3+\frac{1}{n}\right)^{2n}}}$$
$$=\lim_{n\to\infty}\left(\frac{\sqrt[n]{8}}{\left(3+\frac{1}{n}\right)^{2}}\right)$$
$$=\frac{1}{9}<1$$
$$\Rightarrow \sum_{n=1}^{\infty}\frac{8}{\left(3+\frac{1}{n}\right)^{2n}} \text{ converges}$$

(5) converges

(6) $\dfrac{1}{n^{1+n}} \geq 0$ for all $n \geq 2$;

$$\lim_{n\to\infty}\sqrt[n]{\frac{1}{n^{1+n}}}$$
$$=\lim_{n\to\infty}\left(\frac{\sqrt[n]{1}}{n^{1/n+1}}\right)$$
$$=\lim_{n\to\infty}\left(\frac{\sqrt[n]{1}}{n\sqrt[n]{n}}\right)$$
$$=0<1$$
$$\Rightarrow \sum_{n=2}^{\infty}\frac{1}{n^{1+n}} \text{ converges}$$

EX 10

(1) converges by the alternating convergence Test

(2) converges $\Rightarrow$ converges by Alternating Series Test since; $a_n=\dfrac{1}{n \cdot 3^n}>0$ for all $n \geq 1$; $\quad n \geq 1 \Rightarrow n+1 \geq n \Rightarrow 3^{n+1} \geq 3^n$ $\Rightarrow (n+1)3^{n+1} \geq n3^n$

$$\Rightarrow \frac{1}{(n+1)3^{n+1}} \leq \frac{1}{n3^n} \Rightarrow a_{n+1} \leq a_n;$$

$$\lim_{n\to\infty} a_n = \lim_{n\to\infty}\frac{1}{n \cdot 3^n}=0.$$

(3) Converges by the alternating convergence Test

(4) diverges $\Rightarrow$ diverges by n^{th} Term Test for Divergence since $\displaystyle\lim_{n\to\infty}\frac{2^n}{n^2}=\infty \Rightarrow$

$$\lim_{n\to\infty}(-1)^{n+1}\frac{2^n}{n^2}=\text{does not exist}$$

(5) diverges by the nth- Term Test

(6) converges by the Alternating Series Test since $f(x)=\dfrac{\ln x}{x} \Rightarrow f'(x)=\dfrac{1-\ln x}{x^2}<0$ when $x>e \Rightarrow f(x)$ is decreasing $\Rightarrow a_n \geq a_{n+1}$; also $a_n \geq 0$ for $n \geq 1$ and

$$\lim_{n\to\infty} a_n = \lim_{n\to\infty}\frac{\ln n}{n}=\lim_{n\to\infty}\frac{\left(\frac{1}{n}\right)}{1}=0.$$

EX 11

(1) converges absolutely since

$\displaystyle\sum_{n=1}^{\infty}|a_n|=\sum_{n=1}^{\infty}\left(\frac{1}{10}\right)^n$ a convergent geometric series

(2) converges conditionally since

$\dfrac{1}{\sqrt{n}}>\dfrac{1}{\sqrt{n+1}}>0$ and $\displaystyle\lim_{n\to\infty}\frac{1}{\sqrt{n}}=0 \Rightarrow$ convergence; but $\displaystyle\sum_{n=1}^{\infty}|a_n|=\sum_{n=1}^{\infty}\frac{1}{n^{1/2}}$ in a divergent p-series

(3) converges absolutely since

$\displaystyle\sum_{n=1}^{\infty}|a_n|=\sum_{n=1}^{\infty}\frac{n}{n^3+1}$ and $\dfrac{n}{n^3+1}<\dfrac{1}{n^2}$ which is the nth-term of a converging p-series

(4) converges conditionally since

$\dfrac{1}{n+3}>\dfrac{1}{(n+1)+3}>0$ and $\displaystyle\lim_{n\to\infty}\frac{1}{n+3}=0 \Rightarrow$ convergence;

but $\displaystyle\sum_{n=1}^{\infty}|a_n|=\sum_{n=1}^{\infty}\frac{1}{n+3}$ diverges because $\frac{1}{n+3} \geq \frac{1}{4n}$ and $\sum_{n=1}^{\infty}\frac{1}{n}$ is a divergent series

(5) diverges by the nth-Term Test since

$$\lim_{n\to\infty}\frac{3+n}{5+n}=1 \neq 0$$

(6) diverges by the nth- Term Test since

$$\lim_{n\to\infty} a_n = \lim_{n\to\infty} 10^{1/n} = 1 \neq 0$$

EX 12

(1) $|error| < \left|(-1)^6\left(\frac{1}{5}\right)\right| = 0.2$

(2) $|error| < \left|(-1)^6\left(\frac{1}{10^5}\right)\right| = 0.00001$

(3) $|error| < \left|(-1)^6\frac{(0.01)^5}{5}\right| = 2\times 10^{-11}$

(4) $|error| < |(-1)^4 t^4| = t^4 < 1$

EX 13

(1) $|error| < 0.001$
$\Rightarrow a_{n+1} < 0.001$
$\Rightarrow \frac{1}{(n+1)^2+3} < 0.001$
$\Rightarrow (n+1)^2+3 > 1000$
$\Rightarrow n > -1+\sqrt{997} \approx 30.575$
$\Rightarrow n \geq 31$

(2) $|error| < 0.001 \Rightarrow a_{n+1} < 0.001$
$\Rightarrow \frac{n+1}{(n+1)^2+1} < 0.001$
$\Rightarrow (n+1)^2+1 > 1000(n+1)$
$\Rightarrow n > \frac{998+\sqrt{998^2+4(998)}}{2} \approx 998.9999$
$\Rightarrow n \geq 999$

(3)
$|error| < 0.001 \Rightarrow a_{n+1} < 0.001$
$\Rightarrow \frac{1}{((n+1)+3\sqrt{n+1})^3} < 0.001$
$\Rightarrow (n+1+3\sqrt{n+1})^3 > 1000$
$\Rightarrow (\sqrt{n+1})^2 + 3\sqrt{n+1} - 10 > 0$
$\Rightarrow \sqrt{n+1} > \frac{-3+\sqrt{9+40}}{2} = 2$
$\Rightarrow n > 3$
$\Rightarrow n \geq 4$

8.3 Power series

EX 1

(1) $-1 < x \leqq 1$

(2) $\lim_{n\to\infty}(n+1)(x-3) = \infty$ unless $x=3$.

(3) $-1 \leq x \leq 1$

(4) Since $\lim_{n\to\infty}\left|\frac{a_{n+1}}{a_n}\right| = |x|$,

the series converges

if $|x| < 1$. We must the end-points: When x =1, we get the divergent harmonic series; x =-1 yields the convergent alternating harmonic series.

(5) $-\infty < x < \infty$

(6) $\lim_{n\to\infty}\left|\frac{x+1}{n+1}\right| = 0$ for all $x\neq-1$; since the given

series converges to 0 if x=-1, it therefore converges for all x.

(7) $1 \leqq x \leqq 3$

(8) $\left|\frac{a_{n+1}}{a_n}\right| = \left|\frac{(n+1)!x^{n+1}}{n!x^n}\right| = (n+1)|x|$

$\to\infty$ unless x=0. The series diverges for all values of x except x=0.

EX 2

(1) $P_3(x)=1+2x+2x^2+\frac{4}{3}x^3$

(2) $f(x)=\ln x, f'(x)=\frac{1}{x}, f''(x)=-\frac{1}{x^2}$,

$f'''(x)=\frac{2}{x^3}; f(1)=\ln 1=0, f'(1)=1$,

$f''(1)=-1, f'''(1)=2$,

$P_3(x)=(x-1)-\frac{1}{2}(x-1)^2+\frac{1}{3}(x-1)^3$

(3) $P_3(x)$

$=\frac{1}{2}-\frac{1}{4}(x-2)+\frac{1}{8}(x-2)^2-\frac{1}{16}(x-2)^3$

(4) $f(x)=\sin x, f'(x)=\cos x, f''(x)=-\sin x$,

$f'''(x)=-\cos x; f(\frac{\pi}{4})=\cos\frac{\pi}{4}=\frac{\sqrt{2}}{2}$,

$f''(\frac{\pi}{4})=\cos\frac{\pi}{4}=\frac{\sqrt{2}}{2}$,

$f''(\frac{\pi}{4})=-\sin\frac{\pi}{4}=-\frac{\sqrt{2}}{2}$,

$f'''(\frac{\pi}{4})=-\cos\frac{\pi}{4}=-\frac{\sqrt{2}}{2}$,

$P_3(x)=\frac{\sqrt{2}}{2}+\frac{\sqrt{2}}{2}(x-\frac{\pi}{4})-$

$\frac{\sqrt{2}}{4}(x-\frac{\pi}{4})^2-\frac{\sqrt{2}}{12}(x-\frac{\pi}{4})^3$

(5) $P_3(x)$

$=2+\frac{1}{4}(x-4)-\frac{1}{64}(x-4)^2+\frac{1}{512}(x-4)^3$

(6)

$f(x)=\tan x, f'(x)=\sec^2 x, f''(x)=2\sec^2 x\tan x$,

$f'''(x)=2\sec^4 x+4\sec^2 x\tan^2 x; f(\frac{\pi}{4})=\tan\frac{\pi}{4}=1$,

$f'(\frac{\pi}{4})=\sec^2(\frac{\pi}{4})=2, f''(\frac{\pi}{4})=2\sec^2(\frac{\pi}{4})\tan(\frac{\pi}{4})=4$

$f'''(\frac{\pi}{4})=2\sec^4(\frac{\pi}{4})+4\sec^2(\frac{\pi}{4})\tan^2(\frac{\pi}{4})=16$

$P_3(x)=1+2(x-\frac{\pi}{4})+2(x-\frac{\pi}{4})^2+\frac{8}{3}(x-\frac{\pi}{4})^3$

EX 3

(1) $1-x+\frac{1}{2}x^2-\frac{1}{6}x^3+\cdots=\sum_{n=0}^{\infty}\frac{(-1)^n}{n!}x^n$

(2) $f(x)=xe^x, f'(x)=xe^x+e^x$,

$f''(x)=xe^x+2e^x, f'''(x)=xe^x+3e^x$

$\Rightarrow ...f^{(k)}(x)=xe^x+ke^x; f(0)=(0)e^{(0)}=0$

$f'(0)=1, f''(0)=2, f'''(0)=3, ..., f^{(k)}(0)=k$

$\Rightarrow x+x^2+\frac{1}{2}x^3+\cdots=\sum_{n=1}^{\infty}\frac{1}{(n-1)!}x^n$

(3) $1-x+x^2-x^3+\cdots=\sum_{n=0}^{\infty}(-x)^n$

(4) $f(x)=x^4-2x^3-5x+4 \Rightarrow f'(x)=4x^3-6x^2-5$,

$f''(x)=12x^2-12x, f'''(x)=24x-12, f^{(4)}(x)=24$

$\Rightarrow f^{(n)}(x)=0$ if $n\ge 5; f(0)=4, f'(0)=-5, f''(0)=0$

$f'''(0)=-12,\ f^{(4)}(0)=24, f^{(n)}(0)=0$ if $n\ge 5$

$\Rightarrow x^4-2x^3-5x+4=4-5x-\frac{12}{3!}x^3+\frac{24}{4!}x^4$

EX 4

(1) $8+10(x-2)+\frac{12}{2!}(x-2)^2+\frac{6}{3!}(x-2)^3$

(2) $f(x)=x^{-2}$

$\Rightarrow f'(x)=-2x^{-3}, f''(x)=3!x^{-4}, f'''(x)=-4!x^{-5}$

$\Rightarrow f^{(n)}(x)=(-1)^n(n+1)!x^{-n-2}$;

$f(1)=1, f'(1)=-2, f''(1)=3!, f'''(1)=-4!$,

$f^{(n)}(1)=(-1)^n(n+1)! \Rightarrow \frac{1}{x^2}$

$=1-2(x-1)+3(x-1)^2-4(x-1)^3+\cdots$

$=\sum_{n=0}^{\infty}(-1)^n(n+1)(x-1)^n$

(3) $e^2+e^2(x-2)+\frac{e^2}{2}(x-2)^2+\frac{e^2}{3!}(x-2)^3+\cdots$

$=\sum_{n=0}^{\infty}\frac{e^2}{n!}(x-2)^n$

(4)

$f(x)=2^x \Rightarrow f'(x)=2^x\ln 2, f''(x)=2^x(\ln 2)^2, f'''(x)=2^x(\ln 2)^3$

$\Rightarrow f^{(n)}(x)=2^x(\ln 2)^n; f(1)=2, f'(1)=2\ln 2, f''(1)=2(\ln 2)^2$,

$f'''(1)=2(\ln 2)^3, ..., f^{(n)}(1)=2(\ln 2)^n$

$\Rightarrow 2^x=2+(2\ln 2)(x-1)+\frac{2(\ln 2)^2}{2}(x-1)^2+\frac{2(\ln 2)^3}{3!}(x-1)^3$

$+\cdots=\sum_{n=0}^{\infty}\frac{2(\ln 2)^n(x-1)^n}{n!}$

EX 5

(1) $P_1(x)=f(0)+f'(0)x=-4+f'(0)x$

$P_1(\frac{1}{2})=-4+f'(0)\cdot\frac{1}{2}=-3$

$f'(0)\cdot\frac{1}{2}=1$

$f'(0)=2$

(2) $P_3(x)=-4+2x+(-\frac{2}{3})\cdot\frac{x^2}{2!}+\frac{1}{3}\cdot\frac{x^3}{3!}$

$=-4+2x-\frac{1}{3}x^2+\frac{1}{18}x^3$

(3) Let $Q_n(x)$ denote the Taylor polynomial of degree n for h about x=0.

$h'(x)=f(2x), h''(x)=2f'(2x),$

$h'''(x)=4f''(2x)$

$h'(0)=f(0)=-4, h''(0)=2f'(0)=4,$

$h'''(0)=4f''(0)=-\frac{8}{3}$

$Q_3(x)=7-4x+4\cdot\frac{x^2}{2!}-\frac{8}{3}\cdot\frac{x^3}{3!}$

$=7-4x+2x^2-\frac{4}{9}x^3$

EX 6

(1) $\left|\frac{x^{2n+3}}{2n+5}\cdot\frac{2n+3}{x^{2n+1}}\right|=\left(\frac{2n+3}{2n+5}\right)\cdot x^2$

$\lim_{n\to\infty}\left(\frac{2n+3}{2n+5}\right)\cdot x^2=x^2$

$x^2<1 \Rightarrow -1<x<1$

The series converges when $-1<x<1$.

When x = -1, the series is

$-\frac{1}{3}+\frac{1}{5}-\frac{1}{7}+\frac{1}{9}-\cdots$ This series converges by the Alternating Series Test.

When x=1, the series is

$\frac{1}{3}-\frac{1}{5}+\frac{1}{7}-\frac{1}{9}+\cdots$ This series converges by the Alternating Series Test. Therefore, the interval of convergence is $-1\le x\le 1$.

(2) $\left|g\left(\frac{1}{2}\right)-\frac{17}{120}\right|<\frac{\left(\frac{1}{2}\right)^5}{7}=\frac{1}{224}<\frac{1}{200}$

(3)

$g'(x)=\frac{1}{3}-\frac{3}{5}x^2+\frac{5}{7}x^4-\cdots+(-1)^n\left(\frac{2n+1}{2n+3}\right)x^{2n}-\cdots$

EX 7

(1) The power series is geometric with ratio (x +1).

The series converges if and only if $|x+1|<1$.

Therefore, the interval of convergence is $-2<x<0$.

or

$\lim_{n\to\infty}\left|\frac{(x+1)^{n+1}}{(x+1)^n}\right|=|x+1|<1$ when $-2<x<0$

At x=-2, the series is $\sum_{n=0}^{\infty}(-1)^n$, which diverges since the terms do not converge to 0. At x=0, the series is $\sum_{n=0}^{\infty}1$, which similarly diverges. Therefore, the interval of convergence is $-2<x<0$.

(2) Since the series is geometric,

$f(x)=\sum_{n=0}^{\infty}(x+1)^n$

$=\frac{1}{1-(x+1)}$

$=-\frac{1}{x}$ for $-2<x<0$.

(3) $g\left(-\frac{1}{2}\right)=\int_{-1}^{-\frac{1}{2}}-\frac{1}{x}dx=\ln 2$

(4) $h(x)=f(x^2-1)=1+x^2+x^4+\cdots+x^{2n}+\cdots$

$$h\left(\frac{1}{2}\right)=f\left(-\frac{3}{4}\right)=\frac{1}{1-\frac{1}{4}}=\frac{4}{3}$$

EX 8

We first make a table of the derivatives, evaluated at $x=0$ and giving us the coefficients.

n	$f^{(n)}(x)$	$f^{(n)}(0)$	$c_n=\frac{f^{(n)}(0)}{n!}$
0	$\cos\left(x+\frac{\pi}{4}\right)$	$\frac{\sqrt{2}}{2}$	$\frac{\sqrt{2}}{2}$
1	$-\sin\left(x+\frac{\pi}{4}\right)$	$-\frac{\sqrt{2}}{2}$	$-\frac{\sqrt{2}}{2}$
2	$-\cos\left(x+\frac{\pi}{4}\right)$	$-\frac{\sqrt{2}}{2}$	$-\frac{\sqrt{2}}{2\cdot 2!}$
3	$\sin\left(x+\frac{\pi}{4}\right)$	$\frac{\sqrt{2}}{2}$	$\frac{\sqrt{2}}{2\cdot 3!}$

Thus

$$P_3(x)=\frac{\sqrt{2}}{2}-\frac{\sqrt{2}}{2}x-\frac{\sqrt{2}}{2\cdot 2!}x^2+\frac{\sqrt{2}}{2\cdot 3!}x^3.$$

Since this is not an alternating series for $x-0.1$, we must use the Lagrange error bound:

$$R_3(x)<\max\left|\frac{f^{(4)}(c)}{4!}x^4\right|,$$

where $x=0.1$ and $0<c<0.1$.

Note that $f^{(4)}(c)=\cos\left(c+\frac{\pi}{4}\right)$ is decreasing on the interval $0<c<0.1$, so its maximum value occurs at $c=0$. Hence:

$$R_3(x)<\max\left|\frac{\cos\left(c+\frac{\pi}{4}\right)}{4!}x^4\right|=\frac{\cos\left(0+\frac{\pi}{4}\right)}{4!}0.1^4$$

$$=\frac{\sqrt{2}}{2\cdot 4!\cdot 10^4}$$

EX 9

(1) $\sin x=x-\frac{x^3}{3!}+\frac{x^5}{5!}-\frac{x^7}{7!}+\cdots$

$\sin(x^2)=x^2-\frac{x^6}{3!}+\frac{x^{10}}{5!}-\frac{x^{14}}{7!}+\cdots$

(2) $\cos x=1-\frac{x^2}{2!}+\frac{x^4}{4!}-\frac{x^6}{6!}+\cdots$

$f(x)=1+\frac{x^2}{2}+\frac{x^4}{4!}-\frac{121x^6}{6!}+\cdots$

(3) $\frac{f^{(6)}(0)}{6!}$ is the coefficient of x^6 in the Taylor series for f about $x=0$. Therefore $f^{(6)}(0)=-121$.

(4) The graph of $y=|f^{(5)}x|$ indicates that $\max_{0\le x\le\frac{1}{4}}|f^{(5)}(x)|<40$.

Therefore

$$\left|P_4\left(\frac{1}{4}\right)-f\left(\frac{1}{4}\right)\right|\le\max_{0\le x\le\frac{1}{4}}|f^{(5)}(x)|/5!$$

$$\cdot\left(\frac{1}{4}\right)^5<\frac{40}{120\cdot 4^5}=\frac{1}{3072}<\frac{1}{3000}.$$

EX 10

$f(x)=\ln(1+x)$; $\quad f(0)=0$;

$f'(x)=-\frac{1}{1+x}$; $\quad f'(0)=1$;

$f''(x)=-\frac{1}{(1+x)^2}$; $\quad f''(0)=-1$;

$f'''(x)=\frac{2}{(1+x)^3}$; $\quad f'''(0)=2!$;

$\vdots$

$f^{(n)}(x)=\frac{(-1)^{n-1}(n-1)!}{(1+x)^n}$; $\quad f^{(n)}(0)=(-1)^{n-1}(n-1)!$

$f^{(n+1)}(x)=\frac{(-1)^n\cdot n!}{(1+x)^{n+1}}$

Then

$\ln(1+x)$
$=x-\frac{x^2}{2}+\frac{x^3}{3}-\frac{x^4}{4}+\cdots+(-1)^{n-1}\cdot\frac{x^n}{n}+R_n(x),$

where the Lagrange error bound is

$$R_n(x)<\max\left|\frac{(-1)^n}{(1+c)^{n+1}}\cdot\frac{x^{n+1}}{n+1}\right|.$$

EX 11

(1) $e^x=1+x+\frac{x^2}{2!}+\cdots=\sum_{n=0}^{\infty}\frac{x^n}{n!}$
$\Rightarrow e^{-5x}=1+(-5x)+\frac{(-5x)^2}{2!}+\cdots$
$=1-5x+\frac{5^2x^2}{2!}-\frac{5^3x^3}{3!}+\cdots$
$=\sum_{n=0}^{\infty}\frac{(-1)^n5^nx^n}{n!}$

(2) $\sin x=x-\frac{x^3}{3!}+\frac{x^5}{3!}-\cdots=\sum_{n=0}^{\infty}\frac{(-1)^nx^{2n+1}}{(2n+1)!}\Rightarrow$

$\sin\frac{\pi x}{2}=\frac{\pi x}{2}-\frac{(\frac{\pi x}{2})^3}{3!}+\frac{(\frac{\pi x}{2})^5}{5!}-\frac{(\frac{\pi x}{2})^7}{7!}+\cdots$
$=\sum_{n=0}^{\infty}\frac{(-1)^n\pi^{2n+1}x^{2n+1}}{2^{2n+1}(2n+1)!}$

(3) $\ln(1+x)=\sum_{n=1}^{\infty}\frac{(-1)^{n-1}x^n}{n}\Rightarrow\ln(1+x^2)$
$=\sum_{n=1}^{\infty}\frac{(-1)^{n-1}(x^2)^n}{n}=\sum_{n=1}^{\infty}\frac{(-1)^{n-1}x^{2n}}{n}$
$=x^2-\frac{x^4}{2}+\frac{x^6}{3}-\frac{x^8}{4}+\cdots$

(4)

$\frac{1}{1-x}=\sum_{n=0}^{\infty}x^n\Rightarrow\frac{1}{2-x}=\frac{1}{2}\frac{1}{1-\frac{1}{2}x}=\frac{1}{2}\sum_{n=0}^{\infty}(\frac{1}{2}x)^n$
$=\frac{1}{2}\sum_{n=0}^{\infty}(\frac{1}{2})^nx^n=\frac{1}{2}+\frac{1}{4}x+\frac{1}{8}x^2+\frac{1}{16}x^3+\cdots$

EX 12

(1) $e^x=\sum_{n=0}^{\infty}\frac{x^n}{n!}\Rightarrow xe^x=x(\sum_{n=0}^{\infty}\frac{x^n}{n!})=\sum_{n=0}^{\infty}\frac{x^{n+1}}{n!}$
$=x+x^2+\frac{x^3}{2!}+\frac{x^4}{3!}+\frac{x^5}{4!}+\cdots$

(2) $\cos x=\sum_{n=0}^{\infty}\frac{(-1)^nx^{2n}}{(2n)!}\Rightarrow x\cos\pi x$
$=x\sum_{n=0}^{\infty}\frac{(-1)^n(\pi x)^{2n}}{(2n)!}=\sum_{n=0}^{\infty}\frac{(-1)^n\pi^{2n}x^{2n+1}}{(2n)!}$
$=x-\frac{\pi^2x^3}{2!}+\frac{\pi^4x^5}{4!}-\frac{\pi^6x^7}{6!}+\cdots$

(3) $x\ln(1+2x)=x\sum_{n=1}^{\infty}\frac{(-1)^{n-1}(2x)^n}{n}$
$=\sum_{n=1}^{\infty}\frac{(-1)^{n-1}2^nx^{n+1}}{n}$
$=2x^2-\frac{2^2x^3}{2}+\frac{2^3x^4}{3}-\frac{2^4x^5}{4}+\cdots$

(4)

$\tan^{-1}x=x-\frac{1}{3}x^3+\frac{1}{5}x^5-\frac{1}{7}x^7+\cdots\Rightarrow x\tan^{-1}x^2$
$=x\left(x^2-\frac{1}{3}(x^2)^3+\frac{1}{5}(x^2)^5-\frac{1}{7}(x^2)^7+\cdots\right)$
$=x^3-\frac{1}{3}x^7+\frac{1}{5}x^{11}-\frac{1}{7}x^{15}+\cdots$
$=\sum_{n=1}^{\infty}\frac{(-1)^{n+1}x^{4n-1}}{2n-1}$

EX 13

(1)

$\int_0^{0.2}\sin x^2dx=\int_0^{0.2}\left(x^2-\frac{x^6}{3!}+\frac{x^{10}}{5!}-\cdots\right)dx$
$=\left[\frac{x^3}{3}-\frac{x^7}{7\cdot3!}+\cdots\right]_0^{0.2}\approx\left[\frac{x^3}{3}\right]_0^{0.2}\approx0.00267\, with\, error$
$|E|\le\frac{(0.2)^7}{7\cdot3!}\approx0.0000003$

(2)

$\int_0^{0.2}\frac{e^{-x}-1}{x}dx$
$=\int_0^{0.2}\frac{1}{x}\left(1-x+\frac{x^2}{2!}-\frac{x^3}{3!}+\frac{x^4}{4!}-\cdots-1\right)dx$
$=\int_0^{0.2}\left(-1+\frac{x}{2}-\frac{x^2}{6}+\frac{x^3}{24}-\cdots\right)dx$
$=\left[-x+\frac{x^2}{4}-\frac{x^3}{18}+\cdots\right]_0^{0.2}$
$\approx-0.19044\ with\ error\,|E|\le\frac{(0.2)^4}{96}\approx0.00002$

EX 14

(1) $\int_0^{0.1} \frac{\sin x}{x} dx = \int_0^{0.1} \left(1 - \frac{x^2}{3!} + \frac{x^4}{5!} - \frac{x^6}{7!} + \cdots\right) dx$

$= \left[x - \frac{x^3}{3\cdot 3!} + \frac{x^5}{5\cdot 5!} - \frac{x^7}{7\cdot 7!} + \cdots\right]_0^{0.1}$

$\approx \left[x - \frac{x^3}{3\cdot 3!} + \frac{x^5}{5\cdot 5!}\right]_0^{0.1}$

$\approx 0.0999444611,\ |E| \le \frac{(0.1)^7}{7\cdot 7!} \approx 2.8 \times 10^{-12}$

(2) $\int_0^{0.1} \exp(-x^2)\, dx$

$= \int_0^{0.1} \left(1 - x^2 + \frac{x^4}{2!} - \frac{x^6}{3!} + \frac{x^8}{4!} - \cdots\right) dx$

$= \left[x - \frac{x^3}{3} + \frac{x^5}{10} + \frac{x^7}{42} + \cdots\right]_0^{0.1}$

$\approx \left[x - \frac{x^3}{3} + \frac{x^5}{10} + \frac{x^7}{42}\right]_0^{0.1}$

$\approx 0.0996676643,\ |E| \le \frac{(0.1)^9}{216} \approx 4.6 \times 10^{-12}$

(3) $\int_0^1 \left(\frac{1-\cos x}{x^2}\right) dx$

$= \int_0^1 \left(\frac{1}{2} - \frac{x^2}{4!} + \frac{x^4}{6!} - \frac{x^6}{8!} + \frac{x^8}{10!} - \cdots\right) dx$

$\approx \left[\frac{x}{2} - \frac{x^3}{3\cdot 4!} + \frac{x^5}{5\cdot 6!} - \frac{x^7}{7\cdot 8!} + \frac{x^9}{9\cdot 10!}\right]_0^1$

$\approx 0.4863853764,\ |E| \le \frac{1}{11\cdot 12!} \approx 1.9 \times 10^{-10}$

EX 15

(1) $\frac{1}{x^2}(e^x - (1+x))$

$= \frac{1}{x^2}\left(\left(1 + x + \frac{x^2}{2} + \frac{x^3}{3!} + \cdots\right) - 1 - x\right)$

$= \frac{1}{2} + \frac{x}{3!} + \frac{x^2}{4!} + \cdots \Rightarrow \lim_{x\to 0} \frac{e^x - (1+x)}{x^2}$

$= \lim_{x\to 0}\left(\frac{1}{2} + \frac{x}{3!} + \frac{x^2}{4!} + \cdots\right) = \frac{1}{2}$

(2) $\frac{1}{t^4}\left(1 - \cos t - \frac{t^2}{2}\right)$

$= \frac{1}{t^4}\left[1 - \frac{t^2}{2} - \left(1 - \frac{t^2}{2} + \frac{t^4}{4!} - \frac{t^6}{6!} + \cdots\right)\right]$

$= -\frac{1}{4!} + \frac{t^2}{6!} - \frac{t^4}{8!} + \cdots \Rightarrow \lim_{t\to 0} \frac{1 - \cos t - \left(\frac{t^2}{2}\right)}{t^4}$

$= \lim_{t\to 0}\left(-\frac{1}{4!} + \frac{t^2}{6!} - \frac{t^4}{8!} + \cdots\right) = -\frac{1}{24}$

(3) $\frac{1}{y^3}(y - \tan^{-1} y) = \frac{1}{y^3}\left[y - \left(y - \frac{y^3}{3} + \frac{y^5}{5} - \cdots\right)\right]$

$= \frac{1}{3} - \frac{y^2}{5} + \frac{y^4}{7} - \cdots \Rightarrow \lim_{y\to 0} \frac{y - \tan^{-1} y}{y^3}$

$= \lim_{y\to 0}\left(\frac{1}{3} - \frac{y^2}{5} + \frac{y^4}{7} - \cdots\right) = \frac{1}{3}$

(4) $\frac{\ln(1+x^2)}{1-\cos x} = \frac{\left(x^2 - \frac{x^4}{2} + \frac{x^6}{3} - \cdots\right)}{1 - \left(1 - \frac{x^2}{2!} + \frac{x^4}{4!} - \cdots\right)}$

$= \frac{\left(1 - \frac{x^2}{2} + \frac{x^4}{3} - \cdots\right)}{\left(\frac{1}{2!} - \frac{x^2}{4!} + \cdots\right)} \Rightarrow \lim_{x\to 0} \frac{\ln(1+x^2)}{1 - \cos x}$

$= 2$

EX 16

(1) $x^3 - \frac{x^6}{2} + \frac{x^9}{3} + \cdots + (-1)^{n+1} \cdot \frac{x^{3n}}{n} + \cdots$

(2) The interval of convergence is centered at $x = 0$. At $x = -1$, the series is

$-1 - \frac{1}{2} - \frac{1}{3} - \frac{1}{4} - \cdots - \frac{1}{n} - \cdots$, which diverges

because the harmonic series diverges. At $x = 1$,

the series is

$1 - \frac{1}{2} + \frac{1}{3} - \frac{1}{4} + \cdots + (-1)^{n+1} \cdot \frac{1}{n} \cdots$,

the alternating harmonic series, which converges.

Therefore the interval of convergence is $-1 < x \le 1$.

(3) The Maclaurin series for $f'(x), f'(t^2)$, and $g(x)$ are

$$f'(x): \sum_{n=1}^{\infty}(-1)^{n+1}\cdot 3x^{3n-1}$$
$$=3x^2-3x^5+3x^8-3x^{11}+\cdots$$

$$f'(t^2): \sum_{n=1}^{\infty}(-1)^{n+1}\cdot 3t^{6n-2}$$
$$=3t^4-3t^{10}+3t^{16}-3t^{22}+\cdots$$

$$g(x): \sum_{n=1}^{\infty}(-1)^{n+1}\cdot\frac{3x^{6n-1}}{6n-1}$$
$$=\frac{3x^5}{5}-\frac{3x^{11}}{11}+\frac{3x^{17}}{17}-\frac{3x^{23}}{23}+\cdots$$

Thus $g(1)\approx\frac{3}{5}-\frac{3}{11}=\frac{18}{55}$.

(4) The Maclaurin series for g evaluated at $x=1$ is alternating, and the terms decrease in absolute value to 0.

Thus $\left|g(1)-\frac{18}{55}\right|<\frac{3\cdot 1^{17}}{17}=\frac{3}{17}<\frac{1}{5}$.

EX 17

(1) e^{-x^2}

$$=1+\frac{(-x^2)}{1!}+\frac{(-x^2)^2}{2!}+\frac{(-x^2)^3}{3!}+\cdots+\frac{(-x^2)^n}{n!}+\cdots$$
$$=1-x^2+\frac{x^4}{2}-\frac{x^6}{6}+\cdots+\frac{(-1)^n x^{2n}}{n!}$$

(2) $\frac{1-x^2-f(x)}{x^4}=-\frac{1}{2}+\frac{x^2}{6}+\sum_{n=4}^{\infty}\frac{(-1)^{n+1}x^{2n-4}}{n!}$

Thus, $\lim_{x\to 0}\left(\frac{1-x^2-f(x)}{x^4}\right)=-\frac{1}{2}$.

(3) $\int_0^x e^{-t^2}dt$

$$=\int_0^x\left(1-t^2+\frac{t^4}{2}-\frac{t^6}{6}+\cdots+\frac{(-1)^n t^{2n}}{n!}+\cdots\right)dt$$
$$=x-\frac{x^3}{3}+\frac{x^5}{10}-\frac{x^7}{42}+\cdots$$

Using the first two terms of this series, we estimate that

$$\int_0^{1/2}e^{-t^2}dt\approx\frac{1}{2}-\left(\frac{1}{3}\right)\left(\frac{1}{8}\right)=\frac{11}{24}.$$

(4) $\left|\int_0^{1/2}e^{-t^2}dt-\frac{11}{24}\right|<\left(\frac{1}{2}\right)^5\cdot\frac{1}{10}=\frac{1}{320}<\frac{1}{200}$,

since $\int_0^{1/2}e^{-t^2}dt=\sum_{n=0}^{\infty}\frac{(-1)^n\left(\frac{1}{2}\right)^{2n+1}}{n!(2n+1)}$, which is an alternating series with individual terms that decrease in absolute value to 0.

EX 18

(1)

$$\frac{1}{1-u}=1+u+u^2+\cdots+\cdots$$
$$\frac{1}{1+x^2}=1-x^2+x^4-x^6+\cdots+(-x^2)^n+\cdots$$
$$\frac{2x}{1+x^2}=2x-2x^3+2x^5-2x^7+\cdots+(-1)^n 2x^{2n+1}+\cdots$$

(2) No, the series does not converge when $x=1$ because when $x=1$, the terms of the series do not converge to 0.

(3) $\ln(1+x^2)=\int_0^x\frac{2t}{1+t^2}dt$

$$=\int_0^x(2t-2t^3+2t^5-2t^7+\cdots)dt$$
$$=x^2-\frac{1}{2}x^4+\frac{1}{3}x^6-\frac{1}{4}x^8+\cdots$$

(4)

$$\ln\left(\frac{5}{4}\right)=\ln\left(1+\frac{1}{4}\right)$$
$$=\left(\frac{1}{2}\right)^2-\frac{1}{2}\left(\frac{1}{2}\right)^4+\frac{1}{3}\left(\frac{1}{2}\right)^6-\frac{1}{4}\left(\frac{1}{2}\right)^8+\cdots$$

Let $A=\left(\frac{1}{2}\right)^2-\left(\frac{1}{2}\right)\left(\frac{1}{2}\right)^4=\frac{7}{32}$.

Since the series is a converging alternating series and the absolute values of the individual terms decrease to 0,

$$\left|A-\ln\left(\frac{5}{4}\right)\right|<\left|\frac{1}{3}\left(\frac{1}{2}\right)^6\right|=\frac{1}{3}\cdot\frac{1}{64}<\frac{1}{100}.$$

CHAPTER 09.

Parametric Equations and Polar Coordinates

9.1 Parametric Equations

EX 1

(1) $x=-\sqrt{t}, y=t, t\ge 0 \Rightarrow x=\sqrt{y}$
or $y=x^2, x\le 0$

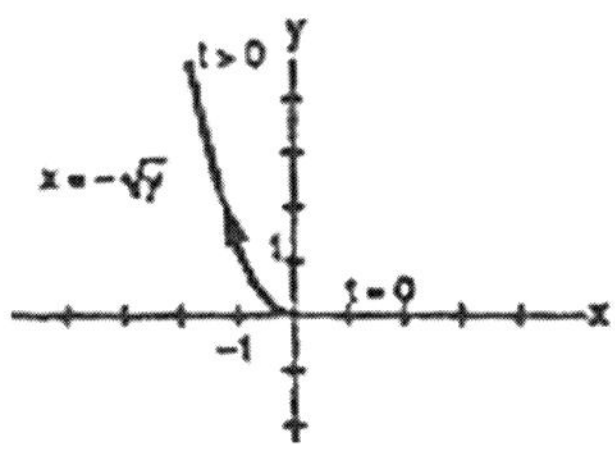

(2) $x=3-3t, y=2t, 0\le t\le 1 \Rightarrow \dfrac{y}{2}=t$
$\Rightarrow x=3-3\left(\dfrac{y}{2}\right) \Rightarrow 2x=6-3y$
$\Rightarrow y=2-\dfrac{2}{3}x, 0\le x\le 3$

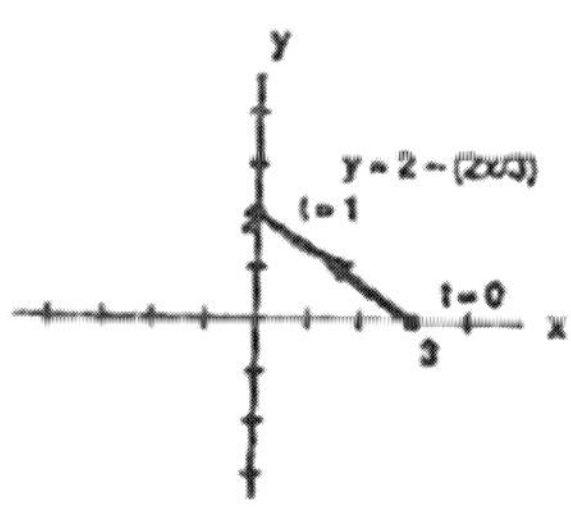

(3) $x=\cos 2t, y=\sin 2t, 0\le t\le \pi$
$\Rightarrow \cos^2 2t+\sin^2 2t=1 \Rightarrow x^2+y^2=1$

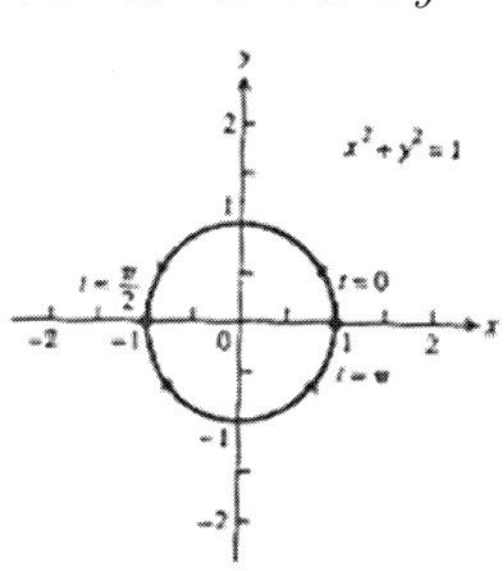

(4) $x=4\cos t, y=2\sin t, 0\le t\le 2\pi$
$\Rightarrow \dfrac{16\cos^2 t}{16}+\dfrac{4\sin^2 t}{4}=1 \Rightarrow \dfrac{x^2}{16}+\dfrac{y^2}{4}=1$

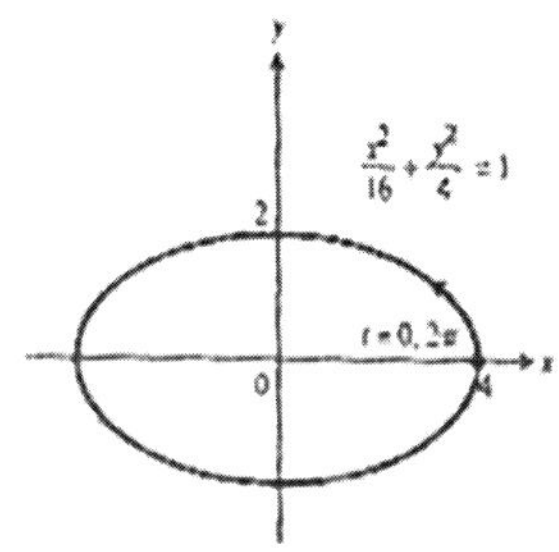

(5) $x=\sin t, y=\cos 2t, -\dfrac{\pi}{2}\le t\le \dfrac{\pi}{2}$
$\Rightarrow y=\cos 2t=1-2\sin^2 t \Rightarrow y=1-2x^2$

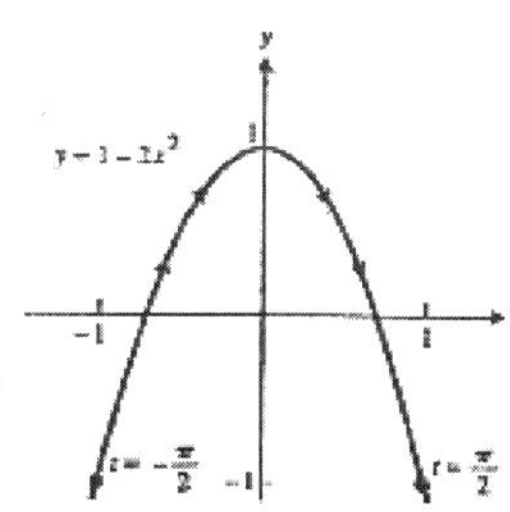

(6) $x=1+\sin t, y=\cos t-2, 0\le t\le \pi$
$\Rightarrow \sin^2 t+\cos^2 t=1 \Rightarrow (x-1)^2+(y+2)^2=1$

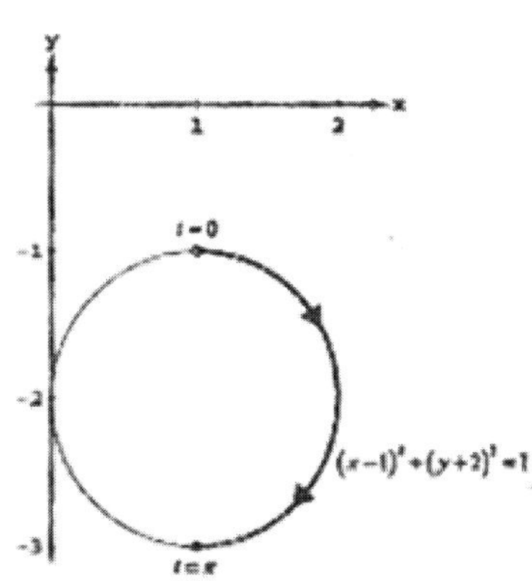

(7) $x=t, y=\sqrt{1-t^2}, -1 \le t \le 0$
$\Rightarrow y=\sqrt{1-x^2}$

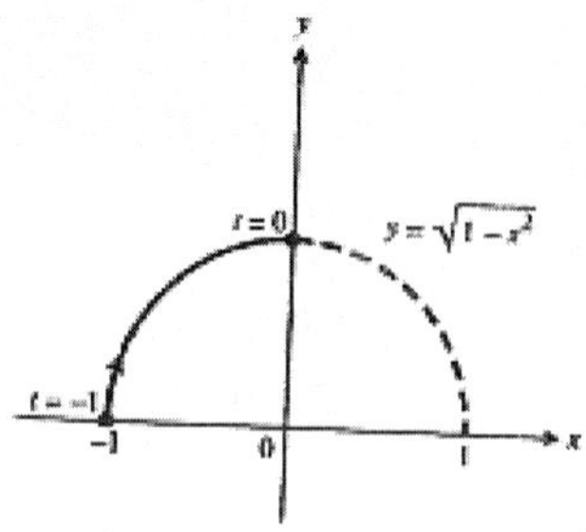

(8) $x=\sec^2 t-1, y=\tan t, -\dfrac{\pi}{2}<t<\dfrac{\pi}{2}$
$\Rightarrow \sec^2 t-1=\tan^2 t \Rightarrow x=y^2$

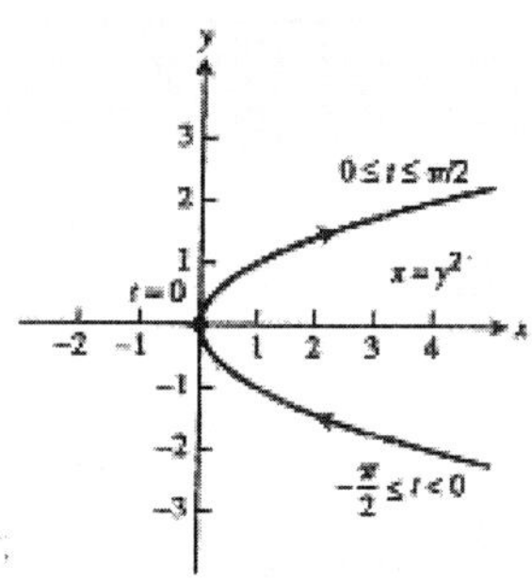

EX 2

(1) $y=-x+2\sqrt{2}, \quad \dfrac{d^2y}{dx^2}=-\sqrt{2}$

(2) $t=\dfrac{\pi}{4}$

$\Rightarrow x=4\sin\dfrac{\pi}{4}=2\sqrt{2}, y=2\cos\dfrac{\pi}{4}=\sqrt{2};$

$\dfrac{dx}{dt}=4\cos t, \dfrac{dy}{dt}=-2\sin t$

$\Rightarrow \dfrac{dy}{dx}=\dfrac{dy/dt}{dx/dt}=\dfrac{-2\sin t}{4\cos t}=-\dfrac{1}{2}\tan t$

$\Rightarrow \left.\dfrac{dy}{dx}\right|_{t=\frac{\pi}{4}}=-\dfrac{1}{2}\tan\dfrac{\pi}{4}=-\dfrac{1}{2};$

tangent line is $y-\sqrt{2}=-\dfrac{1}{2}(x-2\sqrt{2})$ or

$y=-\dfrac{1}{2}x+2\sqrt{2};$

$\dfrac{dy'}{dt}=-\dfrac{1}{2}\sec^2 t$

$\Rightarrow \dfrac{d^2y}{dx^2}=\dfrac{dy'/dt}{dx/dt}=\dfrac{-\frac{1}{2}\sec^2 t}{4\cos t}=-\dfrac{1}{8\cos^3 t}$

$\Rightarrow \left.\dfrac{d^2y}{dx^2}\right|_{t=\frac{\pi}{4}}=-\dfrac{\sqrt{2}}{4}$

(3) $y=x+\dfrac{1}{4}, \quad \dfrac{d^2y}{dx^2}=-2$

(4) $t=\dfrac{\pi}{6}$

$\Rightarrow x=\sec\dfrac{\pi}{6}=\dfrac{2}{\sqrt{3}}, y=\tan\dfrac{\pi}{6}=\dfrac{1}{\sqrt{3}};$

$\dfrac{dx}{dt}=\sec t\tan t, \dfrac{dy}{dt}=\sec^2 t$

$\Rightarrow \dfrac{dy}{dx}=\dfrac{dy/dt}{dx/dt}=\dfrac{\sec^2 t}{\sec t\tan t}=\csc t$

$\Rightarrow \left.\dfrac{dy}{dx}\right|_{t=\frac{\pi}{6}}=\csc\dfrac{\pi}{6}=2;$

tangent line is $y-\dfrac{1}{\sqrt{3}}=2\left(x-\dfrac{2}{\sqrt{3}}\right)$ or

$y=2x-\sqrt{3}; \quad \dfrac{dy'}{dt}=-\csc t\cot t$

$\Rightarrow \dfrac{d^2y}{dx^2}=\dfrac{dy'/dt}{dx/dt}=\dfrac{-\csc t\cot t}{\sec t\tan t}=-\cot^3 t$

$\Rightarrow \left.\dfrac{d^2y}{dx^2}\right|_{t=\frac{\pi}{6}}=-3\sqrt{3}$

(5) $y=x-4, \quad \dfrac{d^2y}{dx^2}=\dfrac{1}{2}$

(6) $t=1$

$\Rightarrow x=1, y=-2; \dfrac{dx}{dt}=-\dfrac{1}{t^2}, \dfrac{dy}{dt}=\dfrac{1}{t}$

$\Rightarrow \dfrac{dy}{dx}=\dfrac{\left(\frac{1}{t}\right)}{\left(-\frac{1}{t^2}\right)}=-t$

$\Rightarrow \left.\dfrac{dy}{dx}\right|_{t=1}=-1;$

tangent line is $y-(-2)=-1(x-1)$ or

$$y=-x-1;\frac{dy'}{dt}=-1\Rightarrow\frac{d^2y}{dx^2}=\frac{-1}{\left(-\frac{1}{t^2}\right)}=t^2$$

$$\Rightarrow\left.\frac{d^2y}{dx^2}\right|_{t=1}=1$$

(7) $y=2,\ \frac{d^2y}{dx^2}=-1$

(8) $t=0$

$\Rightarrow x=0+e^0=1, y=1-e^0=0;$

$\frac{dx}{dt}=1+e^t, \frac{dy}{dt}=-e^t$

$\Rightarrow\frac{dy}{dx}=\frac{-e^t}{1+e^t}$

$\Rightarrow\left.\frac{dy}{dx}\right|_{t=0}=\frac{-e^0}{1+e^0}=-\frac{1}{2}$

tangent line is

$y=-\frac{1}{2}x+\frac{1}{2};\frac{dy'}{dt}=\frac{-e^t}{(1+e^t)^2}$

$\Rightarrow\frac{d^2y}{dx^2}=\frac{-e^t}{(1+e^t)^3}$

$\Rightarrow\left.\frac{d^2y}{dx^2}\right|_{t=0}\frac{-e^0}{(1+e^0)^3}=-\frac{1}{8}$

EX 3

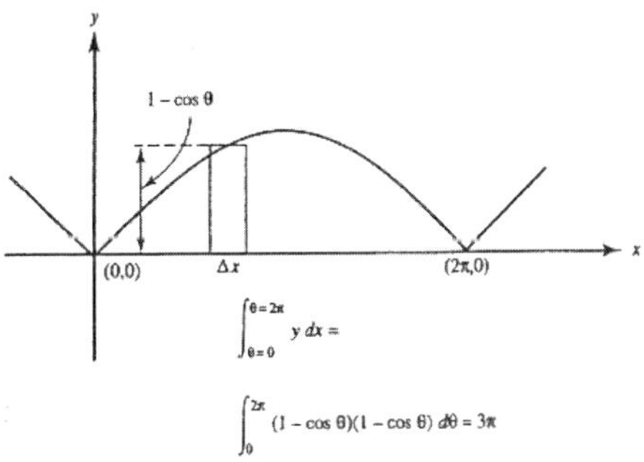

EX 4

By symmetry, the enclosed area is 4 times the area beneath the curve in the first quadrant where $0\le t\le\pi/2$.

$$A=4\int_0^1 ydx=4\int_{\pi/2}^0 \sin^3t(-3\cos^2t\ \sin t)dt$$
$$=4\int_0^{\pi/2}\sin^3t\cdot 3\cos^2t\ \sin t\ dt$$
$$=12\int_0^{\pi/2}\left(\frac{1-\cos2t}{2}\right)^2\left(\frac{1+\cos2t}{2}\right)dt$$
$$=\frac{3}{2}\int_0^{\pi/2}(1-2\cos2t+\cos^2 2t)(1+\cos2t)dt$$
$$=\frac{3}{2}\int_0^{\frac{\pi}{2}}(1-\cos2t-\cos^2 2t+\cos^3 2t)dt$$
$$=\frac{3}{2}\left[\int_0^{\pi/2}(1-\cos2t)dt-\int_0^{\pi/2}\cos^2 2t\,dt+\int_0^{\pi/2}\cos^3 2t\,dt\right]$$
$$=\frac{3}{2}\left[\left(t-\frac{1}{2}\sin2t\right)-\frac{1}{2}\left(t+\frac{1}{4}\sin2t\right)+\frac{1}{2}\left(\sin2t-\frac{1}{3}\sin^3 2t\right)\right]_0^{\pi/2}$$
$$=\frac{3}{2}\left[\left(\frac{\pi}{2}-0-0-0\right)-\frac{1}{2}\left(\frac{\pi}{2}+0-0-0\right)+\frac{1}{2}(0-0-0+0)\right]\Big\}$$

EX 5

(1) $\frac{1}{27}(13^{3/2}-8)$

(2) We find

$$\frac{dx}{dt}=-r\sin t,\quad \frac{dy}{dt}=r\cos t$$

and

$$\left(\frac{dx}{dt}\right)^2+\left(\frac{dy}{dt}\right)^2=r^2(\sin^2t+\cos^2t)=r^2.$$

so

$$L=\int_0^{2\pi}\sqrt{r^2}\,dt=r[t]_0^{2\pi}=2\pi r.$$

(3) $\sqrt{2}\left(1-\frac{1}{e^{\pi}}\right)$

(4) Because of the curve's symmetry with respect to the coordinate axes, its length is four times the length of the first-quadrant portion. We have

$x=\cos^3t\quad y=\sin^3t$

$$\left(\frac{dx}{dt}\right)^2=[3\cos^2t(-\sin t)]^2=9\cos^4t\sin^2t$$

$$\left(\frac{dy}{dt}\right)^2=[3\sin^2t(\cos t)]^2=9\sin^4t\cos^2t$$

$$\sqrt{\left(\frac{dx}{dt}\right)^2+\left(\frac{dy}{dt}\right)^2}=\sqrt{9\cos^2t\sin^2t(\cos^2t+\sin^2t)}$$
$$=\sqrt{9\cos^2t\sin^2t}$$
$$=3|\cos t\sin t|$$
$$=3\cos t\ \sin t.$$

Therefore,

Length of first-quadrant portion

$$=\int_0^{\pi/2}3\cos t\sin t\,dt$$
$$=\frac{3}{2}\int_0^{\pi/2}\sin 2t\,dt$$
$$=-\frac{3}{4}\cos 2t\Big]_0^{\pi/2}=\frac{3}{2}.$$

The length of the astroid is four times this:

$4(3/2)=6.$

9.2 Polar coordinates

EX 1

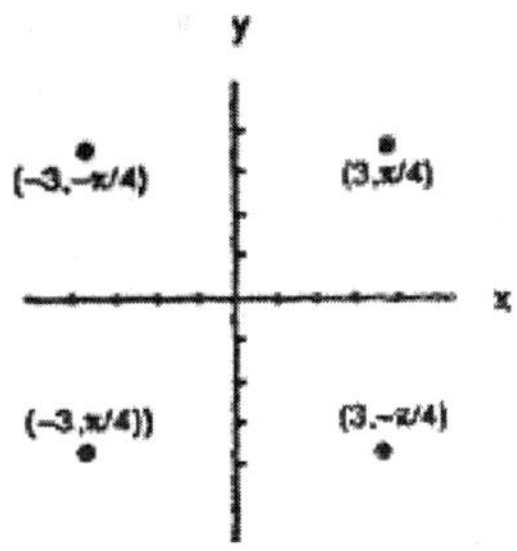

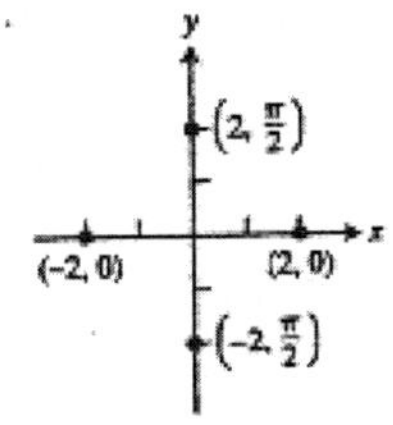

EX 2

(1) $(4,\frac{\pi}{3})$　　(2) $(2,-\frac{\pi}{6})$

(3) $(4\sqrt{2},\frac{5\pi}{4})$　　(4) $(2,\frac{\pi}{6})$

(5) $(3\sqrt{2},\frac{3\pi}{4})$　　(6) $(4,\frac{2\pi}{3})$

EX 3

(1) $x=\sqrt{2}\ \cos\frac{\pi}{4}=1,\ y=\sqrt{2}\sin\frac{\pi}{4}=1$

$\Rightarrow$ Cartesian coordinates are (1, 1)

(2) $x=1\cos 0=1,\ y=1\ \sin 0=0$

$\Rightarrow$ Cartesian coordinates are (1, 0)

(3) $x=0\ \cos\ \frac{\pi}{2}\ =0,\ y=0\ \sin\ \frac{\pi}{2}\ =0$

$\Rightarrow$ Cartesian coordinates are (0, 0)

(4) $x=-\sqrt{2}\ \cos\left(\frac{\pi}{4}\right)=-1,$

$y=-\sqrt{2}\ \sin\left(\frac{\pi}{4}\right)=-1$

$\Rightarrow$ Cartesian coordinates are (-1, -1)

(5) $x=-3\cos\frac{5\pi}{6}=\frac{3\sqrt{3}}{2},$

$y=-3\sin\frac{5\pi}{6}=-\frac{3}{2}$

$\Rightarrow$ Cartesian coordinates are $\left(\frac{3\sqrt{3}}{2},\ -\frac{3}{2}\right)$

(6) $x=5\cos\left(\tan^{-1}\frac{4}{3}\right)=3,$

$y=5\sin\left(\tan^{-1}\frac{4}{3}\right)=4$

$\Rightarrow$ Cartesian coordinates are (3, 4)

(7) $x=-1\cos 7\pi=1,\ y=-1\sin 7\pi=0$

$\Rightarrow$ Cartesian coordinates are (1, 0)

(8) $x=2\sqrt{3}\cos\frac{2\pi}{3}=-\sqrt{3},$

$y=2\sqrt{3}\sin\frac{2\pi}{3}=3$

$\Rightarrow$ Cartesian coordinates are $(-\sqrt{3},3)$

EX 4

(1) $r\cos\theta=7$

(2) $r\cos\theta=r\sin\theta,\ \theta=\frac{\pi}{4}$

(3) $r^2(\cos^2\theta-\sin^2\theta)=1,\ r^2\cos 2\theta=1$

(4) $4r^2\cos^2\theta+9r^2\sin^2\theta=36$

(5) $r\sin^2\theta=4\cos\theta$

(6) $x^2+y^2-4y+4=4,\ r^2=4r\sin\theta$

(7) $r^2=6r\cos\theta-2r\sin\theta-6$

EX 5

(1) $x=2$ (2) $y=0$

(3) $y=4$ (4) $x+y=1$

(5) $r\sin\theta-2r\cos\theta=5,\ y-2x=5$

(6) $r=\frac{\cos\theta}{\sin\theta}\frac{1}{\sin\theta}$

$\Rightarrow r\sin^2\theta=\cos\theta$

$\Rightarrow r^2\sin^2\theta=r\cos\theta$

$\Rightarrow y^2=x$

(7) $y=e^x$

(8) $x^2+y^2=-4x$ (9) $x^2+y^2=8y$

(10) $r^2=2r\cos\theta+2r\sin\theta,\ x^2+y^2=2x+2y.$

EX 6

(1)

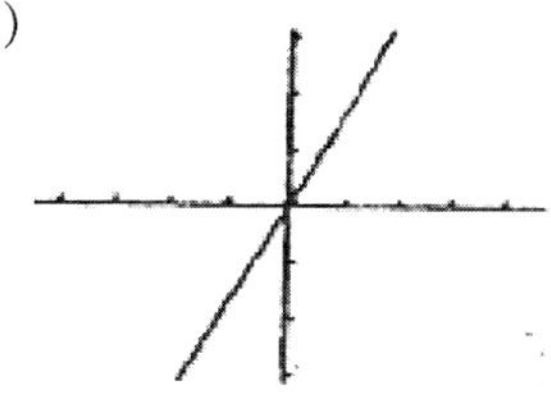

(2)

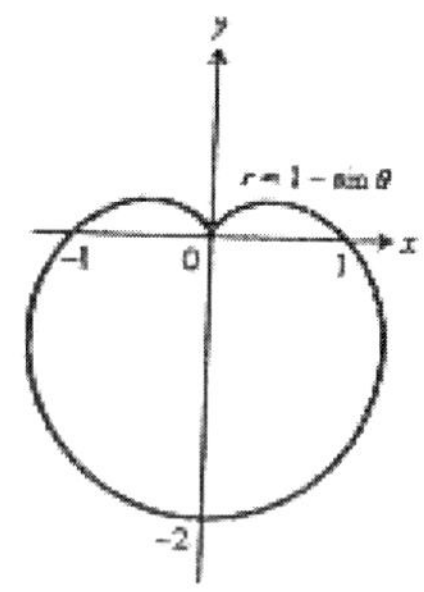

$1-\sin(-\theta)=1+\sin\theta\neq r$ and $1-\sin(\pi-\theta)=1-\sin\theta\neq -r\Rightarrow$ not symmetric about the x-axis; $1-\sin(\pi-\theta)=1-\sin\theta=r\Rightarrow$ symmetric about the y-axis; therefore not symmetric about the origin

(3)

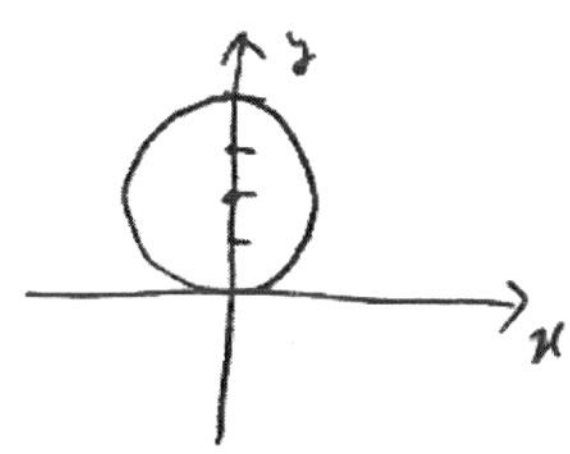

(4)

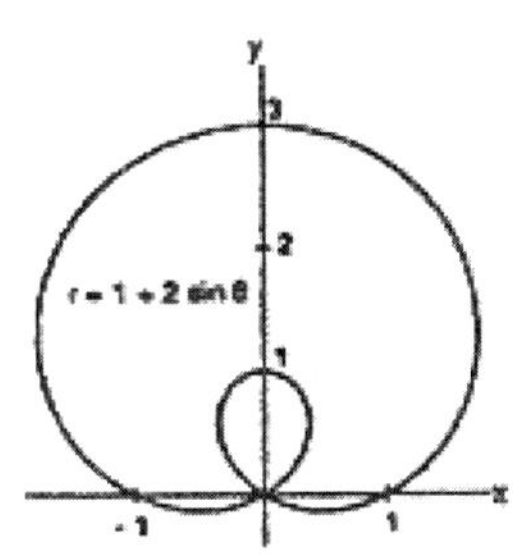

$1+2\sin(-\theta)=1-2\sin\theta\neq r$ and $1+2\sin(\pi-\theta)=1+2\sin\theta\neq -r\Rightarrow$ not symmetric about the x-axis; $1+2\sin(\pi-\theta)=1+2\sin\theta=r\Rightarrow$ symmetric about the y-axis; therefore not symmetric about the origin

(5)

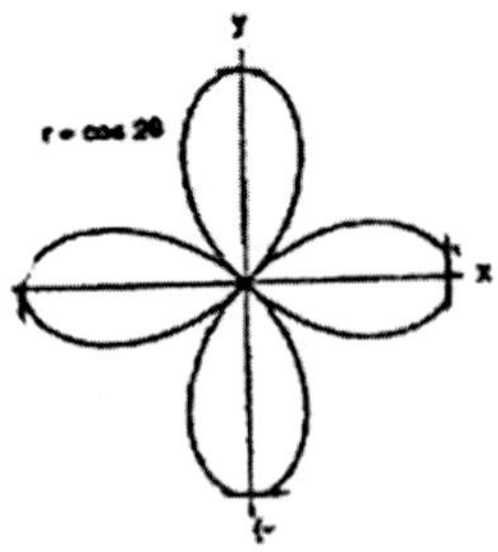

(6)

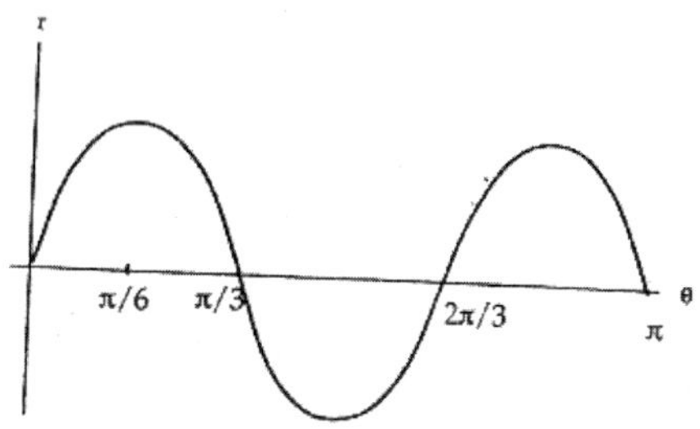

Since $\sin 3(\pi-\theta)=\sin 3\theta$, the graph is symmetric about the y-axis, Now from the Cartesian graph of $r=2\sin 3\theta$, we see that r increases from 0 to 2 as θ increases from 0 to $\frac{\pi}{6}$ Then, as θ increases from $\frac{\pi}{6}$ to $\frac{\pi}{3}$, r decreases from 2 to 0. This completes the right loop of the curve.

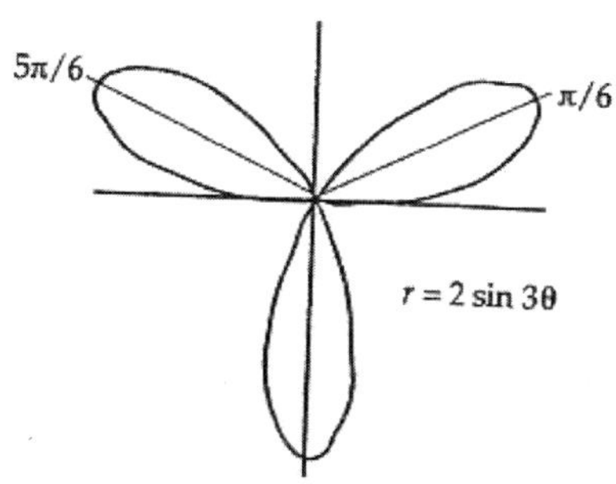

As θ increases from $\frac{\pi}{3}$ to $\frac{2\pi}{3}$, r is negative, first decreasing to -2 and then increasing to 0, thus tracing the lower loop. For $\frac{2\pi}{3}<\theta<\pi, r$ is positive and increasing to 2 and then decreasing to 0 to form the left loop of the curve. The curve is called a three-leaved rose.

EX 7

(1) Use $r=2(1+\cos\theta), x=r\cos\theta, y=r\sin\theta$ and $r'=-2\sin\theta$; *then*

$$\frac{dy}{dx}=\frac{dy/d\theta}{dx/d\theta}=\frac{(-2\sin\theta)\sin\theta+2(1+\cos\theta)(\cos\theta)}{(-2\sin\theta)\cos\theta-2(1+\cos\theta)(\sin\theta)}.$$

At $\theta=\frac{\pi}{6}, \frac{dy}{dx}=-1,$

(2) Since the cardioid is symmetric to $\theta=0$, we need consider only the upper half of the curve for part, The tangent is horizontal where $\frac{dy}{d\theta}=0$ (provided $\frac{dx}{d\theta}\neq 0$.) Since $\frac{dy}{d\theta}$ factors into $2(2\cos\theta-1)(\cos\theta+1)$, which equals 0 for $\cos\theta=\frac{1}{2}$ or -1, $\theta=\frac{\pi}{3}$ or π. From part (1), $\frac{dx}{d\theta}\neq 0$ at $\frac{\pi}{3}$, but $\frac{dx}{d\theta}$ does equal 0 at π. Therefore, the tangent is horizontal only at $\frac{\pi}{3}$(and, by symmetry, at $\frac{5\pi}{3}$).

EX 8

(1) $-3=r(\theta)\cos\theta=(3\theta+\sin\theta)\cos\theta$
$\theta=2.01692$
$y=r(\theta)\sin(\theta)=6.272$

(2) $y=r(\theta)\sin\theta=(3\theta+\sin\theta)\sin\theta$

$$\left.\frac{dy}{dt}\right|_{\theta=2\pi/3}=\left[\frac{dy}{d\theta}\cdot\frac{d\theta}{dt}\right]_{\theta=2\pi/3}=-2.819$$

The y-coordinate of the particle is decreasing at a rate of 2.819.

EX 9

(1) $x = r\cos\theta$ and $y = r\sin\theta$

$$\frac{dr}{d\theta} = 2\sin\theta$$
$$\frac{dx}{d\theta} = \frac{dr}{d\theta}\cos\theta - r\sin\theta = 4\sin\theta\cos\theta - \sin\theta$$
$$\frac{dy}{d\theta} = \frac{dr}{d\theta}\sin\theta + r\cos\theta$$
$$= 2\sin^2\theta + (1-2\cos\theta)\cos\theta$$

(2) When $\theta = \frac{\pi}{2}$, we have $x=0$, $y=1$.

$$\left.\frac{dy}{dx}\right|_{\theta=\frac{\pi}{2}} = \left.\frac{dy/d\theta}{dx/d\theta}\right|_{\theta=\frac{\pi}{2}} = -2$$

The tangent line is given by $y = 1-2x$.

EX 10

(1) $A = \int_0^{\pi} \frac{1}{2}\theta^2 d\theta = \left[\frac{1}{6}\theta^3\right]_0^{\pi} = \frac{\pi^3}{6}$

(2) $$A = \int_{\pi/4}^{\pi/2} \frac{1}{2}(2\sin\theta)^2 d\theta$$
$$= 2\int_{\pi/4}^{\pi/2} \sin^2\theta\, d\theta$$
$$= 2\int_{\pi/4}^{\pi/2} \frac{1-\cos 2\theta}{2} d\theta$$
$$= \int_{\pi/4}^{\pi/2} (1-\cos 2\theta) d\theta$$
$$= \left[\theta - \frac{1}{2}\sin 2\theta\right]_{\pi/4}^{\pi/2}$$
$$= \left(\frac{\pi}{2} - 0\right) - \left(\frac{\pi}{4} - \frac{1}{2}\right) = \frac{\pi}{4} + \frac{1}{2}$$

(3) $$A = 2\int_0^{\pi/4} \frac{1}{2}\cos^2 2\theta\, d\theta$$
$$= \int_0^{\pi/4} \frac{1+\cos 4\theta}{2} d\theta$$
$$= \frac{1}{2}\left[\theta + \frac{\sin 4\theta}{4}\right]_0^{\pi/4}$$
$$= \frac{\pi}{8}$$

(4) $$A = \int_{-\pi/6}^{\pi/6} \frac{1}{2}(\cos 3\theta)^2 d\theta$$
$$= \frac{1}{2}\int_{-\pi/6}^{\pi/6} \cos^2 3\theta\, d\theta$$
$$= \frac{1}{2}\int_{-\pi/6}^{\pi/6} \frac{1+\cos 6\theta}{2} d\theta$$
$$= \frac{1}{4}\int_{-\pi/6}^{\pi/6} (1+\cos 6\theta) d\theta$$
$$= \frac{1}{4}\left[\theta + \frac{1}{6}\sin 6\theta\right]_{-\pi/6}^{\pi/6}$$
$$= \frac{1}{4}\left(\frac{\pi}{6}+0\right) - \frac{1}{4}\left(-\frac{\pi}{6}+0\right)$$
$$= \frac{\pi}{12}$$

EX 11

(1)

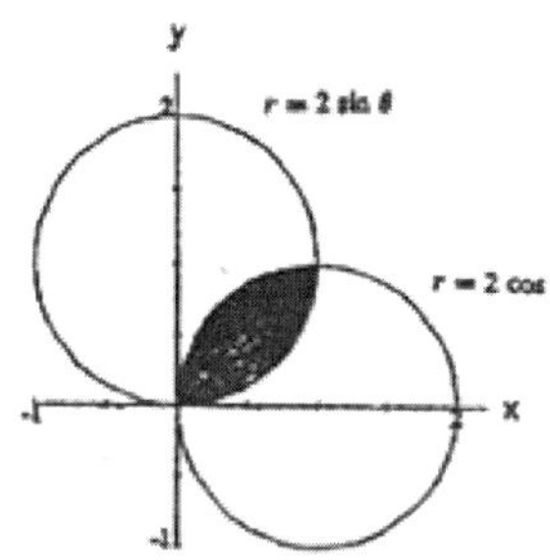

$r = 2\cos\theta$ and $r = 2\sin\theta \Rightarrow 2\cos\theta = 2\sin\theta$

$\Rightarrow \cos\theta = \sin\theta \Rightarrow \theta = \frac{\pi}{4}$; therefore

$$A = 2\int_0^{\pi/4} \frac{1}{2}(2\sin\theta)^2 d\theta = \int_0^{\pi/4} 4\sin^2\theta\, d\theta$$
$$= \int_0^{\pi/4} 4\left(\frac{1 \quad \mathrm{co}32\theta}{2}\right) d\theta$$
$$= \int_0^{\pi/4} (2 - 2\cos 2\theta)\, d\theta$$
$$= [2\theta - \sin 2\theta]_0^{\pi/4} = \frac{\pi}{2} - 1$$

(2)

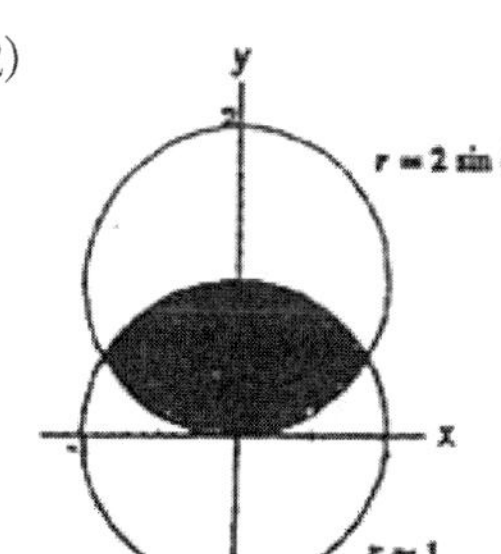

$r=1$ and $r=2\sin\theta \Rightarrow 2\sin\theta = 1 \Rightarrow \sin\theta = \frac{1}{2}$

$\Rightarrow \theta = \frac{\pi}{6}$ or $\frac{5\pi}{6}$; therefore

$$A=\pi(1)^2-\int_{\pi/6}^{5\pi/6}\frac{1}{2}[(2\sin\theta)^2-1^2]d\theta$$
$$=\pi-\int_{\pi/6}^{5\pi/6}(2\sin^2\theta-\frac{1}{2})d\theta$$
$$=\pi-\int_{\pi/6}^{5\pi/6}(1-\cos 2\theta-\frac{1}{2})\theta$$
$$=\pi-\int_{\pi/6}^{5\pi/6}\left(\frac{1}{2}-\cos 2\theta\right)d\theta$$
$$=\pi-\left[\frac{1}{2}\theta-\frac{\sin 2\theta}{2}\right]_{\pi/6}^{5\pi/6}$$
$$=\pi-\left(\frac{5\pi}{12}-\frac{1}{2}\sin\frac{5\pi}{3}\right)+\left(\frac{\pi}{12}-\frac{1}{2}\sin\frac{\pi}{3}\right)$$
$$=\frac{4\pi-3\sqrt{3}}{6}$$

(3)

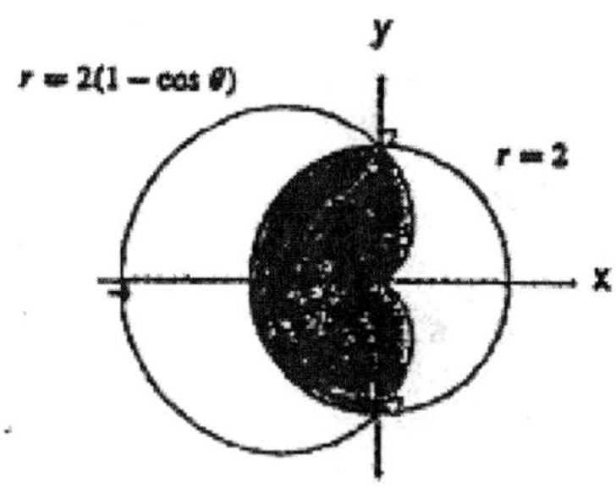

$r=2$ and $r=2(1-\cos\theta) \Rightarrow 2=2(1-\cos\theta)$

$\Rightarrow \cos\theta = 0 \Rightarrow \theta = \pm\frac{\pi}{2}$; therefore

$$A=2\int_{0}^{\pi/2}\frac{1}{2}[2(1-\cos\theta)]^2d\theta+\frac{1}{2}$$

(area of the circle)

$$=\int_{0}^{\pi/2}4(1-2\cos\theta+\cos^2\theta)d\theta+\frac{\pi}{2}(2)^2$$
$$=\int_{0}^{\pi/2}4\left(1-2\cos\theta+\frac{1+\cos 2t\theta}{2}\right)d\theta+2\pi$$
$$=\int_{0}^{\pi/2}(4-8\cos\theta+2+2\cos 2\theta)d\theta+2\pi$$
$$=[6\theta-8\sin\theta+\sin 2\theta]_0^{\pi/2}+2\pi$$
$$=5\pi-8$$

(4)

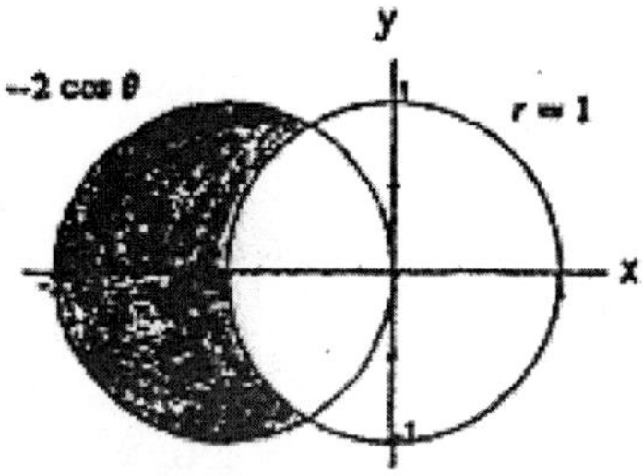

$r=1$ and $r=-2\cos\theta \Rightarrow 1=-2\cos\theta$

$\Rightarrow \cos\theta = -\frac{1}{2} \Rightarrow \theta = \frac{2\pi}{3}$ in quadrant II;

therefore

$$A=2\int_{2\pi/3}^{\pi}\frac{1}{2}[(-2\cos\theta)^2-1^2]\,d\theta$$
$$=\int_{2\pi/3}^{\pi}(4\cos^2\theta-1)d\theta$$
$$=\int_{2\pi/3}^{\pi}[2(1+\cos 2\theta)-1]d\theta$$
$$=\int_{2\pi/3}^{\pi}(1+2\cos 2\theta)d\theta$$
$$=[\theta+\sin 2\theta]_{2\pi/3}^{\pi}$$
$$=\frac{\pi}{3}+\frac{\sqrt{3}}{2}$$

(5)

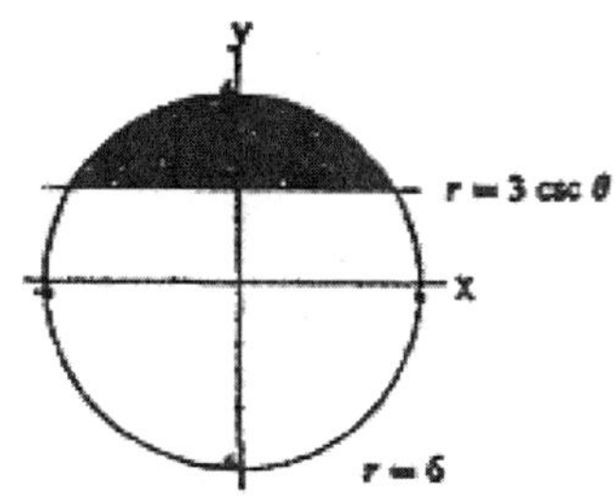

$r=6$ and $r=3\csc\theta \Rightarrow 6\sin\theta = 3 \Rightarrow \sin\theta = \frac{1}{2}$

$\Rightarrow \theta = \frac{\pi}{6}$ or $\frac{5\pi}{6}$;

therefore $A=\int_{\pi/6}^{5\pi/6}\frac{1}{2}(6^2-9\csc^2\theta)d\theta$

$$=\int_{\pi/6}^{5\pi/6}(18-\frac{9}{2}\csc^2\theta)d\theta$$
$$=[18\theta+\frac{9}{2}\cot\theta]_{\pi/6}^{5\pi/6}$$
$$=\left(15\pi-\frac{9}{2}\sqrt{3}\right)-\left(3\pi+\frac{9}{2}\sqrt{3}\right)$$
$$=12\pi-9\sqrt{3}$$

(6)

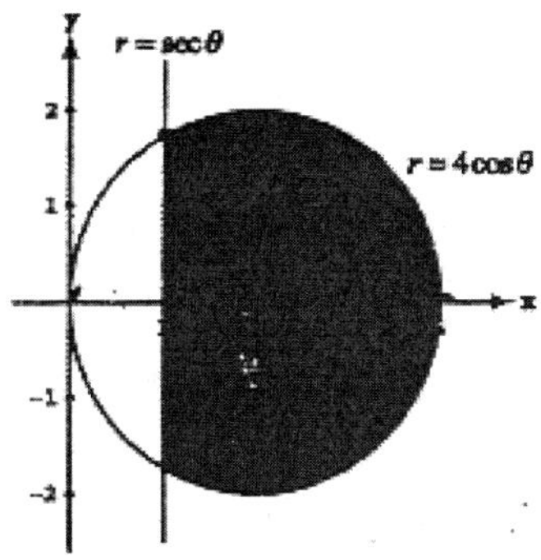

$r=\sec\theta$ and $r=4\cos\theta \Rightarrow 4\cos\theta = \sec\theta$

$\Rightarrow \cos^2\theta = \dfrac{1}{4}$

$\Rightarrow \theta = \dfrac{\pi}{3}, \dfrac{2\pi}{3}, \dfrac{4\pi}{3}$ or $\dfrac{5\pi}{3}$; therefore

$$\begin{aligned} A &= 2\int_0^{\pi/3} \frac{1}{2}(16\cos^2\theta - \sec^2\theta)d\theta \\ &= \int_0^{\pi/3}(8+8\cos 2\theta - \sec^2\theta)d\theta \\ &= [8\theta + 4\sin 2\theta - \tan\theta]_0^{\pi/3} \\ &= (\frac{8\pi}{3} + 2\sqrt{3} - \sqrt{3}) - (0+0-0) \\ &= \frac{8\pi}{3} + \sqrt{3} \end{aligned}$$

EX 12

(1) Area = $3^2\pi \times \dfrac{2}{3} + \dfrac{1}{2}\displaystyle\int_{\pi/6}^{5\pi/6}(4-2\sin\theta)^2 d\theta$

$= 24.709$ (or 24.708)

(2) $x = r\cos\theta \Rightarrow x(\theta) = (4-2\sin\theta)\cos\theta$
$x(t) = (4-2\sin(t^2))\cos(t^2)$
$x(t) = -1$ *when* $t = 1.428$ (or 1.427)

(3) $y = r\sin\theta \Rightarrow y(\theta) = (4-2\sin\theta)\sin\theta$
$y(t) = (4-2\sin(t^2))\sin(t^2)$

Position vector $= < x(t), y(t) >$

$= < (4-2\sin(t^2)\cos(t^2), (4-2\sin(t^2))\sin(t^2) >$

$v(1.5) = < x'(1.5), y'(1.5) >$
$= < -8.072, -1.673 >$ (or $< -8.072, -1.672 >$)

EX 13

(1) Area $= \dfrac{2}{3}\pi(2)^2 + \dfrac{1}{2}\displaystyle\int_{2\pi/3}^{4\pi/3}(3+2\cos\theta)^2 d\theta$
$= 10.370$

(2) $\left.\dfrac{dr}{dt}\right|_{\theta=\pi/3} = \left.\dfrac{dr}{d\theta}\right|_{\theta=\pi/3} = -1.732$

The particle is moving closer to the origin, since $\dfrac{dr}{dt} < 0$ and $r > 0$ when $\theta = \dfrac{\pi}{3}$.

(3) $y = r\sin\theta = (3+2\cos\theta)\sin\theta$

$\dfrac{dy}{d\theta} = -2\sin\theta\sin\theta + (3+2\cos\theta)\cos\theta$

$\left.\dfrac{dy}{dt}\right|_{\theta=\pi/3} = \left.\dfrac{dy}{d\theta}\right|_{\theta=\pi/3} = 0.5$

The particle is moving away from the x-axis, since $\dfrac{dy}{dt} > 0$ and $y > 0$ when $\theta = \dfrac{\pi}{3}$.

CHAPTER 10. Vectors and the Geometry of Space

10.1 Vectors

EX 1

(1)

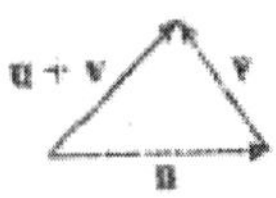

(2)

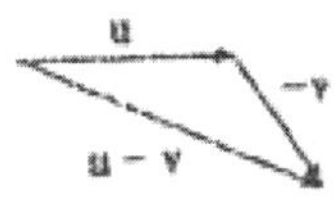

(3)

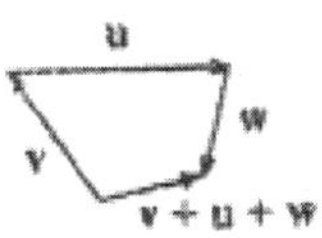

(4)

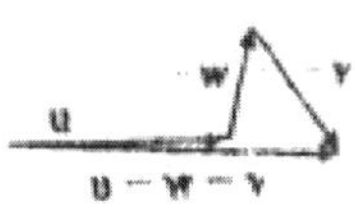

EX 2

(1)

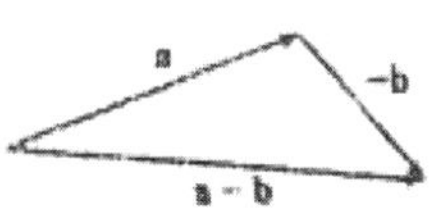

(2)

(3)

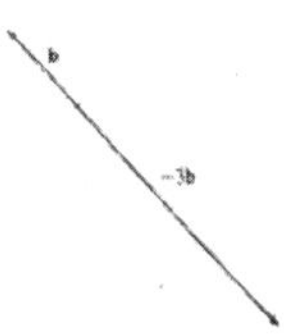

(4)

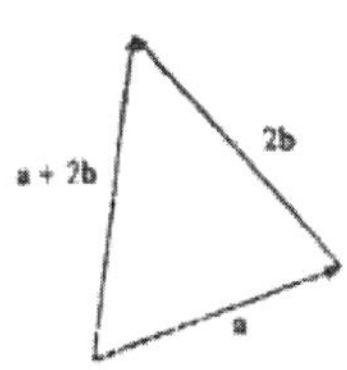

EX 3

(1) (a)

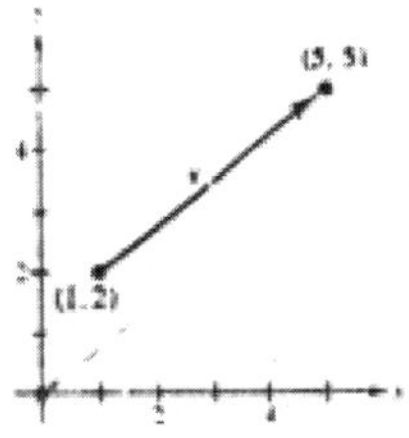

(b) $\mathrm{v} = \langle 5-1,\ 5-2 \rangle = \langle 4, 3 \rangle$

© $|\mathrm{v}| = \sqrt{16+9} = 5$

(2) (a)

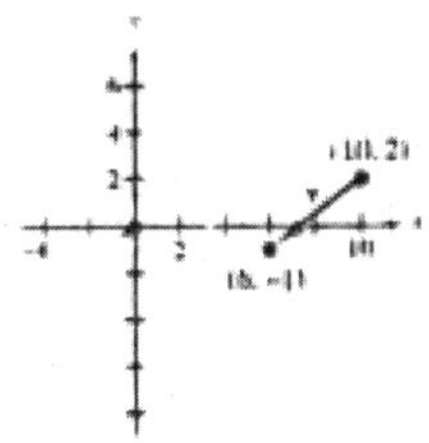

(b) $\mathrm{v} = \langle 6-10,\ -1-2 \rangle = \langle -4, -3 \rangle$

(c) $|\mathrm{v}| = \sqrt{16+9} = 5$

(3) (a)

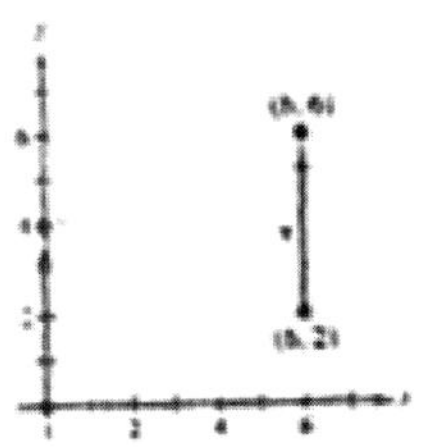

(b) $\mathrm{v} = \langle 6-6,\ 6-2 \rangle = \langle 0, 4 \rangle$

(c) $|\mathrm{v}| = \sqrt{0^2+4^2} = 4$

(4) (a)

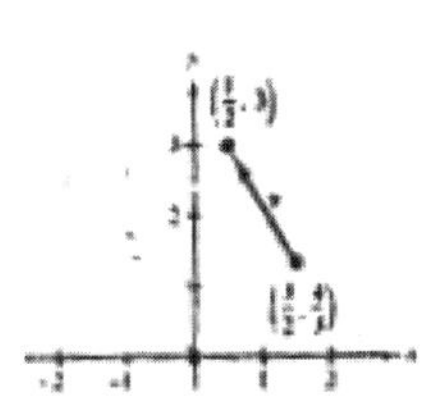

(b) $\mathrm{v} = \left\langle \frac{1}{2} - \frac{3}{2},\ 3 - \frac{4}{3} \right\rangle = \left\langle -1, \frac{5}{3} \right\rangle$

(c) $|\mathrm{v}| = \sqrt{(-1)^2 + \left(\frac{5}{3}\right)^2} = \frac{\sqrt{34}}{3}$

EX 4

(1) $\frac{2}{3}u = \frac{2}{3}\langle -3, -8\rangle = \left\langle -2, -\frac{16}{3}\right\rangle$

(2) $v - u = \langle 8, 25\rangle - \langle -3, -8\rangle = \langle 11, 33\rangle$

(3) $2u + 5v = 2\langle -3, -8\rangle + 5\langle 8, 25\rangle = \langle 34, 109\rangle$

EX 5

(1) $\|u\| = \sqrt{3^2 + 12^2} = \sqrt{153}$

$v = \frac{u}{\|u\|} = \frac{\langle 3, 12\rangle}{\sqrt{153}} = \left\langle \frac{3}{\sqrt{153}}, \frac{12}{\sqrt{153}}\right\rangle$

$= \left\langle \frac{\sqrt{17}}{17}, \frac{4\sqrt{17}}{17}\right\rangle$ unit vector

(2) $\|u\| = \sqrt{5^2 + 15^2} = \sqrt{250} = 5\sqrt{10}$

$v = \frac{u}{\|u\|} = \frac{\langle 5, 15\rangle}{5\sqrt{10}} = \left\langle \frac{1}{\sqrt{10}}, \frac{3}{\sqrt{10}}\right\rangle$

unit vector

(3) $\|u\| = \sqrt{\left(\frac{3}{2}\right)^2 + \left(\frac{5}{2}\right)^2} = \frac{\sqrt{34}}{2}$

$v = \frac{u}{\|u\|} = \frac{\langle (3/2), (5/2)\rangle}{\sqrt{34}/2} = \left\langle \frac{3}{\sqrt{34}}, \frac{5}{\sqrt{34}}\right\rangle$

$= \left\langle \frac{3\sqrt{34}}{34}, \frac{3\sqrt{34}}{34}\right\rangle$ unit vector

(4) $\|u\| = \sqrt{(-6.2)^2 + (3.4)^?} = \sqrt{50} - 5\sqrt{2}$

$v = \frac{u}{\|u\|} = \frac{\langle -6.2, 3.4\rangle}{5\sqrt{2}} = \left\langle \frac{-1.24}{\sqrt{2}}, \frac{0.68}{\sqrt{2}}\right\rangle$

unit vector

EX 6

(1) $v = 3[(\cos 0^\circ)i + (\sin 0^\circ)j]$

$= 3i = \langle 3, 0\rangle$

(2) $v = 5[(\cos 120^\circ)i + (\sin 120^\circ)j]$

$= -\frac{5}{2}i + \frac{5\sqrt{3}}{2}j = \left\langle -\frac{5}{2}, \frac{5\sqrt{3}}{2}\right\rangle$

(3) $v = 2[(\cos 150^\circ)i + (\sin 150^\circ)j]$

$= -\sqrt{3}i + j = \langle -\sqrt{3}, 1\rangle$

(4) $v = (\cos 3.5^\circ)i + (\sin 3.5^\circ)j$

$= 0.99813i + 0.0610j = \langle 0.9981, 0.0610\rangle$

EX 7

$u = 400i$(plane)

$v = 50(\cos 135^\circ i + \sin 135^\circ j) = -25\sqrt{2}i + 25\sqrt{2}j$ (wind)

$u + v = (400 - 25\sqrt{2})i + 25\sqrt{2}j \approx 364.64i + 35.36j$

$\tan\theta = \frac{35.36}{364.64} \Rightarrow \theta \approx 5.54^\circ$

Direction North of East: $\approx N84.46^\circ E$

Speed: ≈ 336.35 mph

EX 8

$F_1 = \langle -|F_1|\cos 30^\circ, |F_1|\sin 30^\circ\rangle$

$= \left\langle -\frac{\sqrt{3}}{2}|F_1|, \frac{1}{2}|F_1|\right\rangle$,

$F_2 = \langle |F_2|\cos 45^\circ, |F_2|\sin 45^\circ\rangle$

$= \left\langle \frac{1}{\sqrt{2}}|F_2|, \frac{1}{\sqrt{2}}|F_2|\right\rangle$, and

$w = \langle 0, -100\rangle$. *Since* $F_1 + F_2$

$= \langle 0, 100\rangle$

$\Rightarrow \left\langle -\frac{\sqrt{3}}{2}|F_1| + \frac{1}{\sqrt{2}}|F_2|, \frac{1}{2}|F_1| + \frac{1}{\sqrt{2}}|F_2|\right\rangle$

$= \langle 0, 100\rangle$

$\Rightarrow -\frac{\sqrt{3}}{2}|F_1| + \frac{1}{\sqrt{2}}|F_2|$

$= 0$ and $\frac{1}{2}|F_1| + \frac{1}{\sqrt{2}}|F_2| = 100$

Solving the first equation for $|F_2|$ results in:

$|F_2| = \frac{\sqrt{6}}{2}|F_1|$

Substituting this result into the second equation gives us:

$\frac{1}{2}|F_1| + \frac{1}{\sqrt{2}}\left(\frac{\sqrt{6}}{2}|F_1|\right) = 100 \Rightarrow |F_1|$

$= \frac{200}{1 + \sqrt{3}} \approx 73.205\,N$

$\Rightarrow |F_2| = \dfrac{100\sqrt{6}}{1+\sqrt{3}} \approx 89.658\,N$

$\Rightarrow F_1 \approx \langle -63.397, 36.603 \rangle$

and $F_2 = \approx \langle 63.397, 63.397 \rangle$

EX 9

(1) $u = 10i$

(2) $v = 20\cos 60i + 20\sin 60j$

$= 10i + 10\sqrt{3}j$

(3) $w = u + v = 30i + 10\sqrt{3}j$

(4) $|w| = \sqrt{30^2 + (10\sqrt{3})^2} = \sqrt{1200} = 20\sqrt{3}$

$\tan\theta = \dfrac{10\sqrt{3}}{30} = \dfrac{1}{\sqrt{3}}, \ \theta = 30°$

10.2 Space Coordinates and Vectors in Space

EX 10

(1)

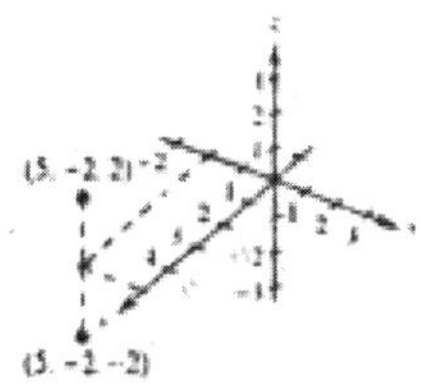

(2)

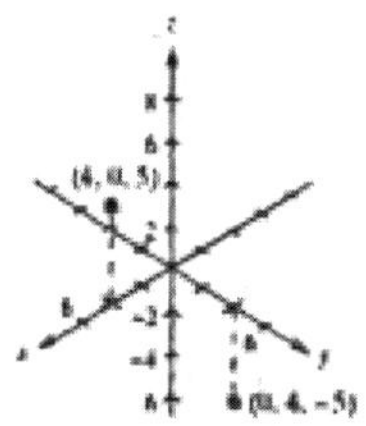

EX 11

(1) $|P_1P_2| = \sqrt{(3-1)^2 + (3-1)^2 + (0-1)^2}$
$= \sqrt{9} = 3$

(2) $|P_1P_2| = \sqrt{(2+1)^2 + (5-1)^2 + (0-5)^2}$
$= \sqrt{50} = 5\sqrt{2}$

(3) $|P_1P_2| = \sqrt{(4-1)^2 + (-2-4)^2 + (7-5)^2}$
$= \sqrt{49} = 7$

(4) $|P_1P_2| = \sqrt{(2-3)^2 + (3-4)^2 + (4-5)^2} = \sqrt{3}$

EX 12

(1) Completing squares in the equation

$x^2 + y^2 + z^2 - 2x - 4y + 8z = 15$ gives

$(x^2 - 2x + 1) + (y^2 - 4y + 4) + (z^2 + 8z + 16)$

$= 15 + 1 - 4 + 16$

$\Rightarrow (x-1)^2 + (y-2)^2 + (z+4)^2 = 36,$

which we recognize as an equation of a sphere with center $(1,2,-4)$ and radius 6.

(2) Completing squares in the equation gives

$(x^2 + 8x + 16) + (y^2 - 6y + 9) + (z^2 + 2z + 1)$

$= -17 + 16 + 9 + 1$

$\Rightarrow (x+4)^2 + (y-3)^2 + (z+1)^2 = 9,$

which we recognize as an equation of a sphere with center $(-4, 3, -1)$ and radius 3.

(3) Completing squares in the equation

$2x^2 - 8x + 2y^2 + 2z^2 + 24z = 1$ gives

$2(x^2 - 4x + 4) + 2y^2 + 2(x^2 + 12z + 36)$

$= 1 + 8 + 72$

$\Rightarrow 2(x-2)^2 + 2y^2 + 2(z+6)^2 = 81$

$\Rightarrow (x-2)^2 + y^2 + (z+6)^2 = \dfrac{81}{2}$, which we recognize as an equation of a

sphere with center $(2, 0, -6)$ and radius

$\sqrt{\dfrac{81}{2}} = 9/\sqrt{2}.$

(4) Completing squares in the equation

$3x^2 + 3y^2 - 6y + 3z^2 - 12z = 10$ gives

$3x^2 + 3(y^2 - 2y + 1) + 3(z^2 - 4z + 4)$

$= 10 + 3 + 12$

$\Rightarrow 3x^2 + 3(y-1)^2 + 3(z-2)^2 = 25$

$\Rightarrow x^2+(y-1)^2+(z-2)^2=\frac{25}{3}$,

which we recognize as an equation of a sphere with center $(0,1,2)$ and radius $\sqrt{\frac{25}{3}}=5/\sqrt{3}$.

EX 13

(1) An equation of the sphere with center $(1, -4, 3)$ and radius 5 is

$(x-1)^2+[y-(-4)]^2+(x-3)^2=5^2$ or

$(x-1)^2+(y+4)^2+(z-3)^2=25$.

The intersection of this sphere with the $xz-$plane is the set of points on the sphere whose $y-$coordinate is 0. Putting $y=0$ into the equation, we have $(x-1)^2+4^2+(z-3)^2=25,\ y=0$ or $(x-1)^2+(z-3)^3=9,\ y=0$, which represents a circle in the $xz-$plane with center $(1, 0, 3)$ and radius3.

(2) An equation of the sphere with center $(2, -6, 4)$ and radius 5 is

$(x-2)^2+[y-(-6)]^2+(z-4)^2=5^2$ or

$(x-2)^2+(y+6)^2+(z-4)^2=25$.

The intersection of this sphere with the $xy-$plane is the set of points on the sphere whose $z-$coordinate is 0. Putting $z=0$ into the equation, we have

$(x-2)^2+(y+6)^2=9,\ z=0$

which represents a circle in the $xy-$plane with center $(2, -6, 0)$ and radius 3. To find the intersection with the $xz-$plane, we set $y=0$:$(x-2)^2+(z-4)^2=-11$. Since no points satisfy this equation, the sphere does not intersect the $xz-$plane. (Also note that the distance from the center of the sphere to the $xz-$plane is greater than the radius of the sphere.)

To find the intersection with the $yz-$plane, we set $x=0$: $(y+6)^2+(z-4)^2=21,\ x=0$, a circle in the $yz-$plane with center $(0, -6, 4)$ and radius $\sqrt{21}$.

(3) The radius of the sphere is the distance between $(4, 3, -1)$ and

$(3, 8, 1)$: r

$=\sqrt{(3-4)^2+(8-3)^2+[1-(-1)]^2}$

$=\sqrt{30}$

Thus, an equation of the sphere is

$(x-3)^2+(y-8)^2+(z-1)^2=30$.

(4) If the sphere passes through the origin, the radius of the sphere must be the distance form the origin to the point $(1,2,3)$: $r=\sqrt{(1-0)^2+(2-0)^2+(3-0)^2}$ $=\sqrt{14}$. Then an equation of the sphere is $(x-1)^2+(y-2)^2+(z-3)^2=14$.

EX 14

(1) $\overrightarrow{P_1P_2}=(3,\ 4,\ -5),\ |\overrightarrow{P_1P_2}|=5\sqrt{2}$

$\Rightarrow$ unit vector is $\left(\frac{3}{5\sqrt{2}}, \frac{4}{5\sqrt{2}}, \frac{-5}{5\sqrt{2}}\right)$

(2) $\overrightarrow{P_1P_2}=3\mathrm{i}-6\mathrm{j}+2\mathrm{k},\ |\overrightarrow{P_1P_2}|=7$

$\Rightarrow$ unit vector is $\frac{3}{7}\mathrm{i}-\frac{6}{7}\mathrm{j}+\frac{2}{7}\mathrm{k}$

(3) $\overrightarrow{P_1P_2}=-\mathrm{i}-\mathrm{j}-\mathrm{k},\ |\overrightarrow{P_1P_2}|=\sqrt{3}$

$\Rightarrow$ unit vector is $-\frac{1}{\sqrt{3}}\mathrm{i}-\frac{1}{\sqrt{3}}\mathrm{j}-\frac{1}{\sqrt{3}}\mathrm{k}$

(4) $\overrightarrow{P_1P_2}=(2,\ -2,\ -2),\ |\overrightarrow{P_1P_2}|=2\sqrt{3}$

$\Rightarrow$ unit vector is $\left(\frac{1}{\sqrt{3}}, -\frac{1}{\sqrt{3}}, -\frac{1}{\sqrt{3}}\right)$

10.3 The Dot product of Two Vectors

EX 15

(1) (a) $u \cdot v = \langle 5, -1\rangle \cdot \langle -3, 2\rangle$

$= 5(-3) + (-1)(2) = -17$

(b) $\|u\|^2 = u \cdot u = 26$

(c) $(u \cdot v)v = (-17)\langle -3, 2\rangle = \langle 51, -34\rangle$

(d) $u \cdot (2v) = 2(u \cdot v) = 2(-17) = -34$

(2) $u = \langle 2, -3, 4\rangle$, $v = \langle 0, 6, 5\rangle$

(a) $u \cdot v = 2(0) + (-3)(6) + (4)(5) = 2$

(b) $\|u\|^2 = 29$

(c) $(u \cdot v)v = 2\langle 0, 6, 5\rangle = \langle 0, 12, 10\rangle$

(d) $u \cdot (2v) = 2(u \cdot v) = 2(2) = 4$

(3) $u = 2i - j + k$, $v = i - k$

(a) $u \cdot v = 2(1) + (-1)(0) + 1(-1) = 1$

(b) $\|u\|^2 = 6$

(c) $(u \cdot v)v = v = i - k$

(d) $u \cdot (2v) = 2(u \cdot v) = 2$

EX 16

(1) $u = 3i + j$, $v = -2i + 4j$

$$\cos\theta = \frac{u \cdot v}{\|u\|\,\|v\|} = \frac{-2}{\sqrt{10}\sqrt{20}} = \frac{-1}{5\sqrt{2}}$$

$$\theta = \arccos\left(\frac{1}{5\sqrt{2}}\right) = 98.1^\circ$$

(2) $u = \cos\left(\frac{\pi}{6}\right)i + \sin\left(\frac{\pi}{6}\right)j = \frac{\sqrt{3}}{2}i + \frac{1}{2}j$

$$v = \cos\left(\frac{3\pi}{4}\right)i + \sin\left(\frac{3\pi}{4}\right)j = -\frac{\sqrt{2}}{2}i + \frac{\sqrt{2}}{2}j$$

$$\cos\theta = \frac{u \cdot v}{\|u\|\|v\|}$$

$$= \frac{\sqrt{3}}{2}\left(-\frac{\sqrt{2}}{2}\right) + \frac{1}{2}\left(\frac{\sqrt{2}}{2}\right) = \frac{\sqrt{2}}{4}(1 - \sqrt{3})$$

$$\theta = \arccos\left[\frac{\sqrt{2}}{4}(1 - \sqrt{3})\right] = 105^\circ$$

(3) $u = \langle 1, 1, 1\rangle$, $v = \langle 2, 1, -1\rangle$

$$\cos\theta = \frac{u \cdot v}{\|u\|\|v\|} = \frac{2}{\sqrt{3}\sqrt{6}} = \frac{\sqrt{2}}{3}$$

$$\theta = \arccos\frac{\sqrt{2}}{3} = 61.9^\circ$$

(4) $u = 3i + 2j + k$, $v = 2i - 3j$

$$\cos\theta = \frac{u \cdot v}{\|u\|\|v\|} = \frac{3(2) + 2(-3) + 0}{\|u\|\|v\|} = 0$$

$$\theta = \frac{\pi}{2}, \text{ orthogonal}$$

EX 17

(1) $u = \langle 3, 2, -2\rangle$ $\|u\| = \sqrt{17}$

$$\cos\alpha = \frac{3}{\sqrt{17}} \Rightarrow \alpha \approx 0.7560 \text{ or } 43.3^\circ$$

$$\cos\beta = \frac{2}{\sqrt{17}} \Rightarrow \beta \approx 1.0644 \text{ or } 61.0^\circ$$

$$\cos\gamma = \frac{-2}{\sqrt{17}} \Rightarrow \gamma \approx 2.0772 \text{ or } 119.0^\circ$$

(2) $u = \langle -2, 6, 1\rangle$ $\|u\| = \sqrt{41}$

$$\cos\alpha = \frac{-2}{\sqrt{41}} \Rightarrow \alpha \approx 1.8885 \text{ or } 108.2^\circ$$

$$\cos\beta = \frac{6}{\sqrt{41}} \Rightarrow \beta \approx 0.3567 \text{ or } 20.4^\circ$$

$$\cos\gamma = \frac{1}{\sqrt{41}} \Rightarrow \gamma \approx 1.4140 \text{ or } 81.0^\circ$$

EX 18

(1) $u = \langle 2, 3\rangle$, $v = \langle 5, 1\rangle$

(a) $w_1 = \left(\frac{u \cdot v}{\|v\|^2}\right)v = \frac{13}{26}\langle 5, 1\rangle = \left\langle \frac{5}{2}, \frac{1}{2}\right\rangle$

(b) $w_2 = u - w_1 = \left\langle -\frac{1}{2}, \frac{5}{2}\right\rangle$

(2) $u=\langle 2, -3\rangle$, $v=\langle 3, 2\rangle$

(a) $w_1=\left(\frac{u \cdot v}{\|v\|^2}\right)v=0v=\langle 0, 0\rangle$

(b) $w_2=u-w_1=\langle 2, -3\rangle$

(3) $u=\langle 2, 1, 2\rangle$, $v=\langle 0, 3, 4\rangle$

(a) $w_1=\left(\frac{u \cdot v}{\|v\|^2}\right)v$
$=\frac{11}{25}\langle 0,3,4\rangle=\left\langle 0, \frac{33}{25}, \frac{44}{25}\right\rangle$

(b) $w_2=u-w_1=\left\langle 2, -\frac{8}{25}, \frac{6}{25}\right\rangle$

(4) $u=\langle 1,0,4\rangle$, $v=\langle 3, 0, 2\rangle$

(a) $w_1=\left(\frac{u \cdot v}{\|v\|^2}\right)v$
$=\frac{11}{13}\langle 3,0,2\rangle=\left\langle \frac{33}{13}, 0, \frac{22}{13}\right\rangle$

(b) $w_2=u-w_1=\langle 1,0,4\rangle-\left\langle \frac{33}{13}, 0, \frac{22}{13}\right\rangle$
$=\left\langle -\frac{20}{13}, 0, \frac{30}{13}\right\rangle$

EX 19

(1) The displacement vector is
$D=(6-0)i+(12-10)j+(20-8)k$

$=6i+2j+12k$

so, the work does is

$W=F \cdot D$
$=(8i-6j+9k) \cdot (6i+2j+12k)$
$=48-12+108$

$=144$ joules.

(2) Here $|D|=1000\text{m}$, $|F|=1500\text{N}$, and $\theta=30^\circ$.
Thus

$W=F \cdot D=|F||D|\cos\theta$
$=(1500)(1000)\left(\frac{\sqrt{3}}{2}\right)$
$=750{,}000\sqrt{3}$ joules

(3) $W=|F||D|\cos\theta$

$=(140)(4)\cos 20^\circ$

$=560\cos 20^\circ$

$\approx 526\,\text{J}$

(4) $W=|F||D|\cos\theta$

$=(100)(5)\cos 40^\circ$

$=500\cos 40^\circ$

$\approx 383\,\text{J}$

10.4 The Cross Product

EX 20

(1)

$$a\times b=\begin{vmatrix} i & j & k \\ 1 & 1 & -1 \\ 2 & 4 & 6 \end{vmatrix}=\begin{vmatrix} 1 & -1 \\ 4 & 6 \end{vmatrix}i-\begin{vmatrix} 1 & -1 \\ 2 & 6 \end{vmatrix}j+\begin{vmatrix} 1 & 1 \\ 2 & 4 \end{vmatrix}k$$

$=[6-(-4)]i-[6-(-2)]j+(4-2)k=10i-8j+2k$

(2)

$$a\times b=\begin{vmatrix} i & j & k \\ 1 & 3 & -2 \\ -1 & 0 & 5 \end{vmatrix}=\begin{vmatrix} 3 & -2 \\ 0 & 5 \end{vmatrix}i-\begin{vmatrix} 1 & -2 \\ -1 & 5 \end{vmatrix}j+\begin{vmatrix} 1 & 3 \\ -1 & 0 \end{vmatrix}k$$

$=(15-0)i-(5-2)j+[0-(-3)]k=15i-3j+3k$

(3)

$$a\times b=\begin{vmatrix} i & j & k \\ 0 & 1 & 7 \\ 2 & -1 & 4 \end{vmatrix}=\begin{vmatrix} 1 & 7 \\ -1 & 4 \end{vmatrix}i-\begin{vmatrix} 0 & 7 \\ 2 & 4 \end{vmatrix}j+\begin{vmatrix} 0 & 1 \\ 2 & -1 \end{vmatrix}k$$

$=[4-(-7)]i-(0-14)j+(0-2)k=11i+14j-2k$

(4)

$$a\times b=\begin{vmatrix} i & j & k \\ t & \cos t & \sin t \\ 1 & -\sin t & \cos t \end{vmatrix}$$
$$=\begin{vmatrix} \cos t & \sin t \\ -\sin t & \cos t \end{vmatrix}i-\begin{vmatrix} t & \sin t \\ 1 & \cos t \end{vmatrix}j+\begin{vmatrix} t & \cos t \\ 1 & -\sin t \end{vmatrix}k$$

$=[\cos^2 t-(-\sin^2 t)]i-(t\cos t-\sin t)j+(-t\sin t-\cos t)k$

$=i+(\sin t-t\cos t)j+(-t\sin t-\cos t)k$

EX 21

(1) $u=\langle 12, -3, 0\rangle,\ v=\langle -2, 5, 0\rangle$

$$u\times v=\begin{vmatrix} i & j & k \\ 12 & -3 & 0 \\ -2 & 5 & 0\end{vmatrix}=54k=\langle 0,0,54\rangle$$

$$u\cdot(u\times v)=12(0)+(-3)(0)+0(54)$$
$$=0 \Rightarrow u\perp u\times v$$
$$v\cdot(u\times v)=-2(0)+5(0)+0(54)$$
$$=0 \Rightarrow v\perp u\times v$$

(2) $u=i+j+k,\ v=2i+j-k$

$$u\times v=\begin{vmatrix} i & j & k \\ 1 & 1 & 1 \\ 2 & 1 & -1\end{vmatrix}=-2i+3j-k$$
$$=\langle -2,3,-1\rangle$$
$$u\cdot(u\times v)=1(-2)+1(3)+1(-1)$$
$$=0 \Rightarrow u\perp u\times v$$
$$v\cdot(u\times v)=2(-2)+1(3)+(-1)(-1)$$
$$=0 \Rightarrow v\perp u\times v$$
$$(-v)\times u=-(v\times u)=u\times v$$

EX 22

(1) $u=j$
$v=i+k$

$$u\times v=\begin{vmatrix} i & j & k \\ 0 & 1 & 0 \\ 0 & 1 & 1\end{vmatrix}=i$$
$$A=\|u\times v\|=\|i\|=1$$

(2) $u=i+j+k$
$v=j+k$

$$u\times v=\begin{vmatrix} i & j & k \\ 1 & 1 & 1 \\ 0 & 1 & 1\end{vmatrix}=-j+k$$
$$A=\|u\times v\|=\|-j+k\|=\sqrt{2}$$

(3) $u=\langle 3,2,-1\rangle$
$v=\langle 1,2,3\rangle$

$$u\times v=\begin{vmatrix} i & j & k \\ 3 & 2 & -1 \\ 1 & 2 & 3\end{vmatrix}=\langle 8,-10,4\rangle$$
$$A=\|u\times v\|=\|\langle 8,-10,4\rangle\|$$
$$=\sqrt{180}=6\sqrt{5}$$

(4) $u=\langle 2,-1,0\rangle$
$v=\langle -1,2,0\rangle$

$$u\times v=\begin{vmatrix} i & j & k \\ 2 & -1 & 0 \\ -1 & 2 & 0\end{vmatrix}=\langle 0,0,3\rangle$$
$$A=\|u\times v\|=\|\langle 0,0,3\rangle\|$$
$$=3$$

EX 23

(1) $A(0,0,0),\ B(1,2,3),\ C(-3,0,0)$

$$\overrightarrow{AB}=\langle 1,2,3\rangle,\ \overrightarrow{AC}=\langle -3,0,0\rangle$$
$$\overrightarrow{AB}\times\overrightarrow{AC}=\begin{vmatrix} i & j & k \\ 1 & 2 & 3 \\ -3 & 0 & 0\end{vmatrix}$$
$$=-9j+6k$$
$$A=\frac{1}{2}\|\overrightarrow{AB}\times\overrightarrow{AC}\|=\frac{1}{2}\sqrt{117}=\frac{3}{2}\sqrt{13}$$

(2) $A(2,-3,4),\ B(0,1,2),\ C(-1,2,0)$

$$\overrightarrow{AB}=\langle -2,4,-2\rangle,\ \overrightarrow{AC}=\langle -3,5,-4\rangle$$
$$\overrightarrow{AB}\times\overrightarrow{AC}=\begin{vmatrix} i & j & k \\ -2 & 4 & -2 \\ -3 & 5 & -4\end{vmatrix}$$
$$=-6i-2j+2k$$
$$A=\frac{1}{2}\|\overrightarrow{AB}\times\overrightarrow{AC}\|=\frac{1}{2}\sqrt{44}=\sqrt{11}$$

(3) $A(2,-7,3)$, $B(-1,5,8)$, $C(4,6,-1)$

$\overrightarrow{AB}=\langle -3,12,5\rangle$, $\overrightarrow{AC}=\langle 2,13,-4\rangle$

$$\overrightarrow{AB}\times\overrightarrow{AC}=\begin{vmatrix} i & j & k \\ -3 & 12 & 5 \\ 2 & 13 & -4 \end{vmatrix}$$

$$=\langle -113,-2,-63\rangle$$

$$Area=\frac{1}{2}\|\overrightarrow{AB}\times\overrightarrow{AC}\|=\frac{1}{2}\sqrt{16,742}$$

(4) $A(1,2,0)$, $B(-2,1,0)$, $C(0,0,0)$

$\overrightarrow{AB}=\langle -3,-1,0\rangle$, $\overrightarrow{AC}=\langle -1,-2,0\rangle$

$$\overrightarrow{AB}\times\overrightarrow{AC}=\begin{vmatrix} i & j & k \\ -3 & -1 & 0 \\ -1 & -2 & 0 \end{vmatrix}$$

$$=5k$$

$$A=\frac{1}{2}\|\overrightarrow{AB}\times\overrightarrow{AC}\|=\frac{5}{2}$$

EX 24

(1) $u\cdot(v\times w)=\begin{vmatrix} 1 & 1 & 0 \\ 0 & 1 & 1 \\ 1 & 0 & 1 \end{vmatrix}=2$

$V=|u\cdot(v\times w)|=2$

(2) $u\cdot(v\times w)=\begin{vmatrix} 1 & 3 & 1 \\ 0 & 6 & 6 \\ -4 & 0 & -4 \end{vmatrix}=-72$

$V=|u\cdot(v\times w)|=72$

EX 25

(1) $u=\langle 3,0,0\rangle$

$v=\langle 0,5,1\rangle$

$w=\langle 2,0,5\rangle$

$u\cdot(v\times w)=\begin{vmatrix} 3 & 0 & 0 \\ 0 & 5 & 1 \\ 2 & 0 & 5 \end{vmatrix}=75$

$V=|u\cdot(v\times w)|=75$

(2) $u=\langle 1,1,0\rangle$

$v=\langle 1,0,2\rangle$

$w=\langle 0,1,1\rangle$

$u\cdot(v\times w)=\begin{vmatrix} 1 & 1 & 0 \\ 1 & 0 & 2 \\ 0 & 1 & 1 \end{vmatrix}=-3$

$V=|u\cdot(v\times w)|=3$

EX 26

(1) Point: $\langle 0,0,0\rangle$

Direction vector: $v=\langle 1,2,3\rangle$

Direction number: 1, 2, 3

(a) Parametric: $x=t$, $y=2t$, $z=3t$

(b) Synmetric: $x=\dfrac{y}{2}=\dfrac{z}{3}$

(2) Point: $\langle 0,0,0\rangle$

Direction vector: $v=\left\langle -2,\dfrac{5}{2},1\right\rangle$

Direction number: -4, 5, 2

(a) Parametric: $x=-4t$, $y=5t$, $z=2t$

(b) Synmetric: $\dfrac{x}{-4}=\dfrac{y}{5}=\dfrac{z}{2}$

(3) Point: $\langle -2,0,3\rangle$

Direction vector: $v=\langle 2,4,-2\rangle$

Direction number: 2, 4, -2

(a) Parametric:

$x=-2+2t$, $y=4t$, $z=3-2t$

(b) Synmetric: $\dfrac{x+2}{2}=\dfrac{y}{4}=\dfrac{z-3}{-2}$

(4) Point: $\langle -3,0,2\rangle$

Direction vector: $v=\langle 0,6,3\rangle$

Direction number: 0, 2, 1

(a) Parametric:

$x=-3t$, $y=2t$, $z=2+t$

(b) Synmetric: $\dfrac{y}{2}=z-2$, $x=-3$

(5) Point: $\langle 1, 0, 1 \rangle$

Direction vector: $v = 3i - 2j + k$

Direction number: 3, -2, 1

(a) Parametric:

$x = 1 + 3t,\ y = -2t,\ z = 1 + t$

(b) Synmetric: $\frac{x-1}{3} = \frac{y}{-2} = \frac{z-1}{1}$

(6) Point: $\langle -3, 5, 4 \rangle$

Direction number: 3, -2, 1 153

(a) Parametric:

$x = -3 + 3t,\ y = 5 - 2t,\ z = 4 + t$

(b) Synmetric: $\frac{x+3}{3} = \frac{y-5}{-2} = z - 4$

EX 27

(1) Point: $\langle 5, -3, -2 \rangle, \left(\frac{-2}{3}, \frac{2}{3}, 1\right)$

Direction vector: $v = \frac{17}{3}i - \frac{11}{3}j - 3k$

Direction number: 17, -11, -9

(a) Parametric: $x = 5 + 17t$,

$y = -3 - 11t,\ z = -2 - 9t$

(b) Synmetric: $\frac{x-5}{17} = \frac{y+3}{-11} = \frac{z+2}{-9}$

(2) Point: $(2, 0, 2)$, $(1, 4, -3)$

Direction vector: $\langle 1, -4, 5 \rangle$

Direction number: 1, -4, 5

(a) Parametric:

$x = 2 + t,\ y = -4t,\ z = 2 + 5t$

(b) Synmetric: $x - 2 = \frac{y}{-4} = \frac{z-2}{5}$

(3) Point: $(2, 3, 0)$, $(10, 8, 12)$

Direction vector: $\langle 8, 5, 12 \rangle$

Direction number: 8, 5, 12

(a) Parametric:

$x = 2 + 8t,\ y = 3 + 5t,\ z = 12t$

(b) Synmetric: $\frac{x-2}{8} = \frac{y-3}{5} = \frac{z}{12}$

(4) Point: $(0, 0, 25)$, $(10, 10, 0)$

Direction vector: $\langle 10, 10, -25 \rangle$

Direction number: 2, 2, -5

(a) Parametric:

$x = 2t,\ y = 2t,\ z = 25 - 5t$

(b) Synmetric: $\frac{x}{2} = \frac{y}{2} = \frac{z-25}{-5}$

EX 28

(1) Point: $(2, 1, 2)$

$n = i = \langle 1, 0, 0 \rangle$

$$1(x-2) + 0(y-1) + 0(z-2) = 0$$
$$x - 2 = 0$$

(2) Point: $(1, 0, -3)$

$n = k = \langle 0, 0, 1 \rangle$

$$0(x-1) + 0(y-1) + 1[z - (-3)] = 0$$
$$z + 3 = 0$$

(3) Point: $(3, 2, 2)$

Normal vector: $n = 2i + 3j - k$

$$2(x-3) + 3(y-2) - 1(z-2) = 0$$
$$2x + 3y - z = 10$$

(4) Point: $(1, 0, -3)$

Normal vector: $n = -3i + 2k$

$$-3(x-0) + 0(y-0) + 2(z-0) = 0$$
$$-3x + 2z = 0$$

(5) Point: $(0, 0, 6)$

Normal vector: $n = -i + j - 2k$

$$-1(x-0) + 1(y-0) - 2(z-6) = 0$$
$$-x + y - 2z + 12 = 0$$
$$x - y + 2z = 12$$

(6) Point: $(3, 2, 2)$

Normal vector: $n = 4i + j - 3k$

$$4(x-3) + (y-2) - 3(z-2) = 0$$
$$4x + y - 3z = 8$$

EX 29

(1) Let u be the vector from $(0, 0, 0)$ to $(1, 2, 3)$: $u = i + 2j + 3k$

Let v be the vector from $(0, 0, 0)$ to $(-2, 3, 3)$: $v = -2i + 3j + 3k$

Normal vector: $u \times v = \begin{vmatrix} i & j & k \\ 1 & 2 & 3 \\ -2 & 3 & 3 \end{vmatrix}$

$= -3i + (-9)j + 7k$

$-3(x-0) - 9(y-0) + 7(z-0) = 0$

$3x + 9y - 7z = 0$

(2) Let u be the vector from $(2, 3, -2)$ to $(3, 4, 2)$: $\langle 1, 1, 4 \rangle$

Let v be the vector from $(2, 3, -2)$ to $(1, -1, 0)$: $\langle -1, -4, 2 \rangle$

Normal vector:

$u \times v = \begin{vmatrix} i & j & k \\ 1 & 1 & 4 \\ -1 & -4 & 2 \end{vmatrix} = \langle 18, -6, -3 \rangle$

$= -3\langle -6, 2, 1 \rangle$

$-6(x-2) + 2(y-3) + 1(z+2) = 0$

$-6x + 2y + z = -8$

(3) Let u be the vector from $(1, 2, 3)$ to $(3, 2, 1)$: $u = 2i - 2k$

Let v be the vector from $(1, 2, 3)$ to $(-1, -2, 2)$: $v = -2i - 4j - k$

Normal vector:

$\left(\frac{1}{2}u\right) \times (-v) = \begin{vmatrix} i & j & k \\ 1 & 0 & -1 \\ 2 & 4 & 1 \end{vmatrix} = 4i - 3j + 4k$

$4(x-1) - 3(y-2) + 4(z-3) = 0$

$4x - 3y + 4z = 10$

(4) $(1, 2, 3)$, Normal vector: $v = i$, $1(x-1) = 0$, $x = 1$

EX 30

(1) $n_1 = i + j + k$ and $n_2 = i + j$

$\Rightarrow n_1 \times n_2 = \begin{vmatrix} i & j & k \\ 1 & 1 & 1 \\ 1 & 1 & 0 \end{vmatrix} = -i + j$, the direction of the desired line: $(1, 1, -1)$ is on both planes $\Rightarrow$ the desired line is

$x = 1 - t,\ y = 1 + t,\ z = -1$

(2) $n_1 = 3i - 6j - 2k$ and $n_2 = 2i + j - 2k$

$\Rightarrow n_1 \times n_2 = \begin{vmatrix} i & j & k \\ 3 & -6 & -2 \\ 2 & 1 & -2 \end{vmatrix} = 14i + 2j + 15k$. the direction of the desired line: $(1, 0, 0)$ is on both planes $\Rightarrow$ the desired line is

$x = 1 + 14t,\ y = 2t,\ z = 15t$

(3) $n_1 = i - 2j + 4k$ and $n_2 = i + j - 2k$

$\Rightarrow n_1 \times n_2 = \begin{vmatrix} i & j & k \\ 1 & -2 & 4 \\ 1 & 1 & -2 \end{vmatrix} = 6j + 3k$.the direction of the desired line: $(4, 3, 1)$ is on both planes $\Rightarrow$ the desired line is

$x = 4,\ y = 3 + 6t,\ z = 1 + 3t$

(4) $n_1 = 5i - 2j$ and $n_2 = 4j - 5k$

$\Rightarrow n_1 \times n_2 = \begin{vmatrix} i & j & k \\ 5 & -2 & 0 \\ 0 & 4 & -5 \end{vmatrix} = 10i + 25j + 20k$, the direction of the desired line: $(1, -3, 1)$ is on both planes

$\Rightarrow$ the desired line is

$x = 1 + 10t,\ y = -3 + 25t,\ z = 1 + 20t$

EX 31

(1) $S(2, -3, 4)$, $x + 2y + 2z = 13$ and $P(13, 0, 0)$ is on the plane $\Rightarrow \overrightarrow{PS} = -11i - 3j + 4k$ and $n = i + 2j + 2k$

$\Rightarrow d = \left| \overrightarrow{PS} \cdot \frac{n}{|n|} \right| = \left| \frac{-11 - 6 + 8}{\sqrt{1+4+4}} \right| = \left| \frac{-9}{\sqrt{9}} \right|$

$= 3$

(2) S(0, 1, 1), $4y+3z=-12$ and P(0, −3, 0) is on the plane $\Rightarrow \overrightarrow{PS}=4j+k$ and

$n=4j+3k$

$$\Rightarrow d=\left|\overrightarrow{PS}\cdot\frac{n}{|n|}\right|=\left|\frac{16+3}{\sqrt{16+9}}\right|=\frac{19}{5}$$

(3) S(0, −1, 0), $2x+y+2z=4$ and P(2, 0, 0) is on the plane $\Rightarrow \overrightarrow{PS}=-2i-j$ and

$n=2i+j+2k$

$$\Rightarrow d=\left|\overrightarrow{PS}\cdot\frac{n}{|n|}\right|=\left|\frac{-4-1+0}{\sqrt{4+1+4}}\right|=\frac{5}{3}$$

(4) The point P(1, 0, 0) is on the first plane and S(10, 0, 0) is a point on the second plane $\Rightarrow$ $\overrightarrow{PS}=9i$, and $n=i+2j+6k$ is normal to the first plane $\Rightarrow$ the distance form S to the first plane is

$$d=\left|\overrightarrow{PS}\cdot\frac{n}{|n|}\right|=\left|\frac{9}{\sqrt{1+4+36}}\right|=\frac{9}{\sqrt{41}},$$

which is also the distance between the planes.

EX 32

(1) S(0, 0, 12), P(0, 0, 0) and $v=4i-2j+2k$

$$\Rightarrow \overrightarrow{PS}\times v=\begin{vmatrix} i & j & k \\ 0 & 0 & 12 \\ 4 & -2 & 2 \end{vmatrix}$$

$$=24i+48j=24(i+2j)$$

$$\Rightarrow d=\left|\frac{\overrightarrow{PS}\cdot v}{|v|}\right|=\frac{24\sqrt{1+4}}{\sqrt{16+4+4}}=\frac{24\sqrt{5}}{\sqrt{24}}$$

$$=\sqrt{5\cdot 24}=2\sqrt{30}$$

is the distance from S to the line.

(2) S(2, 1, 3), P(2, 1, 3) and $v=2i+6j$

$$\Rightarrow \overrightarrow{PS}\times v=0 \Rightarrow d=\left|\frac{\overrightarrow{PS}\cdot v}{|v|}\right|=\frac{0}{\sqrt{40}}=0$$

is the distance from S to the line. (i.e., the point S lies on the line)

(3) S(3, −1, 4), P(4, 3, −5) and $v=-i+2j+3k$

$$\Rightarrow \overrightarrow{PS}\times v=\begin{vmatrix} i & j & k \\ -1 & -4 & 9 \\ -1 & 2 & 3 \end{vmatrix}$$

$$=-30i-6j-6k$$

$$\Rightarrow d=\left|\frac{\overrightarrow{PS}\cdot v}{|v|}\right|=\frac{\sqrt{900+36+36+4}}{\sqrt{1+4+9}}$$

$$=\frac{\sqrt{972}}{\sqrt{14}}=\frac{\sqrt{486}}{\sqrt{7}}$$

$=\dfrac{\sqrt{81\cdot 6}}{7}=\dfrac{9\sqrt{42}}{7}$ is the distance from S to the line.

(4) S(−1, 4, 3), P(10, −3, 0) and $v=4i+4k$

$$\Rightarrow \overrightarrow{PS}\times v=\begin{vmatrix} i & j & k \\ -11 & 7 & 3 \\ 4 & 0 & 4 \end{vmatrix}$$

$$=28i+56j-28k=28(i+2j-k)$$

$$\Rightarrow d=\left|\frac{\overrightarrow{PS}\cdot v}{|v|}\right|=\frac{28\sqrt{1+4+1}}{4\sqrt{1+1}}=7\sqrt{3}$$

is the distance from S to the line.

EX 33

(1) $n_1=i+j$ and

$n_2=2i+j-2k$

$$\Rightarrow \theta=\cos^{-1}\left(\frac{n_1n_2}{|n_1||n_2|}\right)$$

$$=\cos^{-1}\left(\frac{2+1}{\sqrt{2}\sqrt{9}}\right)$$

$$=\cos^{-1}\left(\frac{1}{\sqrt{2}}\right)$$

$$=\frac{\pi}{4}$$

(2) $n_1=5i+j-k$ and

$n_2=i-2j+3k$

$$\Rightarrow \theta=\cos^{-1}\left(\frac{n_1n_2}{|n_1||n_2|}\right)$$

$$=\cos^{-1}\left(\frac{5-2-3}{\sqrt{27}\sqrt{14}}\right)$$

$$=\cos^{-1}(0)=\frac{\pi}{2}$$

EX 34

(1) The direction vectors for the lines are

$\mathbf{u}=-2\mathbf{i}+\mathbf{j}+\mathbf{k}$.

$\mathbf{v}=-3\mathbf{i}+4\mathbf{j}-\mathbf{k}$

Normal vector:

$$\mathbf{u}\times\mathbf{v}=\begin{vmatrix}\mathbf{i} & \mathbf{j} & \mathbf{k}\\ -2 & 1 & 1\\ -3 & 4 & -1\end{vmatrix}=-5(\mathbf{i}+\mathbf{j}+\mathbf{k})$$

Point of intersection of the lines: $(-1, 5, 1)$

$(x+1)+(y-5)+(z-1)=0$

$x+y+z=5$

(2) The direction of the line is $\mathbf{u}=2\mathbf{i}-\mathbf{j}+\mathbf{k}$.

Choose any point on the line. [(0, 4, 0) for example]. and let v be the vector from (0, 4, 0) to the given point (2, 2, 1):

$\mathbf{v}=2\mathbf{i}-2\mathbf{j}+\mathbf{k}$

Normal vector: $\mathbf{u}\times\mathbf{v}=\begin{vmatrix}\mathbf{i} & \mathbf{j} & \mathbf{k}\\ 2 & -1 & 1\\ 2 & -2 & 1\end{vmatrix}=\mathbf{i}-2\mathbf{k}$

$(x-2)-2(z-1)=0$

$x-2z=0$

(3) Let v be the vector from $(-1, 1, -1)$ to $(2, 2, 1)$: $\mathbf{v}=3\mathbf{i}+\mathbf{j}+2\mathbf{k}$

Let n be a vector normal to the plane $2x-3y+z=3$: $\mathbf{n}=2\mathbf{i}-3\mathbf{j}+\mathbf{k}$

Since v and n both lie in the plane P. the normal vector to P is

$$\mathbf{v}\times\mathbf{n}=\begin{vmatrix}\mathbf{i} & \mathbf{j} & \mathbf{k}\\ 3 & 1 & 2\\ 2 & -3 & 1\end{vmatrix}=7\mathbf{i}+\mathbf{j}-11\mathbf{k}$$

$7(x-2)+1(y-2)-11(z-1)=0$

$7x+y-11z=5$

(4) Let v be the vector from $(3, 2, 1)$ to $(3, 1, -5)$: $\mathbf{v}=-\mathbf{j}-6\mathbf{k}$

Let n be the normal to the given plane: $\mathbf{n}=6\mathbf{i}+7\mathbf{j}+2\mathbf{k}$

Since v and n both lie in the plane P. the normal vector to P is:

$$\mathbf{v}\times\mathbf{n}=\begin{vmatrix}\mathbf{i} & \mathbf{j} & \mathbf{k}\\ 0 & -1 & -6\\ 6 & 7 & 2\end{vmatrix}=40\mathbf{i}-36\mathbf{j}+6\mathbf{k}$$

$$=2(20\mathbf{i}-18\mathbf{j}+3\mathbf{k})$$

$20(x-3)-18(y-2)+3(z-1)=0$

$20x-18y+3z=27$

(5) Let $\mathbf{u}=\mathbf{i}$ and let v be the vector from $(1, -2, -1)$ to $(2, 5, 6)$: $\mathbf{v}=\mathbf{i}+7\mathbf{j}+7\mathbf{k}$

Since u and v both lie in the plane P. the normal vector to P is:

$$\mathbf{u}\times\mathbf{v}=\begin{vmatrix}\mathbf{i} & \mathbf{j} & \mathbf{k}\\ 1 & 0 & 0\\ 1 & 7 & 7\end{vmatrix}=-7\mathbf{j}+7\mathbf{k}=-7(\mathbf{j}-\mathbf{k})$$

$[y-(-2)]-[z-(-1)]=0$

$y-z=-1$

(6) Let $\mathbf{u}=\mathbf{k}$ and let v be the vector from $(4, 2, 1)$ to $(-3, 5, 7)$: $\mathbf{v}=-7\mathbf{i}+3\mathbf{j}+6\mathbf{k}$

Since u and v both lie in the plane P. the normal vector to P is:

$$\mathbf{u}\times\mathbf{v}=\begin{vmatrix}\mathbf{i} & \mathbf{j} & \mathbf{k}\\ 0 & 0 & 1\\ -7 & 3 & 6\end{vmatrix}=-3\mathbf{i}-7\mathbf{j}=-(3\mathbf{i}+7\mathbf{j})$$

$3(x-4)+7(y-2)=0$

$3x+7y=26$

CALCULUS INDEX

Authors

Man Sik Min
Department of Mathematics, Division of Data Science, University of Suwon
PhD of Mathematics, Korea University

Hyeong Chul Jeong
Department of Statistics, Division of Data Science, University of Suwon
PhD of Statistics, Korea University

Hyejung Lee
Center for Statistical Research and Management, Department of Information and Statistical Research, Korea Institute for Health And Social Affairs
PhD of Statistics, Korea University

Calculus

Copyright ⓒ 2022, Hantoo Media
Published on 31 March 2022
Author: Man Sik Min · Hyeong Chul Jeong · Hyejung Lee
Publisher: Junho Kim | HANTEE MEDIA
3F 67 Donggyoro 23-gil, Mapo-gu, Seoul
Registered No. 15-571 (15 May, 2006)
Tel. No. 02)332-7993~4 | Fax No. 02)332-7995

Editors: Eunsu Kim, Chaewon Yoo
Marketers: Hogeun Noh, Jaein Park, Sangwook Choi, Wonkuk Kim, Taeksung Kim
Managers: Jiyoung Kim, Jihee Moon

ISBN 978-89-6421-435-0 (93410)
Price 35,000won
www.hanteemedia.co.kr | hantee@hanteemedia.co.kr

All Rights Reserved.
Authorized this book published by the authors.
No part of this publication may be reproduced, stored in a retrieval system or transmitted in any form or by any means, electronic, mechanical, photocopying, recording, scanning or otherwise, without the permission in writing of the HANTEE MEDIA.

*The books with problems can be exchanged at the place of purchase.